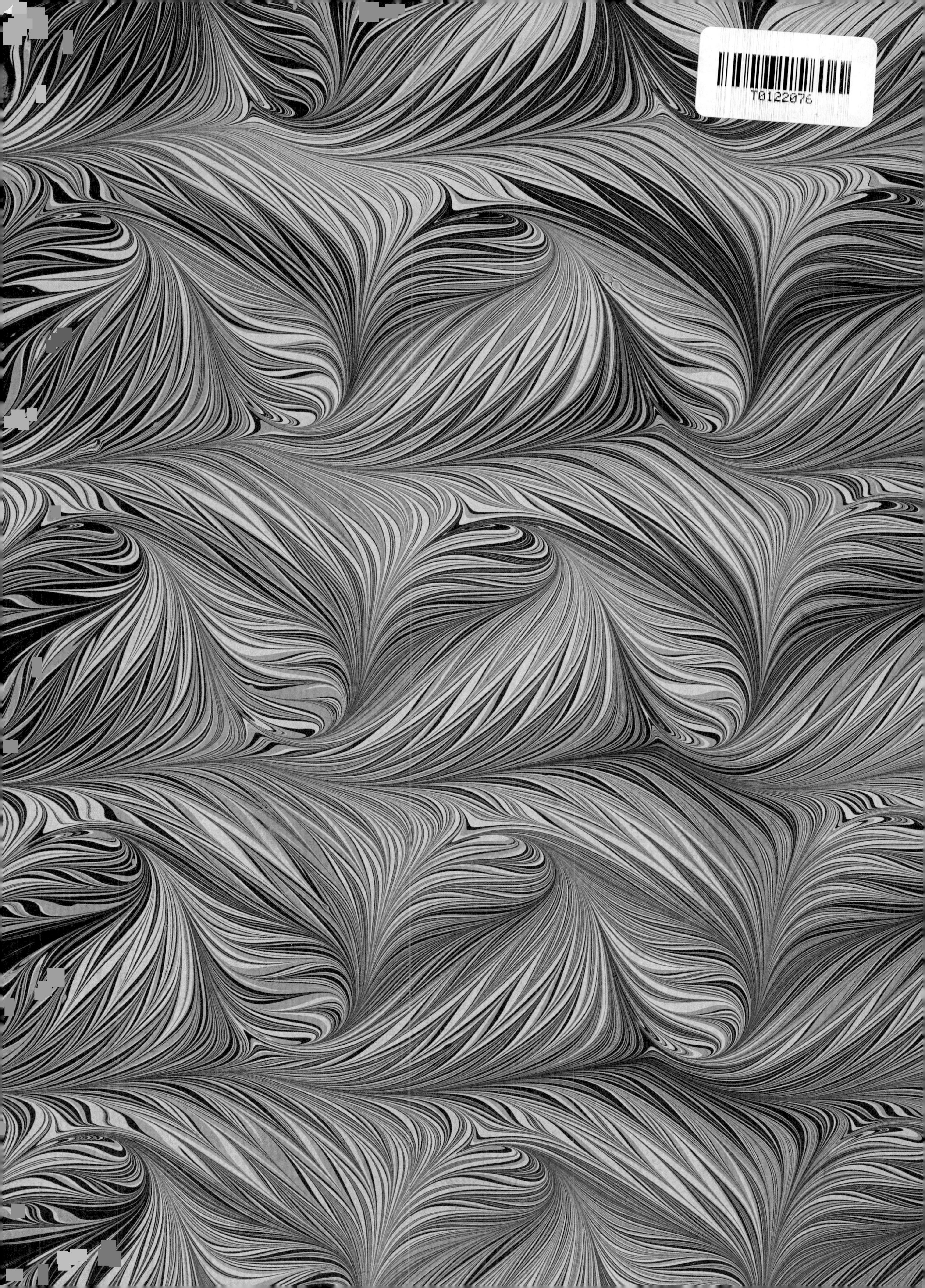
T0122076

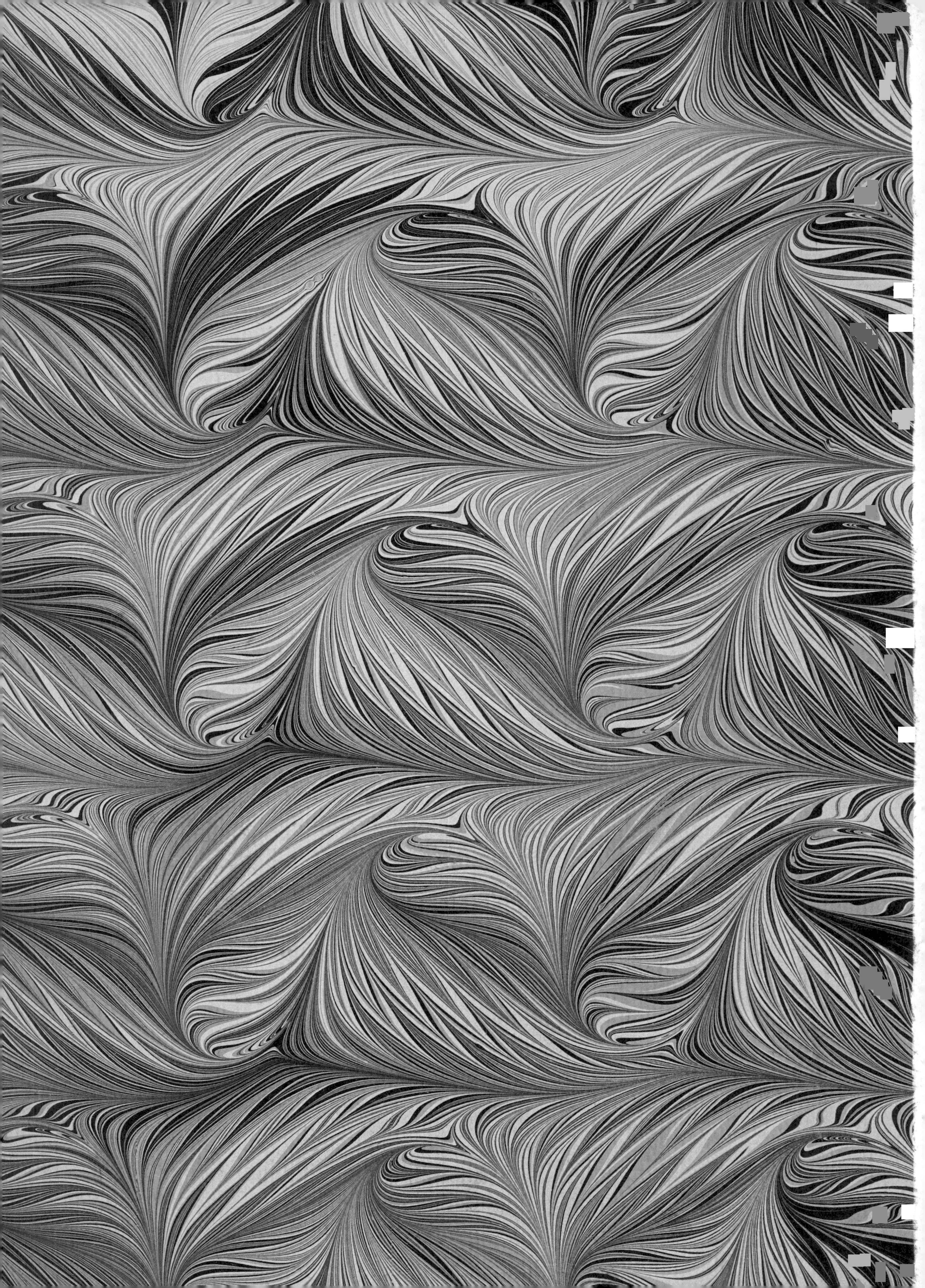

STRATA.

SMITH'S STRATA INDEX *of* SELECTED BRITISH COUNTIES.

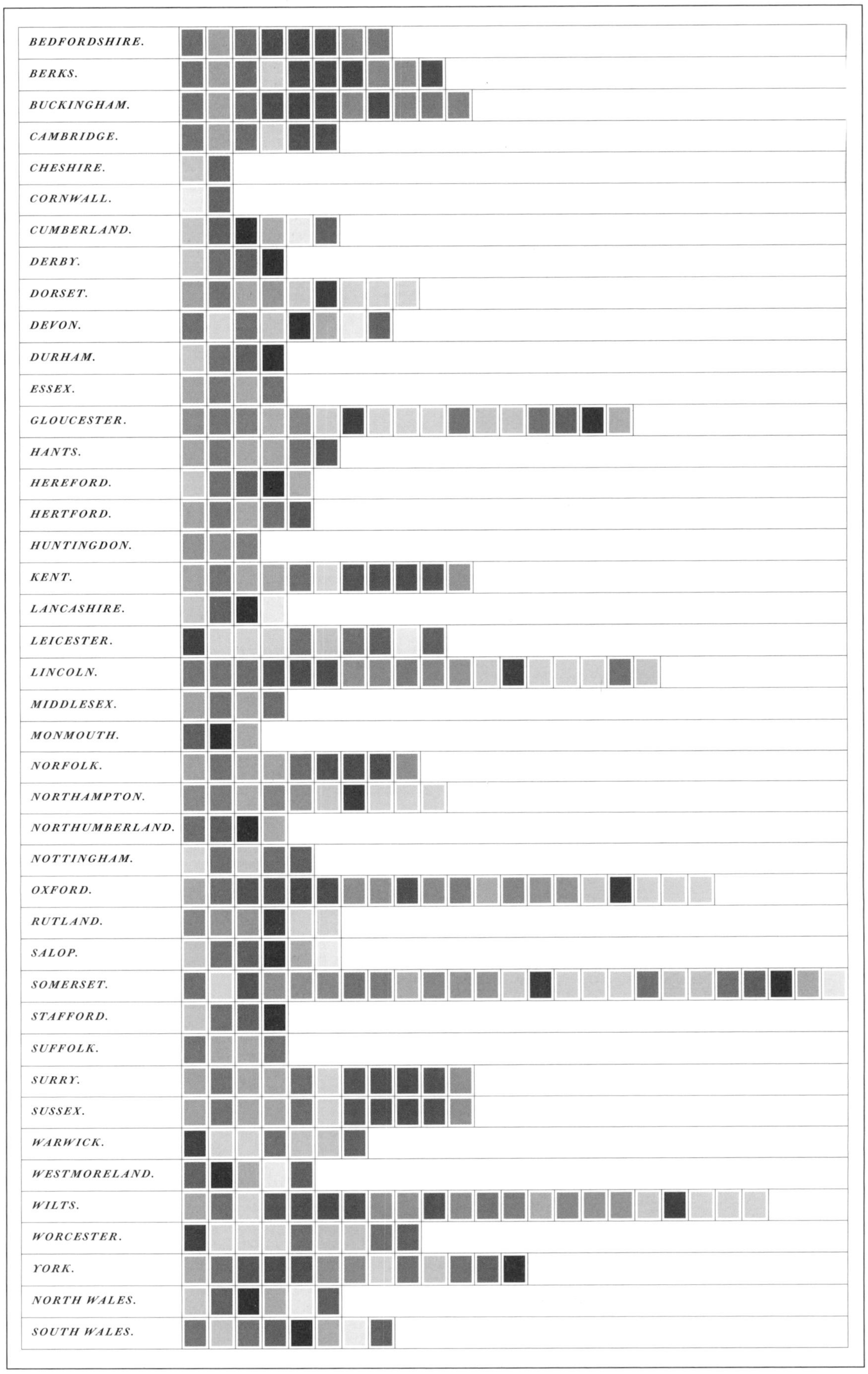

BEDFORDSHIRE.
BERKS.
BUCKINGHAM.
CAMBRIDGE.
CHESHIRE.
CORNWALL.
CUMBERLAND.
DERBY.
DORSET.
DEVON.
DURHAM.
ESSEX.
GLOUCESTER.
HANTS.
HEREFORD.
HERTFORD.
HUNTINGDON.
KENT.
LANCASHIRE.
LEICESTER.
LINCOLN.
MIDDLESEX.
MONMOUTH.
NORFOLK.
NORTHAMPTON.
NORTHUMBERLAND.
NOTTINGHAM.
OXFORD.
RUTLAND.
SALOP.
SOMERSET.
STAFFORD.
SUFFOLK.
SURRY.
SUSSEX.
WARWICK.
WESTMORELAND.
WILTS.
WORCESTER.
YORK.
NORTH WALES.
SOUTH WALES.

STRATA.

William Smith's Geological Maps.

FOREWORD *by* ROBERT MACFARLANE.

THE UNIVERSITY OF CHICAGO PRESS

A DELINEATION *of* THE STRATA OF ENGLAND AND WALES WITH PART OF SCOTLAND.

COMPOSITE MAP.

PAGE 2: A RECONSTRUCTION OF SMITH'S STRATA INDEX OF SELECTED BRITISH COUNTIES.

The strata present in each British county, according to Smith's *Table of British Organized Fossils* (1817). The strata are represented by the colours assigned to them by Smith in his 1815 map shown above and in his 1817 table shown on pp. 10–11. Smith's geological county maps, both published and unpublished, are shown throughout the book in the relevant geographical chapters.

***A DELINEATION OF THE STRATA OF ENGLAND AND WALES, WITH PART OF SCOTLAND*, WILLIAM SMITH, 1815.**

William Smith's seminal 1815 map, coloured to indicate Britain's geological strata, shown as a composite whole (above) and in its separate sheets (opposite). This particular print of the map, from Oxford University Museum of Natural History, is one of the finest examples, and dates from 23 February 1816. It is used as the base map throughout this book.

A DELINEATION *of* THE STRATA OF ENGLAND AND WALES WITH PART OF SCOTLAND.

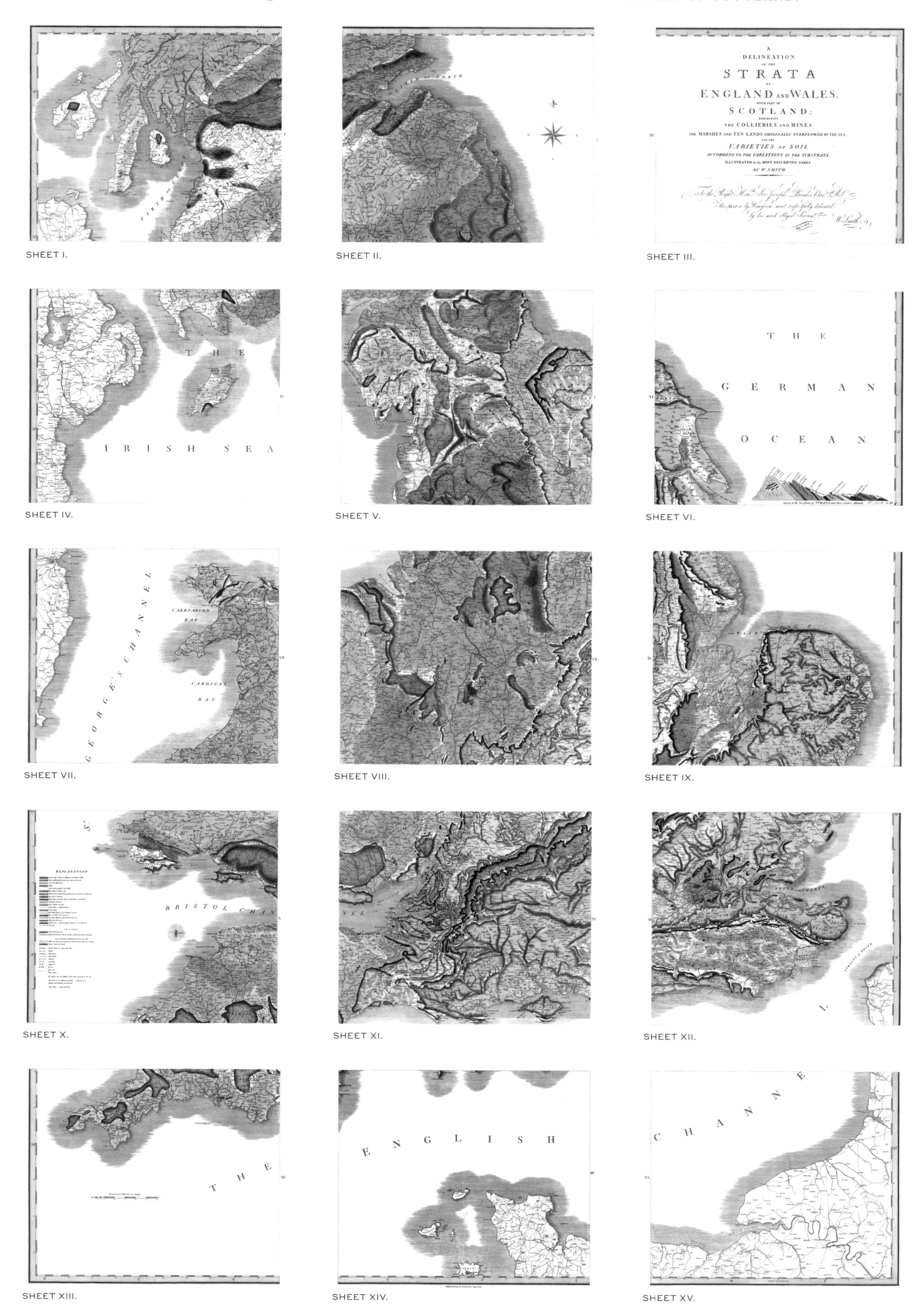

TABLE *of* CONTENTS.

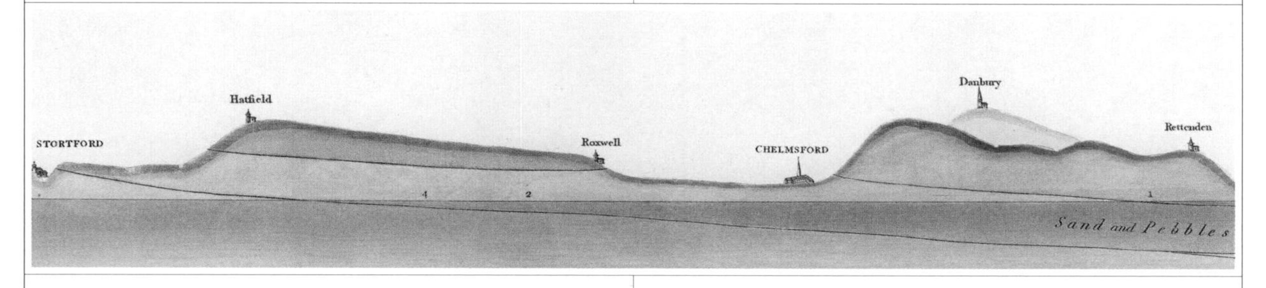
STORTFORD
Hatfield
Roxwell
CHELMSFORD
Danbury
Rettenden
Sand and Pebbles

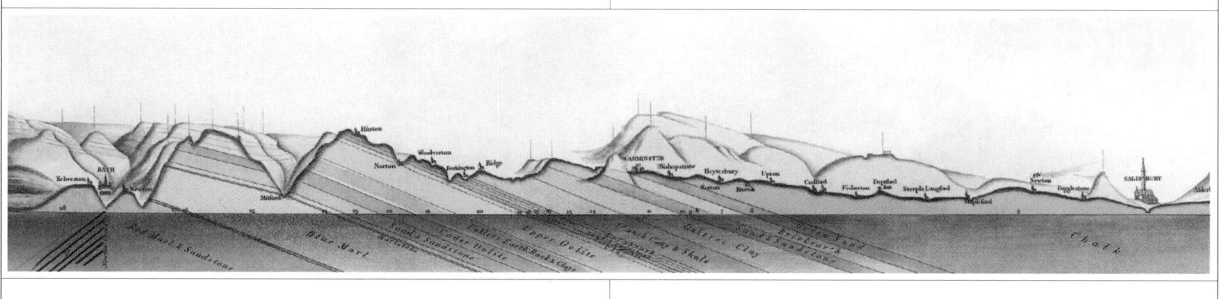
Blue Marl
Chalk

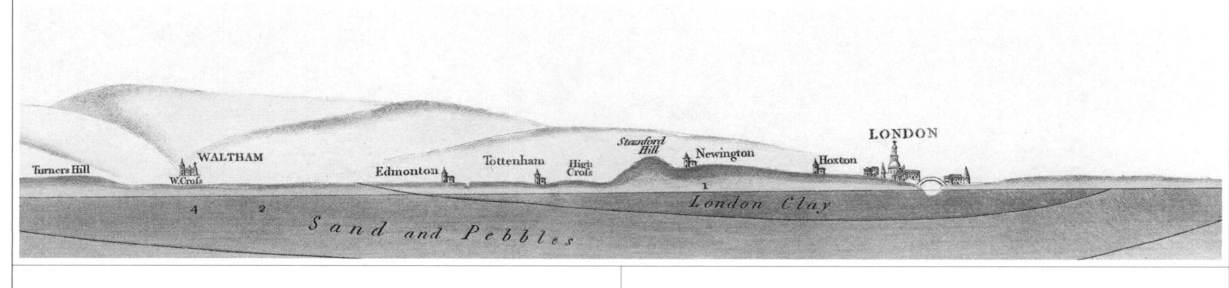
Turners Hill
WALTHAM
W.Cross
Edmonton
Tottenham
High Cross
Stamford Hill
Newington
Hoxton
LONDON
London Clay
Sand and Pebbles

FOSSILS FROM WILLIAM SMITH'S COLLECTION.

This selection of fossils formed part of Smith's large personal collection, which he was forced to sell to the British Museum in 1818. The accompanying catalogue *A Stratigraphical System of Organized Fossils* (1817) indicates that the specimens above were all collected from the 'Under Oolite' (Inferior Oolite) stratum, and those on the page opposite were all collected from the Lias and 'Kelloways Stone' (Kellaways Rock) strata. The final five of these specimens (23–27) were also included in Smith's *Strata Identified by Organized Fossils* (1816–19) alongside illustrations by the artist and minerologist James Sowerby. These can be seen on pp. 150–51.

1. *Acanthothyris spinosa*
2. *Trigonia costata*
3. *Trigonia costata*
4. *Variamussium* cf. *pumilum*
5. *Astarte elegans*
6. Fragment of large *Trichites ploti*
7. *Clypeus ploti*
8. *Pseudomelania* sp.
9. *Pleurotomaria granulata*
10. *Ampullina* sp.
11. *Pyrgotrochus* sp.
12. *Pyrgotrochus conoideus*
13. ?*Nerinea* sp.
14. *Cenoceras excavatus*
15. *Nerinea* sp.
16. *Teloceras calyx*
17. *Pleuromya* aff. *uniformis*
18. *Pleurotomaria* cf. *cognata*
19. *Euagassiceras sauzeanum*
20. *Zugodactylites braunianus*
21. *Isocrinus psilonoti*
22. *Pentacrinites fossilis*
23. *Cadoceras sublaevis*
24. *Ornithella ornithocephala*
25. *Sigaloceras calloviense*
26. *Proplanulites koenigi*
27. *Gryphaea dilobotes*

GEOLOGICAL TABLE *of* BRITISH ORGANIZED FOSSILS.

WHICH IDENTIFY THE COURSES AND CONTINUITY OF THE STRATA *IN THEIR ORDER OF SUPERPOSITION;*

AS ORIGINALLY DISCOVERED BY W. SMITH, *Civil Engineer; WITH REFERENCE TO HIS*

GEOLOGICAL MAP of ENGLAND and WALES.

ORGANIZED FOSSILS *which identify the respective* STRATA.		NAMES OF STRATA *on the shelves of the* GEOLOGICAL COLLECTION.	COLOURS *of* STRATA *on the* MAP.
Volatæ, Restellariæ, Fusus, Cerithia, Nautili, Toredo, Crabs Teeth and bones	*Plains*	*London Clay*	1
		Sand	2
Murices, Turbo, Pentunculus, Cardia, Venus, Ostreæ		*Crag*	3
		Sand	4
Flint, alcyonia, Ostreæ, Echini ... *Plagiostoma*	*Chalk Hills*	*Chalk Upper*	
Terebratulæ, Teeth, Palates ... *Plagiostoma*		*Lower*	5
Funnelform, Alcyonia, Venus, Chama, Pectines, Terebratulæ, echini		*Green Sand*	6
Belemnites, Ammonites	*Clay Vales*	*Brickearth*	7
		Sand	8
Turritella, Ammonites, Trigoniæ, Pecten, Wood		*Portland Rock*	9
		Sand	10
Trochus, Nautilus, Ammonites in Masses, Ostreæ (in a bed), bones		*Oaktree Clay*	11
Various Madreporæ, Melaniæ, Ostreæ, Echini and Spines		*Coral Rag & Pisolite*	12
		Sand	13
Belemnites, Ammonites, Ostreæ		*Clunch Clay & Shale*	14
Ammonites, Ostreæ		*Kelloways Stone*	15
Modiola, Cardia, Ostreæ, Avicula, Terebratulæ	*Stonebrash Hills*	*Cornbrash*	16
		Sand & Sandstone	17
Pectines, Teeth and Bones, Wood		*Forest Marble*	18
Pear Encrinus, Terebratulæ, Ostreæ		*Clay over the Upper Oolite*	19
Madreporæ		*Upper Oolite*	20
Modiolæ, Cardia		*Fuller's Earth & Rock*	21
Madreporæ, Trochi, Nautilus, Ammonites, Pecten		*Under Oolite*	22
Ammonites, Belemnites as in the under oolite		*Sand*	23
Numerous Ammonites		*Marlstone*	24
Belemnites, Ammonites in mass	*Marl Vales*	*Blue Marl*	25
Pentaerini, Numerous Ammonites, Plagiostoma, Ostreæm, Bones		*Lias*	26, 27
		Red Marl	28
Madreporæ, Encrini in masses, Producti		*Redland Limestone*	29
Numerous vegetables, ferns lying over the coal	*Coal Tract*	*Coal Measures*	30
Madreporæ, Encrini in masses, Producti, Trilobites	*Mountainous*	*Mountain Limestone*	31
		Red Rhab & Dunstone	32
		Killas	33
		Granite, Sienite & Gneiss	34

William Smith produced his first table of strata in 1799 alongside his *Geological Map of Bath*. A coloured table of strata also appeared in the memoir accompanying his seminal 1815 map *A Delineation of the Strata of England and Wales with part of Scotland*. The recreated table below (see also pp 164–65) was published in 1817 and offers Smith's most comprehensive account of the succession of strata, their associated fossils and the possible uses of each rock type. The first column lists the fossils that can be used to identify each stratum. The second gives the names of the strata applied by Smith in relation to his own geological collection, which sometimes differ from the names he used in his 1815 memoir; these are listed in the fourth column with notes on their characteristics. Some of the strata names are outdated; for example, Under Oolite is now known as Inferior Oolite. The third column shows the colours applied by Smith to the strata, which accord with the colours he used on the 1815 map, and the fifth column describes the industrial products that can be made from each. The final column identifies the strata found in each county, as noted by Smith in the second edition of the table. Of the forty-two counties listed here, some, such as Cumberland, no longer exist. Smith went on to publish twenty-one dedicated geological county maps.

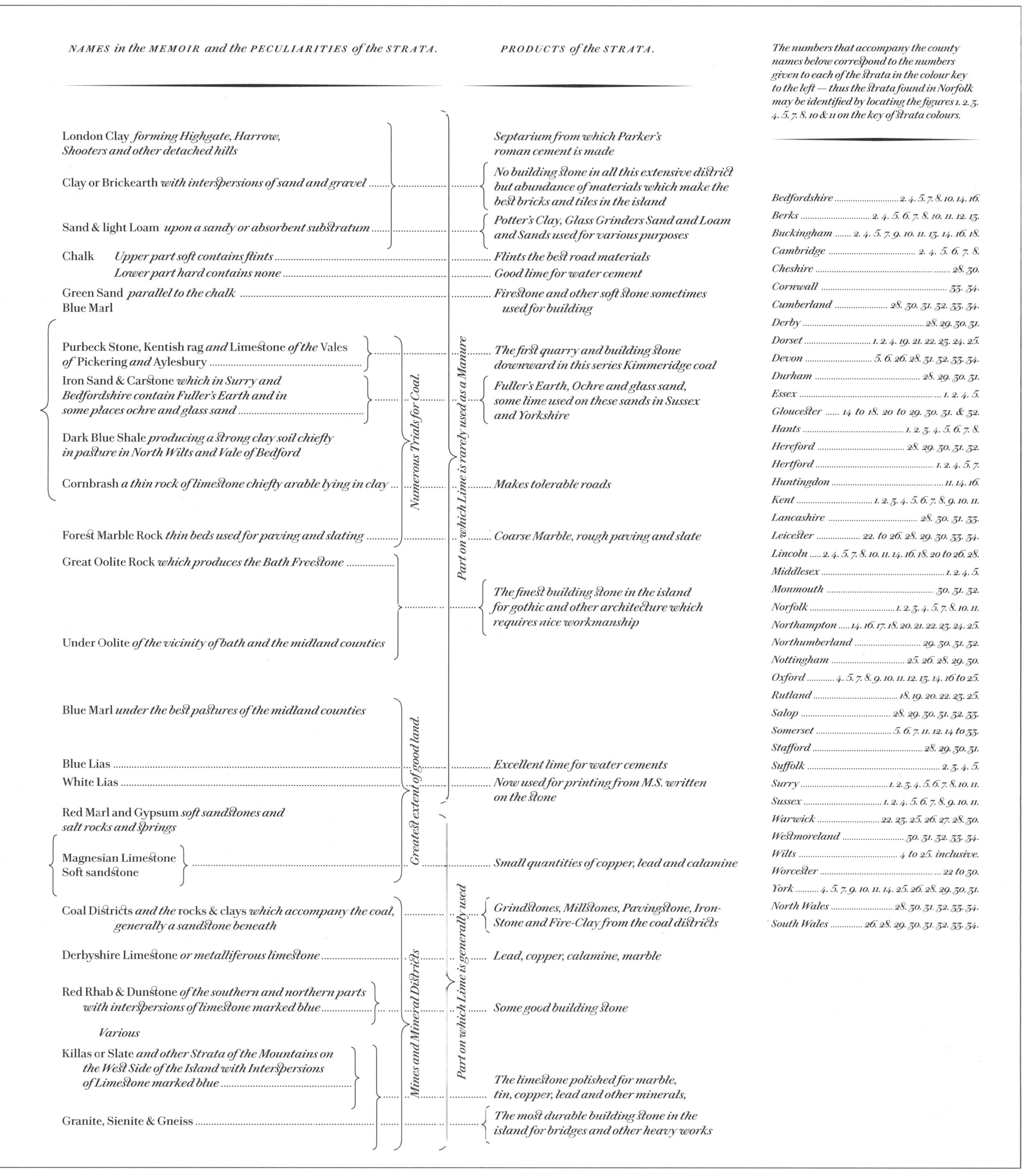

NAMES in the MEMOIR and the PECULIARITIES of the STRATA.		*PRODUCTS of the STRATA.*
London Clay *forming Highgate, Harrow, Shooters and other detached hills*		*Septarium from which Parker's roman cement is made*
Clay or Brickearth *with interspersions of sand and gravel*		*No building stone in all this extensive district but abundance of materials which make the best bricks and tiles in the island*
Sand & light Loam *upon a sandy or absorbent substratum*		*Potter's Clay, Glass Grinders Sand and Loam and Sands used for various purposes*
Chalk *Upper part soft contains flints*		*Flints the best road materials*
Chalk *Lower part hard contains none*		*Good lime for water cement*
Green Sand *parallel to the chalk*		*Firestone and other soft stone sometimes used for building*
Blue Marl		
Purbeck Stone, Kentish rag *and* Limestone *of the* Vales *of* Pickering *and* Aylesbury	*Numerous Trials for Coal.*	*The first quarry and building stone downward in this series Kimmeridge coal*
Iron Sand & Carstone *which in Surry and Bedfordshire contain Fuller's Earth and in some places ochre and glass sand*	*Numerous Trials for Coal.*	*Fuller's Earth, Ochre and glass sand, some lime used on these sands in Sussex and Yorkshire*
Dark Blue Shale *producing a strong clay soil chiefly in pasture in North Wilts and Vale of Bedford*	*Numerous Trials for Coal.*	
Cornbrash *a thin rock of limestone chiefly arable lying in clay*	*Numerous Trials for Coal.*	*Makes tolerable roads*
Forest Marble Rock *thin beds used for paving and slating*	*Numerous Trials for Coal.*	*Coarse Marble, rough paving and slate*
Great Oolite Rock *which produces the Bath Freestone*		*The finest building stone in the island for gothic and other architecture which requires nice workmanship*
Under Oolite *of the vicinity of bath and the midland counties*		
Blue Marl *under the best pastures of the midland counties*	*Greatest extent of good land.*	
Blue Lias	*Greatest extent of good land.*	*Excellent lime for water cements*
White Lias	*Greatest extent of good land.*	*Now used for printing from M.S. written on the stone*
Red Marl and Gypsum *soft sandstones and salt rocks and springs*	*Greatest extent of good land.*	
Magnesian Limestone / Soft sandstone	*Greatest extent of good land.*	*Small quantities of copper, lead and calamine*
Coal Districts *and the* rocks & clays *which accompany the coal, generally a sandstone beneath*	*Mines and Mineral Districts*	*Grindstones, Millstones, Pavingstone, Iron-Stone and Fire-Clay from the coal districts*
Derbyshire Limestone *or metalliferous limestone*	*Mines and Mineral Districts*	*Lead, copper, calamine, marble*
Red Rhab & Dunstone *of the southern and northern parts with interspersions of limestone marked blue*	*Mines and Mineral Districts*	*Some good building stone*
Various		
Killas or Slate *and other Strata of the Mountains on the West Side of the Island with Interspersions of Limestone marked blue*	*Mines and Mineral Districts*	*The limestone polished for marble, tin, copper, lead and other minerals,*
Granite, Sienite & Gneiss	*Mines and Mineral Districts*	*The most durable building stone in the island for bridges and other heavy works*

Part on which Lime is rarely used as a Manure (London Clay to White Lias); *Part on which Lime is generally used* (Red Marl and Gypsum to Granite, Sienite & Gneiss).

The numbers that accompany the county names below correspond to the numbers given to each of the strata in the colour key to the left — thus the strata found in Norfolk may be identified by locating the figures 1. 2. 3. 4. 5. 7. 8. 10 & 11 on the key of strata colours.

Bedfordshire	2. 4. 5. 7. 8. 10. 14. 16.
Berks	2. 4. 5. 6. 7. 8. 10. 11. 12. 13.
Buckingham	2. 4. 5. 7. 9. 10. 11. 13. 14. 16. 18.
Cambridge	2. 4. 5. 6. 7. 8.
Cheshire	28. 30.
Cornwall	33. 34.
Cumberland	28. 30. 31. 32. 33. 34.
Derby	28. 29. 30. 31.
Dorset	1. 2. 4. 19. 21. 22. 23. 24. 25.
Devon	5. 6. 26. 28. 31. 32. 33. 34.
Durham	28. 29. 30. 31.
Essex	1. 2. 4. 5.
Gloucester	14 to 18. 20 to 29. 30. 31. & 32.
Hants	1. 2. 3. 4. 5. 6. 7. 8.
Hereford	28. 29. 30. 31. 32.
Hertford	1. 2. 4. 5. 7.
Huntingdon	11. 14. 16.
Kent	1. 2. 3. 4. 5. 6. 7. 8. 9. 10. 11.
Lancashire	28. 30. 31. 33.
Leicester	22. to 26. 28. 29. 30. 33. 34.
Lincoln	2. 4. 5. 7. 8. 10. 11. 14. 16. 18. 20 to 26. 28.
Middlesex	1. 2. 4. 5.
Monmouth	30. 31. 32.
Norfolk	1. 2. 3. 4. 5. 7. 8. 10. 11.
Northampton	14. 16. 17. 18. 20. 21. 22. 23. 24. 25.
Northumberland	29. 30. 31. 32.
Nottingham	25. 26. 28. 29. 30.
Oxford	4. 5. 7. 8. 9. 10. 11. 12. 13. 14. 16 to 25.
Rutland	18. 19. 20. 22. 23. 25.
Salop	28. 29. 30. 31. 32. 33.
Somerset	5. 6. 7. 11. 12. 14 to 33.
Stafford	28. 29. 30. 31.
Suffolk	2. 3. 4. 5.
Surry	1. 2. 3. 4. 5. 6. 7. 8. 10. 11.
Sussex	1. 2. 4. 5. 6. 7. 8. 9. 10. 11.
Warwick	22. 23. 25. 26. 27. 28. 30.
Westmoreland	30. 31. 32. 33. 34.
Wilts	4 to 25. inclusive.
Worcester	22 to 30.
York	4. 5. 7. 9. 10. 11. 14. 25. 26. 28. 29. 30. 31.
North Wales	28. 30. 31. 32. 33. 34.
South Wales	26. 28. 29. 30. 31. 32. 33. 34.

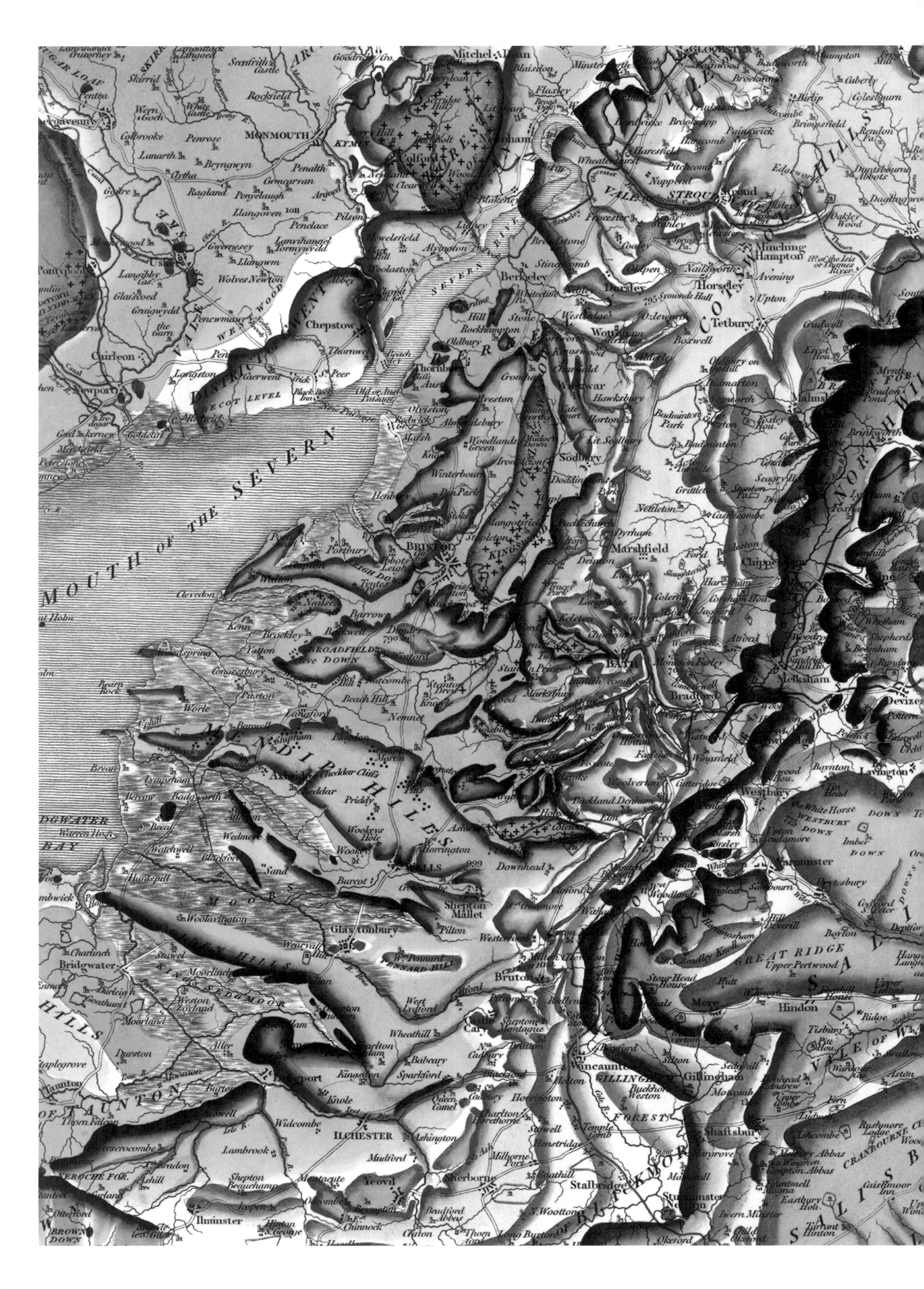

MONMOUTH
Chepstow
Caerleon
Newport
Usk
Pontypool
Ragland
BRISTOL
BATH
Bradford
Melksham
Devizes
Chippenham
Tetbury
Berkeley
Dursley
Sodbury
Marshfield
Thornbury
Clevedon
WELLS
Glastonbury
Shepton Mallet
Bridgwater
ILCHESTER
Yeovil
Sherborne
Wincaunton
Bruton
Gillingham
Shaftsbury
Westbury
Warminster
Frome
Stalbridge
Ilminster
Hindon
Mere
MOUTH OF THE SEVERN
MENDIP HILLS
COTSWOLD HILLS
STROUD
Stroud
VALE OF
DISTRICT OF WENT
SEVERN RIVER
KINGSWOOD
BROADFIELD
GREAT RIDGE
KINGS SEDGMOOR
Cheddar Cliffs
Cheddar
Axbridge
Priddy
Wookey Hole
Burcot
Shipham
Blagdon
Woolavington
Moorlinch
Weston Zoyland
Langport
Kingsdon
Sparkford
Montacute
Mudford
Chinnock
Bradford Abbas
Milborne Port
Henstridge
Templecomb
Stour Head House
Upper Pertwood
Hengestridge
Minchinghampton
Horsley
Nailsworth
Avening
Upton
Badminton Park
Malmsbury
Brinkworth
Nettleton
Castlecombe
Dyrham
Colerne
Corsham Hou.
Monkton Farley
Hinton
Farleigh
Woolverton
Westbury Down
White Horse
Imber Down
Heytesbury
Codford St. Peter
Tisbury
Ashcombe
Rushmore Lodge
Melbury Abbas
Compton Abbas
Iwern Minster
Sturminster Newton
Tarrant Hinton
Eastbury
Ashill
Shepton Beauchamp
Thorn Falcon
Taunton
Durston
Aller
Moorland
Stawel
Charlinch
Brean
Berrow
Lympsham
Badgworth
Wedmore
Blackford
Huntspill
Watchwell
Bleydon
Banwell
Worle
Uphill
Congresbury
Yatton
Kenn
Brockley
Backwell
Barrow
Dundry
Nailsea
Portbury
Abbots Leigh
Portishead
Henbury
Winterbourn
Almondsbury
Olveston
Aust
Oldbury
Rockhampton
Stone
Wickwar
Hawksbury
Horton
Kingswood
Wotton under Edge
Tetbury
Dudmarton
Oldbury on the Hill
Pensford
Stanton Drew
Marksbury
Camerton
Chewton
Ashwick
Downhead
Doulting
Pilton
West Lydford
Wheathill
Babcary
Castle Cary
Shepton Montague
North Cadbury
Queen Camel
Charlton Horethorne
Holton
Horsington
Stowell
Goathill
Pitcombe
Redlynch
Batcombe
Witham
Gloucester
Painswick
Haresfield
Pitchcomb
Frocester
Stinchcomb
Slimbridge
Newnham
Blakeney
Lidney
Alvington
Woolaston
Tidenham
Mitchel Dean
Flaxley
Blaisdon
Colford
Goodrich
Skirrid
Abergavenny
Llanarth
Bryngwyn
Penrose
Penalth
Trelech
Llangovan
Pilston
Hewelsfield
St. Briavels
Penhow
Langstone
Caerwent
St. Peer
Goldcliff
Peterstone
Magor

Northleach
Burford
Witney
Woodstock
Stonesfield
WHICHWOOD FOREST
OXFORD
Thame
VALE OF AYLESBURY
AYLESBURY
Lechlade
Abingdon
Wallingford
Watlington
STOKEN CHURCH HILLS
Slough
West Wycombe
High Wycombe
Great Marlow
CHILTERN FOREST
Henley upon Thames
VALE OF WHITE HORSE
WHITE HORSE HILL
Wantage
Swindon
Wotton Basset
ILSLEY DOWNS
LAMBOURN DOWNS
E.t Ilsley
Lambourn
GORING HEATH
READING
Oakingham
Hungerford
Newbury
Marlborough
MARLBRO'
DOWNS
SAVERNACLE FOREST
VALE OF PEWSEY
WINDSOR
HAMPSHIRE
Kingsclere
Basingstoke
Odiham
Farnham
Ludgarshall
Whitchurch
Andover
Amesbury
Stockbridge
SALISBURY
Wilton
WORTHY DOWN
WINCHESTER
New Alresford
Alton
ALICE HOLT
WOOLMER FOREST
Petersfield
Romsey
Bishops Waltham
SOUTHAMPTON
Fordingbridge
Downton
Broad Chalk

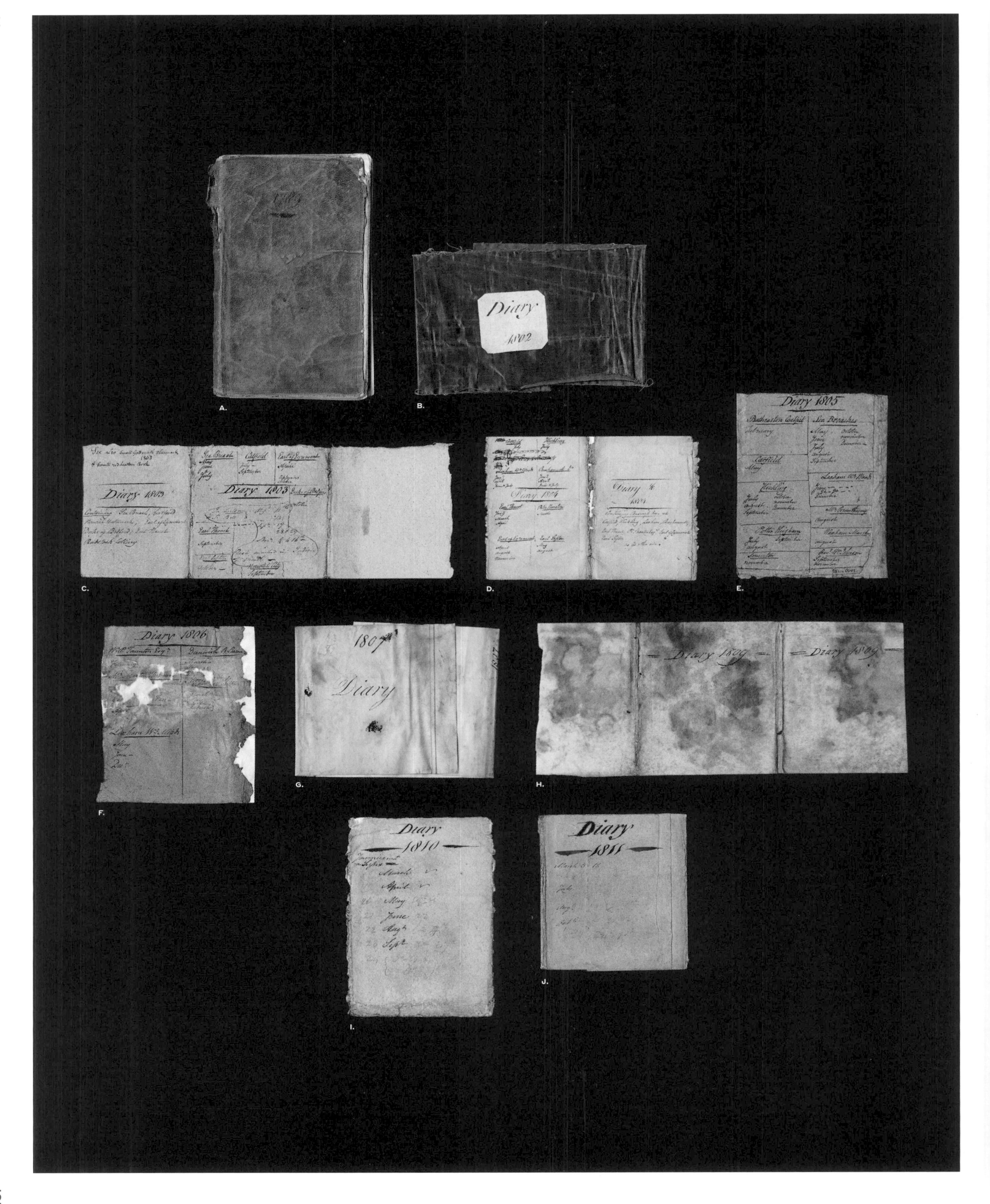

A. Front cover of William Smith's 1789 diary.
B. Front cover of William Smith's 1802 diary.
C. First page of William Smith's 1803 diary.
D. First page of William Smith's 1804 diary.
E. First page of William Smith's 1805 diary.
F. First page of William Smith's 1806 diary.
G. Front cover of William Smith's 1807 diary.
H. Cover of William Smith's 1809 diary.
I. First page of William Smith's 1810 diary.
J. First page of William Smith's 1811 diary.

K. Cover of William Smith's 1812 diary.

L. Cover of William Smith's 1813 diary.

M. First page of William Smith's 1814 diary.

N. First page of William Smith's 1815 diary.

O. Front cover of William Smith's 1816 diary.

P. First page of William Smith's 1817 diary.

Q. First page of William Smith's 1818 diary.

R. First page of William Smith's 1819 diary.

S. First page of William Smith's 1820 diary.

T. First page of William Smith's 1821 diary.

FOREWORD.

We know so little of the worlds beneath our feet. Stand in open country on a clear night, look up, and you will see through troposphere, stratosphere, mesosphere and into outer space. Light will reach your eyes that has travelled for thousands of years across trillions of miles from dying stars. Planets, satellites and galaxies are all within your vision. Look down, though, and you will see as far as the tips of your toes. The Earth's skin stops sight short. What is beneath grass or ground is lost to view and largely lost to knowledge. We stand atop a vast, invisible dominion of soil and rock, rifts and mines, chambers, veins, minerals and groundwater, to the fragile roof of which we are moored by gravity – and yet we give it hardly any thought.

This underworld keeps its secrets well. Only in the past decade, for instance, have scientists begun to comprehend the extent and diversity of the so-called 'deep earth biome': a vast microbial ecosystem, nicknamed the 'subterranean Amazon', that exists within the Earth's crust and contains around three hundred times the biomass of all humans presently alive on the planet. Only in the past twenty years have forest ecologists discovered proof of the 'Wood Wide Web', the immense network of mycorrhizal fungi that connects trees below ground into intercommunicating forests. Globally, the total length of mycorrhizal mycelium in only the top 10 centimetres of soil is enough to span half the width of the galaxy.

Accessing the underworld has always been effortful work. This is true in classical myth as well as in modern science; the many ancient Greek and Roman stories of underworld journeys, from Orpheus to Persephone, all emphasize the hardship involved in gaining access to what lies beneath. But these stories also agree that what one discovers below the surface is often wondrous, and sometimes revelatory.

It is clear that William Smith (1769–1839) was a practical man, not given to melodrama or hyperbole. It is very unlikely that he ever imagined himself as an explorer or pioneer, let alone a figure from myth. But truly, he was a terranaut – a deep-time visionary who taught himself to see down into bedrock and crust, who learnt how to read what Charles Dickens (1812–70) once called 'The Great Stone Book' of Earth history, with strata its pages and fossils its words. It is remarkable, approaching miraculous, that Smith did so in the years around the end of the eighteenth century, when geology scarcely existed as a science, and when the biblical orthodoxy of an Earth whose age could be measured in thousands, not billions, of years was still influential. He achieved this by means of a monumental effort of what archaeologists refer to as 'ground-truthing'; that is to say, by walking and riding the landscape in person over many years – and in this way gathering first-hand local data in order to build up his three-dimensional vision of Britain.

Like many people, I am a cartophile. I collect maps, and books of maps. I learnt to read maps at an early age, and I have continued to do so through a lifetime of mountaineering and hill-walking. Smith's 1815 masterpiece is certainly one of the most beautiful maps that I know,

FIG. 1.

Fig 1. William Smith portrayed in 1837, aged 68. Son of a blacksmith, with only a rudimentary formal education, Smith was largely overlooked by the scientific community for most of his working life. It was not until 1831 that he received recognition for his accomplishments when he was awarded the first Wollaston medal by the Geological Society of London. It was on this occasion that the president, Adam Sedgwick, referred to him as the 'father of English geology'.

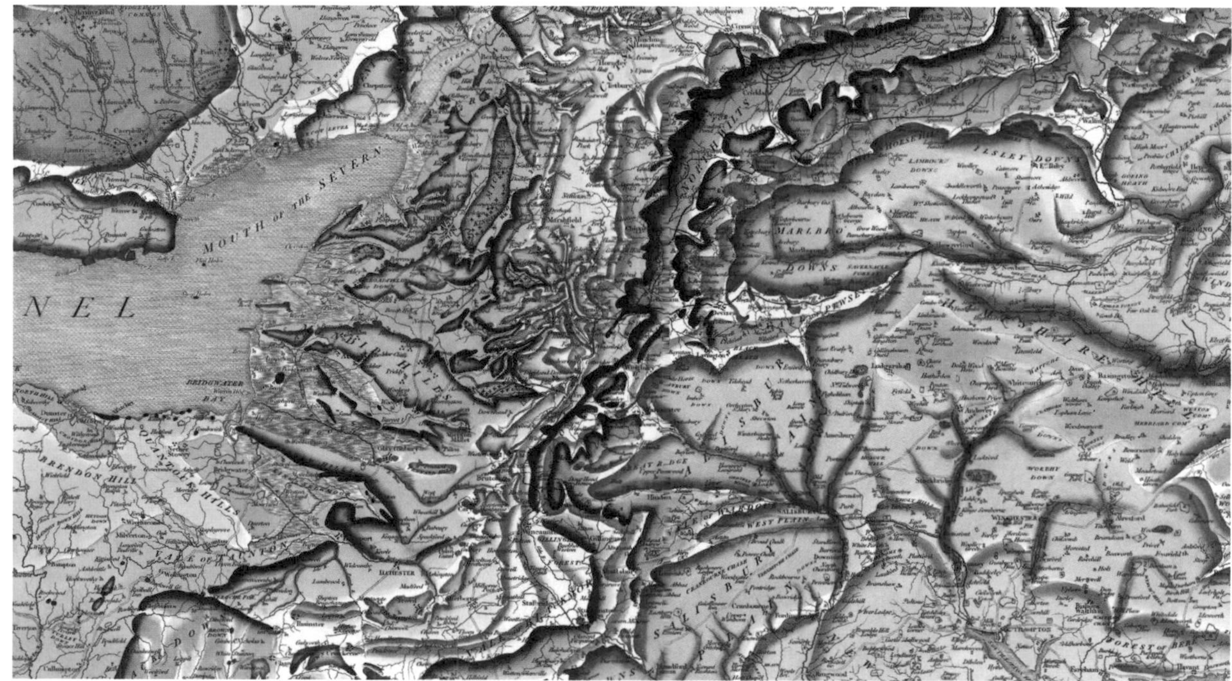

FIG. 2.

Fig. 2.
A section from Sheet 11 of Smith's 1815 geological map, including Wiltshire, Gloucestershire and part of the Bristol Channel. Smith's technicolour markings make visible a rich variety of strata, from the moss-green that indicates Chalk to the bright yellow of Upper Oolite and the peachy pink of Red Marl.

both conceptually and aesthetically. I first saw a full-size version of it in the Earth Sciences Library at Cambridge. Encountered in person, the map is unmistakably a work of art as well as a document of record. The landscape of England has gained a fabulous plumage; become defamiliarized, made brightly strange by its allegiance to what lies beneath. I remember in particular following the moss-green that Smith used to designate the chalk deposits of south-eastern England, on which I've lived and walked for a quarter-century or so. His map lays strikingly bare the continuous extent of the chalk deposit, which runs from the north Norfolk coast south-west all the way to Dorset, and extends eastwards in two huge jaw-lines that we know now as the North Downs and the South Downs. Seeing that great green shape transformed my sense of the land I live on.

Though born chiefly of a pragmatic urgency to exploit the Earth's resources, Smith's map now exists somewhere between artwork, dreamwork and data-set. It gives its readers trilobite-eyes, allowing them to see back into ancient Earth history and glimpse something of how profoundly this buried past shapes the surface world. It reminds us, humblingly, of the inhumanly long periods of geological time which extend both behind and ahead of the present moment, and which our imaginations still find so hard to grasp. The *longues durées* of Earth history 'maketh Pyramids pillars of snow, and all that's past a moment', as Thomas Browne (1605–82) memorably puts it in his 1658 meditation on time and the underworld, *Urne-Buriall.*

Smith was one of the first people to see *in* strata. As such, his work bears much more than a purely historical or aesthetic relevance. Contemporary stratigraphers are currently assessing whether the Earth has entered a new geological epoch, the 'Anthropocene': a period in which human activity has become so forceful in its world-shaping activities that it will leave a long-term signature in the future strata of the planet. In 2009 an Anthropocene Working Group was established by the International Commission on Stratigraphy, tasked with assessing *Homo sapiens*' likely legacy in the rock record. Its current recommendation is that we have entered the Anthropocene epoch, stratigraphically speaking, and that we did so around 1950, when nuclear testing began to disperse radionuclides world-wide, thereby creating an immensely durable 'horizon marker' in the strata.

I have occasionally wondered what it would have been like to be in the field with William Smith; to join this practical, prophetic man on one of his thousands of early morning 'walkings-out', pacing the land with notebook and hammer in hand, peering at the ground, prying at fossils. Smith collected and catalogued ammonites and belemnites, corals and brachiopods, as he unfolded the hidden history of the Earth. Our own epoch is already creating its fossils-to-be. It occurs to me that among the relics that a far-future William Smith might discover, were he to be prospecting the strata of the Anthropocene, would be the trace-fossils of billions of plastic bottles, chicken and swine bones in fabulous abundance, the crushed rubble of our cities, and a curious concentration of Lead-207, the stable isotope at the end of the Uranium-235 decay chain.

Timeline of Events in the Life of William Smith (1769–1839).

1791
In October Smith is sent by Webb to make a survey and valuation of Lady Elizabeth Jones' estate at Stowey in north Somerset.

1792
For more than three years Smith works on Lady Jones' estate, lodging at Rugborne Farm, near High Littleton. The estate he surveys includes coal properties; as he investigates coal seams he takes note of the subterranean geology.

1793
The proposal for a canal to link Bath on the River Avon with Newbury on the River Kennet leads some of the owners of the north Somerset coal mines to consider building a local canal to join it. Smith is engaged to make a survey and take levels along two converging valleys running north-eastward to the Avon. This provides him with an opportunity to confirm his ideas about the regular south-easterly dip of the strata in the county.

1797
Smith makes his first attempt at an order of strata, beginning with Chalk and descending to Limestone, the stratum below Coal Measures.

1798
Smith buys a cottage at Tucking Mill, between Midford and Monkton Combe in Somerset, which remains in his possession until 1819.

1799
Smith's employment as surveyor to the Coal Canal Company is terminated in June. He then works as an adviser on draining and irrigation, and writes his table of strata around Bath with their accompanying fossils. A map of the country around Bath 'coloured geologically in 1799' by Smith is now in the possession of the Geological Society; there are three colours, representing the Bath Oolite, the Lias and the Trias. A geologically coloured map of Somerset, prepared by Smith around the same time, no longer exists.

1800
Smith is employed by Thomas Coke of Holkham in Norfolk, and the following year by the Duke of Bedford at Woburn.

1812
London map engraver and publisher John Cary undertakes to publish Smith's geological map of England and Wales, on a scale of 5 miles to the inch. New plates are engraved to Smith's specification, based on Cary's 1794 map of England and Wales.

1813–1814
Smith adds geological lines to the 1812 map plates. Even after the map's publication he continues to make alterations and additions.

1813
Joseph Townsend publishes *The Character of Moses* in which Smith's stratigraphical system is followed and described.

1814
Smith works on draining the Minsmere marshes in Suffolk, and is arrested twice in London for debt.

1817
Part 3 of *Strata Identified* is published in June. During the next two months Smith also publishes *Geological Section from London to Snowdon*, *Geological Table of British Organized Fossils* and Part 1 of *Stratigraphical System of Organized Fossils*. The last is intended to explain his collection when it is exhibited in the British Museum. Part 2 never appeared. In August he offers the museum a further collection of fossils, which is purchased in 1818.

1818
Dr W. H. Fitton draws attention to Smith's achievements and commends his map. In June Smith issues a lithographed pamphlet stating his claim to the 'Discovery of constancy in the order of superposition and continuity in the course of British Strata with the peculiar mode of identifying them by organized Fossils embedded'.

1819
Eight maps of English counties, coloured geologically by Smith, are published by Cary; these are followed by maps of a further seventeen counties during the next five years. Also in 1819 five large geological sections across different parts of southern England are issued, and Part 4 of *Strata Identified* is published. In June Smith is consigned for debt to a ten-week stay in the King's Bench Prison in London.

1828
Smith moves to Hackness, 8 km (5 miles) from Scarborough, in Yorkshire. While living here he maps on a scale of $6\frac{1}{2}$ inches to the mile the Jurassic rocks of the Hackness or Tabular Hills.

1829
Sir John Vanden Bempde Johnstone, President of the Scarborough Philosophical Society, raises funds for the building of the Rotunda Museum. The museum is designed by William Smith to house a collection of Jurassic geology, and is built the same year.

1831
Smith is selected by the Council of the Geological Society of London as the first recipient of the Wollaston Medal 'in consideration of his being a great original discoverer in English geology'.

1769
William Smith is born on 23 March in Churchill, north Oxfordshire.

1777
Smith's father dies.

1779
Smith's mother remarries; her second husband is Robert Gardner of Churchill, tailor and landlord of the Chequers Inn.

1787
The survey and valuation of Churchill is begun by Edward Webb, and William Smith – now eighteen – is engaged as his assistant. He lives with the Webb family at Stow-on-the-Wold for nearly five years, while he is an apprentice surveyor.

1794
Smith visits London with members of the Canal Committee to give evidence before parliament. The Somersetshire Coal Canal Bill is passed and later in the year Smith accompanies two members of the committee on a tour to the north of England and back to inspect other canals and coal mines. This provides him with evidence for his idea that the strata with which he is already familiar stretch north-eastward to Yorkshire, always dipping to the south-east.

1795
Construction on the Somersetshire Coal Canal begins in the summer, with excavations on two branches designed to meet 2 miles (3.2 km) before the valley of the Avon is reached. In September Smith moves to Cottage Crescent, a mile south-west of Bath.

1796
In January, while at the Swan Inn on the main road to Bath near Dunkerton, Smith writes some notes that show he is already convinced that particular fossils are characteristic of particular strata. During the year, further canal excavations, including a deep one for a caisson, expose sections in the Fuller's Earth clays.

1801
In June, Smith issues a *Prospectus* of a proposed work on the strata of England and Wales. About this time he also creates two maps of England and Wales coloured to show the principal strata. In the summer Coke introduces Smith to Francis, Duke of Bedford, then trying to drain some of his Woburn estates with the assistance of his land steward John Farey. Smith and Farey meet in October 1801 and Smith's results greatly impress Farey with their novelty and economic importance.

1802
Smith rents offices on the east side of Trim Bridge Bath. Farey introduces Smith and his work to Sir Joseph Banks.

1804
Smith takes lease of much larger premises in London, in Buckingham Street, off the Strand. He becomes involved as an adviser to the Batheaston Coal Trial.

1805
Smith visits the site of another intended colliery at Cook's Farm at South Brewham, Somerset.

1808
Members of the Geological Society visit Smith's house in London to see his collection of fossils, arranged on sloping shelves to indicate the strata they are found in.

1815
Smith's completed geological map, *A Delineation of the Strata of England and Wales, with part of Scotland*, is exhibited in London and Smith receives an award of fifty guineas offered for a 'Mineralogical Map' by the Society for the Encouragement of Arts, Manufacture and Commerce. Twenty-three different strata are depicted, in twenty-one different colours. By using a deeper shade at the base of a particular stratum, the general dip is also shown. The map is accompanied by a memoir.

1816
By January 200 copies of the map have been coloured and issued to subscribers. A deep well dug near Swindon enables Smith to get a clearer idea of the succession of strata in the Upper Jurassic, with Oaktree Clay overlaying and Clunch Clay underlying Coral Rag. On copies of his map issued after this Coral Rag was added, using an orange shade in Berkshire, Oxfordshire, Somerset and Wiltshire. Later still, possibly in 1817, he added it in Yorkshire. Also in 1816 Smith's collection of fossils, arranged stratigraphically, is bought for the British Museum and transferred there in June. In June and September the first two parts of his *Strata Identified by Organized Fossils, containing Prints on Coloured Paper of the most characteristic specimens in Each Stratum* are issued.

1820
Cary publishes *A New Geological Map of England and Wales* by Smith, on a much smaller scale, 15 miles to the inch. Compared with the 1815 map it has some additions but fewer strata are shown in separate colours. Certain major errors in Smith's earlier maps (notably the correlation of the Weald Clay with the Kimmeridge and the placing of certain areas of Carboniferous Limestone above the Coal Measures) are very clearly shown on this map. From 1820 onwards Smith resides in the north of England, especially Yorkshire. G. B. Greenough publishes the Geological Society's geological map of England and Wales.

1824–1825
Smith lectures on geology in several Yorkshire towns, assisted by his nephew John Phillips.

1832
Smith's map of the Hackness or Tabular Hills is published. An annuity of £100 a year is granted to him by the government.

1834
Smith moves to Scarborough, Yorkshire.

1835
Trinity College, Dublin, confer on Smith an honorary doctorate of laws (LL.D.).

1838
Smith takes an active part as member of a small committee which tours England and Scotland in search of a suitable freestone for the building of the new Houses of Parliament.

1839
Aged 70, and in good health, Smith catches a chill while staying in Northampton on his way to a meeting of the British Association for the Advancement of Science in Birmingham. He dies within a few days, on 28 August, and is buried in the churchyard of St Peter's Church, Northampton.

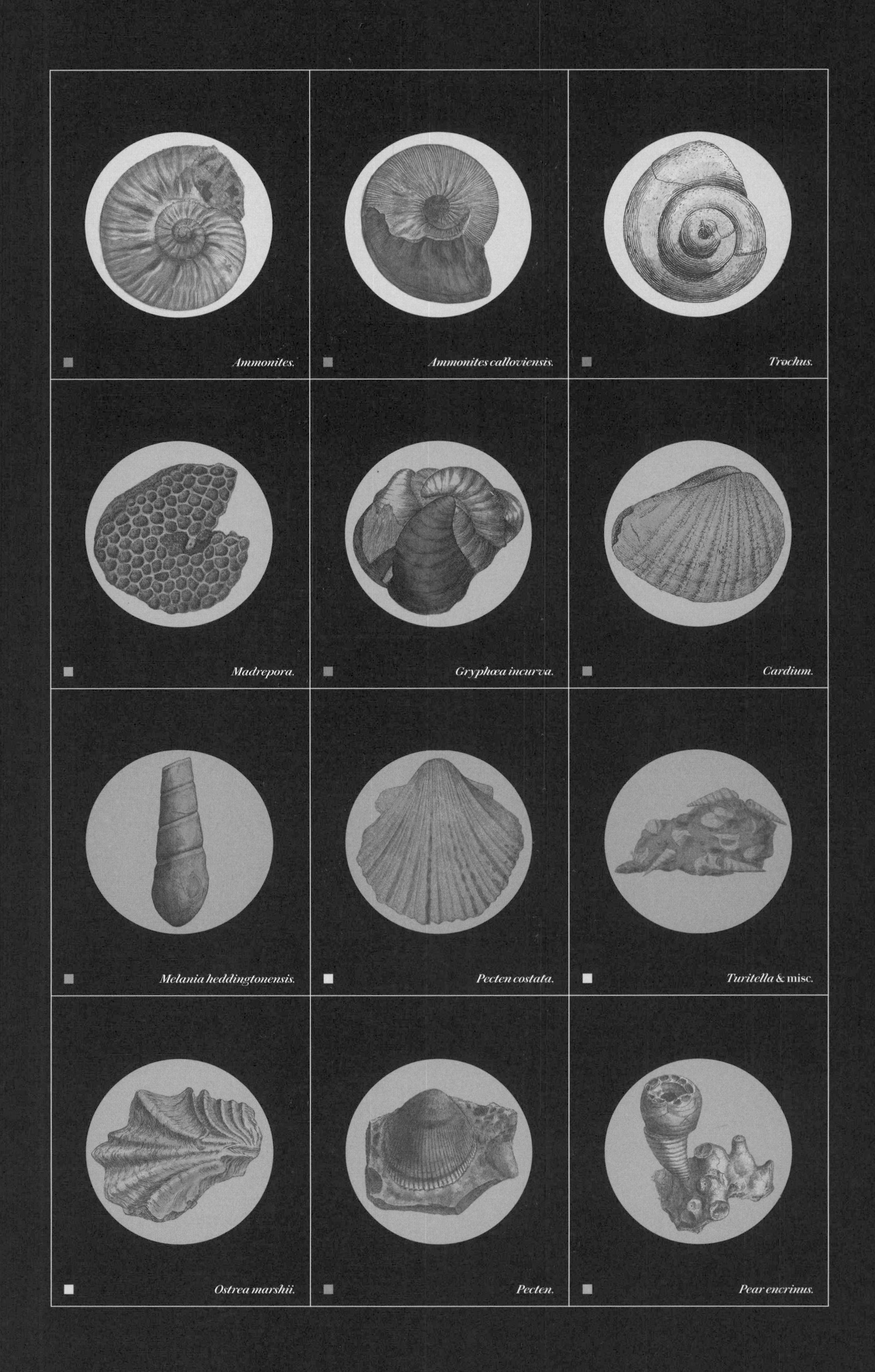
Ammonites.
Ammonites calloviensis.
Trochus.
Madrepora.
Gryphœa incurva.
Cardium.
Melania heddingtonensis.
Pecten costata.
Turitella & misc.
Ostrea marshii.
Pecten.
Pear encrinus.

I	N	T
R	O	D
U	C	T
I	O	N

A Delineation of the Strata of England and Wales, with part of Scotland: exhibiting the collieries and mines, the marshes and fen lands originally overflowed by the sea, and the varieties of soil according to the variations in the substrata, illustrated by the most descriptive names might not be the most concise of titles, but it does accurately describe the pioneering map that William Smith (1769–1839) published in 1815. With a scale of 5 miles to the inch, the map was 8 ft 6 in. high and 6 ft wide (2.5 by 1.7 m), and had to be printed as fifteen separate sheets. Its size was not the only aspect that made the map so remarkable. As the first detailed geological map of the region, it was fundamentally different from the many geological maps that had previously been produced in Britain and Europe. Rather than just another depiction of the geographic distribution of different rocks at the surface, the map also indicated the succession and underground distribution of the sedimentary strata.

Smith's map was laboriously hand coloured in twenty-three different tints to distinguish the different rock types, making it very expensive to produce. With the cheapest version costing 5 guineas, the map was never going to sell in large numbers, so it was not likely to generate much profit, either for Smith or his publisher John Cary (1754–1835). To make matters worse, 1815 was not the most propitious year for publication.

FIG. 1.

Fig. 1.
Smith's 1815 map was issued as fifteen sheets, numbered I to XV from north-west to south-east, plus an index *General Map* showing the area covered by each sheet. The *General Map*, shown here, usually survives only in bound copies of the 1815 map.

Earlier in the summer that Smith's map was published, on Sunday, 18 June, the battle of Waterloo was fought and won by the United Kingdom and its allies, but military success was followed by economic recession. Several days later, when word of the victory eventually reached North Yorkshire, a thirty-year-old Cambridge graduate by the name of Adam Sedgwick (1785–1873) announced the news to the people of his home village of Dent. Sixteen years after that, in 1831, the same Adam Sedgwick, now in his role as President of the Geological Society of London, would proclaim William Smith the 'Father of English Geology'. However, the intervening years were ones of great hardship for William Smith.

FORMATIVE YEARS.

The son of a village blacksmith, William Smith was born in Churchill, Oxfordshire, in 1769. His father died before he was eight, and William was largely brought up on the farm of his yeoman uncle, where like many small boys he collected fossils from the surrounding fields. Little detail of Smith's early life is known, but evidently he was attentive and intelligent and showed an aptitude for drawing and mathematics. Luckily for the young William, his nascent skills were noticed by a land-surveyor, Edward Webb (1751–1828) of Stow-on-the-Wold, who took him on as an assistant in 1787. These were the early years of Britain's Industrial Revolution and a period of rising political turmoil in France. By 1791 Smith was proficient enough to be entrusted with a probate evaluation survey of the estate of Lady Elizabeth Jones (1741–1800) in Somerset. Included in the estate was Sutton Court, a house that had once belonged to John Strachey (1671–1743), an Oxford-educated country squire and landowner who developed an interest in geology and its use in the search for coal.

Smith had the good fortune to be in the right place at the right time. We know that he obtained a copy of Strachey's 1719 geological section through part of the Somerset coalfield. It opened Smith's eyes to the succession of coal-bearing sedimentary rocks and their fossil content. In 1793 Smith was taken on by the Somersetshire Coal Canal Company and experienced for himself rock strata exposed in mines and canal cuttings. These gave Smith the opportunity to check Strachey's geological observations and appreciate their more general implications and potential. Smith began to develop his own geological ideas while earning his living as a peripatetic surveyor and engineer. As such, Smith became one of the foremost practitioners in the army of skilled artisans who were transforming the agricultural and industrial landscape of Britain.

From the mid-1790s, William Smith began single-handedly to gather geological information that would culminate in the 1815 publication of his map. It was a monumental task that involved searching for rock outcrops over thousands of miles of the landscape, travelling on foot or horseback and by stagecoach. Following in Strachey's footsteps, Smith used all this acquired knowledge to map strata both on the surface and below the ground. In order to do this, he developed another of Strachey's innovations, the characterization of strata by their fossils.

Smith's first-hand encounters with stratified sedimentary rocks alerted him to their fossil content. He found that in any one place, successive strata contained distinct associations of fossils. Even in a succession of similar strata, such as limestone or shale, the fossils in the lower, older, strata differed to some extent from those in higher and younger strata. Today, the role of evolution in these changes is understood, but it was not in Smith's day.

Smith assiduously collected series of characteristic fossils, and arranged them in order of occurrence. Known as 'Strata Smith', he would demonstrate to friends and potential clients how fossils could be used to identify the relative age of strata and where they lay in the overall chronology, or stratigraphic succession as it is technically known. In naming the strata he was mapping, Smith used the vernacular names of quarrymen and miners for the different rock types – from 'Killas and slate', through 'Old Red Sandstone', 'Coal', 'Magnesian Limestone' to 'Lias', 'Oolite', 'Chalk', 'Brickearth' and 'London Clay'.

In Smith's hands this understanding enabled the identification and correlation of strata across the country and became known as the 'Smithian method'. It was then incorporated into the general methodology of the mapping of fossil-bearing strata both in Britain and further afield.

GEOLOGICAL STRATA.

Strata, or layers of stratified deposits, underlie much of the British landscape and are best observed in coastal cliffs, quarries and cuttings for road and rail. These sedimentary rocks were originally formed by a variety of geological processes from the erosion of pre-existing materials on the Earth's surface, through transport by wind and water into the sea or low-lying land. The rock materials deposited in this way, mainly sediment but also volcanic products such as lava and ash, form a succession of near horizontal layers.

Often these deposits contain the remains of plants and animals which lived and died in the immediate environment. The three-dimensional

DOUGLAS PALMER.

REGION.	COLOUR.	STRATA.
PLAINS. *Nos. 1–4.*		*LONDON CLAY.*
		CLAY OR BRICKEARTH.
		CRAG.
		Sand.
CHALK HILLS. *Nos. 5–6.*		*UPPER & LOWER CHALK.*
		GREEN SAND.
CLAY VALES. *Nos. 7–15.*		*BRICKEARTH.*
		Sand.
		PORTLAND ROCK.
		Sand.
		OAKTREE CLAY.
		CORAL RAG & PISOLITE.
		Sand.
		CLUNCH CLAY & SHALE.
		KELLOWAYS STONE.
STONEBRASH HILLS. *Nos. 16–24.*		*CORNBRASH.*
		SAND & SANDSTONE.
		FOREST MARBLE.
		CLAY OVER THE UPPER OOLITE.
		UPPER OOLITE.
		FULLER'S EARTH & ROCK.
		UNDER OOLITE.
		Sand.
		MARLSTONE.
MARL VALES. *Nos. 25–28.*		*BLUE MARL.*
		BLUE LIAS.
		WHITE LIAS
		RED MARL.
COAL TRACT. *Nos. 29–30.*		*REDLAND LIMESTONE.*
		COAL MEASURES.
MOUNTAINOUS. *Nos. 31–34.*		*MOUNTAIN LIMESTONE.*
		RED RHAB & DUNSTONE.
		KILLAS.
		GRANITE, SIENITE & GNEISS.

FIG. 2.

Fig. 2.
Smith laid out these strata colours and names in his *Geological Table of British Organized Fossils*, published in 1817 (see pp. 10–11). This early tabulation of the succession of British sedimentary strata with British vernacular names laid the foundation for a stratigraphy and history of geological time, which persists to this day. At the top, the London Clay is the youngest recognized by Smith. At the bottom, the oldest is the Killas strata, today recognized as the Lower Palaeozoic strata of Cambrian to Silurian age. These are underlain by non-stratified Granite, Sienite and Gneiss.

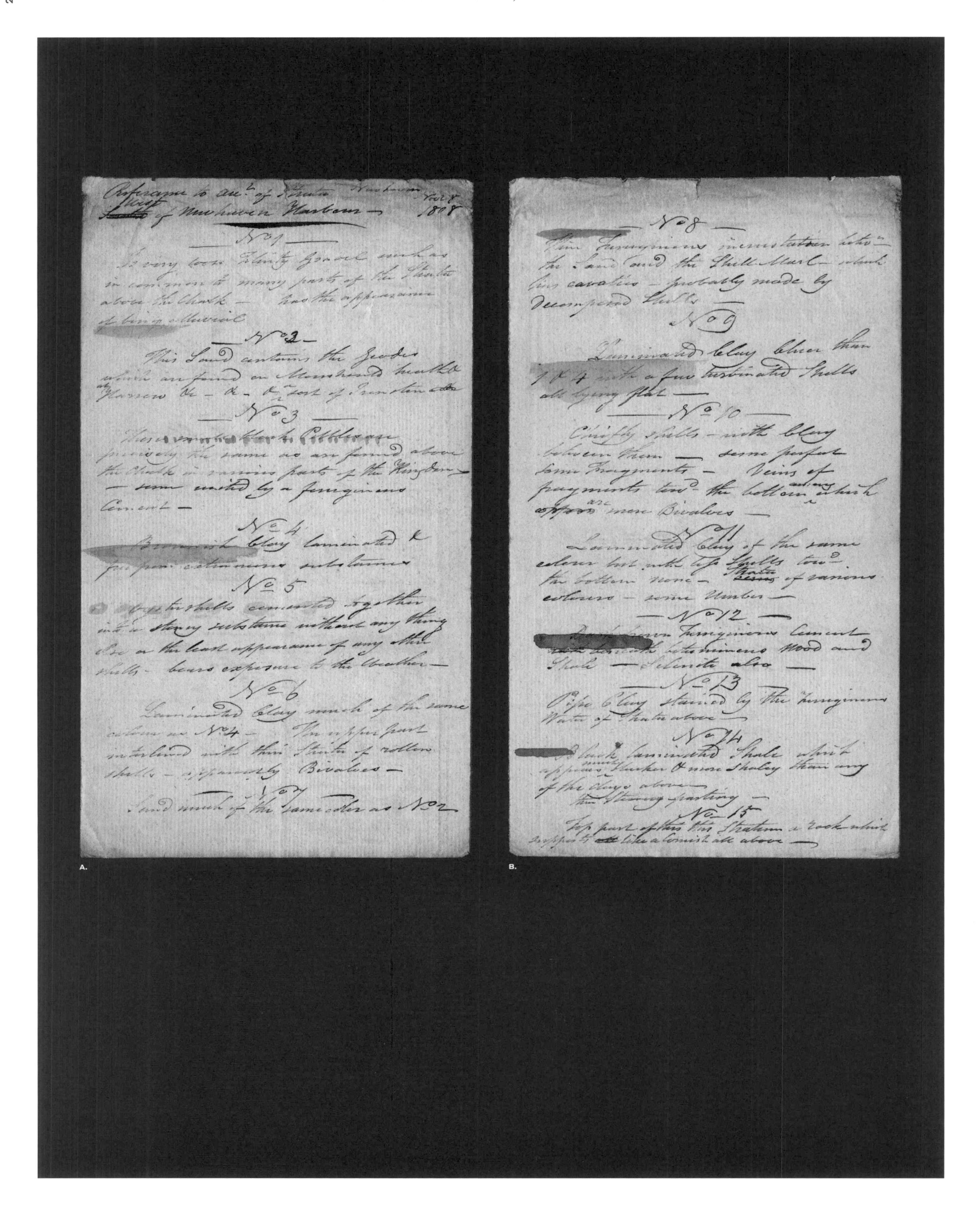

Reference to acc.t of Strata Newhaven Nov.r 8 1808
Cliff of Newhaven Harbour —

— No 1 —
A very loose Flinty Gravel such as is common to many parts of the Strata above the Chalk — has the appearance of being alluvial

— No 2 —
This Sand contains the Geodes which are found on Almondsbury heath at Harrow &c – &c – &c a sort of Ironstone [illegible]

— No 3 —
These [illegible] Pebbles precisely the same as are found above the Chalk in various parts of the Kingdom — some united by a ferruginous Cement —

No 4
Brownish Clay laminated & [illegible] extraneous substances

No 5
Oyster shells cemented together into a stoney substance without any thing else or the least appearance of any other shells – bears exposure to the Weather —

No 6
Laminated Clay much of the same colour as No 4 — The upper part interlined with thin Strata of rotten shells – apparently Bivalves —

No 7
Sand much of the same color as No 2

A.

— No 8 —
Thin ferruginous incrustation betw. the Sand and the Shell Marl which has cavities – probably made by decomposed Shells

No 9
Laminated Clay bluer than No 4 with a few turbinated Shells all lying flat —

— No 10 —
Chiefly shells – with Clay between them — some perfect some fragments – Veins of fragments tow.d the bottom among which are more Bivalves —

No 11
Laminated Clay of the same colour but with less Shells tow.d the bottom none – Strata of various colours – some Umber —

— No 12 —
[illegible] ferruginous Cement [illegible] bituminous Wood and Shale — Selenite also —

— No 13
Pipe Clay stained by the ferruginous Water of Strata above

No 14
Black laminated Shale which appears much thicker & more Shaley than any of the Clays above
the Strong parting —

No 15
Top part of this Stratum a rock which supports all like a Cornish all above —

B.

A. NOTES ON THE STRATA AT CASTLE HILL CLIFF, NEWHAVEN, SUSSEX, 1808. PAGE 1.

Today, the geological importance of this site is recognized by its designation as a protected Local Nature Reserve within a larger Site of Special Scientific Interest (SSSI). However, over 200 years ago, in November 1808, William Smith was the first geologist to examine and measure the well exposed cliff section, and detail its succession of strata in these unpublished notes.

B. NOTES ON THE STRATA AT CASTLE HILL CLIFF, NEWHAVEN, SUSSEX, 1808. PAGE 2.

Smith's manuscript notes distinguish sixteen separate strata. Fifteen lie above the Chalk, which rises from below sea level and forms the thickest stratum (40 ft / 12.2 m in a 127 ft / 38.7 m thick section). He lists and numbers the strata from the youngest to the oldest whereas today this would be reversed. In his brief descriptions Smith also notes when he has seen similar strata elsewhere.

C.

D.

C. NOTES ON THE STRATA AT CASTLE HILL CLIFF, NEWHAVEN, SUSSEX, 1808. PAGE 3.

A decade later Gideon Mantell (1790–1852) also described the Castle Hill Cliff section in his book *The Fossils of the South Downs; or Illustrations of the Geology of Sussex*. His wife Mary Ann Mantell's (1795–1869) painting of the section verified the accuracy of Smith's section. However, as Smith's section was never published his contribution to understanding this area's geology was not appreciated.

D. ILLUSTRATION OF STRATA AT CASTLE HILL CLIFF, NEWHAVEN, SUSSEX, 1808.

Smith's illustration shows a spectacular section of fossiliferous Chalk strata, above which there is a succession of younger sediments. Today, these younger sediments are recognized as Early Paleogene (Eocene age) gravels, sands and clays covered by Ice Age (Late Quaternary) river gravels and clays with flints, and finally post glacial alluvial deposits and peat.

form and scale of the deposits varies from metres to kilometres in thickness and areal extent, and reflects changes in the original environment. Buried over time, stratified deposits and their organic remains then become transformed by physical and chemical processes into layered fossiliferous rocks. When Smith was forming his ideas, the concept of stratification was still being developed.

Understanding of strata can be traced back to the later seventeenth century and the work of the Danish natural philosopher and cleric Nicolaus Steno (1638–1686). From his observations of Italian landscapes in Tuscany, Steno defined the principles of stratigraphic geology in 1669. The fundamental principle is the 'law of superposition', whereby successive surface deposits are laid down with the younger layers on top of older ones. As a result, a sequential history of deposition can be read from the succession of strata.

At this period, the European scholars who were debating the nature of strata were working within a religious tradition that saw Earth's formation in terms of the biblical creation narrative and the great flood of Noah. The latter was thought to have had a global catastrophic effect upon the geology of Earth and life, with rocks and fossil materials seen as products of that history.

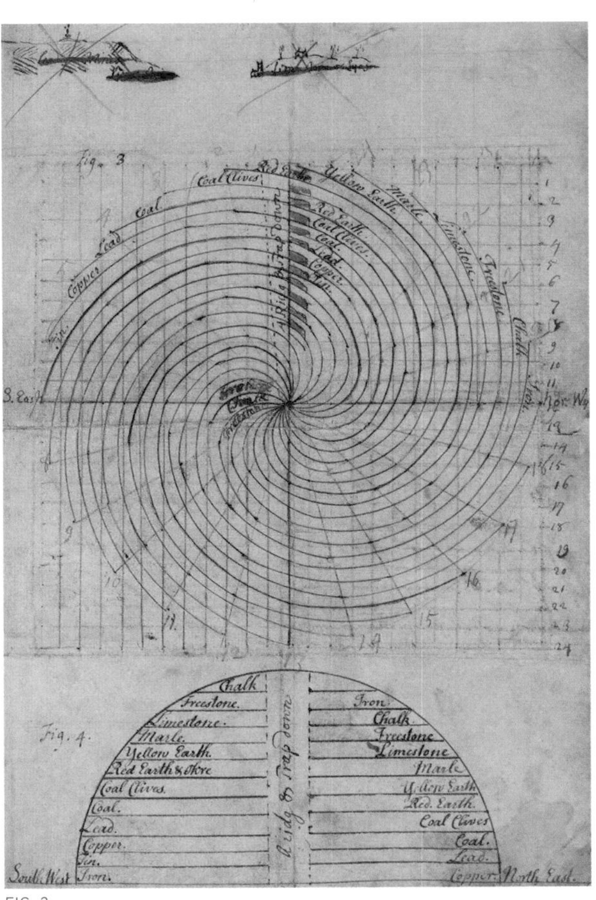

FIG. 3.

Fig. 3.
In 1724 John Strachey created this diagram illustrating an older theory based on the biblical creation story. The theory suggests that on the fourth day, Earth began to rotate and as a result, newly deposited strata were flexed to dip eastwards, extending to Earth's centre. After encountering this theory in the 1790s, Smith believed it for the rest of his life.

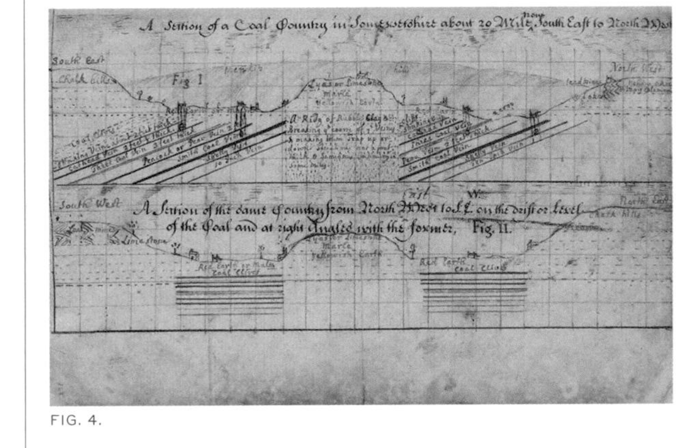

FIG. 4.

Fig. 4.
One of the most important early geological documents: John Strachey's 1719 cross section, which accurately depicts individually named coal veins (seams) dipping from north-west to south-east and displaced by a fault zone. Lying unconformably above are horizontal strata. By 1724 Strachey had added the lower 'strike section' drawn at right angles to the dip of the coal-bearing strata, which is why the strata appear to be horizontal.

JOHN STRACHEY'S INFLUENCE.

It was John Strachey's pioneering investigations of coal-bearing strata in Somerset that were so important for the development of Smith's understanding of strata. Strachey's 1719 cross section of the underground succession and structure of the coal-bearing strata is one of the most important documents in the early history of stratigraphic geology in Britain.

In the cross section Strachey enumerated the succession of coal-bearing strata, their thicknesses and fossils, for instance '*Cockle shells and Fern branches, Peacock or Peaw Veyn, about 2 feet thick*'. From the angle of dip it was possible to calculate the depth at which they would be found at different locations. Above the inclined coal beds he also showed younger strata laid horizontally and therefore unconformably on their eroded surface. In this single diagram Strachey illustrated the fundamental features that characterize the ancient strata underlying so much of the English landscape.

By 1725, Strachey had developed more general and speculative ideas, claiming that the Somerset strata could be traced north-eastwards from the Wessex coast to Northumberland. Furthermore, he stated that they were of global extent: 'All these different *Strata*, as found in any of those Places I have observed myself ... I here likewise have protracted in a globular Projection, supposing the Mass of the terraqueous Globe, to consist of the foregoing, or perhaps of ten thousand other different Minerals' (Strachey, 1727, p. 16).

Biblical texts of the Creation story underpinned his theories. Following previous speculation by English scholars such as the naturalist John Ray (1627–1705) and the antiquarian William Stukeley (1687–1765), Strachey argued that the planet's rotation had deflected all strata from their original position to decline and curve from west to east down to the Earth's centre. William Smith evidently came across these ideas in the late 1790s, as in addition to copies of Strachey's work annotated in Smith's handwriting, there is also a memorandum from 1798 in which Smith refers to Strachey's ideas as 'our theory'. Smith maintained this belief until the end of his life.

Again, the accident of place played a significant role in Smith's geological development. In tracing the geology of the Cotswold Hills from Somerset north-east towards Northumberland, Smith came across strata that seemed to confirm Strachey's theory. They do generally decline (or dip) to the south-east and pass from older to younger in that direction. The same rule also seemed to apply as Smith's mapping of southern England progressed, and again in his 1817 geological section from London to Snowdon. However, Strachey's theory did not stand up when exposed to geological evidence from elsewhere in Britain.

GEOLOGICAL MAPPING BEFORE SMITH.

Economically valuable minerals have been exploited in Europe from ancient times. By the sixteenth century advanced mining technology was being described in scholarly treatises such as the 1556 *De re metallica* by the German Georgius Agricola (1494–1555). The mapping of mineral deposits and their host rocks developed rapidly from the mid-eighteenth century.

In 1746, the French naturalist and mineralogist Jean-Étienne Guettard (1715–1786) published a map of economic mineral distribution in France, England and Wales. A decade later, more sophisticated representations of geology began to appear in Germany, with sections of mineral-bearing rocks constructed by the mineralogists Johann Gottlob Lehmann (1719–1767) and Georg Christian Füchsel (1722–1773). Lehmann proposed a new descriptive and historical grouping of rocks based on three kinds of mountain ranges. First, the high mountains of the Alps had inclined strata, abundant minerals but no fossils. Lying upon these, the second class of mountains, formed during the Noachian Deluge, contained the remains of organisms drowned in the Flood and some minerals such as coal. The third class of mountains had formed since the Deluge, and Lehmann believed that these contained rocks produced by geological events such as volcanic eruptions, storms, landslides and earthquakes.

Füchsel also studied the same mineral-rich rocks of Thuringia, which he illustrated using a novel kind of geological map. With great cartographic skill he depicted oblique-aerial views of the landscape showing the outcrop of strata with their underground relationships seen in cross section. Although not widely known, Füschel's work was as groundbreaking as Strachey's earlier work, which was constrained to two dimensions.

FIG. 5.

Fig. 5.
In 1746 J. E. Guettard drew the distribution of mineral deposits across France and part of England as symbols on a map prepared by P. Buache (1700–1773). The grey shaded area depicts *pierre blanche* (white stone) and is a very rough approximation of the outcrop of Chalk in northern France and southern England.

A FOURFOLD DIVISION OF EARTH HISTORY.

A few years later, in 1759, Giovanni Arduino (1714–1795), an Italian mining engineer, detailed the succession of strata of northern Italy. Describing mountain formation as a long drawn-out process, he proposed a fourfold division of Earth's rock formations into Alluvium, Tertiary, Secondary and Primary, the fundamentals of which were to survive into the mid-nineteenth century.

In the last decades of the eighteenth century, more than fourteen geological maps were published, mostly in Germany but also in France, Switzerland and England. Predominantly economic, the maps depict the geographical, two-dimensional distribution of minerals and rock types with little or no representation of their underground, three-dimensional structure. Whether Smith had access to any of these earlier continental maps is not known, but he may have been shown some by well-to-do friends and supporters such as the Rev. Joseph Townsend (1739–1816) and the President of the Royal Society, Sir Joseph Banks (1743–1820).

THE 'FATHER OF GERMAN GEOLOGY'.

It was another late eighteenth-century German mineralogist who had a considerable impact on the development of geology, one that extended well beyond Germany. Abraham Gottlob Werner (1749–1817) was known as the 'Father of German geology'. His experience and knowledge came from mining in Germany, and his lecturing skills turned the Freiburg Mining Academy into an international centre of geological learning.

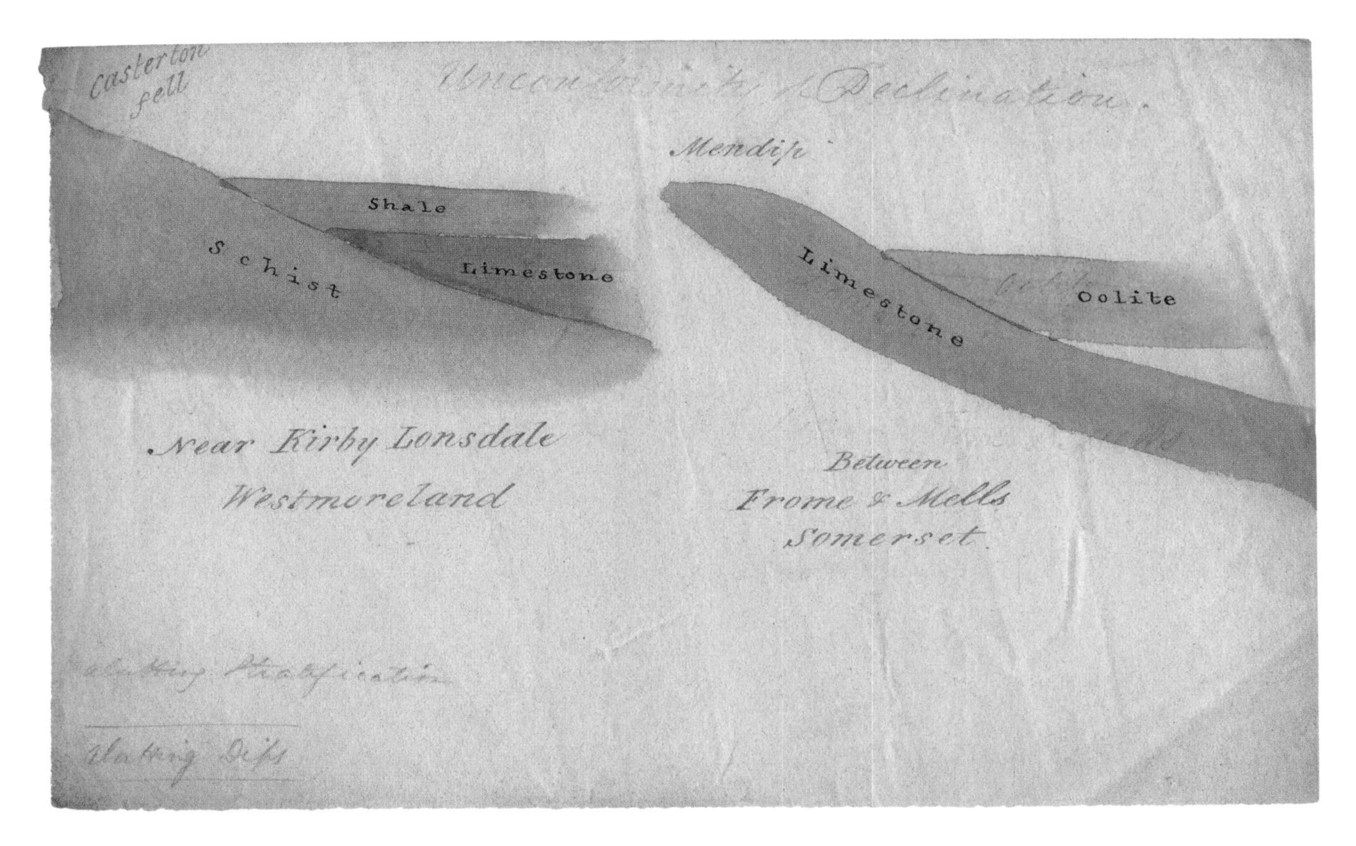

SECTION OF CASTERTON FELL NEAR KIRKBY LONSDALE, WESTMORELAND (LEFT) AND MENDIP HILLS, SOMERSET (RIGHT), C. 1825.

At Casterton Fell, horizontal Limestone and Shale strata (of Carboniferous age) overlap one another on to older 'Schist' rocks (now known as older folded Palaeozoic strata) in what is technically known as an overlapping unconformity. The section between Frome and Mells in Somerset shows declining Limestone strata (Carboniferous in age) unconformably overlain by younger horizontal Oolite Limestone (Jurassic age).

SECTION OF CHARNWOOD FOREST, LEICESTERSHIRE, C. 1825.

This shows Smith's attempt to understand the complex relationships between the hard and ancient schists (Precambrian age volcanic rocks and sedimentary strata) of Charnwood Forest, which are covered unconformably by the Limestone, and Coal Series (Carboniferous age), which in turn, have Red Marl (Triassic age) and Lias strata (Jurassic age) lying unconformably on them.

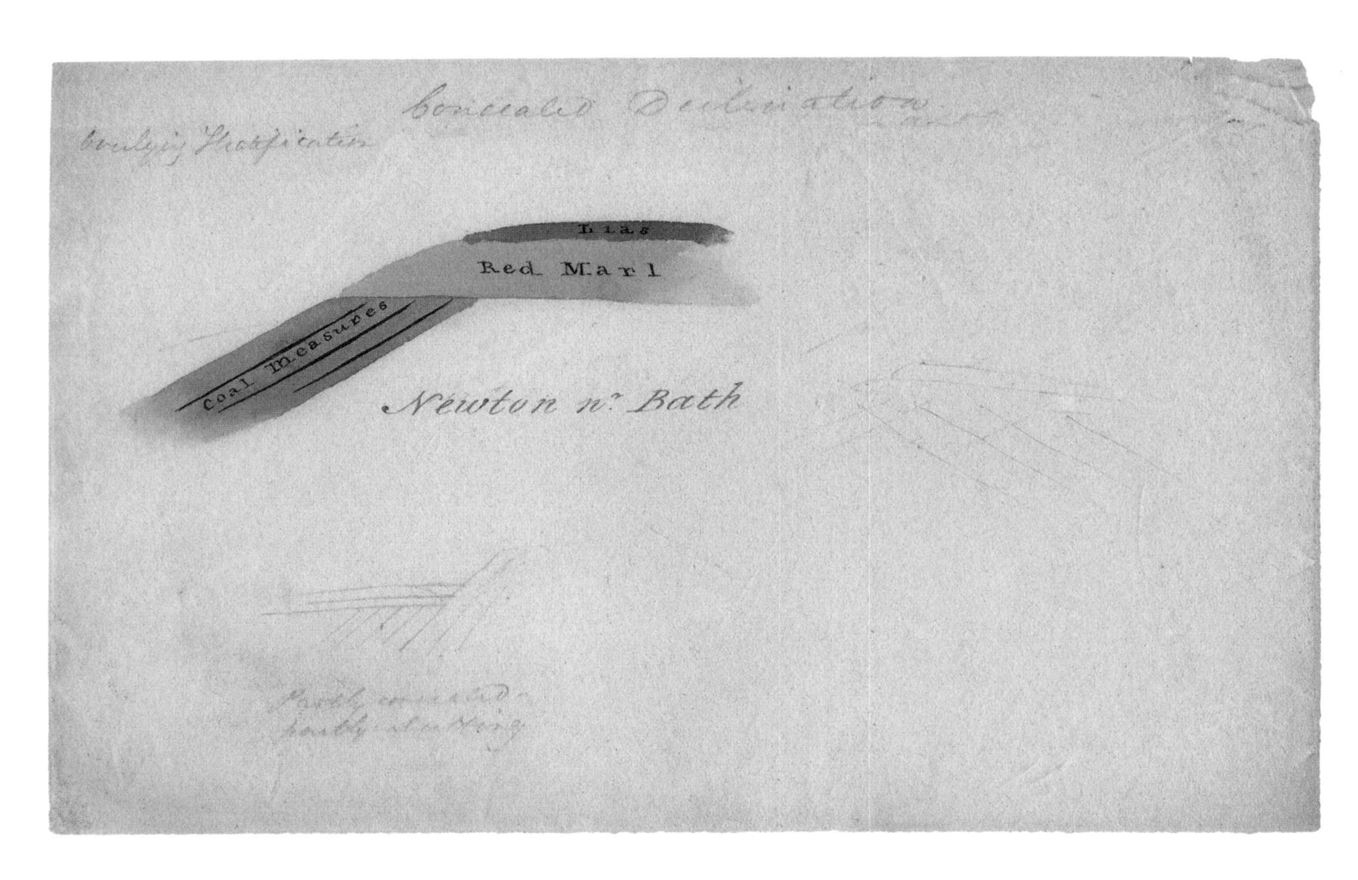

SECTION OF NEWTON NEAR BATH, SHOWING CONCEALED DECLINATION, C. 1825.

Following Strachey, Smith showed how Coal Measure strata (Carboniferous age) may be present below younger, horizontal and unconformable Red Marl and Lias (Triassic and Jurassic age) strata. This is of considerable economic importance and the further implication is that because of the direction of declination of the Coal Measures, older non-coal-bearing strata may also occur below the Red Marl, where a search for coal would be fruitless.

SECTION OF THE CUMBRIAN HILLS AND VALE OF EDEN, C. 1825.

The dislocated (faulted) boundary of the Vale of Eden with the Cumbrian Hills is clearly shown by the displacement of the Limestone Series (Carboniferous) and exposure of the older Schists (Lower Palaeozoic) strata in Hilton Beck. Smith does not seem to fully appreciate the nature of the down-faulted Vale of Eden, but it is nevertheless noted on his section with the pencil annotation 'Unconformity of Elevation', though this is in another hand, possibly Phillips'.

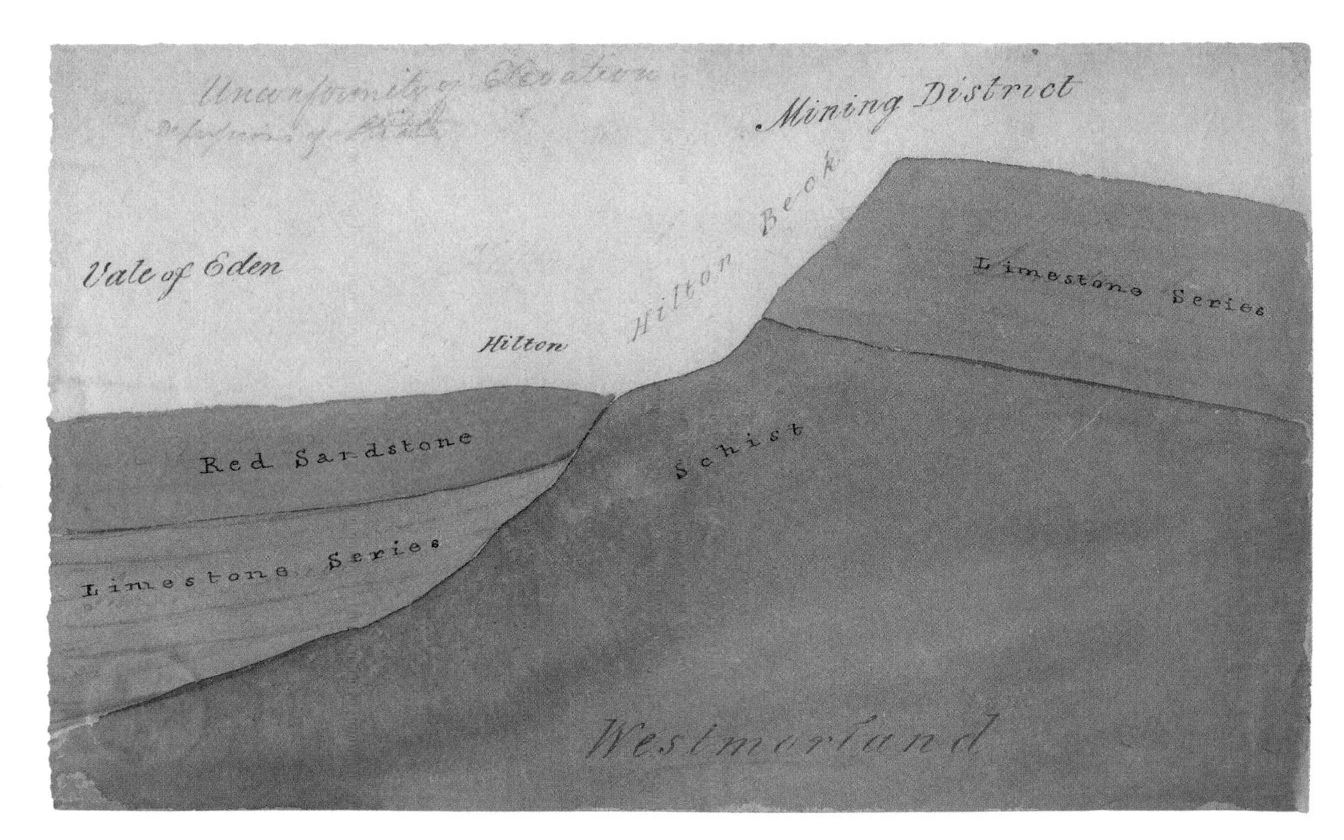

SECTION OF INGLETON, NORTH YORKSHIRE, SHOWING AN UNDULATING DIP, C. 1825.

From Ingleborough Peak in the Yorkshire Dales, the Limestone, Shale and Coal Measure strata (all of Carboniferous age) form the basin-shaped 'undulation' or synclinal fold of the Ingleton coalfield, exploited from the seventeenth century. The Peak Shale has a cap of unnamed strata (Millstone Grit of Carboniferous age) and the limestone is underlain unconformably by unnamed Schist (folded Lower Palaeozoic strata).

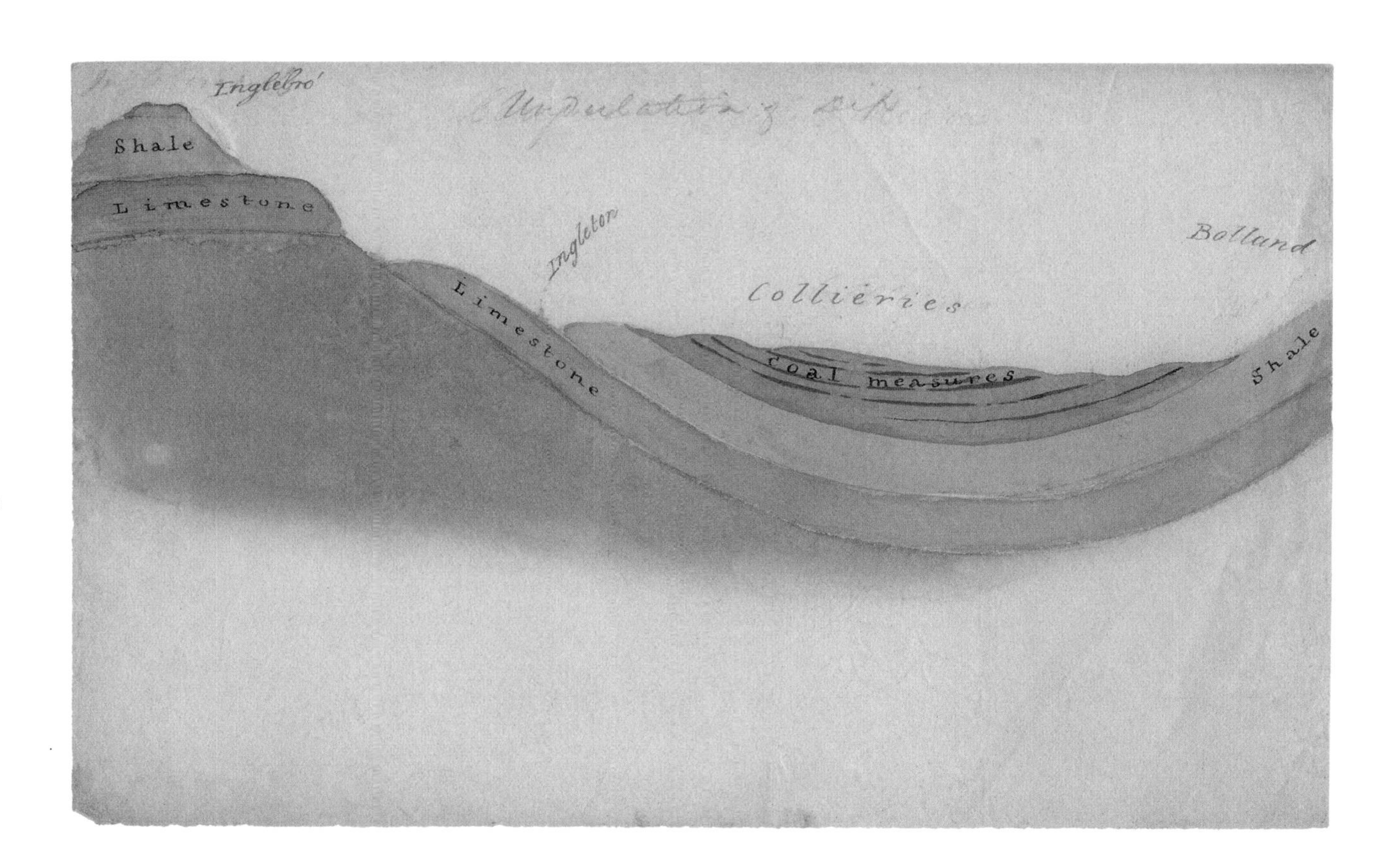

SECTION OF THE NORTH PENNINES, OR THE 'SUMMIT RIDGE OF ENGLAND', C. 1825.

From Cross Fell to Ingleborough in Yorkshire a thick sequence of Limestones with alternations of Shales and Grit (Carboniferous age) are folded into a large syncline and thinner upfold or anticline. These strata are underlain unconformably by Schist (folded Lower Palaeozoic strata) and the peaks are capped by unnamed, brown-coloured outcrops (Millstone Grit, also Carboniferous age).

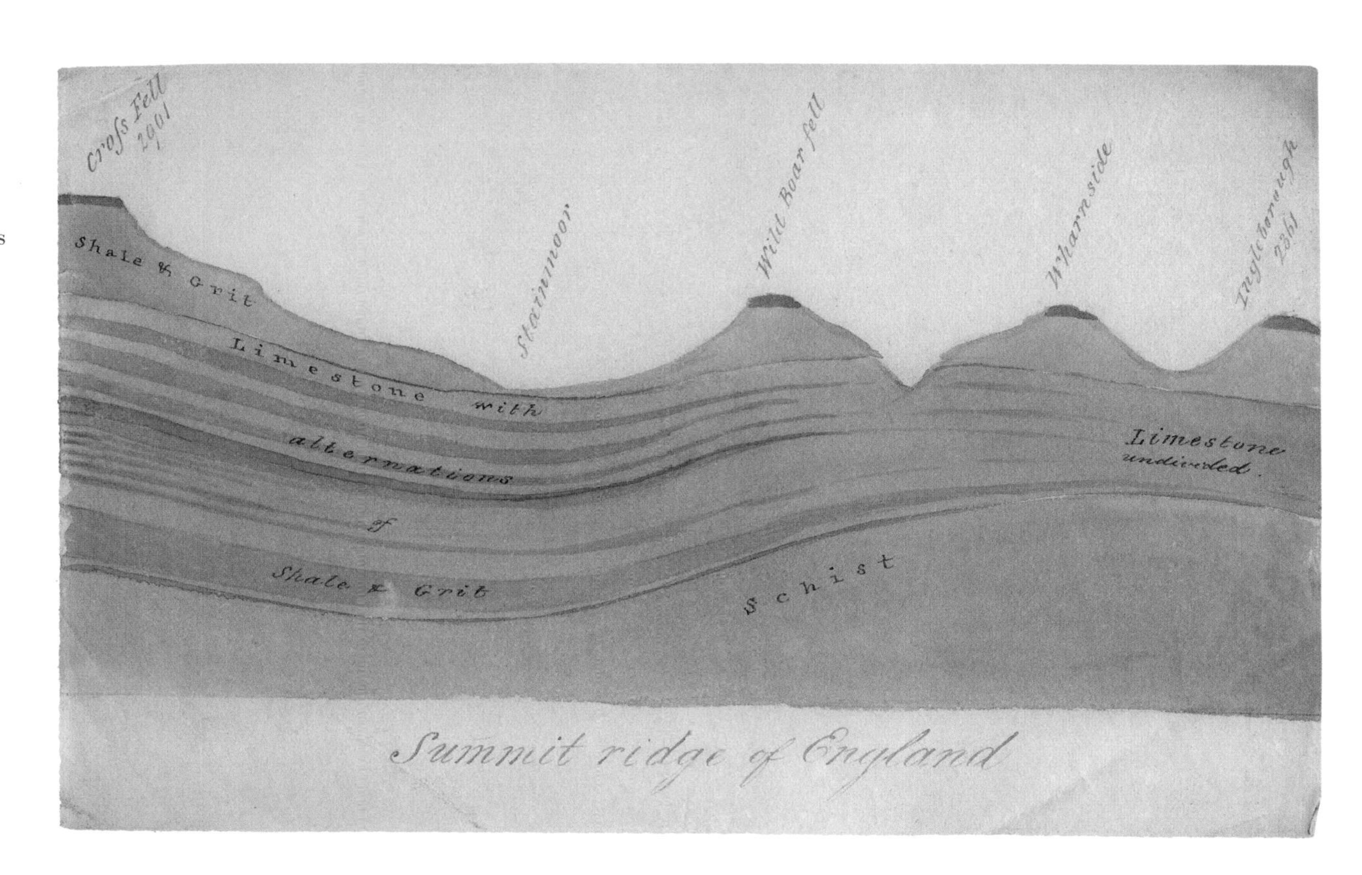

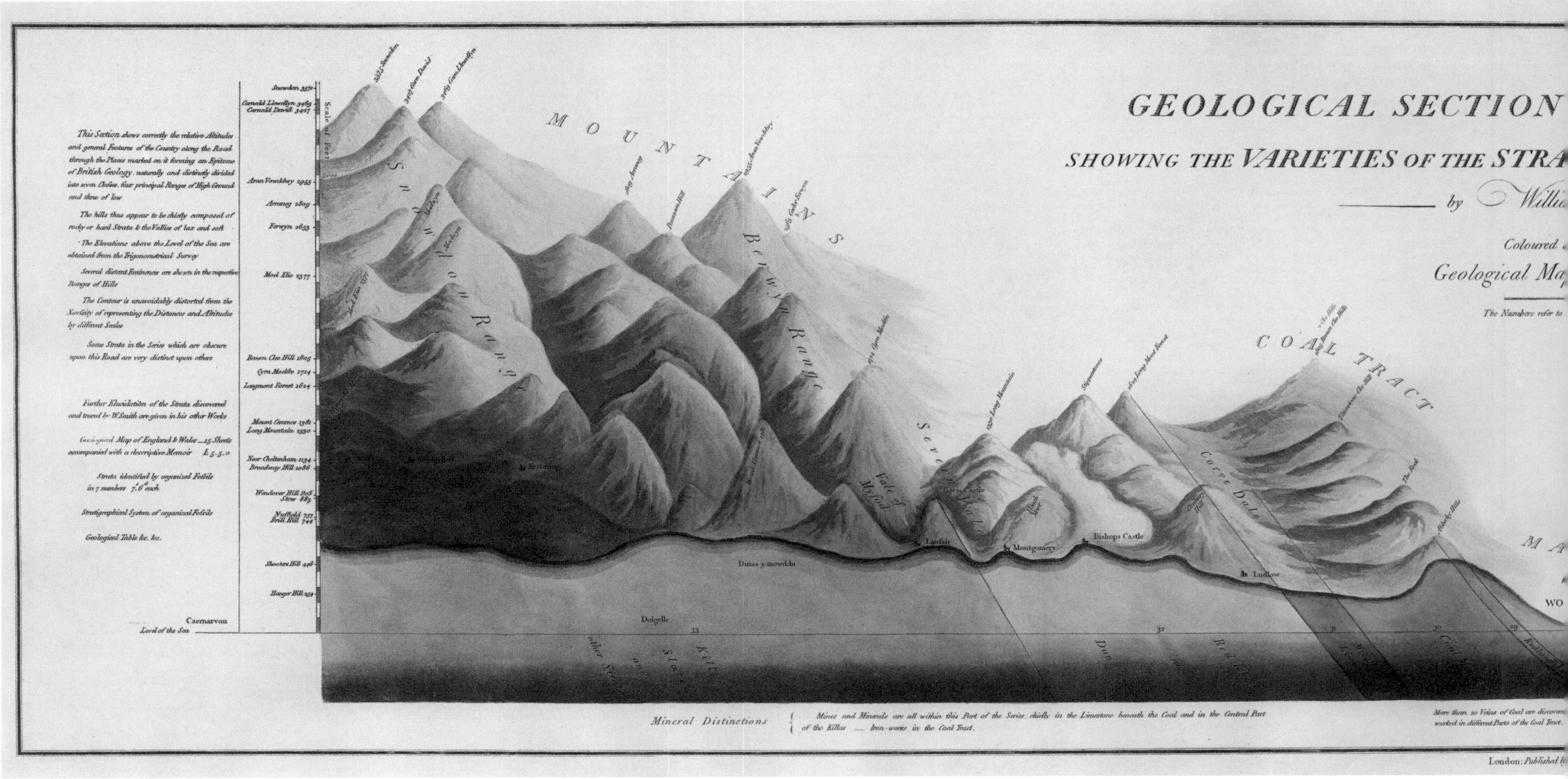

ROCK STRATA DISPLAYED IN *GEOLOGICAL SECTION FROM LONDON TO SNOWDON*, 1817.

Smith's 1817 cross section of rock strata is developed from the much smaller section seen on his 1815 map. It demonstrates his belief that most strata were inclined; the protruding edges or outcrops of each were what he primarily delineated on his maps. The colours of the strata are identical to those used on his maps.

Descriptions of strata given on these pages are extracted from William Smith's *Strata Identified by Organized Fossils* (1816–19) and *A Memoir to the Map and Delineation of the Strata of England and Wales, with Part of Scotland* (1815). They do not all appear in the cross-section above. Smith's close observation of different strata was very wide ranging. It included many comments on topography, drainage, soils, natural vegetation, crops and other economic uses. His purpose was to provide useful information to farmers, land-owners, surveyors, drainers and engineers.

GRANITE, SIENITE, & GNEISS.

The surface of a Granite country, and its great irregularly rounded outlines, bespeak the massive nature of its substrata. Nothing can be more gloomy than the frowning brow of Dartmoor, which overhangs Oakhampton; or more dreary than the rusty-looking surface of the higher parts of the moor, interspersed with deep boggy valleys, and tors or huge masses of rock, against the horizon, instead of trees. Lower situations are drier and better; but large lumps of sterile rock will peep through the barren soil, which seems to be composed of little else than the pulverized fragments of the rock itself. The most remarkable surfaces on these strata, at Malvern, Mountsorrel, and in Anglesea, Cumberland, and the south of Scotland, are small compared to those in Devon and Cornwall; and, but for the singular nature of the rocks, make but little figure in the general features of the country.

KILLAS.

The soil, on shattery fragments of these strata, is good, where tinged with red, and the stone is loose enough to be absorbent, as in some parts of Somerset and Devon. Oaks grow kindly on the steep sides of the woody glens in these strata. And in some parts of Wales, where these rocks are water-tight, such yellow tough Clay may be seen as generally produces oak. The strata of Slate, in the most mountainous parts, usually occupy lower situations than the harder rocks, with which they alternate: some of these, on Snowdon, approach the regular figure of Basalt. The deep narrow valleys of these mountainous districts are filled with fragments of the more rocky heights, abound with large lakes, or are very wet and springy: hence those who have seen the sharp-pointed mountains which pierce the clouds, and the deep bottoms of spungy peat, which hold the water they produce, may account for the humidity of these high situations.

LIMESTONE AND SLATE.

Very hard grey stone, with blue flinty Slate, in almost vertical courses, a blackish soft stone, which resembles some of the accompaniments of Coal, with beds of imperfect Limestone, occupy the space between the Red and Killas, and may be expected to produce a soil as various as the substrata; and, from the same cause, many parts of the surface also rise into the most singular and romantic hills.

RED RHAB & DUNSTONE.

The course of the Red Rhab, as it is called in South Wales, and of the Red and Dunstone of Monmouthshire, may be distinguished by its colour and a peculiar unevenness of surface, abounding with deep narrow dingles, through which small streams of water descend with rapidity. Some parts of its surface soil have a good herbage; the portions in cultivation produce corn of a fine quality; and others, too steep and too high for the plough, are good sheep pastures. So many alternations of Red and Grey Sandstone rocks and much reddish indurated Marl, may be expected to produce a great variety of soil.

MOUNTAIN LIMESTONE.

The Limestone of the Peak of Derby, which rises from beneath the Coal Measures, is a part of the same kind of rock which appears at intermediate distances under the same circumstances, thence up to its termination in the sea, north of Berwick on Tweed. Craven, Richmond and Hexham, are most conspicuous surfaces on this rock. In a part of Westmoreland and Cumberland also, it seems to underlay the coal. It reappears, under the same circumstances, in Flintshire, and may be thence traced to the corresponding point in the sea, in South Wales.

COAL TRACTS.

The mass of strata usually called Coal Measures, is known to be deprived of much of the superficial space which it would occupy by the overlapping of the red earth. In the higher situations to which these strata ascend, in South Wales, the Forest of Dean, Shropshire, Derbyshire, and thence northward to Berwick on Tweed, the surface of the Coal Measures is much alike, and such as cannot be mistaken. These strata, like the red earth, seem to bend over the summit of drainage into Lancashire and other districts.

REDLAND LIMESTONE.

The Magnesian, or Yellow Limestone, which may be very distinctly traced from the neighbourhood of Nottingham, northward to the seaside, at the mouth of the Tyne, is, in some parts of its course, marked with the characters of poverty and rough herbage, unusual to Limestone. The Magnesian Limestone, like that of Mendip, seems to lie in, and belong to, the great stratum of red earth which forms the eastern boundary of Coal.

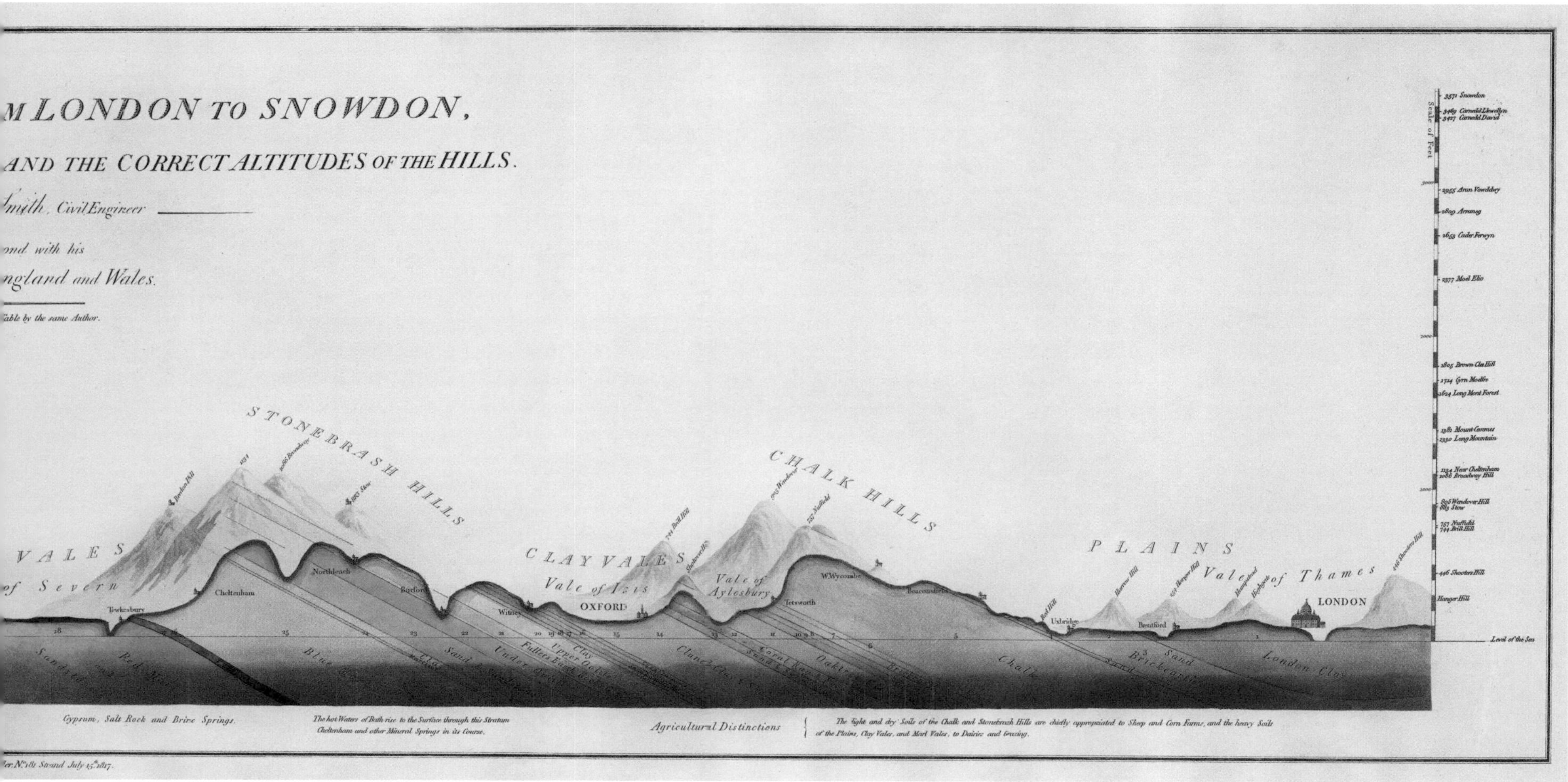

RED MARL.

The Red Marl, red earth, and its beds of soft red Sandstone, and in some places whitish blue beds of indurated Clay, or stone, are first noticed on the south coast of Devon, and in that country and in the vale of Taunton, adjoining, form exceedingly good land, which is thickly covered with fruit-trees. Its surface is narrowed over the Severn, by the sudden rising of the Coal Measures in the Forest of Dean. It extends thence northward by the Severn to Worcester, where an immense expansion of it commences, which extends northward over the greater part of four or five counties. At Nottingham another contraction is occasioned by the Lias and Coal Measures, and its great breadth hence northwards, through the north and south clays, is again contracted by the Fens .... A great breadth of it passes over the summit of drainage between Staffordshire and Shropshire, and expands over part of the latter county, nearly over the whole of Cheshire, and a large portion of the district of Lancashire.

BLUE & WHITE LIAS.

The Blue Lias Limestone, composed of thin beds of stone, imbedded in Clay of the same colour, lies at the edge of the Blue Marl district, and makes a surface so little differing from that of the Blue Marl as to be frequently passed without notice. The White Lias beneath it is still less seen except in particular parts of Somersetshire, where its planes happen to be parallel to the surface of arable land, and fragments of its beds turned up with the plough.

BLUE MARL.

Much of the stiff soil of this stratum is now in pasture. This stratum expands over a great space in the southern part of Somersetshire, where, as in the vale of Gloucester, it produces many of the best orchards .... It spreads wide on each side of the vale of Evesham, through the vale of Red Horse, produces the same kind of pasture in its course through Warwickshire and Leicestershire, and, in a large part of Northamptonshire and Oxfordshire, becomes much interspersed with the stony lands of the stratum above.

UNDER OOLITE.

This stratum is frequently so blended with the one above, in the slopes of the same hill in all the southern parts of its course, as not to be distinguished from it but in maps of larger scale.

UPPER OOLITE.

The vast districts of downs and open fields, recently converted into dry arable lands, inclosed with stone walls, characterize those strata which, on the Cotswold hills, must be well known to the numerous visitors of Bath and Cheltenham. In some parts of Oxfordshire, these walls and similar appearances, are common; but in Northamptonshire, and other districts of these strata, hedges have been more generally planted; but the stone walls, and a similar openness of country and aridity of soil, re-appear in Rutland and in a part of Dorsetshire north of Sherborne. The peculiar dryness of the soil, and vast extent of country without water, will enable the traveller to distinguish the site of this stratum.

CORNBRASH.

The cornbrash is very aptly described by its name, as in the western part of its course, parallel to the strong Clay lands before described, this is almost the only land in tillage. It makes, generally, a good soil. Its course is marked by several considerable market, and other larger towns, which are mostly situated at the places where the outcrop of this stratum crosses the rivers. Being the hardest and best of that stone is, in many parts of its course, much used on the roads.

CLUNCH CLAY & SHALE.

It will be known to the agriculturalist by its wet, tenacious properties and difficulties in tillage. The grazier and skilful land-surveyor well know its pastures by the blue cast of carnation-grass, and other coarse herbage. These are features which may be readily traced through every part of the country which produces it; but its course may be found on a map, by certain districts of low lands, which are frequently subject to inundation. It forms the chief boundary of the great level of the Fens, and continues thence northwards, through the wet and low lands, to the estuary of the Humber.

CORAL RAG & PISOLITE.

The Coral Rag consists chiefly of lumps of Coralline Limestone, which in the quarry are very rough, irregular, and dirty; but where this stone is used as a road material, it wears to a smooth surface. The Pisolite freestone beneath is softer. In some parts it being an Oolite of fine grain is used in building. Coral Rag and Pisolite, with the Sand and Sandstone beneath, make a surface of dry land which, within a generally moist surface of Clay land, is very desirable for tillage.

BRICKEARTH.

The surface of this stratum is frequently so obscured by the loose incumbent stratum of Sand, that in some parts its outcrop may be passed without notice, and its course traced with difficulty. The subsoil is yellower than the soil, retentive, good Brickearth, some of it works freely into tiles and coarse pottery. It often constitutes the base of some of the highest western promontories of Chalk; seldom quits the Chalk hills far enough to occasion any great breadths of Clay land: is often covered with small woods, chiefly of oak in tolerable state of luxuriance. As this Clay keeps up the water of the Chalk and Green Sand, and occasions the first springs at the foot of those hills; the course of it may thus be traced: also, by rushes and other indications of a Clay surface, especially in a district so abounding in Sand.

CHALK.

The Chalk stratum is generally known by the feature which it gives to our island, in the white cliffs of Britain, and long ranges of interior hills. For want of colour in the stratum, it is defined in my map by a green line. The soil is generally brown, but by the sides of hills of various mellow tints, from that to brownish white. Its course through the island from the English Channel to the German Ocean, is from S.W. to N.E. from a point on the western extremity of Dorsetshire, to Flambro' Head in Yorkshire, with a considerable curve to the eastward. It has two singular branches from Hampshire, one through Sussex, and the other through Surry and Kent, which approach the sea in their respective cliffs at Beachy Head and Dover. These are the chain of hills which it forms; its greatest plains are in Wiltshire and Hampshire.

LONDON CLAY.

This thick stratum, from its being the site of the metropolis, and most abundant in its environs, has been called the London Clay. Its course north-eastward to the sea is described in the map. It thence occupies the heights in the hundreds of Essex, and east of Chelmsford and Colchester; extends through the Sokens to the seaside at Walton Nase, and Harwich. The soil is of a mellow brown or umber colour, and the subsoil is generally the same .... The exact boundaries of the soft strata are generally difficult to define, but particularly so in this district, where they alternate with no hard materials in the form of rock.

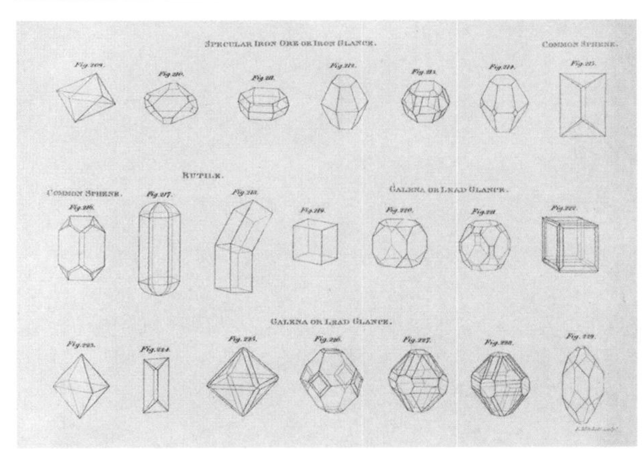

FIG. 6.

Fig. 6. Robert Jameson's illustrations showed how the identification of different minerals could be aided by recognition of their geometrical form.

Charismatic but also dogmatic, Werner published little. Our understanding of his teaching is largely derived from the writings of his pupils, such as the Scottish geologist Robert Jameson (1774–1854), who studied in Freiburg in 1800. In his impressive three-volume *System of Mineralogy* (1804–8) Jameson wrote: 'The illustrious Werner early saw the impossibility of Mineralogy advancing without a determinate language, he therefore made this first object of his attention.' (Jameson, 1804, p. viii–ix)

In 1786 Werner published his *Kurze Klassifikation und Beschreibung der verscheiden Gebirgsarten* (translated into English as *A Short Classification and Description of Various Rocks*), which outlined his ideas about the succession of rocks that form the Earth's solid crust. Building on earlier work, Werner claimed that Earth's rocks originated largely by chemical precipitation from a global primeval ocean. The first 'primitive' rocks thus precipitated included granite, gneiss, slate and basalt. Above them lay more precipitates, the 'transitional' rocks, with some organic remains of newly formed life and deposited by running water. In turn they were covered by the Flötz-Schichten or stratified rocks, also mainly deposited by running water and containing fossil remains, but with some precipitates such as coal and slate.

NEPTUNISTS AND VULCANISTS.

Werner's insistence on the role of a primeval ocean and the chemical precipitation of rocks led to one of the major controversies in the development of geology. His proselytizing acolytes became known as 'Neptunists' and were opposed by 'Vulcanists' (or 'Plutonists') – geologists who believed that heat and igneous activity played a central role in the formation of the Earth's rocks.

'Vulcanism' also had an eminent founding 'father', the Scottish natural philosopher James Hutton (1726–1797). Like Werner, Hutton was a reluctant author, but in his 1788 paper 'Theory of the Earth; or an Investigation of the Laws observable in the Composition, Dissolution, and Restoration of Land upon the Globe', Hutton proposed that Earth's interior heat powered the 'engine' that created new rock such as granite and volcanic basalt. These primary rocks formed land, which was then weathered and eroded, with the products transported and deposited on the seabed. Earth's internal heat also consolidated the deeply buried sediment into rock and uplifted it to form new land, which was then recycled back into the sea.

Most importantly, Hutton argued that the whole geological cycle took an immensely greater length of time than previously envisaged. Hutton's conclusion 'that we find no vestige of a beginning – no prospect of an end' (Hutton, 1788, p. 304) is frequently quoted as seminal for modern geological thought about Earth's great antiquity. In particular, it helped counter claims by theological geologists that the Earth's age could be calculated from biblical sources.

It is no coincidence that as near contemporaries in an age of agricultural and industrial development, Hutton and Smith had common interests, but there were also differences between them. Hutton had the considerable social advantage of a university education. He inherited farmland whose productivity he set about improving. And, from 1764, he was involved with the construction of the Forth and Clyde canal. However, unlike Smith, Hutton was not greatly concerned with geological mapping or fossils.

THE ROLE OF FOSSILS.

For much of the eighteenth century, the development of geological ideas and geological practice took two largely separate paths, one followed by mineralogists and the other by naturalists. Mineralogists, such as Guettard, Lehmann and Werner, were interested in the

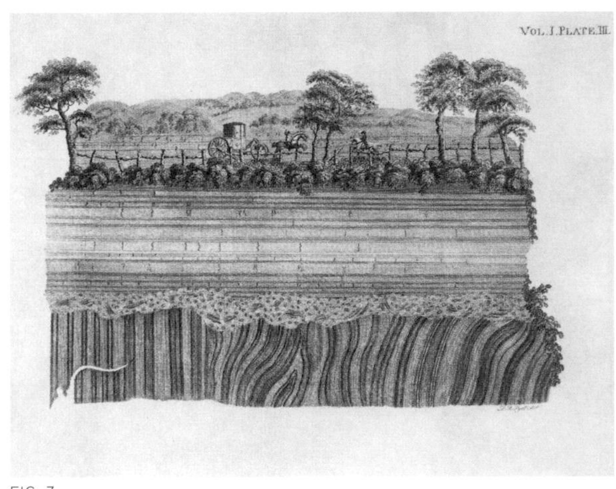

FIG. 7.

Fig. 7. James Hutton came across a dramatic example of an unconformity between lower, vertical strata and higher and younger horizontal strata on the Jed Water, Scotland, in 1787. It was sketched by Hutton's friend John Clerk of Eldin (1728–1812).

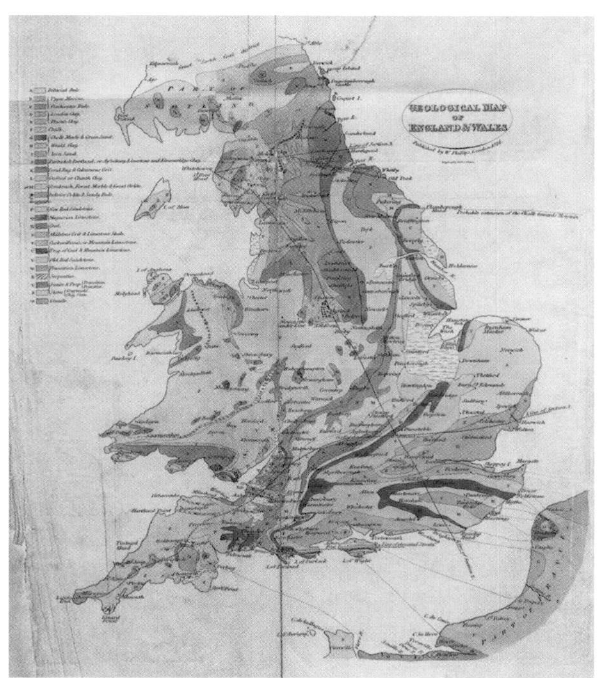

FIG. 8.

Fig. 8.
The best early-nineteenth-century small-scale geological map of England and Wales accompanied the 1822 edition of W. D. Conybeare and W. Phillips's popular *Outlines of the Geology of England and Wales.* With Smith's permission, the authors combined information from his map and that of the Geological Society. It included cross-channel correlation with France.

study of minerals, their formation, economic value and exploitation. Naturalists were more concerned with questions of the fundamental nature of fossils, their preservation and origin. However, few naturalists were looking at fossil distribution within the rock record until William Smith and some of his French contemporaries.

The organic nature of fossils was finally recognized in the late seventeenth century. Many scholars, such as the Sicilian painter and naturalist Agostino Scilla (1629–1700), the Danish-born Nicolaus Steno and the Swiss Johann Jakob Scheuchzer (1672–1733), along with the English naturalists Robert Hooke (1635–1703) and John Woodward (1665–1728), were highly accomplished observers and illustrators of fossils and were influential in the debate over their nature. The accurate drawings of Steno, Scilla and Hooke in particular were pivotal in resolving the very real problems of interpretation presented by the preservation of fossils.

In 1667 Steno illustrated the head of a shark and compared its teeth with fossils known as *glossopetrae* or 'tongue stones'. He argued that the latter were 'petrified' shark's teeth and that their occurrence far from the sea proved the oceans had previously extended far inland. A few years later, in 1670, Scilla's beautiful fossil illustrations led to the same conclusions. Having seen Scilla's book, Woodward bought the Sicilian's fossil collection and used it in developing his own, similar theories, which he published in his two-volume *An Attempt towards a Natural History of the Fossils of England* (1728–29). Woodward's ideas were circulated around Europe following the translation of his book by Scheuchzer, whose own numerous works included *Herbarium diluvianum*, published in 1709. In this Scheuchzer pictured fossils as the remains of organisms drowned by the Noachian Flood.

The idea that fossils were the result of the Flood persisted into the nineteenth century among theological geologists such as the Rev. Dr William Buckland (1784–1856) of the University of Oxford. The title of his 1823 book is revealing – *Reliquiæ Diluvianæ; or, Observations on the Organic Remains ... attesting the Action of the Universal Deluge*. But new information which became available over the next decade on the distribution of fossils within sedimentary strata caused Buckland to realize that the Flood theory could not be maintained.

CATASTROPHISM OR UNIFORMITARIANISM?

James Hutton's cyclical theory of the formation and destruction of rocks was kept alive by one of the new generation of nineteenth-century geologists, Charles Lyell (1797–1875). Lyell's three-volume *Principles of Geology* (1830–33) was enormously influential. In 1831, the ambitious twenty-two-year-old naturalist Charles Darwin (1809–1882) carried the first volume with him on board the *Beagle*. As a Cambridge undergraduate, Darwin had been taught the 'Smithian' principles and practice of geological mapping by the Woodwardian professor of geology, Adam Sedgwick. During the *Beagle* voyage Darwin collected fossils as an aid to identifying and mapping the relative age of strata he encountered.

Another major theme of Lyell's great book was the shaping of Earth at a slow and gradual pace by the same geological processes that operate in the present. Called 'Uniformitarianism' by the Victorian polymath William Whewell (1794–1866), this gradualistic view of Earth's history was contrary to the so-called 'Catastrophist' theory associated with the French naturalist Georges Cuvier (1769–1832).

Despite the ongoing armed conflict with France at the time, Cuvier was acknowledged in Britain as Europe's pre-eminent expert on fossils. It was Cuvier who, in his 1796 study of elephants, both living and fossil, demonstrated that extinctions had occurred. Furthermore, for Cuvier such extinctions were 'catastrophic': 'All these facts ... seem to me to prove the existence of a world previous to ours, destroyed by some kind of catastrophe' (Rudwick, 1997, p. 24). But Cuvier's catastrophism was not a simplistic interpretation of biblical texts, rather he envisaged a series of catastrophic events throughout Earth's history.

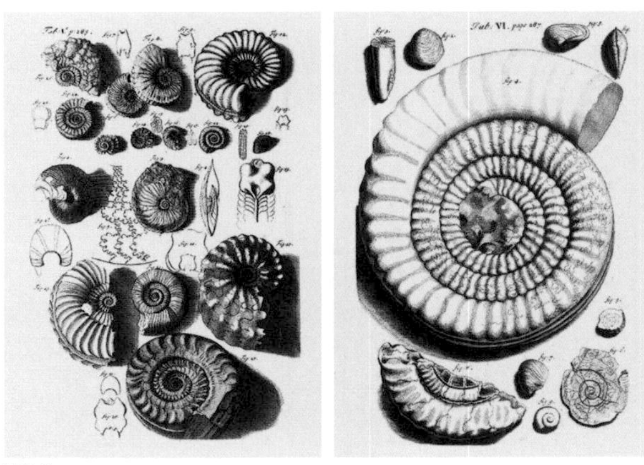

FIG. 9.

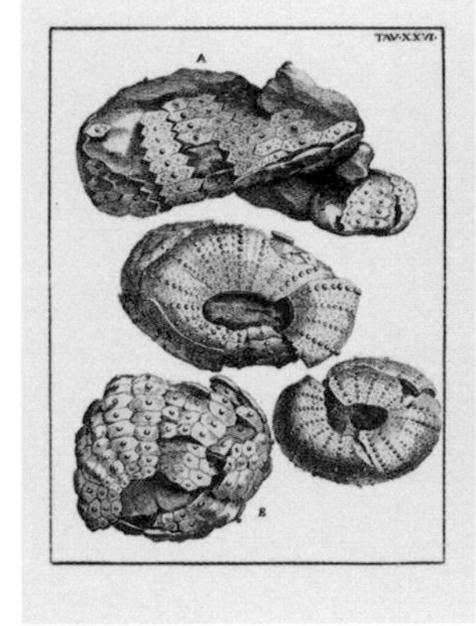

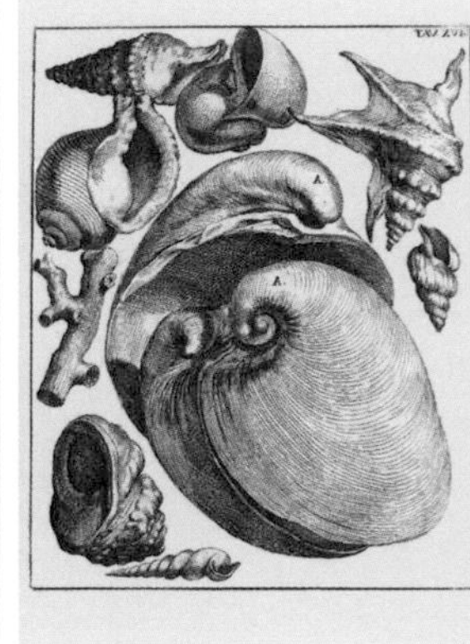

FIG. 10.

Fig. 9.
Some of the earliest scientifically accurate illustrations of fossils were drawn by Robert Hooke in the late seventeenth century. They were designed to help substantiate the claim for their organic origins.

Fig. 10.
Two plates from the Sicilian artist Agostino Scilla's *La vana speculazione disinginnata dal senso* (Vain Speculation Undeceived by Sense), published in 1670. Scilla's skilful representation of fossils was highly influential in supporting the argument for the organic origin of fossils.

PARALLEL ACHIEVEMENTS.

Cuvier also researched the strata and fossils of the Paris basin and mapped them with the mineralogist and naturalist Alexandre Brongniart (1770–1847). Using geological sections, they correlated strata across the region, producing a preliminary map in 1808 and a more detailed one in 1811, the *Carte géognostique des environs de Paris*. Importantly, the strata were characterized by their fossil content and matched with their contemporaneous deposits. Independently, it seems, Cuvier and Brongniart had developed the same method as Smith.

This is perhaps not surprising. Knowledge of the nature and classification of fossils was well established by the end of the eighteenth century, but fossils were still seen in isolation from their sedimentary context. The relationship between individual fossil species and the strata in which they occurred required a well-developed taxonomy. In the nineteenth century in both France and England, the understanding of strata and fossils had reached a stage when it was only a matter of time before some astute observer would 'put two and two together'. Unlike Brongniart and Cuvier, who had made detailed scientific studies of fossils, Smith had not. Luckily, however, he had guidance from naturalists such as James Sowerby (1787–1871) and James Parkinson (1755–1824). This does not detract from Smith's achievements. Cuvier and Brongniart were part of a well-established French scientific elite that was supported by the state, unlike Smith in England.

Although Cuvier and Brongniart were developing the same mapping methodology as Smith, no direct evidence has yet been uncovered to confirm that they knew what Smith was doing or *vice versa*. Though it is perhaps possible that the French learnt of Smith's work through his patron Sir Joseph Banks, who as President of the Royal Society entertained all the leading scientists of Europe at his home in London and was in correspondence with Brongniart.

COMPETING MAPS.

Cuvier and Brongniart's 1808 map of the Paris Basin came at a critical time in the development of the geological sciences. Unfortunately, Smith's map was not then nearly ready for publication, although he had issued a *Prospectus* for it in 1801. It was such an ambitious project that it would not appear for another seven years after the French map, especially as Smith had to earn his living from other work at the same time.

To make matters worse for Smith a new competitor was emerging. The Geological Society of London was founded in 1807, with Georges Bellas Greenough (1778–1855) as Chairman and then President. One of the Society's early concerns was the progress made by the French in the study of fossils, including their stratigraphical distribution and potential in identifying different strata.

On 3 March 1809, the Geological Society set up a Committee of Extraneous Fossils, whose members included Smith's two fossil experts, James Sowerby and James Parkinson. The Society also established a Committee of Maps and Sections for 'the construction of Mineralogical Maps', with the aim of gathering empirical data for a geological map of Britain. Information was to be supplied by a network of correspondents from around the country, mostly well-educated 'gentlemen'.

The base map for their first draft of 1812 proved unsatisfactory. A new base map was not ready until 1814, and the following year Smith got there first with the publication of his map. It took another five years for the Society's map to appear. Its excellent presentation of the geology was due to Thomas Webster (1773–1844), a Scottish geologist, famous for his beautiful illustrations of the geology of the Isle of Wight and for being the first in 1825 to applaud Smith as 'the father of modern English geology'.

It is apparent that awareness of the potential importance of fossils for stratigraphy had been growing. In 1815, William Phillips (1775–1828), in his *An Outline of Mineralogy and Geology* wrote that 'those fossils ... which heretofore were carefully collected as curiosities, now possess a value greater than as *mere* curiosities. They are to the globe what coins are to the history of its inhabitants; they denote the period of revolution; they ascertain at least comparative dates.' (Phillips, 1815, p. 189) Yet it was also clear that the Society's map owed a great deal to Smith's work but failed fully to acknowledge this. By this time, unfortunately, Smith was facing financial ruin. He had already been arrested twice for debt in 1814, but in 1819 he was actually imprisoned for debt.

A NEW LIFE AND OFFERS ABROAD.

Smith's financial woes were primarily the result of unlucky speculation. After forfeiting most of his worldly goods and spending almost ten weeks in debtors' prison in London, Smith, his wife Mary Ann (1790/92–1844) and nephew John Phillips (1800–1874), left the capital for Yorkshire. Smith appears remarkably resilient and undaunted by his experiences and spent the next seven years there working as a surveyor, map-maker and lecturer.

In Yorkshire Smith found an audience thirsty for knowledge about the new science of geology and he soon became a local celebrity. Using his uncle's method of mapping, John Phillips was to help put Yorkshire on the geological map of Britain with his *Illustrations of the Geology of Yorkshire, or a Description of the Strata and Organic Remains of the Yorkshire Coast accompanied by a Geological Map, Sections, and Plates of the Fossil Plants and Animals* (1829–36).

However, Smith's story might have been very different, as he had offers of well-paid work in both Russia and America. The first opportunity for foreign employment in fact arose in 1814. Emperor Alexander I (1777–1825) of Russia wanted to engage an English practical mineralogist to direct coal works in one of the southern

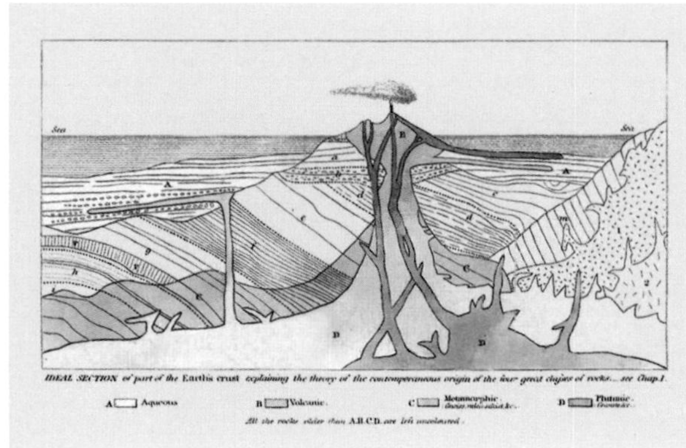

FIG. 11.

FIG. 12.

provinces of his empire. Smith corresponded about the position, which came with a salary of some £800 a year, with the Russian agent Dr Hamel, but ultimately the post was taken in 1817 by J. B. Longmire (1785–1858), an engineer from Whitehaven in Cumberland.

Of greater potential importance for Smith was an offer of employment as canal engineer in North Carolina, which was made in 1819 when his financial woes were at their height. It might seem surprising that Smith's reputation extended as far as North America, but a significant number of British and especially Scottish ex-patriots with an interest in geology were living in America and they were familiar with his work.

Smith's first contact with America had come earlier, when the ex-patriot Scottish businessman and geologist William Maclure (1763–1840) visited him in London on 29 September 1815. Maclure was a successful merchant who had made his home in Virginia in 1796 but continued to travel to Europe, geologizing in Germany and France. When he returned to America, Maclure carried out a geological survey of the states of the Union east of the Mississippi, and by 1809 he had published a memoir and small-scale map (100 miles to the inch) – the first geological map in the Americas. Influenced by the pioneering Scottish geologist, Robert Jameson, Maclure's map was essentially Wernerian and geognostic (that is, related to the study of the distribution of minerals) in its presentation.

At the time of Maclure's visit, Smith was busy supervising the labour-intensive hand-colouring of his great map. It is not known what transpired at the meeting, but it may have been simply a courtesy call by Maclure. Almost four years later, on 5 April 1819, yet another Scottish-born American introduced himself to Smith and on this occasion made him an offer of employment as an engineer. Peter Browne (1764/5–1832)

Fig. 11.
In his *Elements of Geology* (1838) Charles Lyell illustrated the four main categories of rock that comprise Earth's crust – yellow is aqueous (sedimentary rocks deposited in or by water), pale blue is volcanic (rocks formed from lavas), bright blue is metamorphic (older rocks transformed by heat and pressure) and pink is plutonic (rocks cooled from a molten state within the crust).

Fig. 12.
Cuvier and Brongniart's pioneering geological map of the Paris Basin, published in 1811. It has a succession of eleven groups of Tertiary strata, distinguished by their characteristic fossils and rock types. The map was accompanied by a cross section showing the correlation of the strata across region's basinal structure.

practised law in North Carolina and in 1817, on behalf of the state's Commissioners, he sailed to England to engage an engineer, a surveyor and a draughtsman. Such was the demand for first-rate civil engineers in England that Browne could not find one prepared to accept a salary that was thought reasonable in North Carolina. He had to explain to the Commissioners that since the end of the war with France, the British government had turned its attention to improving the country's physical resources and had employed all the best talent.

Browne eventually met Smith in spring 1819 and offered to engage him to 'survey, level, estimate and report on the practicality of overcoming the difficulties of passing the falls in Navigable rivers in North Carolina'. Evidently, Smith must have taken the possibility seriously because two days later he called on Browne at his lodgings in Leicester Square to discuss the proposal. With the help of a large-scale map of North Carolina, Browne explained that the State wanted to achieve the navigation of seven of the principal rivers by canals and locks and to deepen the entrance to the principal harbour, the Roanoke Inlet.

However, 1819 promised to be another busy year for Smith. The first four of his series of geological county maps (for Norfolk, Kent, Wiltshire and Sussex) were published in January, his geological sections were to appear in May, the fourth part of his memoir on *Strata Identified by Organized Fossils* was to be published in June and four more of his county maps (Gloucestershire, Berkshire, Surrey [Surry] and Suffolk) were scheduled for September.

All these publications required close supervision, but they were also costly. Smith must have been tempted by an offer that would potentially solve his financial problems. However, friends who knew of his precarious finances had assured Smith that they would obtain some financial help for him from the Treasury. But none was forthcoming, and by April 1819, Smith was in a court in Westminster and consigned to an insolvent debtors' prison for ten weeks. Following Smith's refusal of the post, Browne eventually employed another Scot, the civil engineer Hamilton Fulton (1781–1834) on a salary of £1,200 a year, which would have been a substantial help to Smith.

Even in his Yorkshire 'exile' Smith was to receive one more American visitor, the London-born George William Featherstonhaugh (1780–1866). Featherstonhaugh had travelled to America in 1806, married an American heiress and become an entrepreneur in agriculture and railway development, and a geologist. In 1828 he came to Britain to study railway construction and visited his childhood home of Scarborough, where he met up with Smith. Apparently, Featherstonhaugh tried to persuade Smith to name strata after their characteristic fossils, but Smith felt that the geologists of the time were not ready to accept such a radical nomenclature.

On his return to America, Featherstonhaugh was appointed in 1834 as the first US Government geologist, with instructions to survey first Arkansas then Wisconsin, Illinois, Georgia and the Carolinas. Recognition of his contributions came in 1835 with his election to the Royal Society of London. Featherstonhaugh retired to England only to be appointed in 1844 as British Consul in France, based in Le Havre. While there he used the name 'William Smith' as a pseudonym to help King Louis Philippe I (1773–1850) escape to England following the revolution of 1848.

LEGACY.

How should we view William Smith and his career today, and what is his place in the history of geology? Thanks to the efforts of many dedicated researchers there is now a much better sense of the sheer scale and range of Smith's achievements. The separate trails

Figs. 13 & 14.
In 1809 William Maclure drew the earliest geological map of America using a base map created by Samuel G. Lewis (left). On this map Maclure distinguished five classes of rock: In brown are Primitive Rocks, in pink are Transition Rocks, in blue are Secondary Rocks, in yellow are Alluvial Rocks and in dark blue is Rock Salt. In 1817 he published an improved version of this map (right) alongside an expanded memoir. On this edition dark blue indicates Old Red Sandstone, and added is a green 'line to the west of which has been found the greatest part of the Salt & Gypsum'.

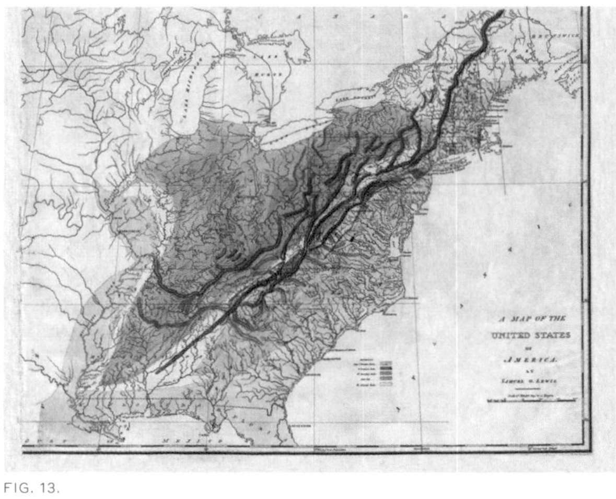

FIG. 13.

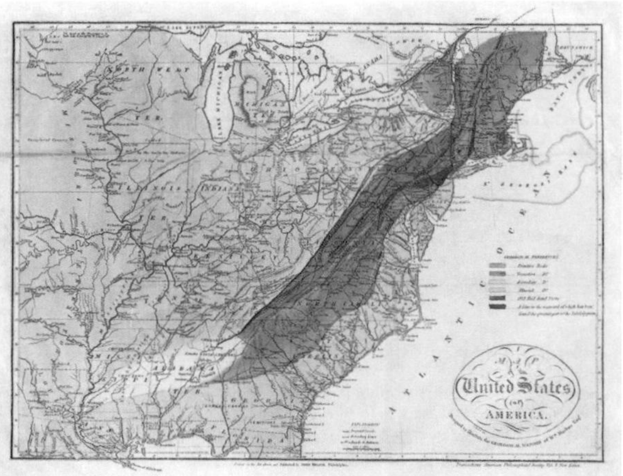

FIG. 14.

of Smith's work as surveyor and engineer of canals and coal mines, and his contributions to land drainage, agriculture and sea-defences, as well as his realization of the importance of fossils, his geological mapping and authorship of several books, have all been studied in detail. This wealth of interest and information surely speaks for itself as his legacy. No other contemporary geologist in Britain or abroad achieved so much in a lifetime spent in pursuit of geological mapping and its practical utility, and so often against the odds. Of particular significance was Smith's mapping and the application of his method to the search for coal. Huge sums of money were wasted in fruitless sinking of exploratory shafts in the hope of finding coal through strata where it was either absent or at too great a depth to be economically extracted.

However, within the rigid social hierarchy of England in the eighteenth century, where breeding and bloodlines counted, William Smith was not generally accepted as a 'gentleman'. This would have presented obstacles to his success and put him at a disadvantage. He was, though, part of a growing group of skilled artisans who were becoming increasingly professional and recognized for their skills. Despite his lack of formal qualifications, Smith came to be acknowledged as a competent and experienced surveyor and engineer. Even so, in the English social game of 'gentlemen and players', Smith would still have been regarded as among the latter.

For rural gentry and aristocracy, such as John Russell (1766–1839), the 6th Duke of Bedford, and Sir Joseph Banks, president of the Royal Society, acceptance of Smith's social status presented no problem, especially as they sought and valued his advice. Through attendance at annual Sheep Shearings and county fairs, Smith advertised his knowledge to the agricultural entrepreneurs of the day. Thomas Coke (1754–1842) of Holkham Hall, Norfolk, appreciated Smith's work and he even celebrated it in a sculpted monumental frieze. Nevertheless, before the advent of any organized or formal technical education in Britain, Smith's achievements in these fields received no official recognition.

To the closed elite that populated the recently founded Geological Society of London, Smith's background was a barrier. Like most gentlemen's clubs of the time it had a selective membership. Although a number of the founders of the Society, such as the Rev. Joseph Townsend, James Parkinson and William Phillips, knew Smith personally and were familiar with his work, it was not sufficient to obtain his election. Even when Smith was earning considerable sums of money as an engineer, it was still not enough for formal recognition by the Society.

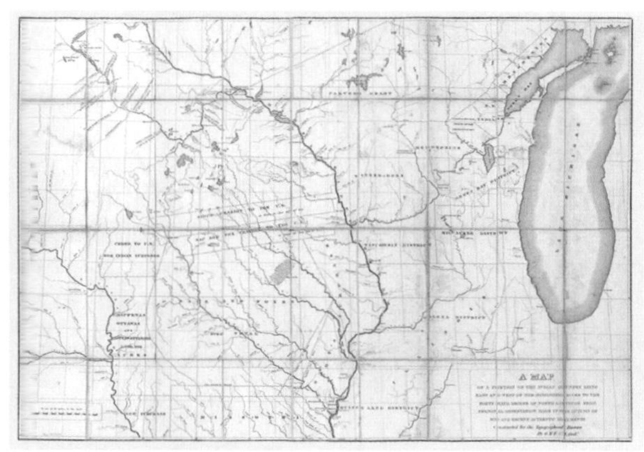

FIG. 15.

Fig. 15.
In 1836 *A map of a portion of the Indian Country lying east and west of the Mississippi River to the Forty Sixth Degree of North Latitude from personal observation made in the autumn of 1835 and recent authentic documents* was published to accompany the report of the first US government geologist, G. W. Featherstonhaugh.

Furthermore, by the 1820s few Fellows of the Society were interested in practical or economic geology. Unlike France and Germany, in Britain there was a much clearer demarcation between 'academic' geology, as imparted by the likes of Adam Sedgwick in Cambridge and William Buckland in Oxford, and 'practical' geology, which was not taught at those universities. The snobbish social attitudes of British university-educated classes towards those working in industry and trade were slow to change.

Then there was the ignominy heaped upon Smith by his imprisonment for debt, which was seriously damaging to a person's reputation. For the well connected such as Marc Isambard Brunel (1769–1849), the renowned French-born engineer and father of Isambard Kingdom Brunel (1806–1859), it was possible for reputations to be restored after such a catastrophe. In 1821, Brunel was, like Smith just before him, in debt and committed to the King's Bench Prison for eighty-eight days. Released following the intervention of the Duke of Wellington, Brunel was eventually knighted by Queen Victoria in 1841. Unfortunately for Smith, the restoration of his reputation and fortunes was not so rapid.

PIONEER.

Before Smith's time the practice of geological mapping had already been developing across Europe. However, until the founding of organizations such as the Geological Society of London in 1807 and the Ordnance Geological Survey in 1835, there was little or no formal method of achieving the systematic production of geological maps that could be extended across larger areas. Furthermore, the majority of the piecemeal geological mapping projects that did occur primarily depicted the distribution of different rock types at the surface.

What Smith did was to make possible predictive geological mapping on a large, regional or national scale. He achieved this through the recognition of the usefulness of palaeontological indicators – fossils – in the characterization of successive sedimentary strata, which overcame the limitations of geological mapping solely by rock type. Smith's expectation was that such strata could be traced from locality to locality across countries and indeed globally. He was not aware, though, that these sedimentary deposits are inherently limited in their extent by their original environments of deposition. Nevertheless, his method was subsequently applied to the correlation of strata over great distances even when the nature of the deposits themselves changed.

The first published contemporary assessment of Smith and his achievements appeared in 1831. At the age of sixty-two, Smith was awarded the first Wollaston Medal by the Geological Society of London in consideration 'of his being a great and original discoverer in English Geology' (Sedgwick, 1831, p. 271). The President of the Geological Society at this time was Adam Sedgwick, Woodwardian professor of geology at the University of Cambridge.

Sedgwick had first met Smith in 1822 in the Lake District. Writing to the poet William Wordsworth (1770–1850), Sedgwick declared that 'my best fossils from Kirkby Moor were procured ... under the guidance of Smith, the "father of English geology", on the day I first became acquainted with him' (Hudson, 1842, p. 35). In the presentation to Smith, Sedgwick took up this personal connection 'I for one, can speak with gratitude of the practical lessons I have received from Mr Smith: it was by tracking his footsteps, with his maps in my hand, through Wiltshire ... that I first learnt the subdivisions of our oolitic series.' With a characteristic rhetorical flourish, Sedgwick concluded by appealing 'to those intelligent men who form the strength and ornament of this Society, whether there was any place for doubt or hesitation ... to place our first honours on the brow of the Father of English Geology.' (Sedgwick, 1831, pp. 278, 279)

There is a clear suggestion here that Sedgwick, one of the new generation of professional geologists, was attempting to make amends for previous slights by the Society and its members on Smith, and the unacknowledged misappropriation of data from his maps. Certainly, the epithet 'Father of English Geology', which actually originated with Thomas Webster in 1825, was linked to Smith from then on.

'SELF-HELP' HERO.

Within a year of Smith's death, public knowledge of his achievements and the difficulties of his life had become sufficiently well established to make him an ideal candidate for Samuel Smiles's great compendium of Victorian heroes, *Self-help*, published in 1859. Smiles (1812–1904) was a Presbyterian Scot whose book was in tune with the ethos of the times. It sold 20,000 copies in the year of publication and a quarter of a million copies by 1904, when Smiles died.

Smith was selected by Smiles as an example of 'patient and laborious effort, and the diligent

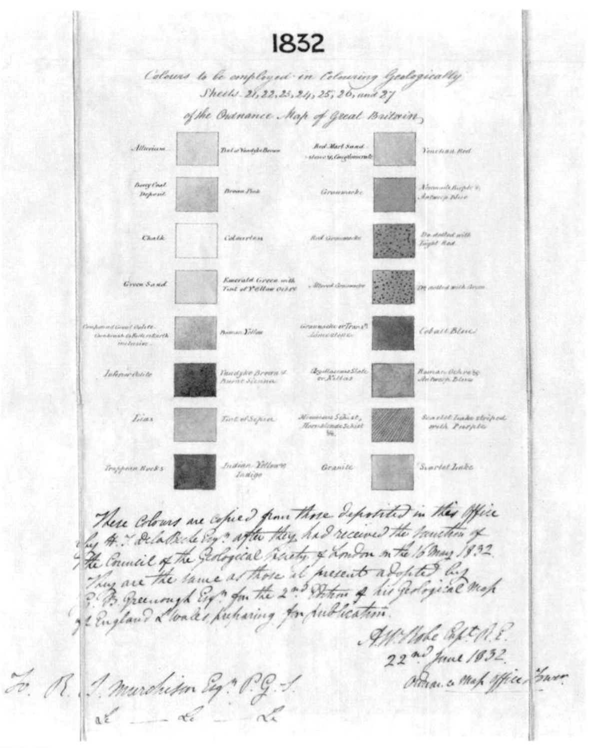

FIG. 16.

FIG. 17.

Fig. 16.
Before the Geological Survey of Great Britain was formally founded in 1835, Henry Thomas De la Beche used Ordnance Survey maps as the base for geological surveys. He secured funding from the Board of Ordnance, on the condition that he establish an agreed scheme of colours with the Geological Society. The first iteration of this colour scheme is shown here.

Fig. 17.
The British Geological Survey began its geological mapping programme at a scale of 1 inch to the mile in the 1830s, using newly available Ordnance Survey maps as their base. Denbigh, North Wales, sheet 79 SE (below) was published in 1850, and Fforest Fawr and Brecon, sheet 42 SW (above) was surveyed in 1845.

cultivation of opportunities' (Smiles, 1859, p. 91). After quoting the 1831 Wollaston Medal citation, Smiles adds that 'it is difficult to speak in terms of too high praise of the first geological map of England', which according to an anonymous obituarist in 1858, 'was a work so masterly in conception and so correct in general outline, that in principle it served as a basis not only for the production of later maps of the British Isles, but for geological maps of all other parts of the world' (Smiles, 1859, p. 97).

However, by the 1820s a new international terminology began to permeate the lexicon of British stratigraphy and gradually displace Smith's vernacular terminology. For instance, in 1822, the coal measures and associated strata were grouped as the 'Carboniferous' system by W. D. Conybeare (1787–1857) and William Phillips, while the Chalk, called the *Terrain Crétace* by the Belgian geologist J. P. J. d'Omalius d'Halloy (1783–1875) was anglicized to 'Cretaceous' system.

In this international race to distinguish, establish and name the successive major divisions of the stratigraphic record, several 'rising stars' of geology in Victorian Britain – Sedgwick, Murchison, Lyell and Smith's nephew, John Phillips – were at the forefront. By mid-century nearly all periods of geological time and the history of life from Cambrian to Quaternary were named, and there was growing international recognition of the global validity of these subdivisions. In 1860 John Phillips took his uncle's use of fossils a stage further to group the history of life into three great eras: the Palaeozoic, meaning 'ancient life', the 'age of fishes'; the Mesozoic, 'middle life', the 'age of reptiles'; and the Cenozoic, 'recent life', the 'age of mammals'.

The story of William Smith's achievements is extraordinary and in many ways tragic, although he seems to have been relatively undaunted by the setbacks. John Phillips, who from 1815 shared much of Smith's life and troubles, was very protective of his uncle's reputation. He admitted as much in his 1844 *Memoirs* and 'purposely softened the darkest outlines of Mr Smith's private and personal fortunes'. In an earlier 1831 letter to Sedgwick, Phillips referred to Smith's 'poverty, disappointment and neglect, forced seclusion from the world of science'. Furthermore, 'these have been heightened by a still more severe and invincible torment, a mad, bad wife'. Strong words, but if anyone would know apart from Smith, it was his nephew. Very little is known about Mary Ann, Smith's wife, apart from her being 'an eccentric little round-faced woman ... about as unsuited for being the partner of a meditative philosopher as she could be', according to William Crawford Williamson (1816–1895). However, even for this thoughtful

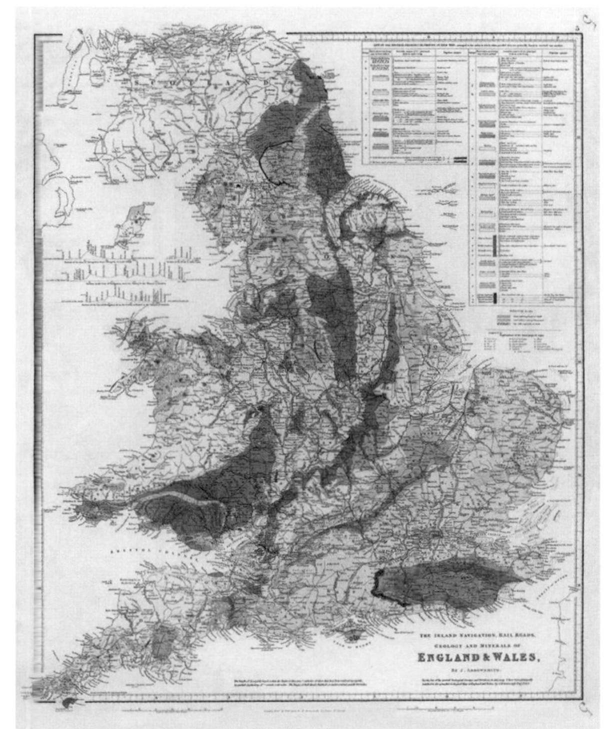

FIG. 18.

Fig. 18.
One of several small-scale geological maps of England and Wales, aimed at a popular market, published in 1834 by John Arrowsmith (1790–1873). On a scale of 1:949,400 and developed from the 1820 maps of Conybeare and Greenough, it was a more professionally produced version and distinguishes some thirty different strata and other rock types.

acquaintance of Smith, the question of 'Who she was, or where he had found her, we never could learn' (Williamson, 1877, p. 65).

It would seem that Smith had the physical and mental resilience to achieve all that he did despite considerable hardship and fairly rough living conditions in pursuit of his aims. Something of an obsessive personality is suggested by his friend the mineral surveyor John Farey (1766–1826), who reported that Smith had to retrieve 'his pecuniary affairs, from the embarrassments that a too ardent zeal in the prosecution of this great and truly *national undertaking*, had brought on him' (Farey, 1815, p. 335). An obituarist for the *Scarborough Herald* portrayed Smith as 'Simple minded, of a heaven born temper, warm hearted, kind, cheerful and instructive.'

Perhaps the last word should go to Smith himself. In 1829, he versified on his empirical view of the distribution of fossils through the succession of strata. It was an attitude that in many ways served him well, as he was not deflected from his main purpose by contemporary arguments about how and why fossils change through time.

Theories that have the earth eroded
May all with safety be exploded
For of the Deluge we have data
Shells in plenty mark the strata,
And though we know not yet awhile
What made them range, what made them pile,
Yet this one thing full well we know –
How to find them ordered so.

(see pages 46–47)

SHEET II.

30 15 1° 45 30 15 0 15

S T

ENGL

S C

THE C

THE MARSHES AND FEN

VA

ACCORDING TO

ILLUSTRA

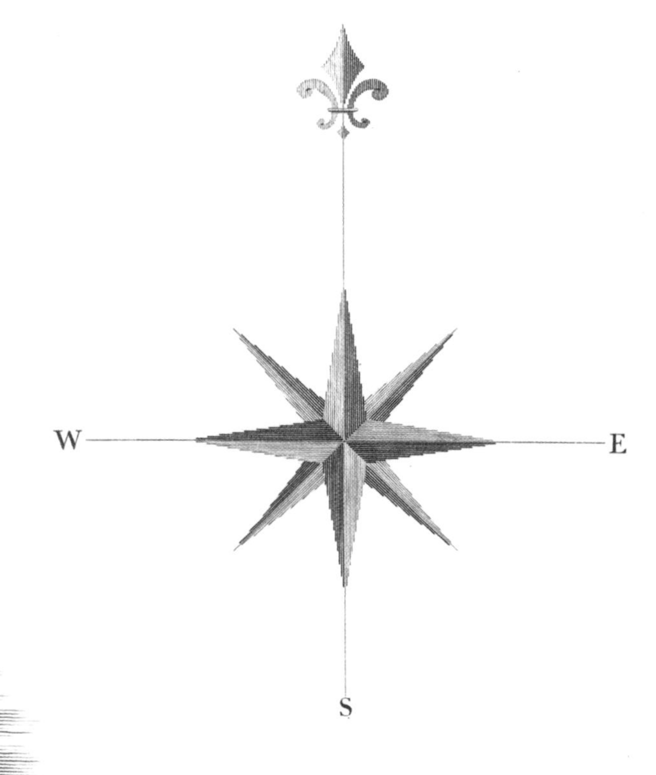

30 45 1° 15 30 45 2° 15

A

ELINEATION

OF THE

RATA

OF

AND AND WALES,

WITH PART OF

OTLAND;

EXHIBITING

LLIERIES AND MINES,

LANDS *ORIGINALLY OVERFLOWED* BY THE SEA,

AND THE

IETIES OF SOIL

E VARIATIONS IN THE SUBSTRATA,

ED by the MOST DESCRIPTIVE NAMES

BY W. SMITH.

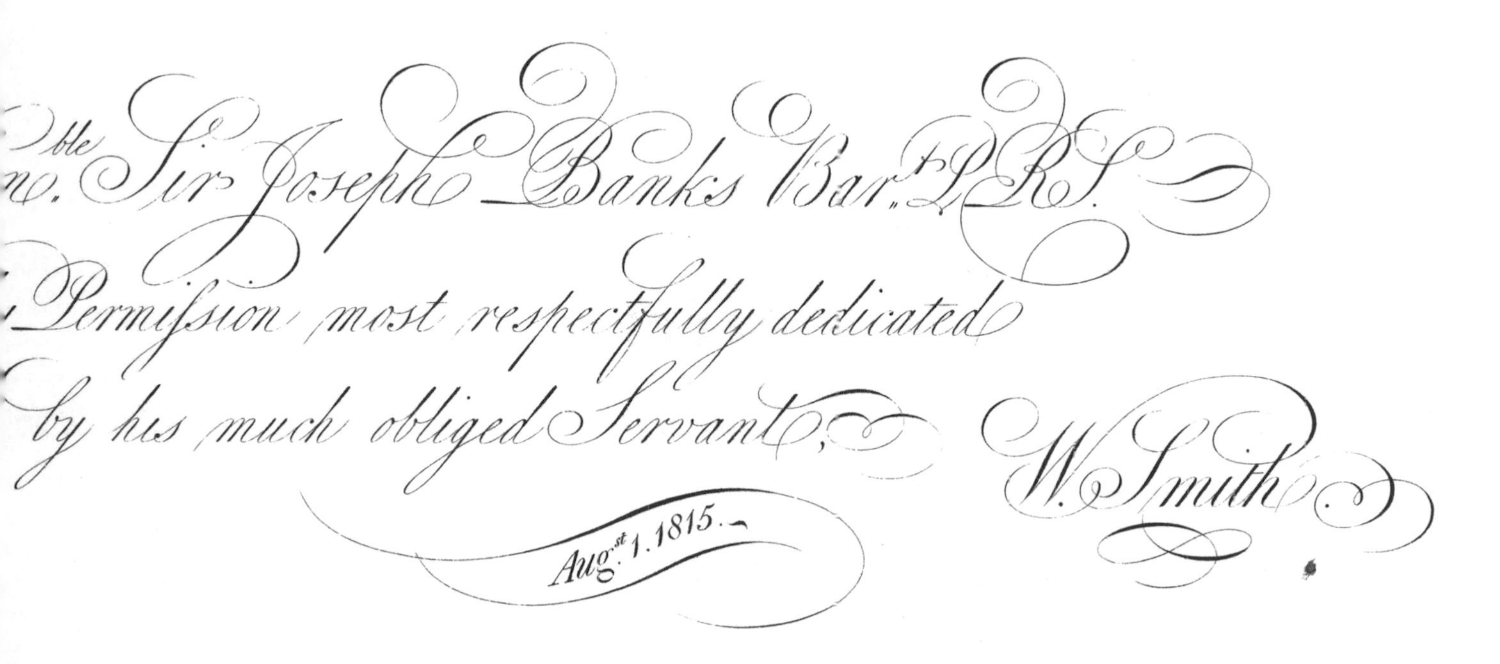

ble
no. Sir Joseph Banks Bar.t P.R.S.
Permission most respectfully dedicated
by his much obliged Servant,
W. Smith
Aug.st 1. 1815.

15 56° 45 30 15 55°

Hartlepool
Wolsingham
WOLSINGHAM COMMON
Bishop Auckland
HAMSTERLY COMMON
TEESDALE FOREST
HARWOOD COMMON
MICKLE FELL
LUNE FOREST
KELTON FELL
Bernard Castle
Staindrop
STAINMOOR
Brough
Kirkby Stephen
Sedgefield
Stockton
Darlington
Redcar
Gisborough
Yarm
CLEVELAND
RICHMOND
ARKENGARTH FOREST
Richmond
Middleton Tyas
Ripon
Bedale
Leyburn
Middleham
Masham
Askrigg
Hawes
Hardrow
COAL PIT MOOR
STAKE FELL
LANGSTERDALE CHACE
INGLEBOROUGH
NETHERDALE FOREST
MASHAMSHIRE CHACE
Thirsk
Helmsley
YORK
Easingwold
Boroughbridge
Knaresborough
Harrowgate
Ripley
Wetherby
Tadcaster
KNARESBORO FOREST
ROMBALDS MOOR
Skipton
Settle
Otley
Keighley
Bingley
Colne
Clitheroe
BOULSWORTH COMMON
Burnley
LEEDS
Bradford
Halifax
Huddersfield
Wakefield
Pontefract
Selby
Rochdale
Bury
Barnsley
Doncaster
AINSTY
FOREST OF GALTRES
SOUTH MOOR

BORDERS AND THE NORTH.

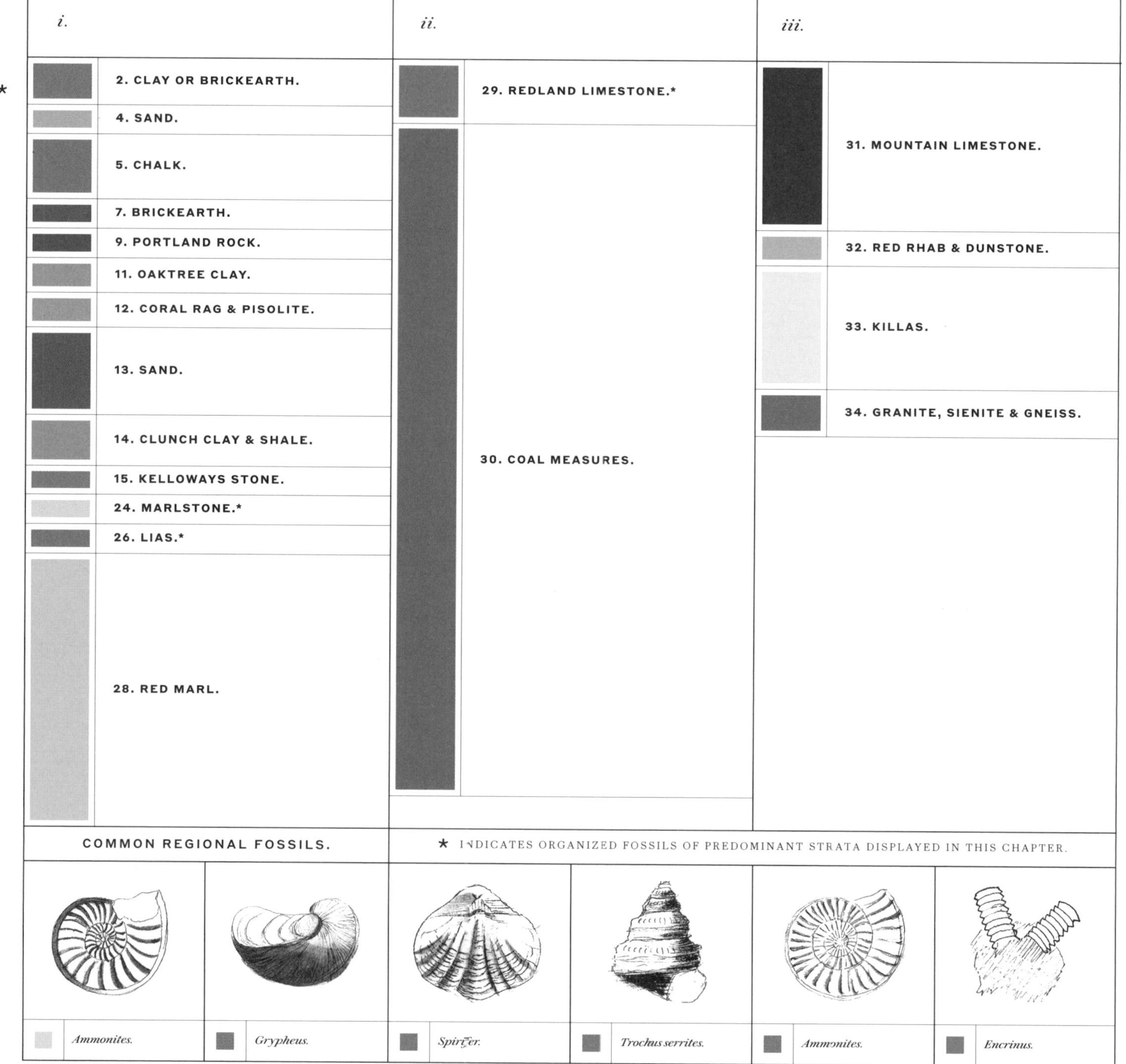

i.
ii.
iii.
2. CLAY OR BRICKEARTH.
4. SAND.
5. CHALK.
7. BRICKEARTH.
9. PORTLAND ROCK.
11. OAKTREE CLAY.
12. CORAL RAG & PISOLITE.
13. SAND.
14. CLUNCH CLAY & SHALE.
15. KELLOWAYS STONE.
24. MARLSTONE.*
26. LIAS.*
28. RED MARL.
29. REDLAND LIMESTONE.*
30. COAL MEASURES.
31. MOUNTAIN LIMESTONE.
32. RED RHAB & DUNSTONE.
33. KILLAS.
34. GRANITE, SIENITE & GNEISS.
COMMON REGIONAL FOSSILS.
* INDICATES ORGANIZED FOSSILS OF PREDOMINANT STRATA DISPLAYED IN THIS CHAPTER.
Ammonites.
Grypheus.
Spirifer.
Trochus serrites.
Ammonites.
Encrinus.

Bar chart of strata indicates the comparative area covered by each stratum in this region.

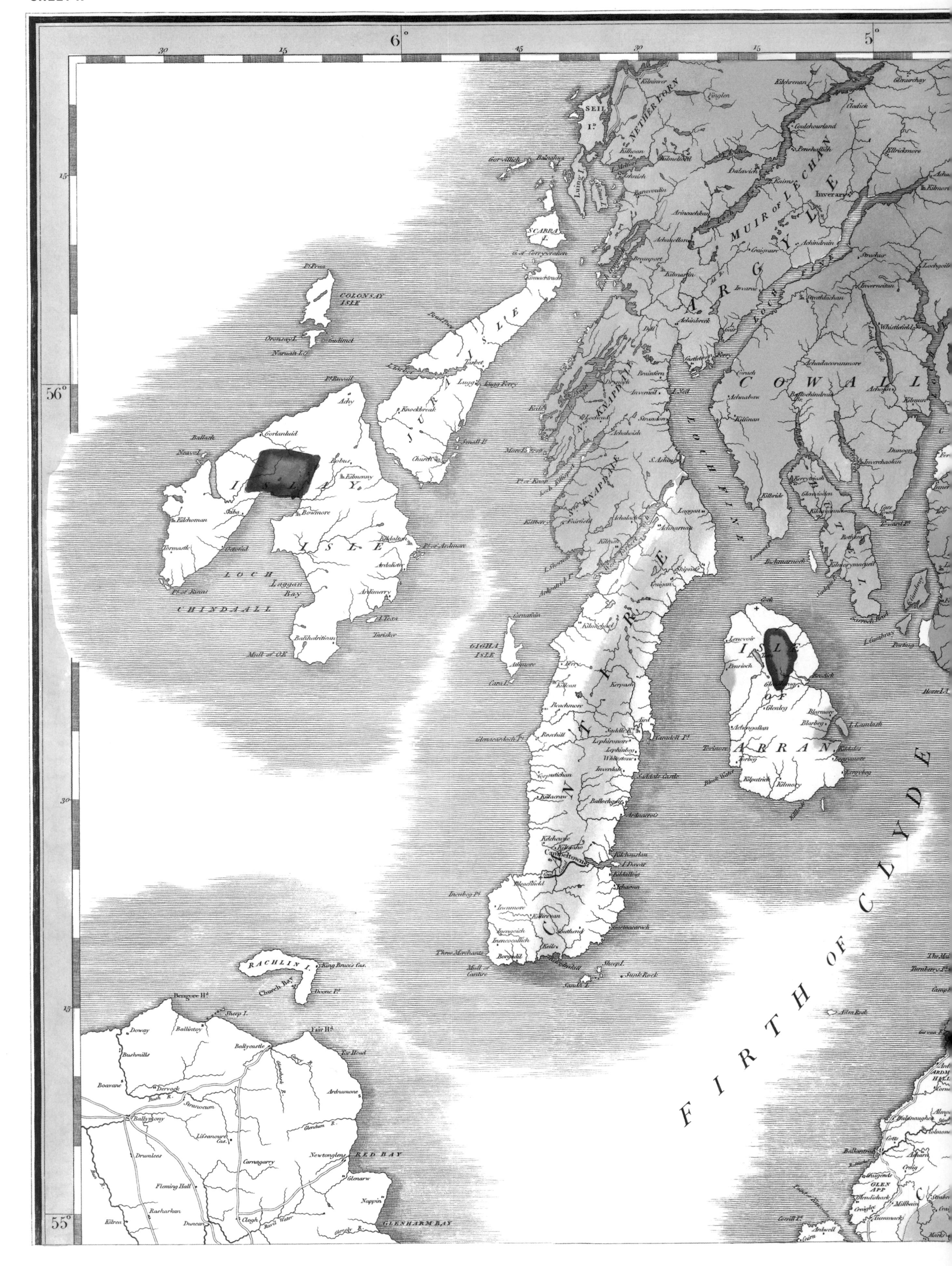
ISLAY ISLE
JURA ISLE
COLONSAY ISLE
GIGHA ISLE
SCARBA I.
SEIL I.
NETHER LORN
MUIR OF LECHAN
ARGYLE
COWALL
LOCH FINE
CANTIRE
ISLE OF ARRAN
FIRTH OF CLYDE
LOCH CHINDAALL
RACHLIN I.
RED BAY
GLENHARM BAY
Inverary
Bowmore
Campbeltown
Ballymony
Ballycastle
Bushmills
Kilchoman
Mull of OE
Mull of Cantire
Fair Hd.
Bengore Hd.
56°
55°
6°
5°

SHEET II.

(see pages 46–47)

(see pages 44–45)

SHEET I.

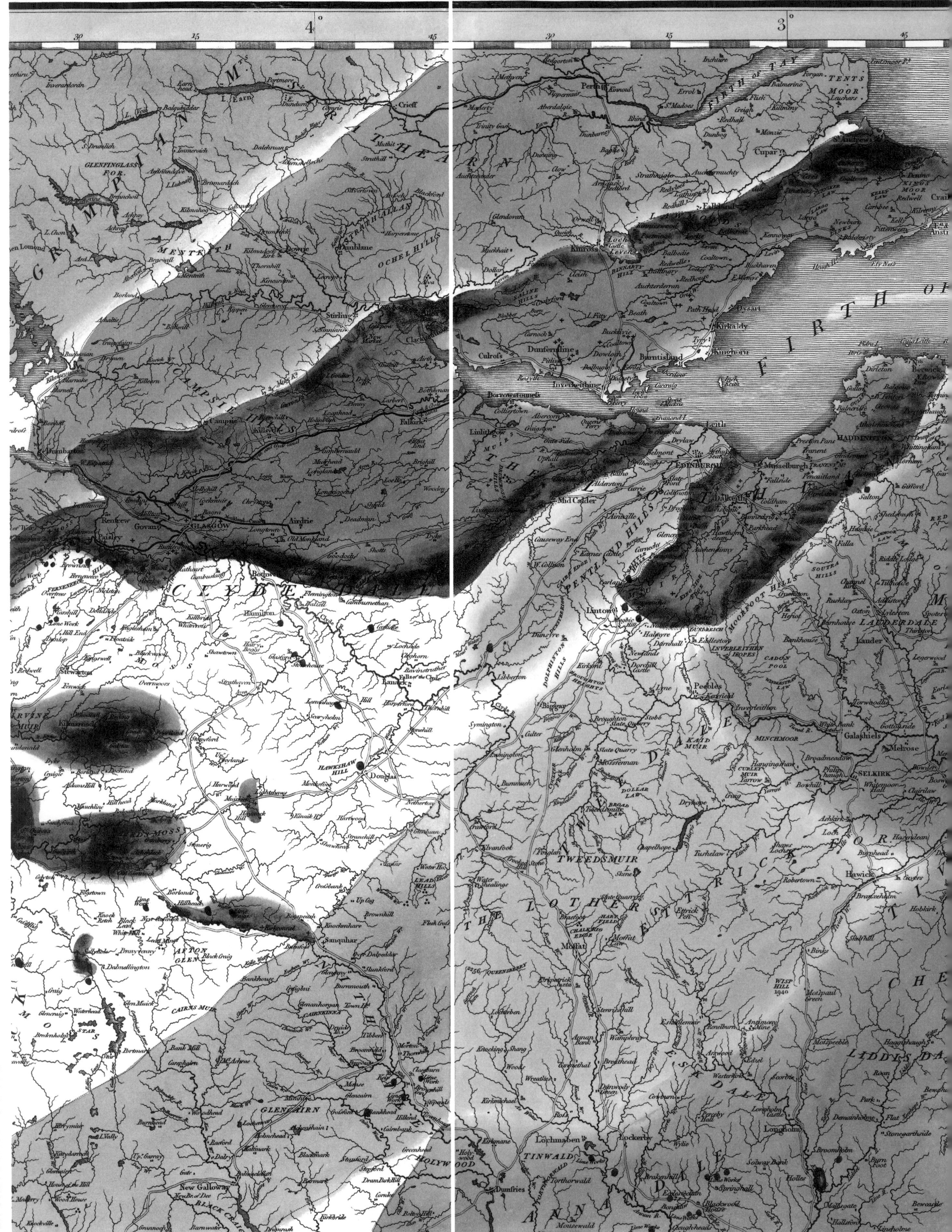

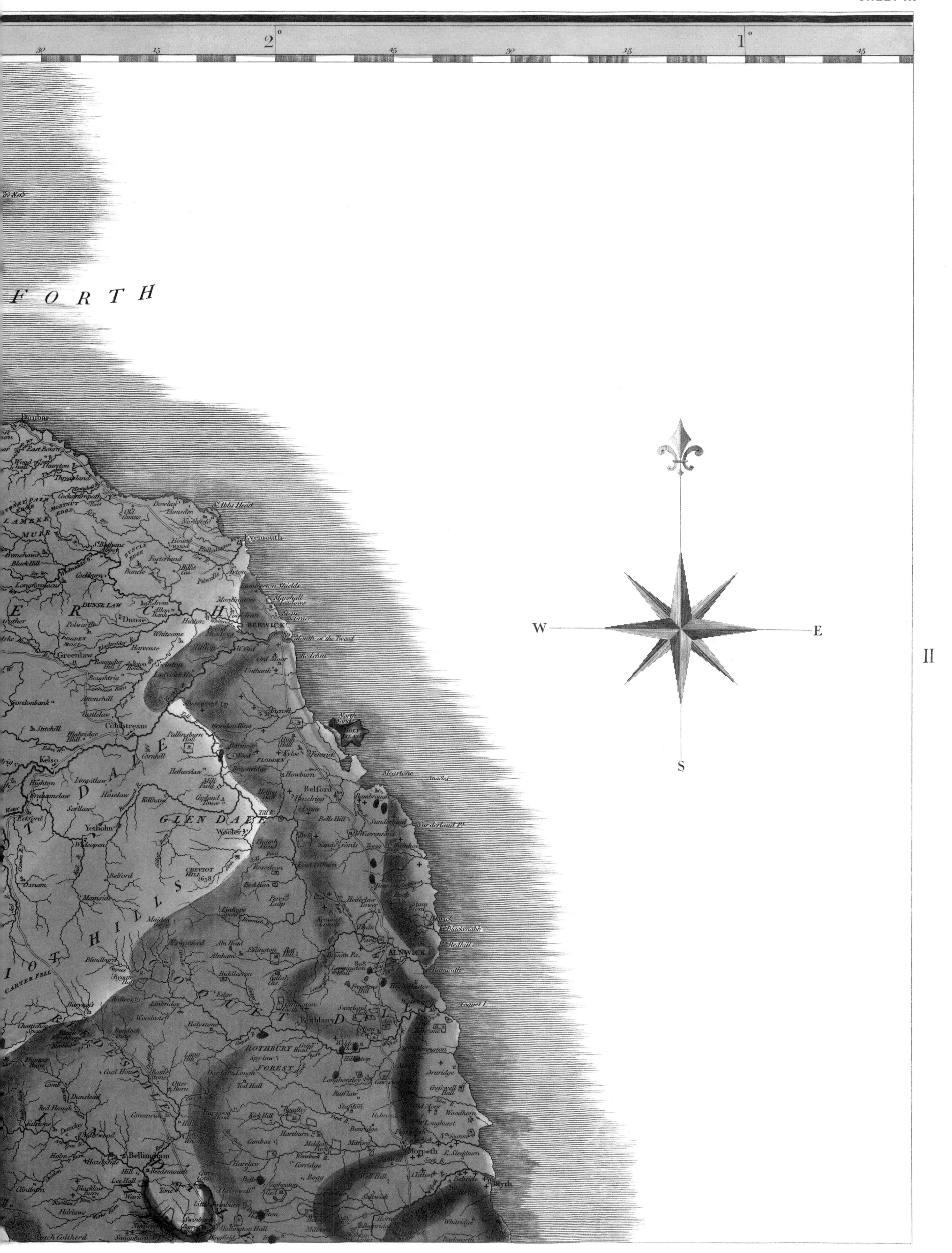
2°
1°
30
15
45
F O R T H
Dunbar
St. Abbs Head
Eyemouth
Dunse
DUNSE LAW
Greenlaw
BERWICK
Mouth of the Tweed
Coldstream
Kelso
Cornhill
HOLY ISLAND
Belford
Yetholm
GLENDALE
Wooler
CHEVIOT HILL
2658
ALNWICK
Alnmouth
Coquet I.
Rothbury
ROTHBURY
FOREST
Morpeth
Bellingham
Blyth
Sunderland Pt.
W
E
S
II

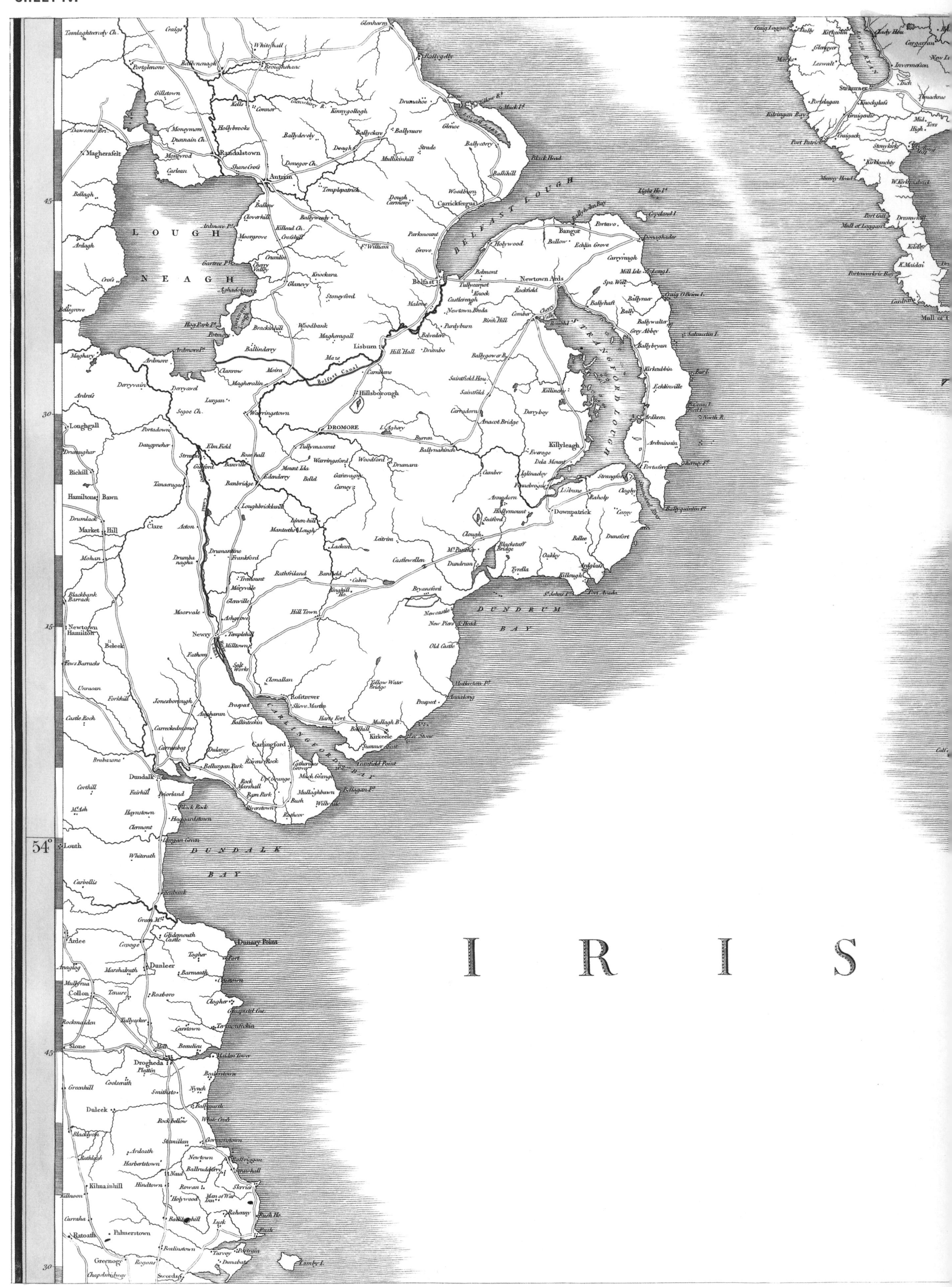
LOUGH NEAGH
BELFAST LOUGH
STRANGFORD LOUGH
DUNDRUM BAY
CARLINGFORD BAY
DUNDALK BAY
IRIS
Belfast
Lisburn
Antrim
Randalstown
Magherafelt
Carrickfergus
Newtown Ards
Bangor
Donaghadee
Hillsborough
DROMORE
Killyleagh
Downpatrick
Portaferry
Newry
Dundalk
Carlingford
Drogheda
Dunleer
Ardee
Louth
Market Hill
Richill
Loughgall
Stranraer
Port Patrick
54°

 (see pages 50–51)

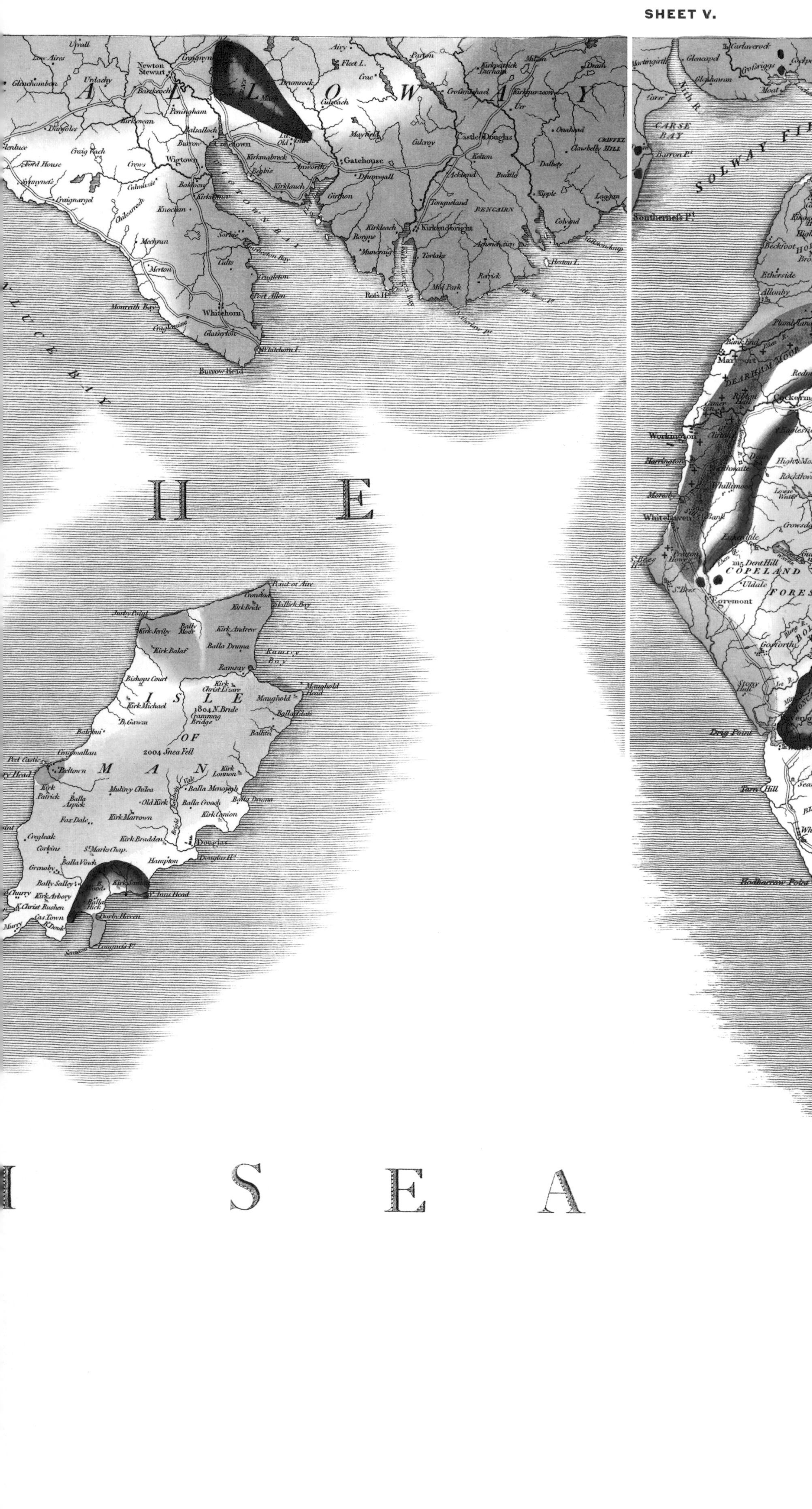

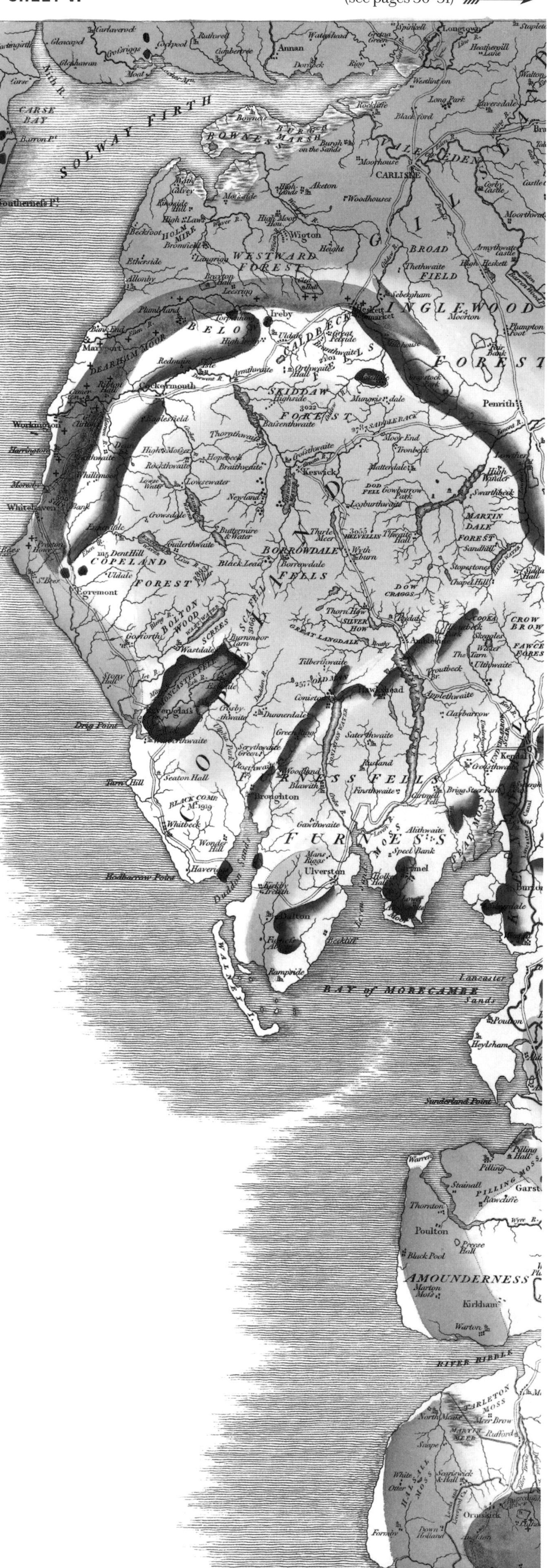

SOLWAY FIRTH
CARSE BAY
Southerness Pt.
BOWNESS MARSH
CARLISLE
Wigton
WESTWARD FOREST
INGLEWOOD FOREST
Brampton
Penrith
Keswick
Cockermouth
Workington
Maryport
Whitehaven
COPELAND FOREST
SKIDDAW FOREST
BORROWDALE FELLS
HELVELLIN
FURNESS FELLS
FURNESS
Ulverston
Drig Point
Hodbarrow Point
Walney
BAY of MORECAMBE
Lancaster
LANCASTER
Heysham
Sunderland Point
Garstang
Poulton
Black Pool
AMOUNDERNESS
Kirkham
RIVER RIBBLE
Preston
Blackburn
Clitheroe
Burnley
TARLETON MOSS
Chorley
Bolton
Bury
Rochdale
Ormskirk
Leigh
Halifax
Huddersfield
Skipton
Settle
Kirkby Lonsdale
Sedbergh
Kendal
Appleby
Brough
Hexham
Allondale
Wolsingham
Stanhope
Richmond
Middleham
Askrigg
Hawes
WEARDALE FOREST
CROSS FELL 2901
MILBOURN FOREST
LUNE FOREST
STAINMOOR
ARKENGARTH FOREST
LANGSTERDALE CHACE
NETHERDALE FOREST
BURNMOOR
BOLLAND FOREST
ROMBALDS MOOR
ROULSWORTH COMMON
ROSENDALE FOREST
CHURCH MOOR
INGLEBOROUGH
WOLSINGHAM COMMON
HUNSTONWORTH COMMON
BULBECK COMMON
GELTSDALE
SPADEHAM WASTE
Annan
Longtown

(see pages 52–53)

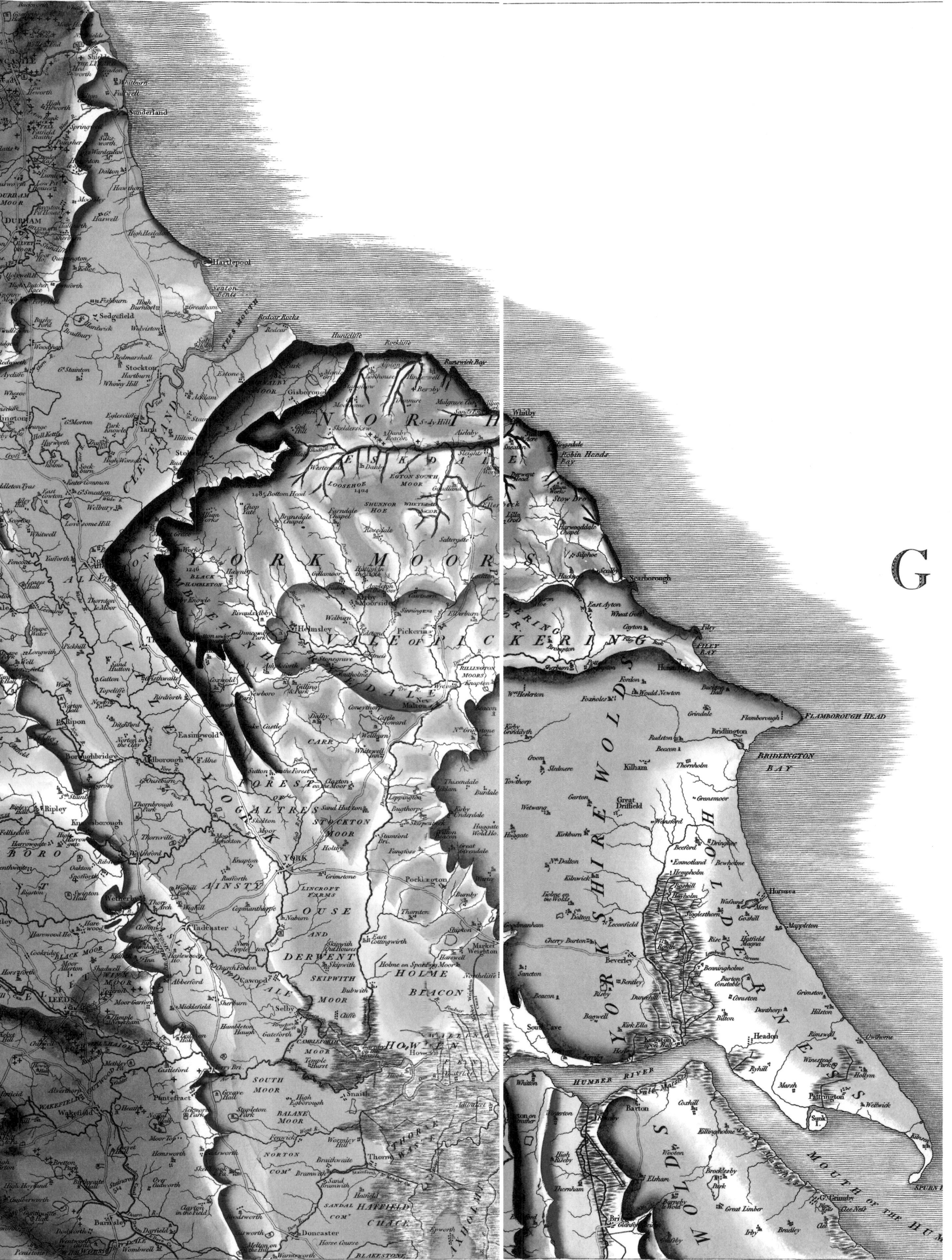

(see pages 50–51)

SHEET V.

T H E

E R M A N

C E A N

45

30

15

54°

45

30

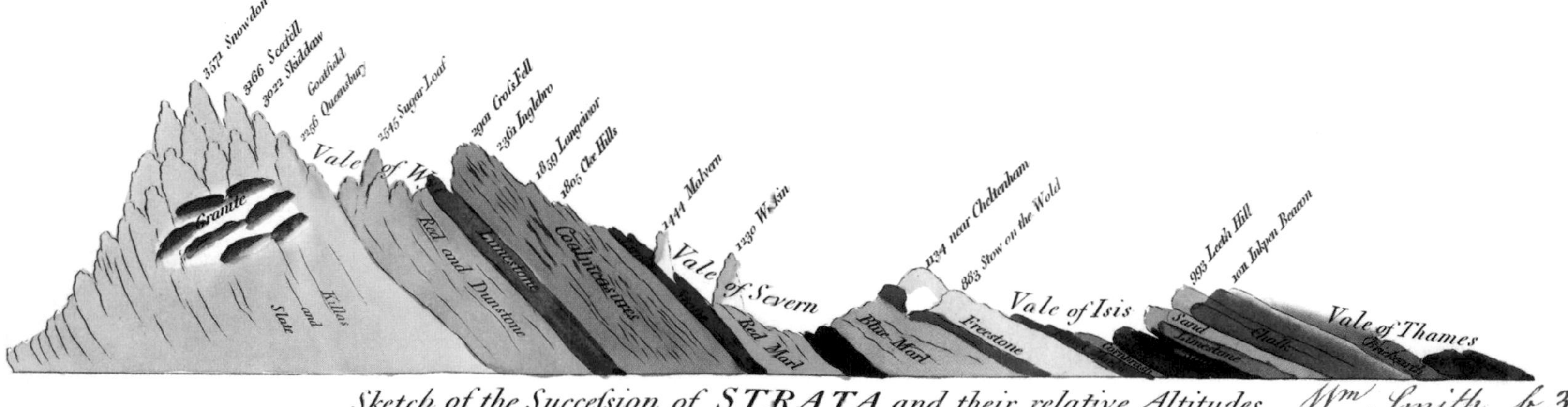

Sketch of the Succeſsion of **STRATA** and their relative Altitudes. Wm. Smith, b. 38

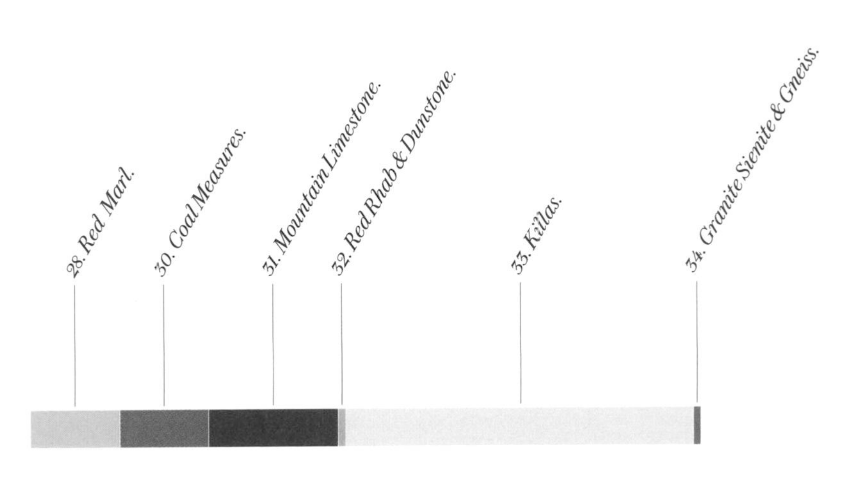

GEOLOGICAL MAP OF WESTMORELAND, 1824.

First published in Part VI of *A New Geological Atlas of England and Wales* with the counties Cumberland, Durham and Northumberland. The upper stratum is composed of Red Marl (28) above Coal Measures and Mountain Limestone (30, 31), Red Rhab and Killas (32, 33), and finally porphyritic granite (34) of Shap Fell and Hause Hill, and also Great Whin Sill 'basalts' (34).

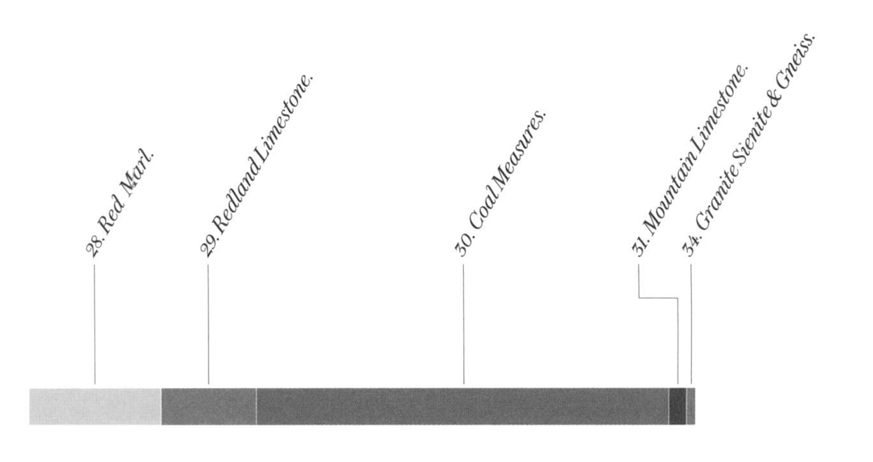

GEOLOGICAL MAP OF DURHAM, 1824.

First published in Part VI of *A New Geological Atlas of England and Wales* with the counties Cumberland, Westmoreland and Northumberland. The upper stratum is composed of Red Marl (28). Within the Coal Measures, Smith recognized Millstone Grit. The lower stratum shown is the Metalliferous Limestone (Mountain Limestone, 31). Igneous rocks are represented by Great Whin Sill 'basalts' (34).

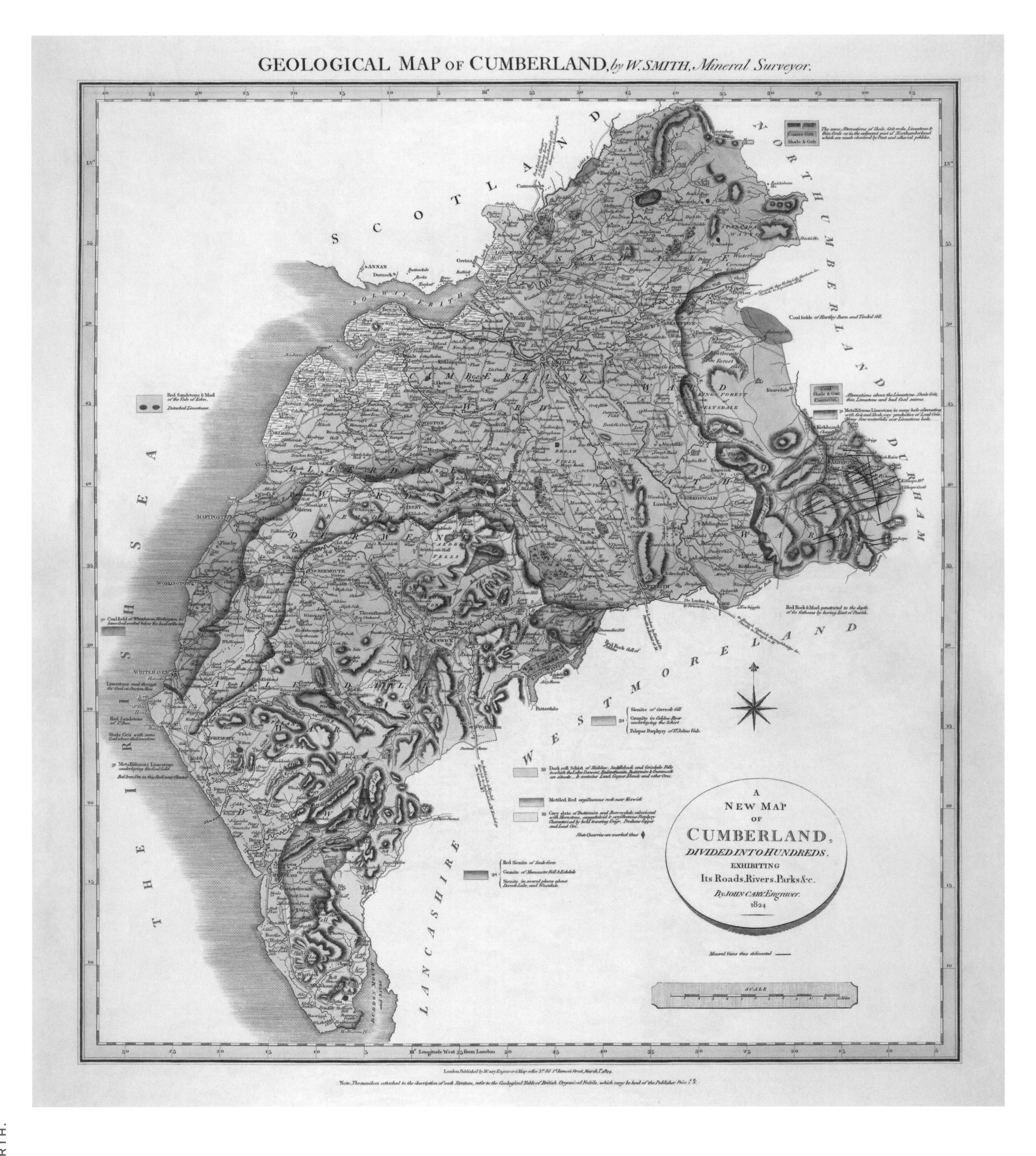

GEOLOGICAL MAP OF CUMBERLAND, 1824.

First published in Part VI of *A New Geological Atlas of England and Wales* with the counties Durham, Westmoreland and Northumberland. The upper stratum is composed of Red Marl (28), Smith included two 'detached' limestones within the Red Marl (possibly considered by him to be Lias, but actually Mountain Limestone) and the lower stratum dark Schist and grey Slate (33). Igneous rocks include Sienite and granites (34) around Eskdale, Ennerdale, Threlkeld and the Calderbeck area.

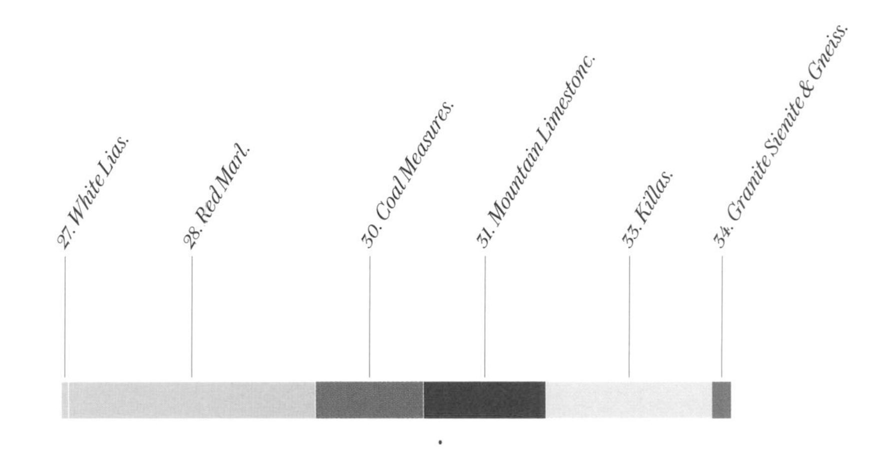

GEOLOGICAL MAP OF NORTHUMBERLAND, *by W. SMITH, Mineral Surveyor.*

A NEW MAP OF NORTHUMBERLAND, DIVIDED INTO WARDS EXHIBITING Its Roads, Rivers, Parks &c. By JOHN CARY, Engraver, 1824.

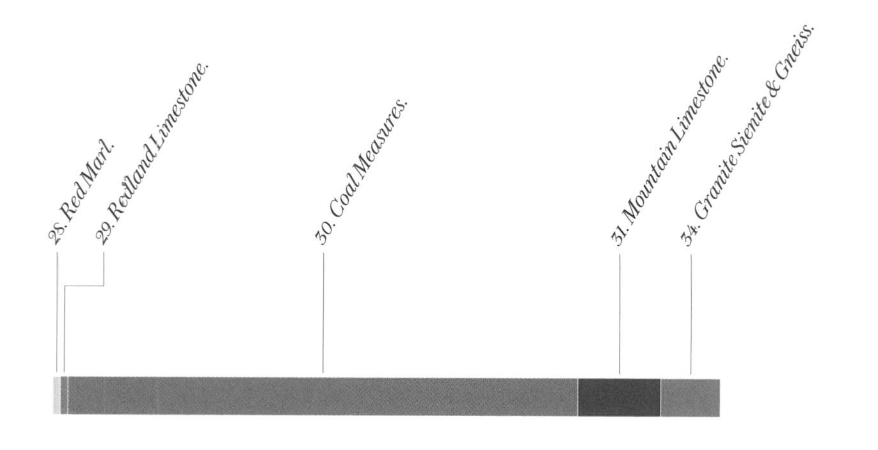

GEOLOGICAL MAP OF NORTHUMBERLAND, 1824.

First published in Part VI of *A New Geological Atlas of England and Wales* with the counties Cumberland, Westmoreland and Durham. The upper stratum is composed of Red Marl (28) above Redland Limestone (29), Coal Measures and Mountain Limestone (30, 31). On the map Smith differentiates Millstone Grit from the Coal Measures. Smith also showed Sienite and granites (34) of the Cheviots and 'basalts' (actually quartz microgabbros) of the Whin Sill.

GEOLOGICAL MAP OF YORKSHIRE, NW, 1821.

First published in Part IV of *A New Geological Atlas of England And Wales*. Smith showed Sand and Grit Freestone (13) over Clunch and Alum Shale (14) to the east, the latter is now known to be Liassic (he did record a very small Lias outcrop at Topcliff). To the west the sequence of strata progresses from Red Marl (28) to Redland Limestone (29) to Coal Measures (30). The lowermost stratum shown on the map is Mountain Limestone (31).

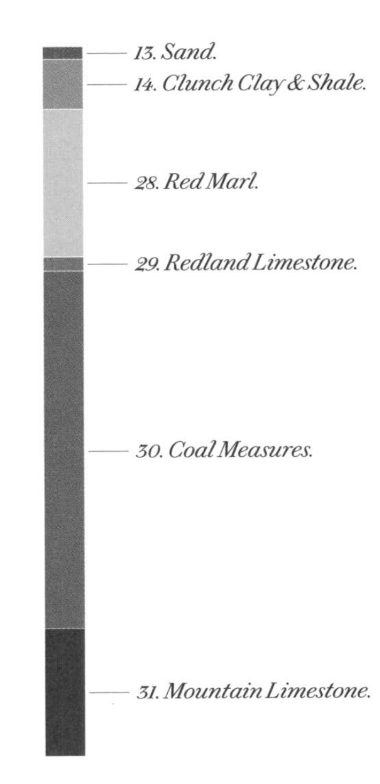

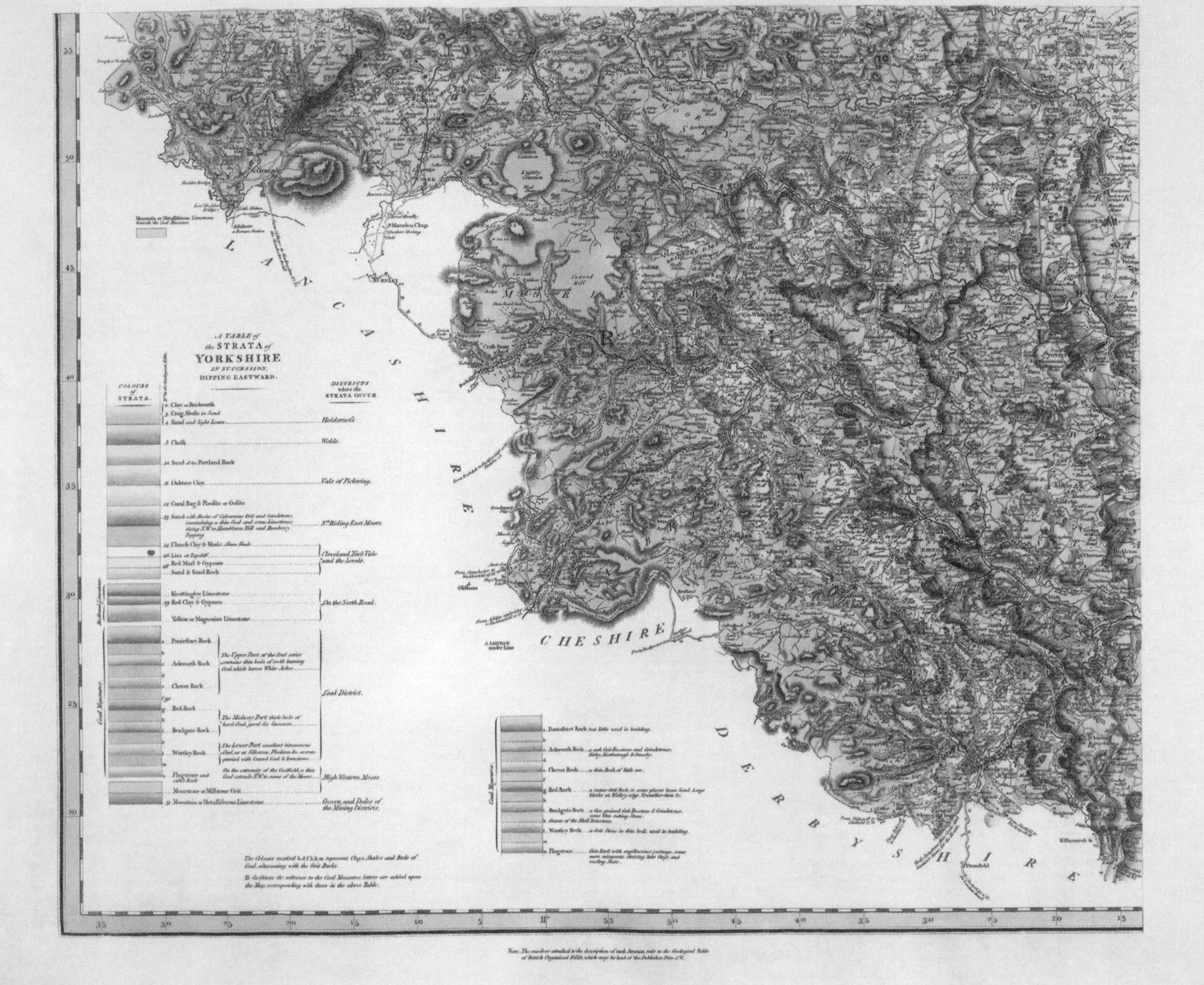

GEOLOGICAL MAP OF YORKSHIRE, SW, 1821.

First published in Part IV of *A New Geological Atlas of England and Wales*. The upper stratum is composed of Red Marl (28) with and the lower stratum Metalliferous (Mountain) Limestone bed (31). This sheet also contains a coloured stratigraphical key: 'A Table of the Strata of Yorkshire in Succession Dipping Eastward'.

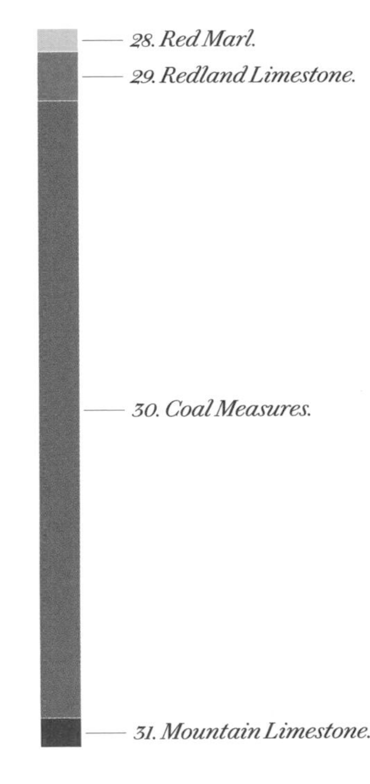

GEOLOGICAL MAP OF YORKSHIRE, NE, 1821.

First published in Part IV of *A New Geological Atlas of England and Wales*. The uppermost strata shown are Brickearth and Sand (2, 4) which overlie the Chalk (5) of the Yorkshire Wolds. On the North York Moors he showed Sand and Grit Freestone (13) over Clunch and Alum Shale (14), this was a mistake as the Clunch is in fact Lias and the sands and grits the equivalent of Oolite strata in southern England.

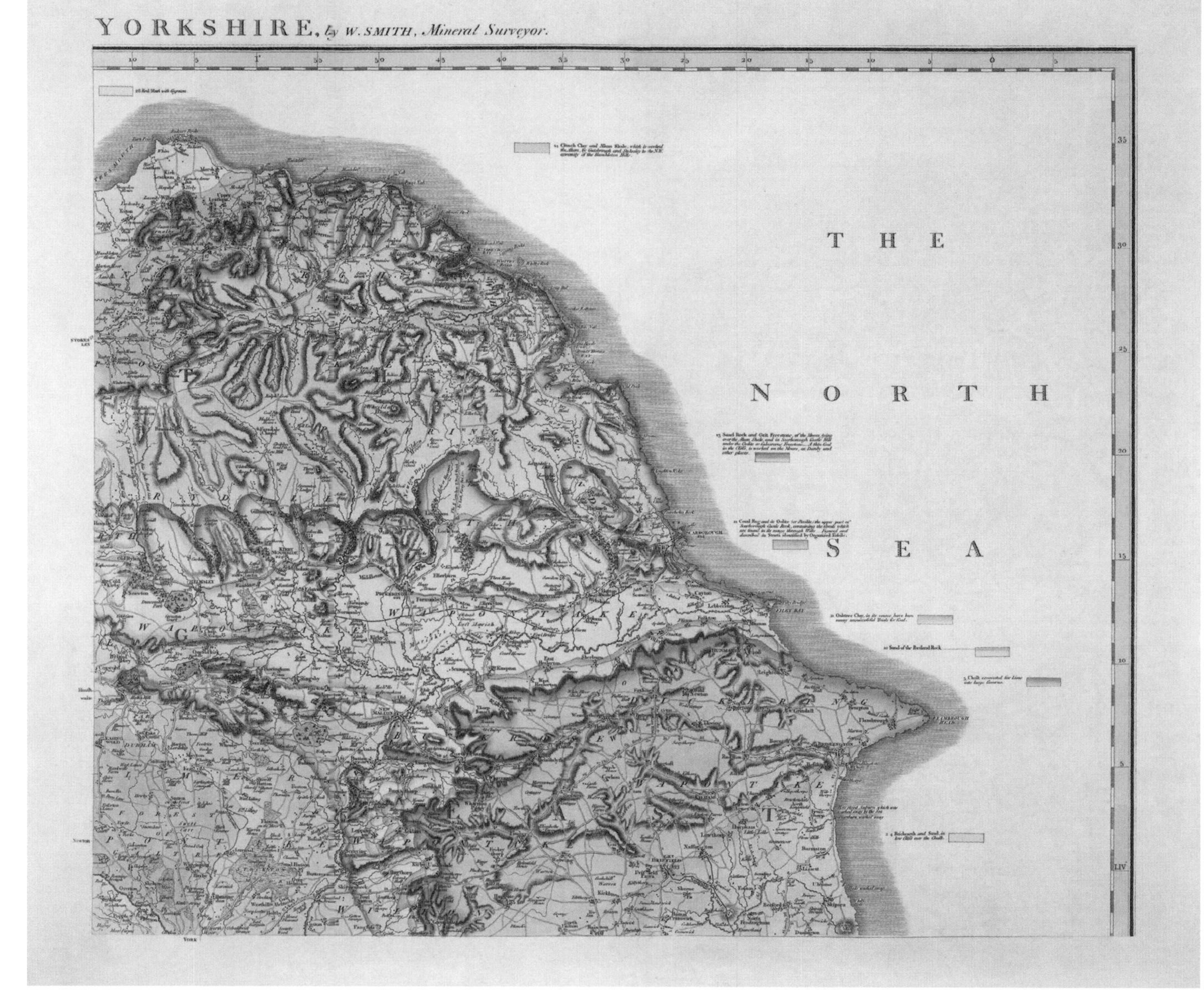

GEOLOGICAL MAP OF YORKSHIRE, SE, 1821.

First published in Part IV of *A New Geological Atlas of England and Wales*. The uppermost strata, Brickearth, Sand and Clay (2, 3, 4) are shown to the east of the map. The sequence to the west shows the Chalk (5), Oaktree Clay (11), Coral Rag Oolite with sands below (12, 13) and the Clunch (14). Below these is the Red Marl (28), Redland Limestone (29) and finally Coal Measures (30).

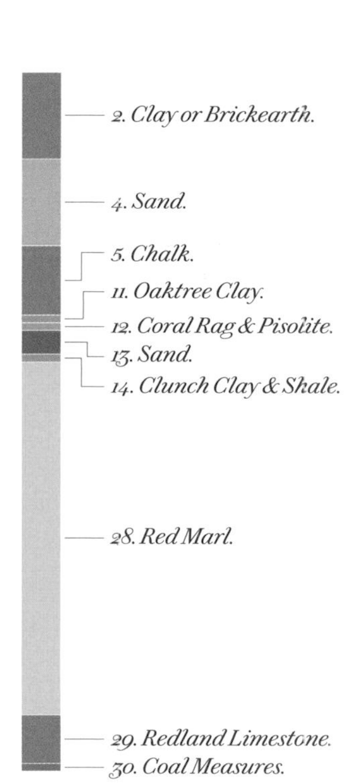

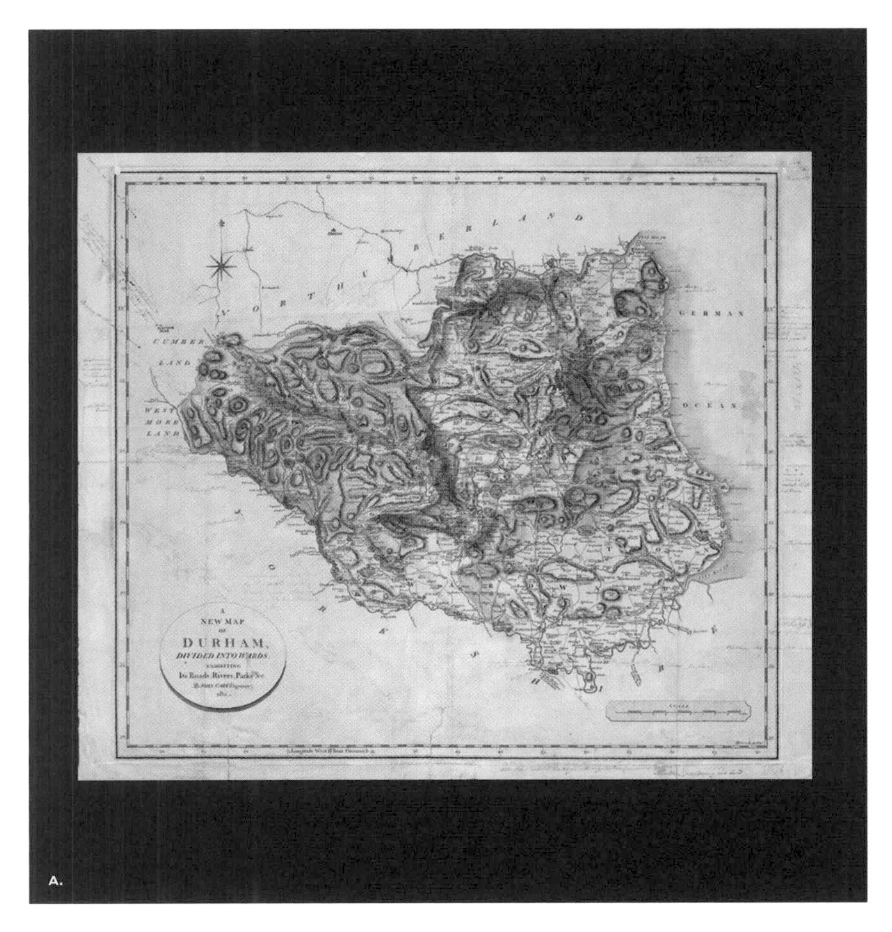

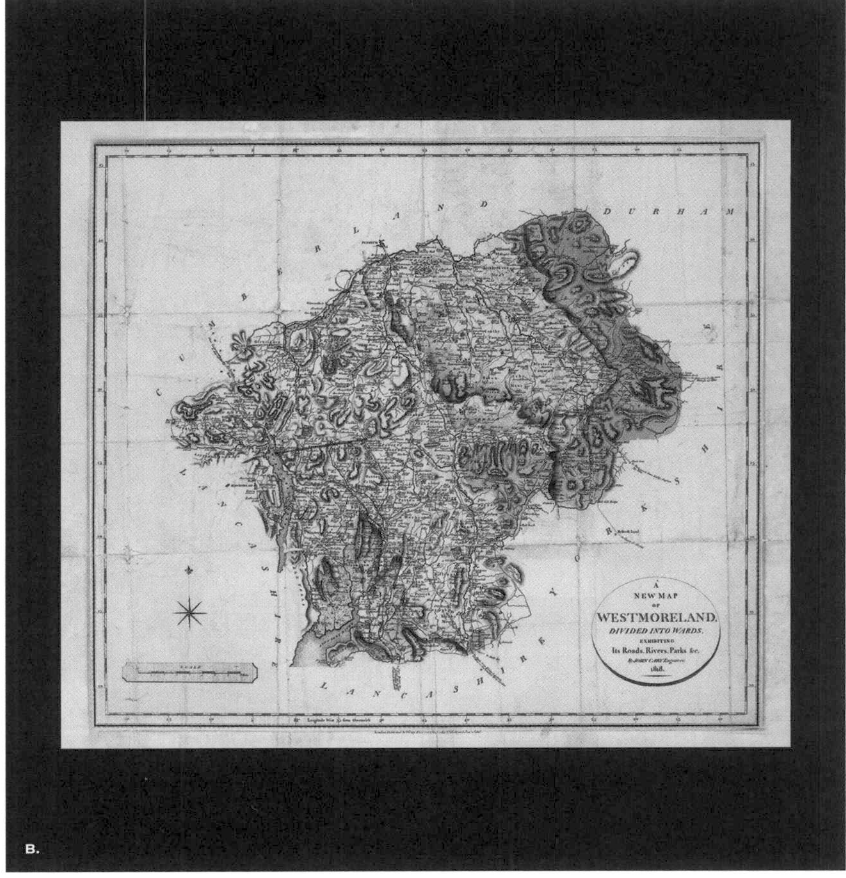

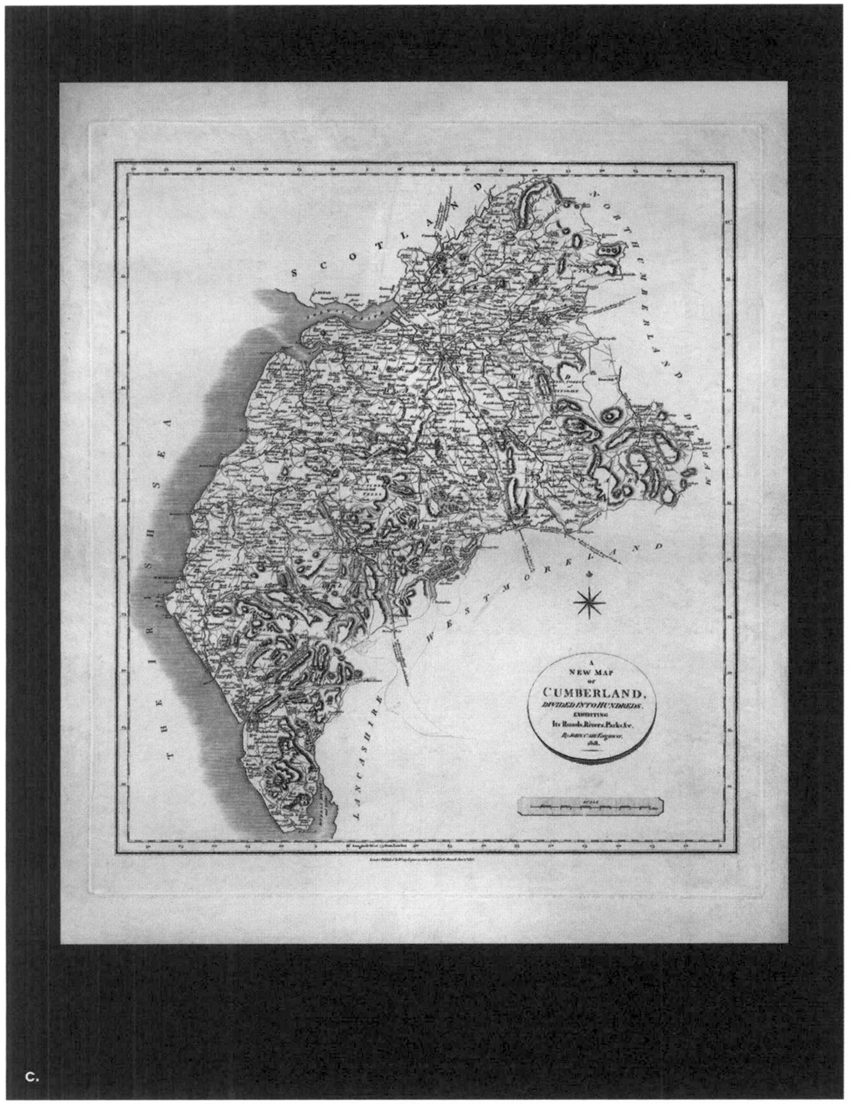

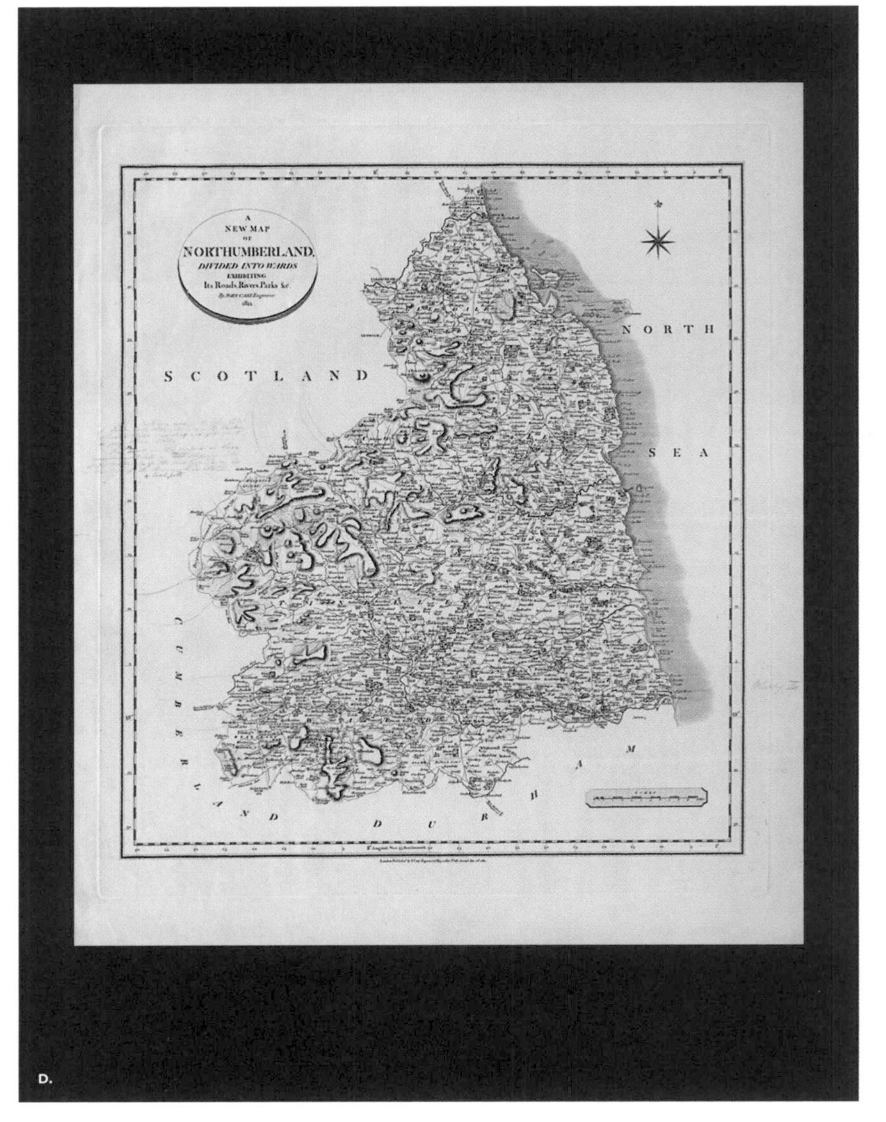

A. UNPUBLISHED GEOLOGICAL MAP OF DURHAM.

The base map is engraved by John Cary, and dated 1811. A draft, working-copy map with extensive annotation in ink and pencil by Smith and geologically coloured to show Smith's Metalliferous Limestone, Millstone Grit and Coal Measures. Smith's completed map of Durham was published in 1824.

B. UNPUBLISHED GEOLOGICAL MAP OF WESTMORELAND.

The base map is engraved by John Cary, and dated 1818. A draft map with Smith's pencil annotations and some colouring of his Metalliferous Limestone in blue and, in the northeast, the overlying Coal Shale and Grit. Smith's completed map of Westmoreland was published in 1824.

C. UNPUBLISHED GEOLOGICAL MAP OF CUMBERLAND.

The base map is engraved by John Cary, and dated 1818. A draft map with a few pencil annotations by Smith. Smith's completed map of Cumberland was published in 1824.

D. UNPUBLISHED GEOLOGICAL MAP OF NORTHUMBERLAND.

The base map is engraved by John Cary, and dated 1811. It has a few pencil annotations by Smith. Smith's completed map of Northumberland was published in 1824.

E. UNPUBLISHED GEOLOGICAL MAP OF LANCASHIRE.

The base map is engraved by John Cary, and dated 1818. A draft map with extensive pencil and ink annotation by Smith, some dating from the 1830s. Colouring shows Smith's Metalliferous Limestone and the Coal Shale and Grit. No completed geological map of Lancashire was published by Smith.

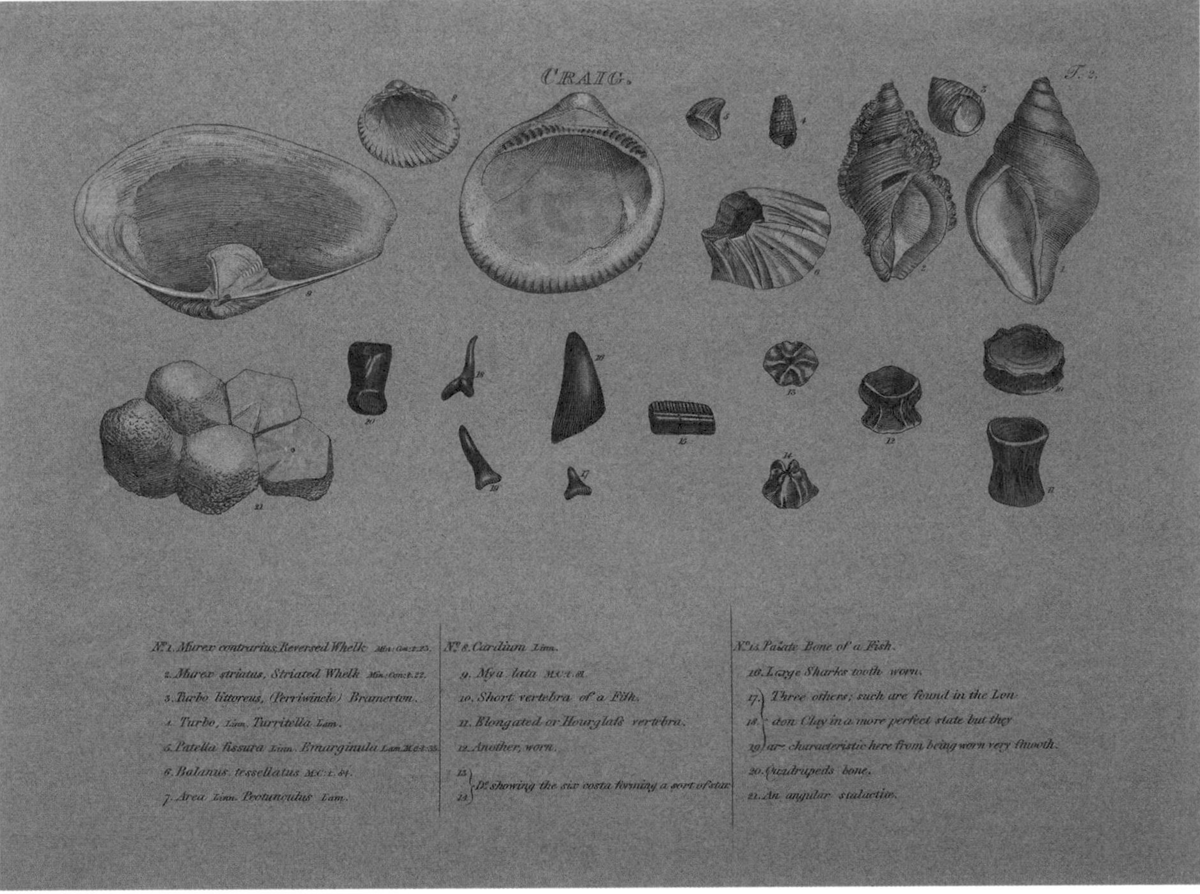

FOSSILS FOUND IN LONDON CLAY [ABOVE].

Smith illustrated fossils from the coast at Sheppey and Bognor, still favourite places for London Clay fossils today. His fossils from Highgate were from a temporary exposure when the tunnel through Highgate Hill in North London collapsed. He includes slightly older fossils from Woolwich and younger ones from Barton, Bracklesham and Hordle Cliff.

FOSSILS FOUND IN CRAG [BELOW].

The title Craig was used by Sowerby in his *Mineral Conchology*; Smith uses Crag throughout his text. Crag fossils are only found in East Anglia and relate to the drainage projects Smith worked on in the area. The fossils illustrated all come from the Red Crag, although the stratigraphic range of some extends to the Coralline Crag below and the Norwich Crag above.

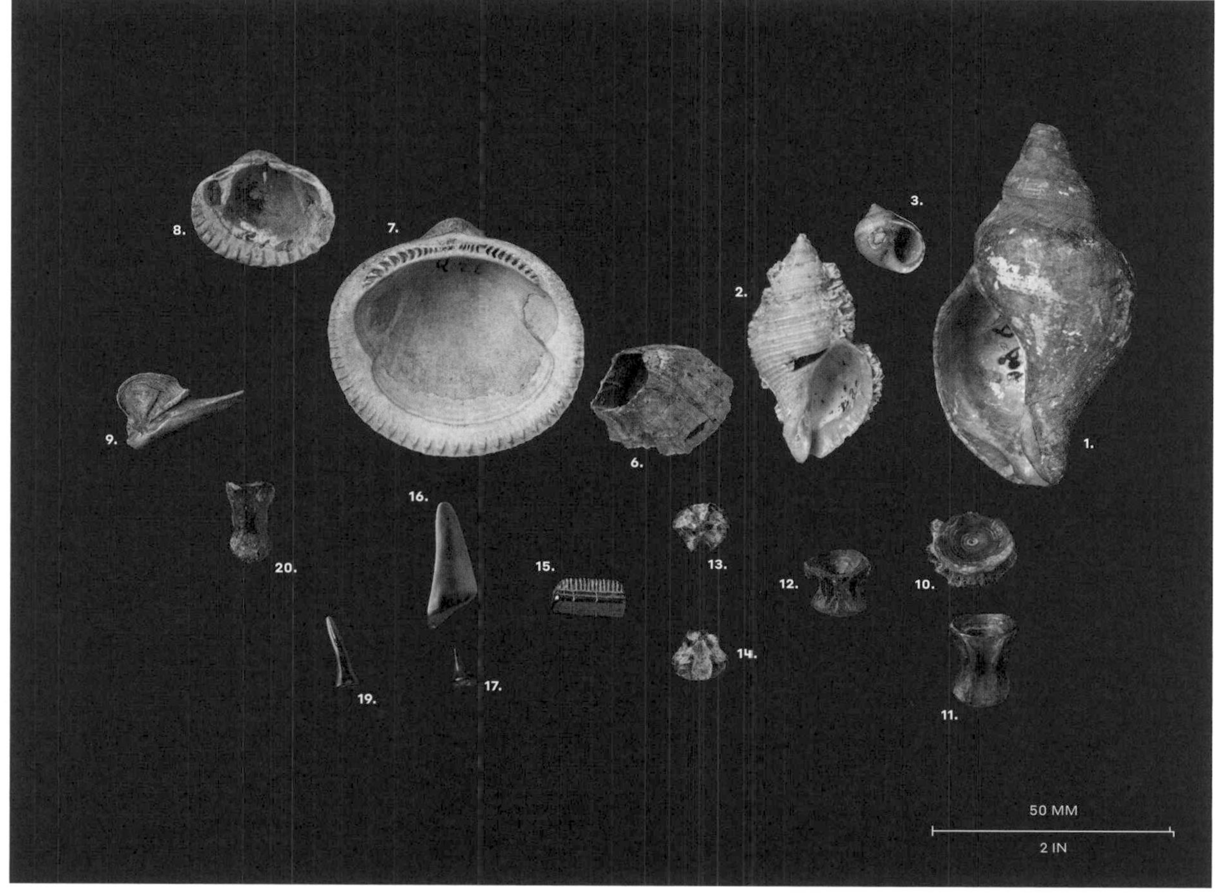

FOSSILS FOUND IN LONDON CLAY [ABOVE].

1. *Viviparus suessoniensis* G1675 Well, Brixton causeway; **2.** *Abra splendens* L1426 Sheppey; **3.** *Glycymeris brevirostris* L1430 Bognor; **6.** *Volutospina denudata* G1563 Bognor; **7.** *Brotia melanioides* G1567 Woolwich; **8.** *Otodus obliquus* P4829 Sheppey; **10.** *Striarca wrigleyi* L1435 Highgate; **13.** *Zanthopsis* sp. I749 Sheppey (substitute).

FOSSILS FOUND IN CRAG [BELOW].

1. *Neptunea angulata* G1546 Alderton; **2.** *Nucella incrassata* G1536 Bramerton; **3.** *Littorina littorea* G1558 Thorpe Common; **6.** *Balanus* sp. I747; **7.** *Glycymeris variabilis* L1409 Tattingstone Park; **8.** *Cerastoderma hostei* L1410 Tattingstone Park; **9.** *Mya arenaria* L1413 Bramerton (substitute); **10–14.** Fish vertebrae P4832, P4839, P75892, P4833 (2 views) Thorpe Common; **8.** Tooth palate ?*Aetobatus* sp. P4834 Tattingstone Park; **16, 17, 19.** Sharks teeth P4835 Stoke Hill, P4836 & P4837 Reading, Ipswich; **20.** Toe bone ?gazelle M1990 Tattingstone Park.

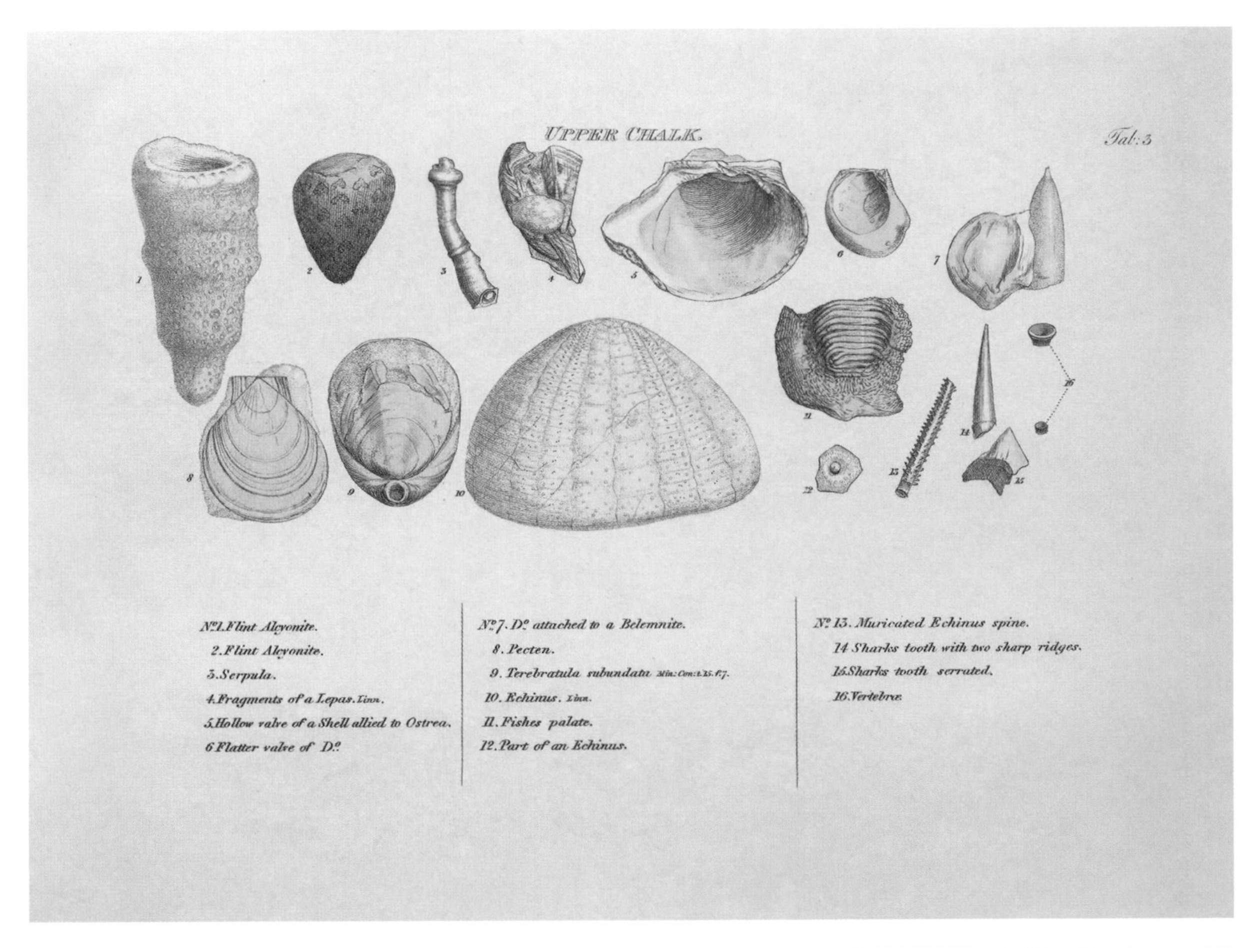

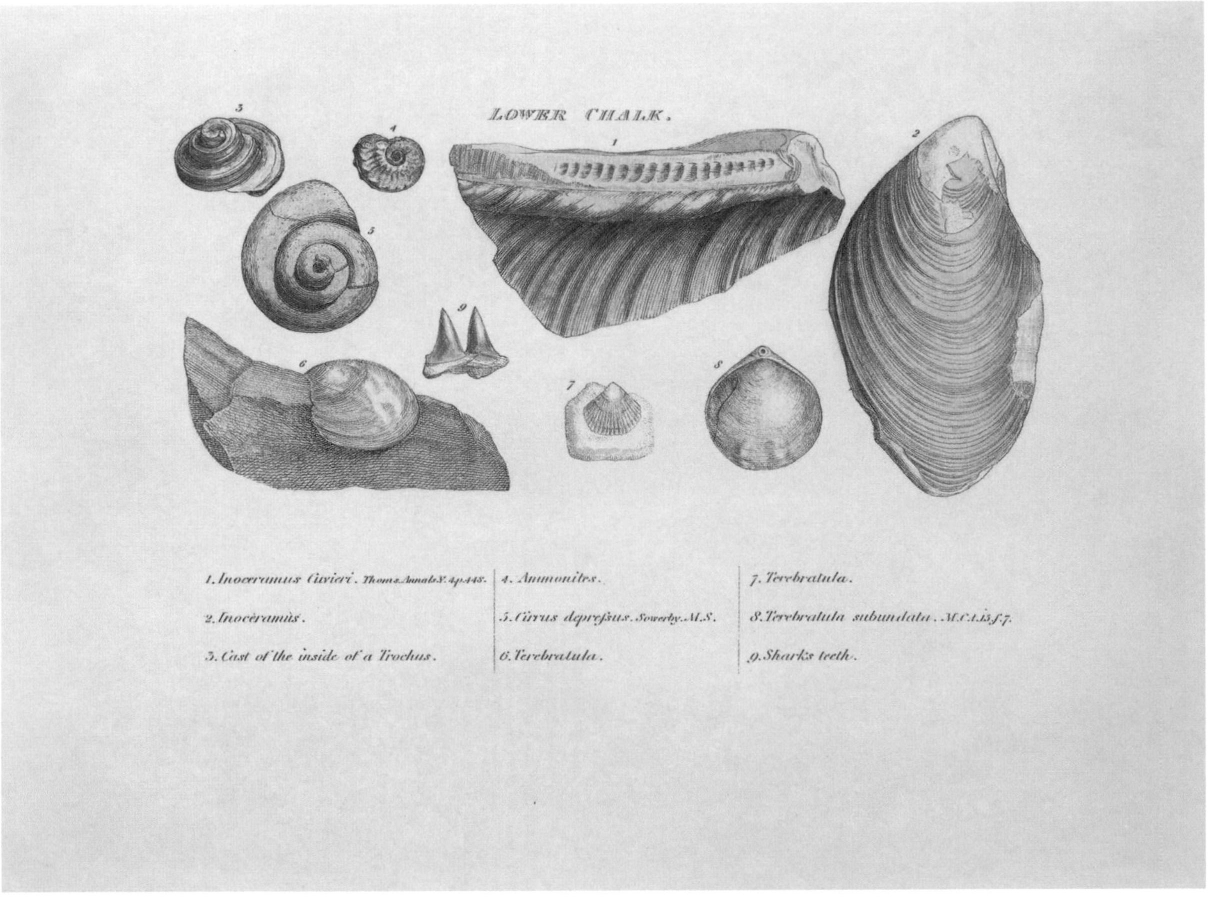

FOSSILS FOUND IN UPPER CHALK [ABOVE].

Although Smith describes the full outcrop of chalk in detail, most of the fossils that he illustrated from the Upper Chalk are from a small area around Norwich, probably collected when Smith was employed by Thomas Coke of Holkham in Norfolk in 1800.

FOSSILS FOUND IN LOWER CHALK [BELOW].

Hard horizons in the Lower Chalk were sometimes used locally as a building stone. Smith mentions Warminster as one such place and most of his Lower Chalk fossils come from here which he will have collected while employed on draining and irrigation projects on estates in the vicinity.

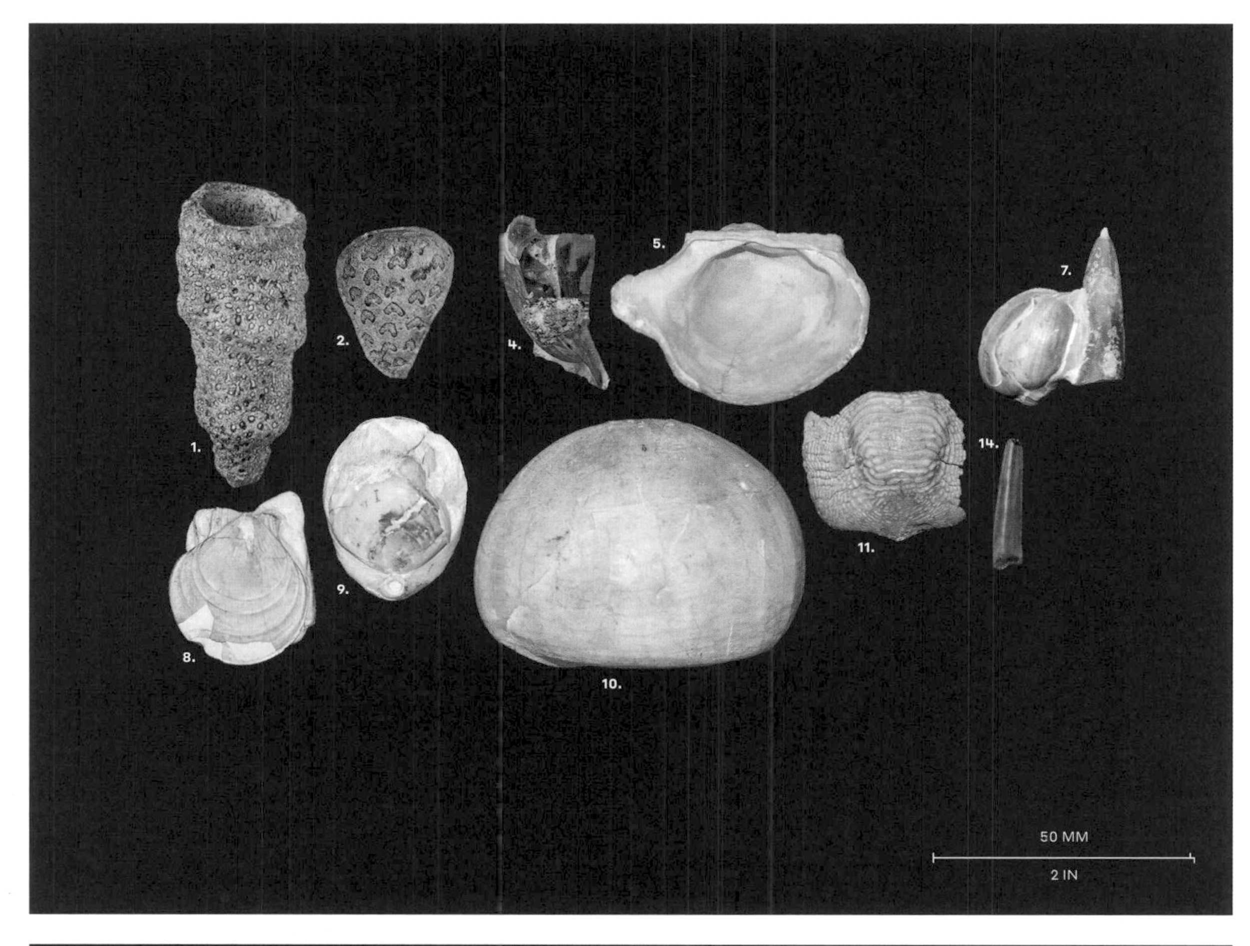

FOSSILS FOUND IN UPPER CHALK [ABOVE].

1. *Sporadoscinia alcyonoides* S9863 Wighton; **2.** *Toulminia catenifer* S9866 Chittern; **4.** *Regioscalpellum maximum* I750 Norwich; **5.** *Pycnodonte vesicularis* L1446 Norwich; **7.** oyster attached to a belemnite *Belemnitella mucronata* L1446 Norwich, **8.** *Mimachlamys mantelliana* L1441 Norwich; **9.** *Concinnithyris subundata* B1392 Norwich; **10.** *Echinocorys scutata* E552 Norwich (substitute); **11.** Fish palate *Ptychodus mammillaris* P4813 Near Warminster; **14.** Fish tooth: *Enchodus* sp. P4816 North of Norwich.

FOSSILS FOUND IN LOWER CHALK [BELOW].

1. *Volviceramus involutus* L1444 Heytesbury; **2.** *Mytiloides labiatus* MB1147 Warminster; **3.** *Bathrotomaria* sp. G1571 Mazen Hill; **4.** *Schloenbachia subtuberculata* C619 Rundaway Hill (substitute); **5.** *Bathrotomaria* sp. G1573 Warminster; **6, 8.** *Gibbithyris semiglobosa* B1387, BF107 Heytesbury; **7.** *Orbirhynchia cuvieri* B1396 Heytesbury (substitute); **9.** Shark tooth P4819 no location (substitute).

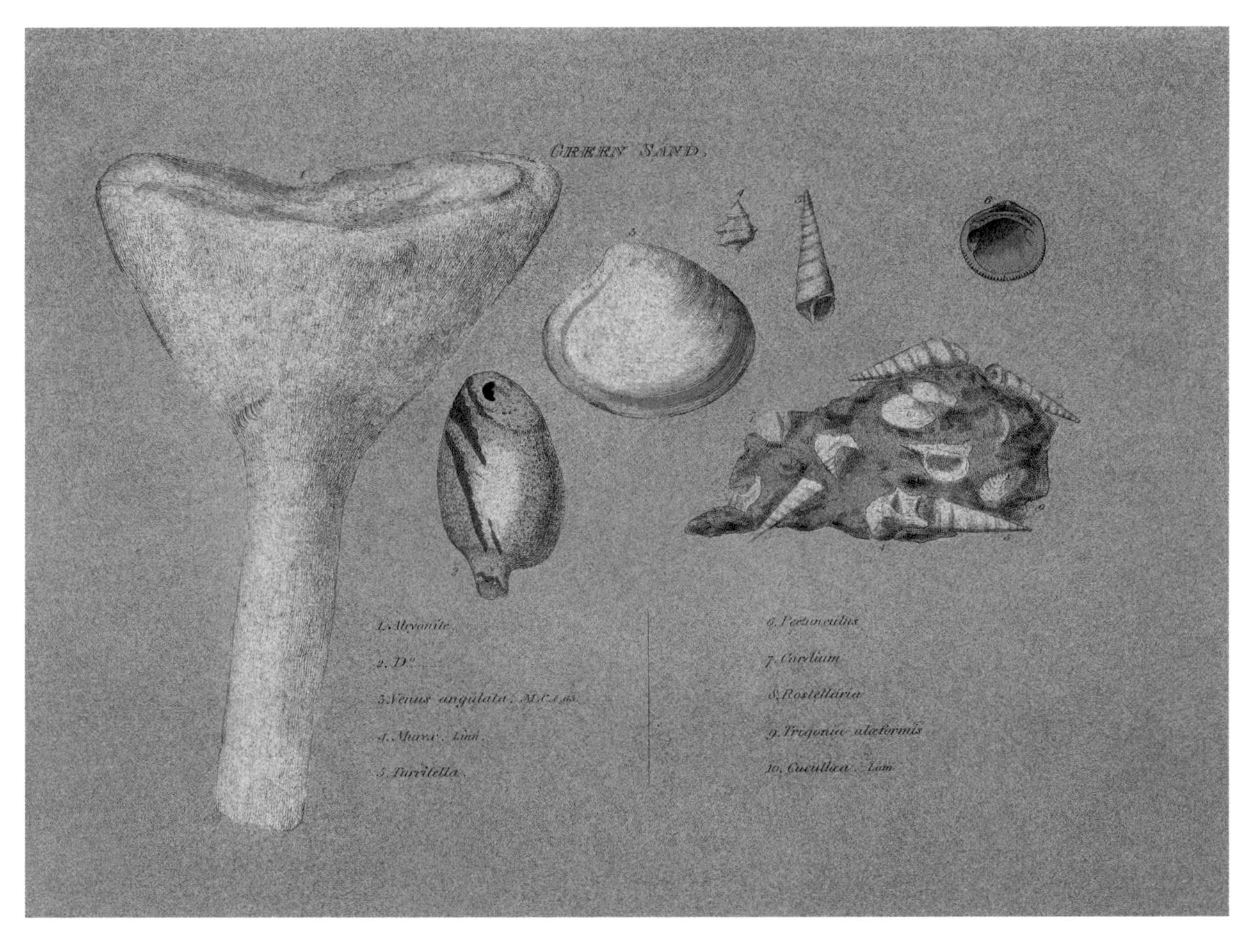

FOSSILS FOUND IN GREEN SAND [ABOVE].

Fossils pictured in the first of the two Green Sand plates mostly come from Blackdown at the edge of an outlier of Upper Greensand in the Blackdown Hills in Devon. In the south-west area, this is Smith's location furthest to the west. The large sponge on this plate is one of the most spectacular fossils in his collection.

FOSSILS FOUND IN GREEN SAND [BELOW].

Nearly all the fossils that Smith illustrates in his two Green Sand plates are from the Upper Greensand, which we now know is the same age as his underlying Brickearth (Gault Clay) with the Greensand being the near-shore facies. On this plate the fossils mostly come from Warminster or nearby Chute, the same locations for his Lower Chalk fossils. The two strata are adjacent in this area.

FOSSILS FOUND IN UPPER GREENSAND [ABOVE].

1. *Pachypoterion compactum* P3030 Warminster; **2.** *Siphonia tulipa* P3018 Pewsey; **3.** *Epicyprina angulata* L1455 Blackdown; **4.** *Cretaceomurex calcar* G1584 Blackdown; **5.** *Torquesia granulata* G1581 Blackdown; **6.** *Glycymeris sublaevis* L1448 Blackdown; **7–10.** *'Mactra' angulata, Drepanocheilus calcaratus, Pterotrigonia* cf. *aliformis*, ?*Idonearca* sp. – all on block G1583 Blackdown.

FOSSILS FOUND IN UPPER GREENSAND [BELOW].

1. *Rotularia concava* A121 Warminster; **2.** *Nummogaultina fittoni* G1580 Rundaway Hill; **3.** *Merklinia scabra* L1470 Chute Farm; **4.** *Dereta pectita* B1407 Chute Farm; **6.** *Cyclothyris latissima* B1406 Chute Farm; **7.** *Amphidonte obliquatum* L1462 Alfred's Tower; **8.** *Neithea gibbosa* L1468 Warminster; **9.** *'Chlamys'* aff. *subacuta* L1473 Chute Farm; **10.** *Amphidonte obliquatum* L1464 Stourhead; **11.** *Salenia petalifera* E476 Warminster; **12.** *Discoides subuculus* E487 Warminster; **13.** *Catopygus columbarius* E485 Chute Farm; **14.** *Holaster laevis* E483 Chute Farm.

I. APPRENTICE.

An account of Smith's early surveying endeavours in Somerset; an edifying plan of the Somersetshire Coal Canal; the unfortunate outcome of the caisson experiment; a period of consultations on slippage and subsidence; contemplations on the natural order of the strata; geological depictions of Bath on circular maps in 1800.

In 1791 William Smith (1769–1839), aged twenty-two, was directed by his employer Edward Webb (1751–1828) to survey an estate near High Littleton in Somerset belonging to Lady Elizabeth Jones (1741–1800). Smith made the journey from Webb's house in Stow-on-the-Wold, Gloucestershire, down to Stowey in Somerset on foot, a distance of over 128 km (80 miles). After reaching Stowey, Smith lodged at a farmhouse called Rugbourne. In his *Memoirs of William Smith*, John Phillips (1800–1874), his nephew, wrote that Smith's diet there included milk, while his host, an 'honourable and hospitable farmer preferred the rich cider of that fertile vale', apparently imbibing four hogsheads per annum – nearly 220 gallons (1,000 litres) (Phillips, 1844, p. 145).

Smith had been an assistant to Webb, a talented surveyor, from the age of eighteen. Under Webb's instruction, Smith became a competent surveyor, a skill which was to serve him well in later life. Before arriving in Somerset, he had worked mainly in Oxfordshire and Gloucestershire. Phillips tells his readers that in 1788 Smith had visited the famous 'Salperton' (Sapperton) tunnel on the Thames & Severn Canal (Phillips, 1844, p. 5). This tunnel, which in 1789 was the longest canal tunnel built in the United Kingdom, traversed three strata which were to prove of great importance to Smith in later life: the Inferior Oolite, the Fuller's Earth and the Upper Oolite. Tunnelling problems at Sapperton associated with the unstable Fuller's Earth foreshadowed those that were to haunt Smith later during his work on the caisson lock on the Somersetshire Coal Canal.

TAKING THE LEVELS.

Smith's first task in Somerset in 1791 was to make a map of High Littleton, which is his earliest surviving map (Henry, 2016). Lady Elizabeth Jones was a shareholder in the Mearns Colliery, which was very close to his lodgings at Rugbourne,

FIG. 1.

Fig. 1.
A sepia aquatint by Samuel Ireland, published in *Picturesque Views on the River Thames* (1791), of the entrance to Sapperton Canal Tunnel in Gloucestershire.

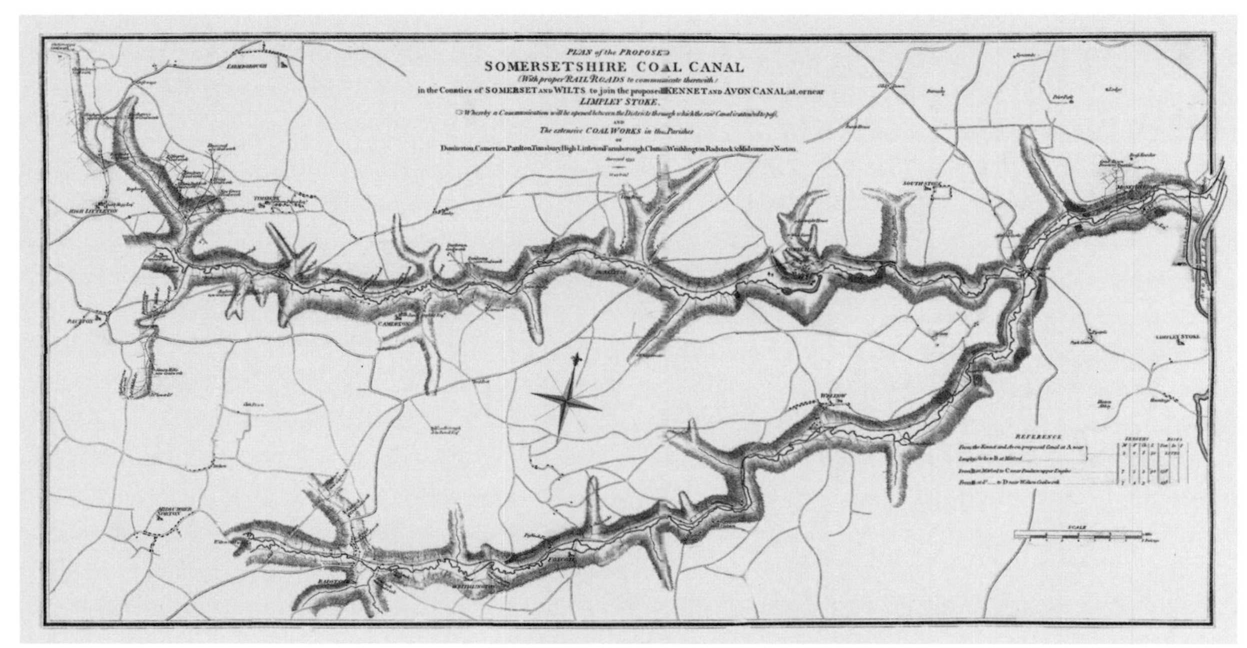

FIG. 2.

Fig. 2.
John Rennie's *Plan of the Proposed Somersetshire Coal Canal.* William Smith conducted the survey for the map in 1793, and John Cary engraved it the following year. The plan was presented along with the parliamentary bill that enabled the construction of the Kennet & Avon Canal and branch canals.

and his next job took him underground to make a survey of the colliery and to value her investment (5⁄16 share) in it. In the late eighteenth century there were many collieries in the vicinity of High Littleton and neighbouring villages. Coal had been mined from outcrops and shallow bell pits in the Somerset coalfield certainly since the fifteenth century and probably as far back as Roman times. Roads in the area were poor and coal was difficult to transport to centres of population, causing costs to rise.

In late 1792 there was talk of building a two-branched canal following adjacent valleys to transport coal out of the area, possibly connecting to the then proposed Kennet & Avon Canal. In early 1793 potential shareholders commissioned the engineer John Rennie (1761–1821), who was also engineer on the Kennet & Avon, to write a report and produce an initial survey for the new canal.

It is here that William Smith enters the story. His work in High Littleton must have impressed Lady Jones and other nearby landowners, including Jacob Mogg, Samborn Palmer and James Stephens, who were all to become shareholders in the canal. It is likely that they recommended Smith to Rennie, for Smith was subsequently engaged to make the survey for the proposed canal in 1793. In his diary entry for 30 January 1794 Smith notes 'finished drawing canal plans on vellum and inking the lines' and the entry for 2 February reads 'examined impressions of plates engraved for canal plans'. The engraver of this map was John Cary (1754–1835), who was later to engrave and publish many of Smith's geological maps.

In 1794 the Canal Bill required to authorize the building of the canal received Royal Assent and in 1795 Smith was instructed to prepare a plan of proposed deviations from the original submission. The major change to the northern branch was to locate the canal at a higher level, with no locks as far east as Combe Hay, and then to lower the level for the rest of the canal by a flight of locks or some other method. For the southern branch the approach was similar, this time incorporating an inclined tramway from Twinhoe to Midford. The alterations were accepted and work began in July 1795.

From a geological perspective the new line of the canal along the two valleys was significant for Smith, as for much of its course the canal maintained a constant level, allowing him to make comparisons in both branches. Smith observed that the strata were gently inclined to the east or south-east; Phillips comments that Smith used the analogy of 'superposed slices of bread and butter' to describe this. His theory was proved correct by 'the levelling processes executed in two parallel valleys, for in each of the levelled lines the strata of "red ground," "lias," and "freestone" ... came down in an eastern direction and sunk below the level, and yielded place to the next in succession' (Phillips, 1844, p. 8). This observation became fundamental to Smith's concept of the strata and their order, as was his use of the fossils contained within them. In a thoughtful memorandum penned in the Swan Inn at Dunkerton, he remarked on the 'wonderful order and regularity with which Nature has disposed of these singular productions [fossils], and assigned to each class its peculiar stratum' (Cox, 1942, p. 12). Smith collected numerous fossils in the area of the canal, many of which survive today at the Natural History Museum, London.

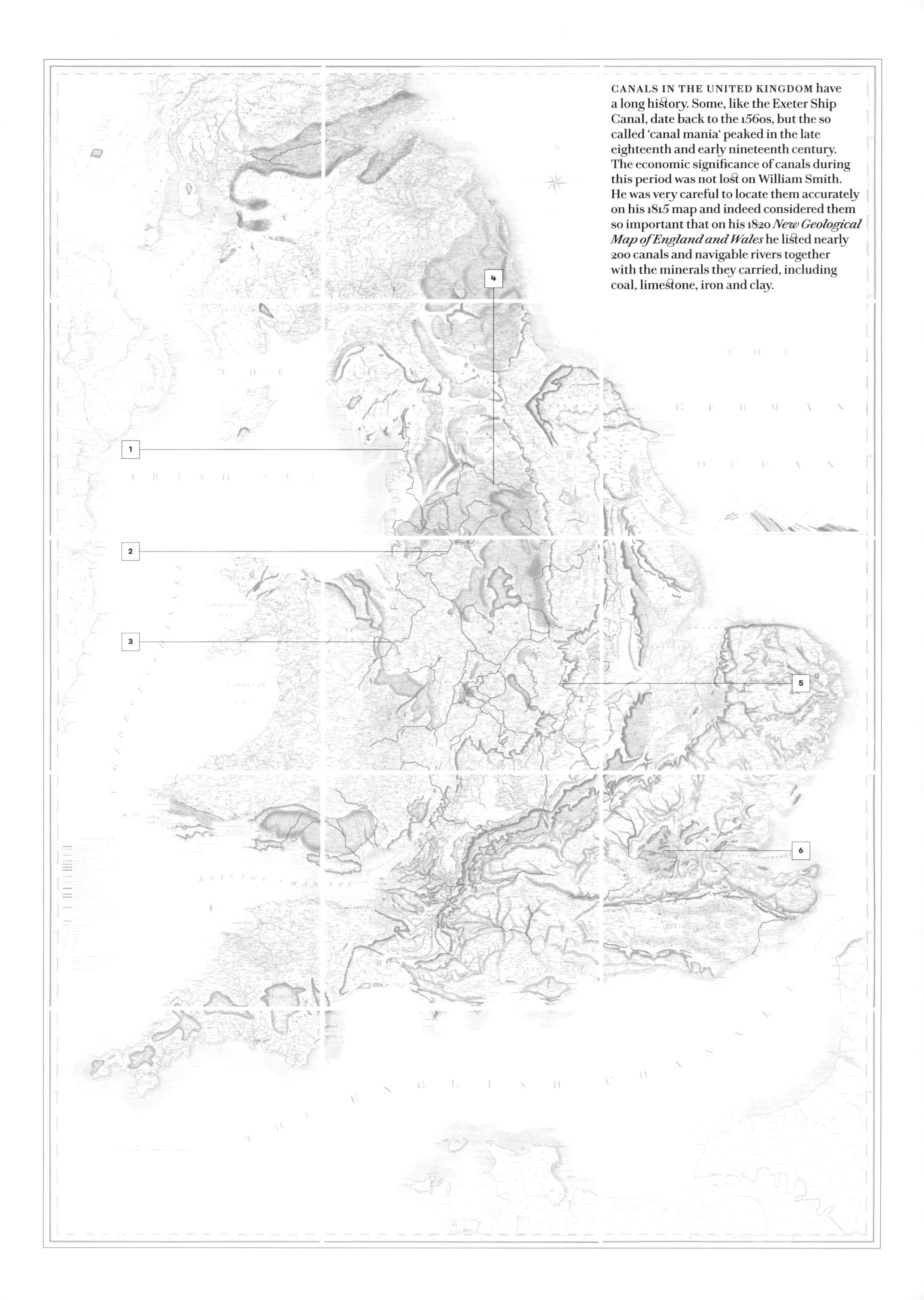

CANALS IN THE UNITED KINGDOM have a long history. Some, like the Exeter Ship Canal, date back to the 1560s, but the so called 'canal mania' peaked in the late eighteenth and early nineteenth century. The economic significance of canals during this period was not lost on William Smith. He was very careful to locate them accurately on his 1815 map and indeed considered them so important that on his 1820 *New Geological Map of England and Wales* he listed nearly 200 canals and navigable rivers together with the minerals they carried, including coal, limestone, iron and clay.

1.

LANCASTER CANAL.
Robert Whitworth, John Rennie. 1792–1826.

The Lancaster Canal is one of the country's few coastal canals, linking Preston to Kendal. It was mainly used to carry coal south and Limestone north.

2.

DUKE OF BRIDGEWATER CANAL.
John Gilbert, James Brindley. 1759–1761.

The Bridgewater Canal, built to transport coal from Worsley to Manchester, was the forerunner to the period of 'canal mania'.

3.

ELLESMERE CANAL.
Thomas Telford. 1791–1805.

In 1791 a canal was proposed to link the River Mersey and the River Severn, via the River Dee. However the canal was never completed in its entirety.

4.

LEEDS & LIVERPOOL CANAL.
John Longbotham, James Brindley. 1770–1816.

The Leeds & Liverpool Canal is the longest single waterway canal in Britain, linking east and west across the Pennines.

5.

GRAND UNION CANAL.
James Barnes. 1793–1814.

Now known as the Old Grand Union Canal, this was built to connect the Grand Junction Canal with the Leicestershire and Northamptonshire Union Canal.

6.

REGENT'S CANAL
James Morgan. 1812–1816.

The Regent's Canal Company was formed in 1812 to cut a new canal in London from the Paddington spur of the Grand Junction Canal to Limehouse.

CANALS IN EXISTENCE IN 1815.

SCOTLAND.

Aberdeen *(coal, iron).*
Aberdeenshire *(granite).*
Forth & Clyde *(coal, ironstone).*

THE NORTH.

Barnsley *(coal, flagstone).*
Bolton & Bury *(coal).*
Bradford *(coal, ironstone, limestone).*
Duke of Bridgewater *(coal, stone).*
Caistor *(gravel).*
Dearne & Dove *(coal, iron, lime, limestone).*
Driffield *(coal).*
Ellesmere *(coal, lime, limestone, iron, lead).*
Foss Dyke *(coal).*
Gresley's *(coal).*
Huddersfield *(coal, lime).*
Hull & Leven *(coal, lime).*
Lancaster *(coal, limestone).*
Leeds & Liverpool *(coal, iron, limestone).*
Manchester Bolton & Bury *(coal).*
Market Weighton *(coal).*
Newcastle-under-Lyme *(lime, coal).*
Newcastle-under-Lyme Junction *(coal, iron, lime).*
Peak Forest *(coal, limestone).*
Ramsden's *(coal, flags, limestone).*
Ripon *(coal, lime, stone, lead).*
Rochdale *(coal, pavingstone).*
Sankey *(coal, slates).*
Stainforth & Keadby *(coal, iron, gypsum).*
Sheffield *(coal, iron, stone, lime).*
Wyrley & Essington *(coal, iron, limestone).*

CENTRAL ENGLAND.

Ashby de la Zouch *(coal, lime).*
Birmingham *(coal, iron, lime).*
Birmingham & Fazeley *(coal, iron).*
Chester *(coal, lime, shale).*
Chesterfield *(coal, iron, lead, lime).*
Coventry *(coal, lime, stone).*
Cromford *(coal, iron, lead, limestone).*
Derby *(coal, iron).*
Donnington Wood *(coal, ironstone, lime, stone).*
Droitwich *(salt, coal).*
Dudley *(coal, iron, lime).*
Erewash *(coal, iron).*
Grand Junction *(coal, flints, lime, limestone).*
Grantham *(coal, lime).*
Ketley *(coal, iron).*
Leicester *(coal, lime).*
Leicestershire & Nottinghamshire Union *(coal, Mount Sorrel stone).*
Leominster *(coal, iron, lime)*
Nottingham *(coal, lime).*
Nutbrook *(coal).*
Oakham *(coal).*
Stafford & Worcestershire *(coal, limestone, pottery).*
Stourbridge *(coal, ironstone, fire clay).*
Stafford *(coal, lime, paving stone).*
Trent & Mersey *(coal, salt, gypsum, potters clay).*
Warwick & Birmingham *(coal, lime).*
Warwick & Napton *(coal, lime).*
Worcester & Birmingham *(coal, lime).*

WALES.

Brecon & Abergavenny *(coal, iron, lime).*
Cyfarthfa *(coal, iron, ore).*
Glamorganshire *(coal, iron, lime).*
Kidwelly *(coal, culm).*
Monmouthshire *(coal, iron).*
Montgomeryshire *(coal, iron, lead, limestone).*
Neath *(coal, iron, lime).*
Swansea *(coal, gulm, iron, stone, copper-ore).*

THE WEST.

Bude & Stratton *(sand).*
St Columb *(coal, china stone, sea sand).*
Coombe Hill *(coal).*
Grand Western *(coal, lime).*
Hereford & Gloucester *(coal).*
Ilchester & Langport *(coal).*
Kennet & Avon *(coal, freestone).*
North Wilts *(coal, stone, flags).*
Polbrook *(coal, stone).*
Somerset *(coal, freestone, fuller's earth).*
Stover *(coal, lime, sea sand, potters clay).*
Stroudwater *(coal).*
Tavistock *(slate, copper-ore, coal, lime).*

THE SOUTH.

Andover *(coal).*
Arundel *(coal).*
Basingstoke *(coal).*
Croydon *(coal, flint, firestone, fuller's earth).*
Grand Surrey *(coal, gravel).*
Oxford *(coal, flints).*
Shorncliff & Rye *(defence sea beach).*
Thames & Severn *(coal, stone).*
Wilts & Berks *(coal, flags, freestone).*

Fig. 3.
An illustration of Robert Weldon's Hydrostatick, or caisson lock, modified from the original patent drawing. Taken from John Billingsley, *General View of the Agriculture of the County of Somerset* (1798).

Geology aside, practical problems remained for Smith in building the canal, particularly the question of how the summit level of the canal at Combe Hay was to be lowered by over 40 m (135 ft) to the valley floor. Before work began, in August 1794 Smith, accompanied by Samborn Palmer (1758–1814) and Richard Perkins (1753–1821), both local owners of coal mines with an interest in the canal, was sent by the Canal Committee on a trip around England to view various canals. On their way home via Shropshire (Torrens, 2003, p. xxiii) they viewed a half-sized version of a potentially revolutionary invention which, it was claimed, could deliver a boat to a lower (or higher) level without any loss of water, something not possible in a regular lock. This was called a 'Hydrostatick' or caisson lock and its inventor was Robert Weldon (?1768–1804).

The full-scale lock that was built consisted of a huge stone cistern, 25 m (80 ft) long, 6 m (20 ft) wide and 18 m (60 ft) deep, filled with water and containing a large box or caisson with recessed water-tight doors at either end. The caisson, partially filled with water, was capable of receiving a 30-tonne boat. After the boat entered the caisson, the doors were closed and a seal maintained by hydrostatic pressure. With adequate ballast, neutral buoyancy was achieved and the caisson could be easily lowered down the cistern using control rods. The lower door was then opened to release the boat via a tunnel. To achieve the required drop on the Somersetshire Coal Canal, three caissons would have been needed; in the event only one was ever built. A caisson 'experiment' took place in late 1795–early 1796 and from the start there were problems with water leaking from the cistern and tunnel. Some of the trials were successful and others were spectacularly less so (in one the Canal Committee nearly suffocated when they became trapped in the submerged caisson). Eventually, by August 1799, the Committee decided that further expenditure was unjustified and other methods were explored. The solution finally arrived at was an expensive flight of twenty-two locks gradually descending down Rowley Bottom.

CLIMATE AND CAREER CHANGE.

The weather in 1799 was atrocious: unseasonably bad, cold early in the year and very wet throughout the summer. Some areas of the country had only eight days without rain from June to November. It was also the year that saw Smith and Weldon out of a job. Probably as a result of the lack of success with the caisson experiment, Smith ceased to be the Canal Surveyor on 5 April. Robert Weldon died in 1804, his obituary noting that 'the failure of the

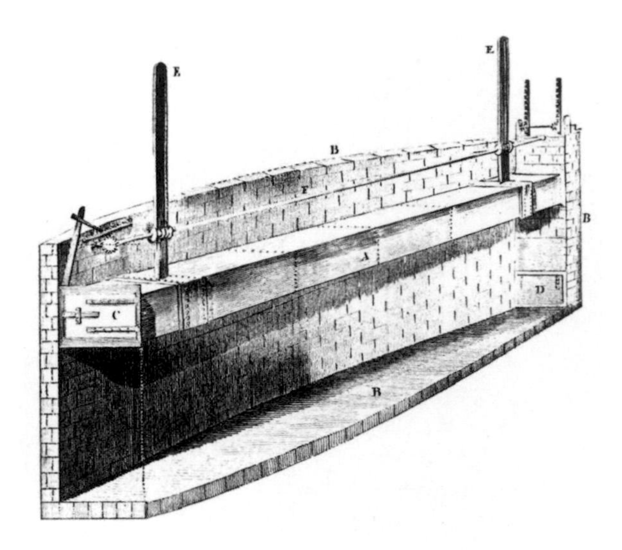

FIG. 3.

masonry ... and not the failure of the invention, had very much injured his health and spirits'.

The caisson did not fail because of the weather but it certainly did not help. The primary reason for its collapse was its geological location in the unstable Fuller's Earth. Fuller's Earth often contains an abundance of 'swelling clays', rich in the mineral montmorillonite. These clays absorb water and swell when wet and conversely shrink when dry. Smith's friend John Farey (1766–1826) wrote some time later that when the water in the cistern was drawn off to allow some alterations to be made, the walls bulged so much that 'the whole was rendered unsafe and useless' (quoted in Torrens, 1975). The unstable nature of the Fuller's Earth is evident in many localities around Bath, often leading to landslips.

But out of disaster arose opportunity. The prevalence of landslips in a very wet year afforded Smith another career path, this time advising on slippage and subsidence. Smith's friend the Rev. Richard Warner (1763–1857) had a house on the east side of Lyncombe Vale (Torrens, 2003, p. 161) which had previously been called Hanging Lands House because of its precarious situation on the steep slope under the scarp (Peach, 1884, p. 8). Warner enlarged the house and it became known as Warner's Cottage. Because of its location and the fact that it was on the Fuller's Earth, the house had experienced slippage and subsidence, which Smith sought to correct. Smith realized that springs in the Fuller's Earth immediately beneath the Upper Oolite limestone were often the cause of the problems. In an uncompleted preface to his proposed 1801 publication Smith says 'Many of these slips, containing more than an acre of ground (in consequence of the late rainy seasons), have put on appearances rather alarming to some of the possessors of land and houses in the neighborhood. I have seen new buildings in the neighborhood of Bath which have been cracked from top to bottom in consequence of these movements' (Cox, 1942, p. 87). He goes on to chastise house owners for choosing a charming view over a house foundation on solid ground.

Smith also described another example of a building in a dangerous state due to slippage at Combe Grove. In this case the building was on the Upper Oolite, but he says 'I am of opinion, that the whole of this beautiful place, in a few years, must inevitably fall a sacrifice to the irresistible pressure of the rocks moving down upon it'. In what may be a description of the same building, Smith recounts that he put sticks into the cracks in the structure in order to measure how fast they were enlarging. His solution was to tunnel into the hill and intercept the springs, which apparently halted any further damage. By coincidence, Smith later acquired a quarry at Kingham Field, close to Combe Grove, but it proved to be a financially disastrous venture which ultimately led to his imprisonment in the King's Bench Prison for debt.

NATURAL ORDER OF THE STRATA.

In 1799 Smith, now a journeyman, was having to earn a rather mundane living advising on slippage, subsidence and later on land drainage, but he also had time for more contemplative matters. Since his arrival in Somerset he had been developing the concept of an order of the strata, initially in relation to the coals and coal measures in the Mearns Colliery and later on observing the sequence of strata he encountered along the line of the Somersetshire Coal Canal. Phillips noted that Smith had developed a general law that the 'same strata were found always in the same order of superposition and contained the same peculiar fossils' (Phillips, 1844, p. 28). The Rev. Benjamin Richardson (1758–1832), a collector of fossils and a friend of Smith, was apparently 'astonished and incredulous' at this assertion, and together with another friend, the Rev. Joseph Townsend (1739–1816), they determined to make field examinations to test it. Needless to say, their investigations confirmed Smith's hypothesis. Later in 1799, after the three dined together, it was proposed that Smith should dictate a table of the strata according to their order of succession, starting with the Chalk, and number them in a continuous sequence down to the coal. Townsend penned the list and Richardson provided the names of the fossils. Thus, the first *Table of Strata* was completed; each member of the triumvirate took a copy and it seems that others were widely distributed. Over subsequent years Smith refined the table.

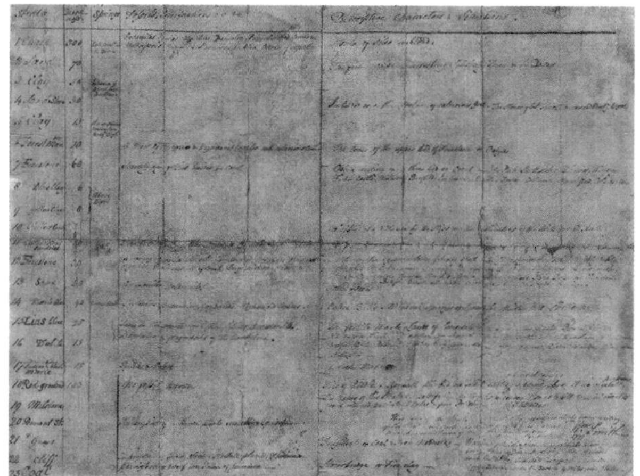

FIG. 4.

FIG. 5.

It does not seem to have taken long for word of Smith's discoveries to get around. Richard Warner in his *The History of Bath* (1801, p. 394) gives a general view of the strata of Bath and refers his readers to 'a more scientific and particular account of them ... written expressly on the subject by the very ingenious Mr. Smith of Midford, near Bath, which, we understand, will shortly be given to the world'. He then goes on to describe a number of strata from the Forest Marble, Oolite, Fuller's Earth, another Oolite, Lyas (Lias) and Coal. Later, in 1811, in his *A New Guide through Bath* he has a section entitled 'Fossilogical Phoenomena' (p. 174), noting that 'this little memoir was drawn up not so much for the information of the scientific, as for the gratification of the curious and the amusement of the idle'. Notwithstanding, he then described in some detail an almost complete succession of strata down to the coal, which he undoubtedly plagiarized from the original table the three men drew up. To his credit, Warner was a good friend of Smith and indeed later tried to secure a job for him in Russia, so perhaps Smith was not too distressed by this lack of acknowledgment.

Smith might have been comfortable with the unacknowledged distribution of his ideas but his other steadfast friend, Benjamin Richardson,

Fig. 4.
The original *Table of Strata near Bath*, dictated by William Smith and written by the Rev. Joseph Townsend in 1799.

Fig. 5.
Tucking Mill House, Somerset, residence of William Smith. A commemorative plaque originally placed on the mill buildings in 1889 was incorrectly transferred to Tucking Mill Cottage in 1932.

***AREA AROUND BATH*, C.1799.**

William Smith's early manuscript map of the Bath district from around 1800 shows the outline of the Oolitic Limestone with the colour yellow. The original map scale was 1½ inch to 1 mile.

AREA AROUND BATH, C. 1799.

This is a slightly later version of the map shown opposite. Although damaged, this map shows much more geological detail, including a sub-division of the Oolitic Limestone, as well as the strata of the Lias and the Red Ground.

was not. In May 1801 he wrote Smith a somewhat enigmatic letter with a veiled warning that perhaps Townsend might be considering publishing Smith's work for his own benefit. Richardson urged Smith speedily to commit his ideas to print, advising him that he needed to do much more work on the Chalk before doing so. This fired Smith into action and on 1 June 1801 he published his *Prospectus* for a work entitled: *Accurate Delineations and Descriptions of the Natural Order of the Various Strata that are Found in Different Parts of England and Wales: with Practical Observations Thereon.* It set out a plan for a spectacularly ambitious project, including a sequential description of all the strata, together with a coloured map showing them, as well as a section. Smith extolled the virtues of the work at great length, with its potential benefits to agriculture, mining and a whole list of other trades and professions. While he recognized that 'the complete history of all the minutiae of Strata would be an endless labour', he said that if the prospectus were well received and supported, he intended to continue the task and would complete the book by November 1802. Unfortunately, his magnum opus was not to be published as a complete work, for his publisher, John Debrett (1753–1822), went bankrupt. It was clear, however, that Smith already had a grand vision of how the strata should be depicted on geological maps, as cross sections and in a descriptive table of strata. His idea may have been clear, but the resources needed to complete such a task would have overwhelmed even the stout-hearted. In the prospectus, Smith quotes from Alexander Pope's *An Essay on Man*, a poem describing the natural order that God has decreed for man: 'All Nature is but Art, unknown to thee', which ends 'One truth is clear, Whatever is, is right.'

FIG. 7.

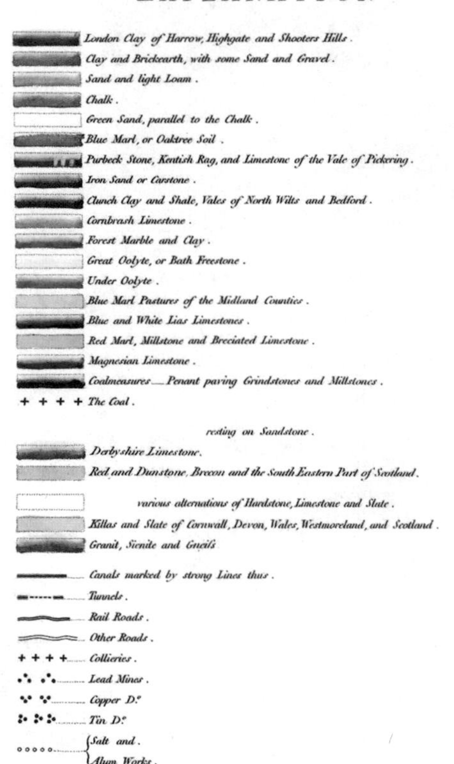

FIG. 6.

Fig. 6.
The legend to William Smith's map *A Delineation of the Strata of England and Wales, with part of Scotland*, published in 1815. The legend indicates the colours used for each strata, and also shows the sequential order of strata, many of which derive from his early work around Bath.

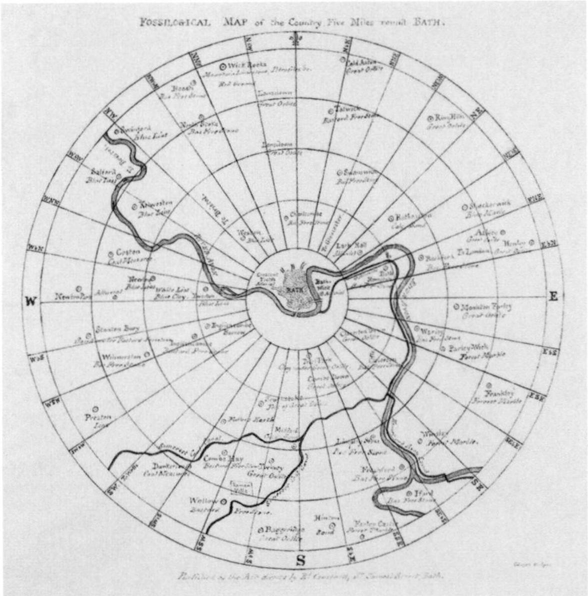

Fig. 7.
Richard Warner's *Fossilogical Map of the Country, five miles around Bath*, published in 1811. The map notes the strata found in various locations, and is clearly influenced by Smith's work.

MAPPING IN CIRCLES.

Smith's first geological map still exists in the library of the Geological Society of London. It was among a number of items donated by him when the Society's first Wollaston Medal was awarded to him in 1831. The original map is almost illegible because of fading and discoloured varnish, but over the years a number of facsimiles have been made. First impressions are that it seems a little odd: the base map is circular with Bath in the centre together with the surrounding countryside in five radiating 1-mile circles at a scale $1\frac{1}{2}$ inches to a mile. The base map was published by Taylor & Meyler in 1799, and numerous copies, including pirated ones, were sold to visitors to the city. Smith had coloured the map geologically, showing the Oolitic limestone, Red Ground and Lias. Although simple, the depiction of the geology is fairly accurate, and for the first time Smith's use of graded tints for outcrops is evident.

A second circular map, probably from around 1800 and like the first poorly legible and also torn, on closer inspection reveals a lot more geological detail. Smith has now clearly differentiated the Upper Oolite limestone from the Inferior Oolite limestone. Furthermore, when this map is compared with the fossil localities from Richard

Warner's 1811 'Fossilogical Map of the Country Five Miles around Bath', a more detailed pattern of the geology emerges. There can be little doubt that the fossil locations shown on Warner's map were from Smith. This, combined with the detail on the circular Bath map, provides an accurate reflection of Smith's understanding of the geology at that time.

LOST SOMERSETSHIRE.

Somerset was the county which, geologically, Smith knew best. As a continuation of his circular Bath maps, it was recorded (Phillips, 1844 p. 27) that he geologically coloured at least part of the county on a 1 inch to 1 mile map, Day and Masters' County Survey of 1782. Smith also mentioned the map in a letter (dated 26 June 1805) to Richard Crawshay (1739–1810), a South Wales ironmaster. He relates how, at the Woburn Sheep Shearing, he met the Duke of Clarence (later King William IV) and 'happened to have my large map of Somersetshire with me, which I have lately completed, as a specimen of what may be done upon all the county maps in the kingdom.' There is no record of what the future 'Sailor King' thought of Smith's map. Smith went on to make many excellent county maps, including for Gloucestershire and Wiltshire, both adjacent to Somerset, but sadly the original Somerset map itself has never been found. It is possible that it could have perished in a fire at his engraver's works in 1820, or that it was seized by bailiffs when Smith was in debt.

This could have been the end of the story for the map, but for what appears to have been a slipshod, but fortuitous, mistake on the part of John Cary's sons, George and John. Shortly after the fire, John Cary retired and his sons took over his business. It was during their tenure that they continued to publish *A New Map of Somersetshire divided into hundreds, exhibiting its roads, Rivers, Parks &c*, a series dating back to the first decade of the nineteenth century. For a number of reissued copies dated to between 1821 and 1831 the brothers appear to have inadvertently used an existing copper plate on which the line work for another map was engraved. The geological lines seem to suggest that this may be a preliminary version, albeit very incomplete, of Smith's long-lost map of the county.

An 1829 version of one of these maps from the Hugh Torrens Collection was digitally scanned at a high resolution. Then, with the aid of modern image processing, all hand-colouring of the hundreds, roads and other boundaries was removed and the resulting map loaded into a GIS (Geographic Information System). The map was spatially adjusted to fit both with Smith's 1815 map and also modern geological data sources. The original line work was digitized and geologically interpreted in order to re-create, as far as possible, Smith's original map. It was most important to honour all of Smith's geological lines, but at the same time not to over-interpret them based on modern knowledge. In some parts of the map it is easy to see what Smith wanted, especially when overlaid on the 1815 map; in other areas it is evident that Smith wished to add more detailed geology. When the interpretation was finalized the image was processed to give the outcrops their distinct Smithian appearance using his colours with graded tinting, and legend tablets were added to the finished map.

FAREWELL TO SOMERSET.

Smith left Somerset for London in 1803; in June 1804 he wrote that his fossils had been removed from Bath and were on the road to London (Torrens, 2003, xxviii). Later in the same letter he says 'my house at 15 Buckingham St Adelphi is very roomy and well calculated for the purpose of fitting up a Collection which should form a good representation of the Strata' (draft to C. J. Harford, OUMNH). His new home in the metropolis was a far cry from his early lodgings at Rugbourne, where over a decade earlier his ideas first began to crystallize.

Smith had achieved much during his sojourn in the county of Somerset. His experience at the Mearns coalfield had given him an understanding of strata, their attitude – dip and strike – and the importance of faulting. The canal afforded an opportunity to further develop his concepts, which resulted in his *Table of Strata* and, most importantly of all, led to geologically coloured maps, first in Somerset and later covering much of the United Kingdom.

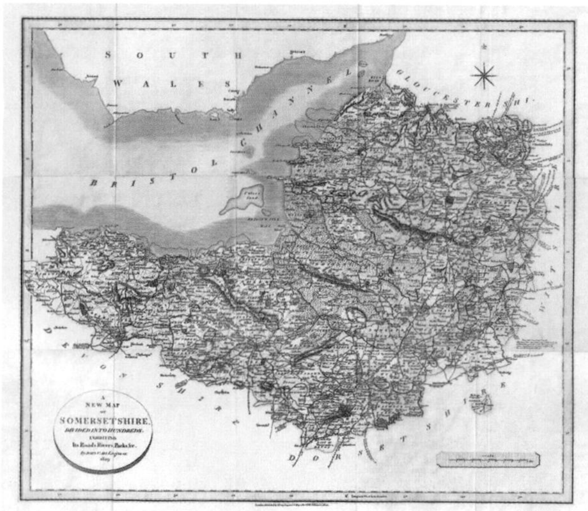

FIG. 8.

Fig. 8.
A New Map of Somersetshire divided into hundred, exhibiting its roads, Rivers, Parks &c, dated 1829. This map has no legend tablets or names of strata, although there are numerous dashed lines showing Smith's geological boundaries between strata. It is from these lines that a modern interpretation of how Smith's lost Somersetshire map may have looked has been created (see p. 215).

A.
B.
C.
D.
E.
F.
G.
H.

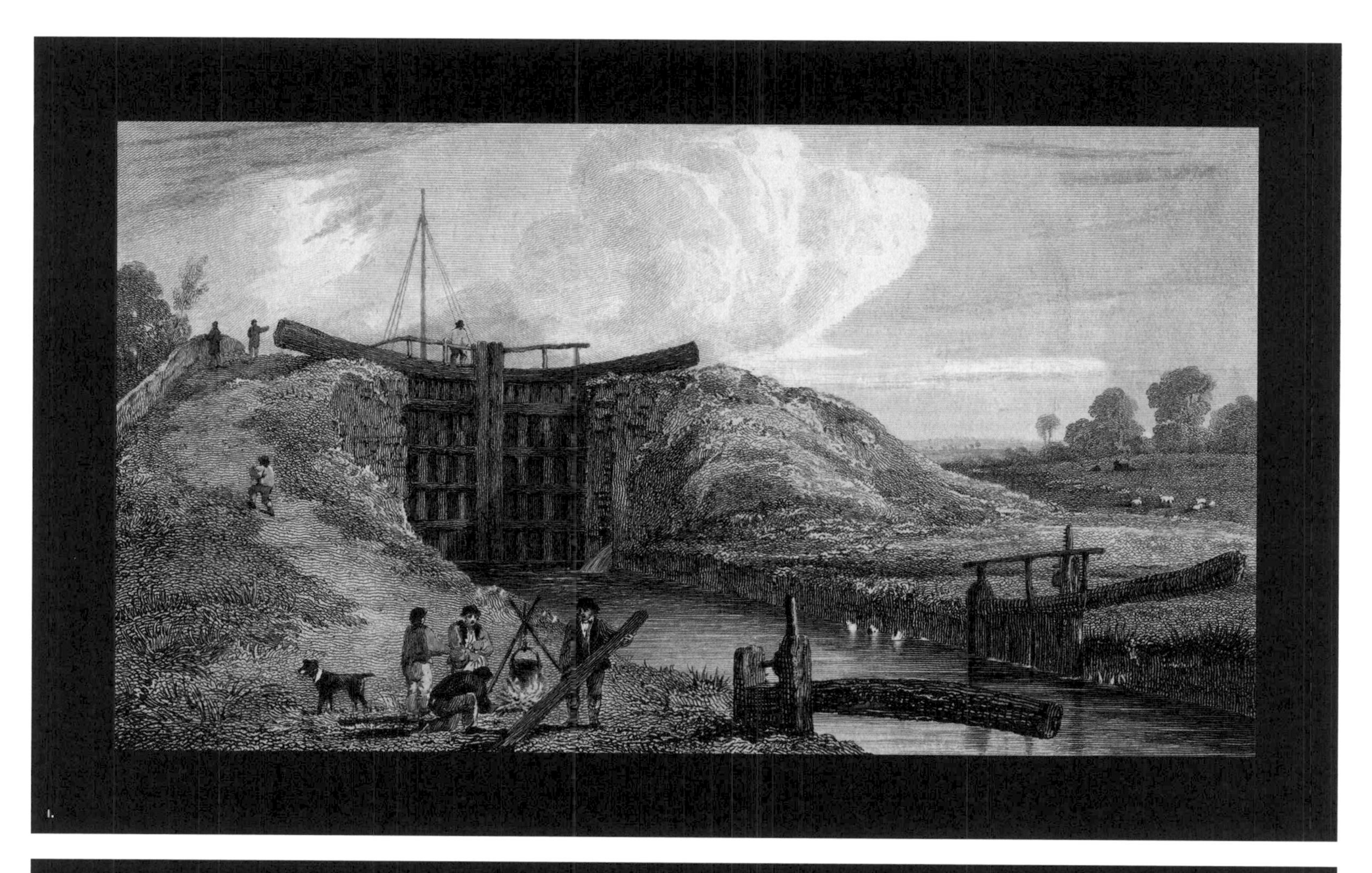

I.

J.

A. **PARK VILLAGE EAST, REGENT'S CANAL, 1827.**
Regent's Canal, connecting the Paddington Canal with the Thames at Limehouse, was opened in 1820.

B. **MACCLESFIELD BRIDGE, REGENT'S PARK, 1827.**
The bridge was renamed from North Gate Bridge, in honour of Lord Macclesfield (1755–1842), the Chairman of the Regent's Canal Company who steered the company through a financial crisis before the canal was completed.

C. **WEST ENTRANCE TO ISLINGTON TUNNEL, REGENT'S CANAL, 1822.**
The Tunnel is 878 metres (960 yds) in length, making it the longest canal tunnel in London. It was completed in 1819.

D. **THE CITY BASIN, REGENT'S CANAL, 1822.**
The canal was built as part of noted architect John Nash's (1752–1835) plan to redevelop a large area of central north London.

E. **THE CITY BASIN, REGENT'S CANAL, 1827.**
The City Basin quickly became Regent's Canal's busiest, acting as a distribution centre for goods into London.

F. **PANCRAS GAS WORKS, REGENT'S CANAL, 1826.**
Many industrial premises, including factories, gasworks and warehouses, were established on the canal's banks and basins, to take advantage of the cheap transportation.

G. **ENTRANCE FROM THE THAMES, REGENT'S CANAL, C. 1825.**
Both canal narrowboats and seagoing yachts could be found at this important connection point between the Thames and the canal system.

H. **THE LIMEHOUSE DOCK, REGENT'S CANAL, C. 1825.**
The canal meant that Limehouse Dock experienced huge commercial success as it played an important role in supplying coal across London.

I. **DUDGROVE DOUBLE LOCK ABOVE LECHLADE, THAMES & SEVERN CANAL, 1814.**
Dudgrove, about 1 kilometre ($\frac{3}{4}$ mile) from the Thames, was the second pound lock on the canal from the point where it connected with the river.

J. **RAILWAY GALLERY UNDER THE CANAL NEAR HOLSDEN GREEN, MIDDLESEX, 1837.**
The Birmingham, Bristol and Thames Junction Railway passed under the Paddington Arm of the Grand Union Canal at Holsden Green (later Harlesden).

A.

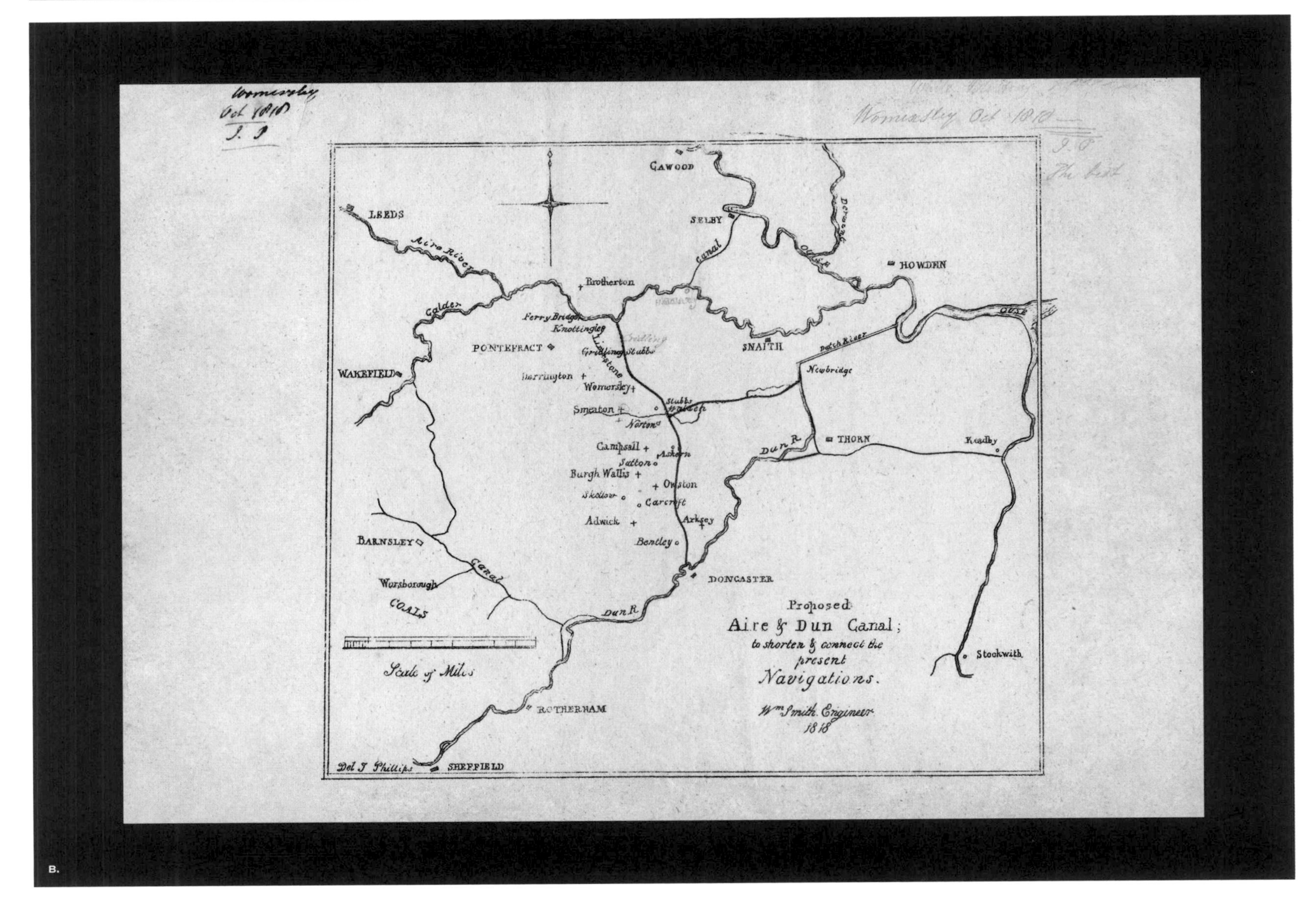

B.

A. PLAN OF THE NAVIGABLE RIVERS AND CANALS CONNECTED WITH THE PROPOSED AIR & DON CANAL, 1818.

Smith's map shows the route of the proposed Air[e] & Don Canal from Knottingley south towards Doncaster. Smith planned to combine canal construction with land drainage. The map was lithographed by his nephew John Phillips.

B. PROPOSED AIRE & DUN CANAL; TO SHORTEN & CONNECT THE PRESENT NAVIGATIONS, 1818.

The Aire & Don Canal was never built due to opposition from the existing Aire and Calder Navigation company. They proposed their own route from Haddlesey to the Dutch River, ultimately leading to the establishment of the new town and docks at Goole.

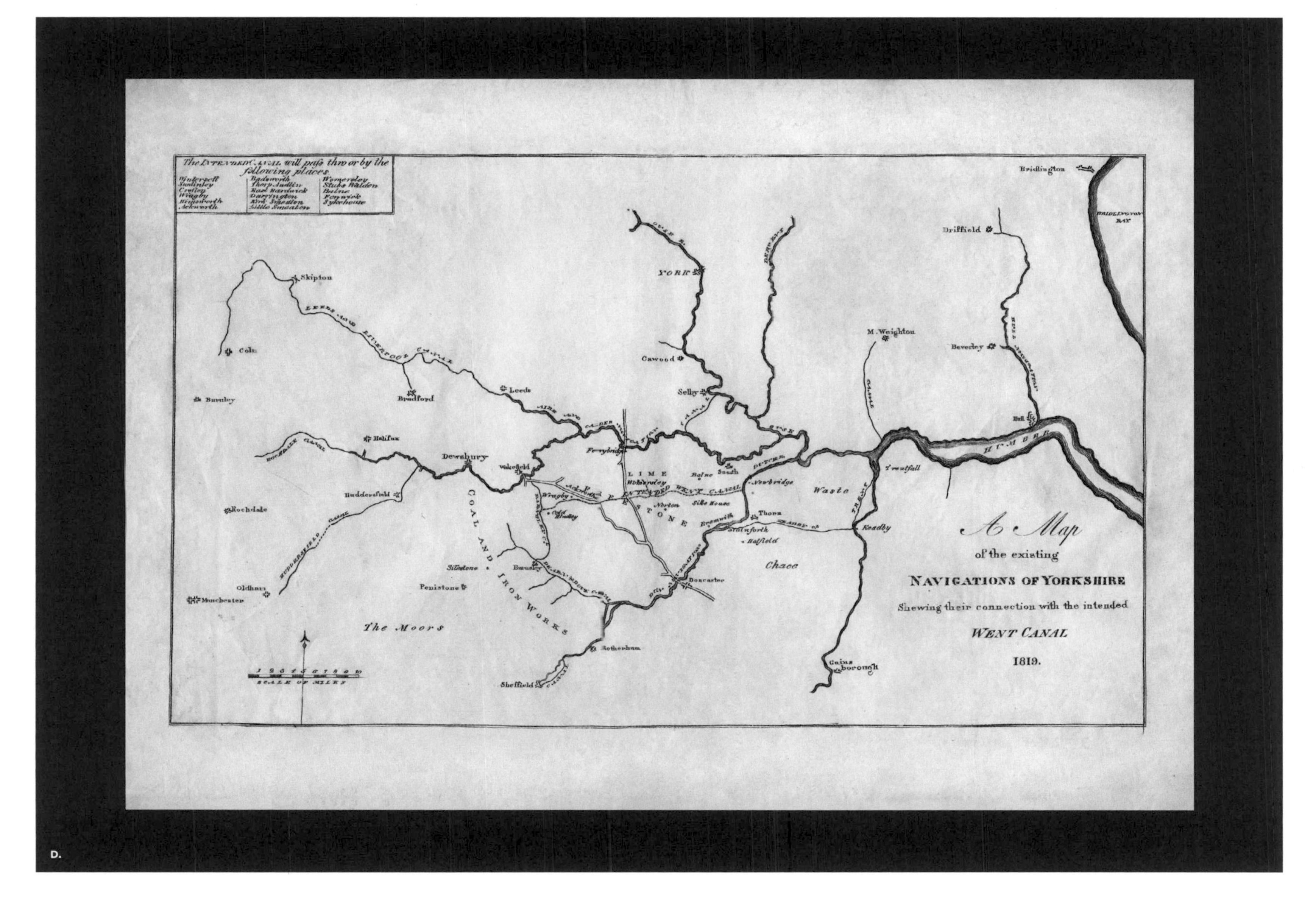

C. A MAP OF THE NAVIGABLE RIVERS & CANALS IN THE VICINITY OF THE WOMERSLEY RAIL ROAD, 1821.

A map by E. Taylor with sketches of possible routes for the proposed Womersley Rail Road and its connection to canals in the area. The route of the rail road eventually ran from Wentbridge through Womersley, and then north-east to Heck.

D. A MAP OF THE EXISTING NAVIGATIONS OF YORKSHIRE SHOWING THE CONNECTIONS WITH THE INTENDED WENT CANAL, 1819.

Smith's map shows the proposed Went Canal running from Cold Hindley on the Barnsley Canal to Newbridge on the Dutch River. Existing canals in East Yorkshire are also shown. Like the Aire & Don Canal, this canal also remained unimplemented.

Stroud
Bisley
Cirencester
Barnsley Park
Daglingworth
Oakley Wood
Pilson
Hewelsfield
Alvington
Woolaston
Minching Hampton
Nailsworth
Avening
Horsley
Berkeley
Dursley
Coaley
Symonds Hall
Upton
Tetbury
Chepstow
Rockhampton
Oldbury
Kingswood
Boxwell
Cricklade
Hawksbury
Horton
Badminton Park
Badminton
Malmsbury
Bradon Pond
Brinkworth
Lydiard Park
Wotton Basset
Sodbury
Olviston
Almondsbury
Woodlands Green
Marsh
Winterbourn
Nettleton
Castlecombe
Marshfield
Dyrham
Bristol
Stapleton
Portbury
Chippenham
Corsham Hou.
Calstone
Yatesbury
Cherhill
Winter Basset
Broad Hinton
West Kennet
Barrow
Dundry
Kelston
Bath
Melksham
Devizes
Bradford
Broadfield Down
Winford
Stanton Prior
Marksbury
Beach Hill
Nemnet
Potterne
Eastwell
Urchfont
Marden
Wilsford
Lavington
Black Heath
Bayntou
Westbury
Woolverton
Farleigh
Wingsfield
Cheddar Cliffs
Priddy
Wookeys Hole
Wookey
Downhead
Shepton Mallet
Warminster
Heytesbury
Upton Scudamore
Westbury Down
White Horse
Imber
Tilshead
Orcheston S. Mary
Shrewton
Winterbourne Stoke
Glastonbury
Pilton
Westercombe
Bruton
Great Ridge
Upper Pertwood
Hanging Langford
Stour Head House
Mere
Hindon
Fonthill House
Ridge
Tisbury
Vale of Wardour
Wardour
West Lydford
Wheathill
Babcary
Sparkford
Kingsdon
Knole
Queen Camel
Cadbury
Horsington
Wincaunton
Gillingham
Buckhorn Weston
Motcomb
Shaftsbury
Henstridge
Milborne Port
Charlton Horethorne
Ashington
Ilchester
Mudford
Stowell
Temple Comb
Ludwell
Ashcombe
Rushmore Lodge
Cranbourne Chase
Woodyates Inn
Pentridge
Bower Chalk
Broad Chalk
Swallowcliffe
Aston
Melbury Abbas
Compton Abbas
Boyton
Deptford
Stapleford
Barford
Groveley Wood
Groveley Lodge
Salisbury

'The canals are added to this map, for the purposes of showing how the heavy articles of subterraneous produce may be best conveyed from their native sites in the strata to their places of consumption. It may thus be seen what parts of the kingdom have benefitted the most by canals, and where they are still wanting, and from whence the heavy articles of tonnage (which alone can render them profitable) may be the most readily obtained.'

SMITH'S *MEMOIR*, 1815.

DETAIL FROM SHEET XI, SHOWING THE SOMERSETSHIRE COAL CANAL.

The canal, which is to the south of Bath, consisted of two branches. The northern branch was from Paulton to Limpley Stoke where it joined the Kennet & Avon Canal. The southern branch from Radstock to Midford was never commercially successful and was subsequently replaced by a tramway and then a railway. The Somerset coalfields served by the canal are shown in grey with crosses.

Wirksworth
Waterfall
Ipstones Edge
Winkhill
Wetley Rocks
Okeover
Sandy Brook
Kirk Ireton
Ashborn
Oakmoor
Cheadle
Hollington
Alveton
Wooton
Mayfield
Clifton
Belper
Haslewood
Holbrook
Weston under Wood
Duffield
Horsley
Kidsley Park
Loscoe
Underwood
Eastwood
Nuthall
Bulwell
Woodborough
Oxton
Southwell
Lay Gross
Budworth
Newstead Abbey
Annesley Wood Ho.
Selstone
Pentrich
Birchwood
Butterley Hall
Darley Moor
Brailsford
Kedleston
Quarndon
Windyworth Dale
Heath
Beamhurst
Fulford
Waste
Longford
Mackworth
DERBY
Dale Abbey
Stanley
Ilkeston
NOTTINGHAM
Radcliffe
Shelford
East Bridgford
Holme Pierpoint
Basingfield
Bingham
Uttoxeter
Doveridge
Hare Hill
Sutton on the Hill
Holbrook Ho.
Ockbrook
Sandiacre
Risley
Bramcote Hills
Chilwell
Wilford
West Bridgford
Clifton
Clipstone
Cropwell
Day Hills
Sandon
Chartley Pa.
Loxley Hall
Park
Sudbury
Dalbury
Hilton
Alvaston
Sinfin Moor
Elvaston
Derby Can.
Ruddington
Bradmore
Stanton on the Wolds
Moorton
Tutbury
Eggington Heath
Twyford
Chellaston
Trent R.
Cavendish Br.
Woodlands
Needwood Forest
Stanton
Castle Donington & Pa.
Kegworth
Shirleywich
Bagots Park
Thorney Lane
Dabble Heath
Abbots Bromley
STAFFORD
Ingestry
Great Haywood
Shugborough
Millford
Wolseley Park
Acton
Rugeley
Blithford
Hall Park
Bromley Hurst
Sinai Park
Burton
Winchill
Tatenhill
Trent and Mersey Canal
Melbourn
Bredon
Diseworth
Langley
Hatherne
Stanton
Hartshorn
Drinkley
Blackfordby
Worthington
Lime Works
Sheepshead
Prestwould
Hoton
Wimeswould
Dalby on the Woulds
Burton on the Woulds
Walton on the Woulds
Sex or Segs Hill
Ragdale
Wolds
Ashby de la Zouch
Coleorton
Charnwood
Thrinkston
Rail Road
Loughborough
Quorndon
Beau Manor
Woodhouse
Mount Sorrel
Wood Houses
Coton Park
Wichnor
Penkridge
Huntington
Cannock
Hednesford
Beaudesart
Redmoor
Elmhurst
Grand Trunk Ca.
Heath
Mease R.
Stretton in the Field
Measham
Normanton on the Heath
Whitwick
Bardon Hill 853
Ravenstone
Forest
Ulverscroft Abbey
Swithland
Elford
Clifton Campvil
Heather
Stanton under Bardon
Markfield
Newtown Linford
Bradgate Park
Groby Pool
LITCHFIELD
Wyrley Bank
Fisherwick Pa.
Statford
Gopsal
Bagworth
Thornton
Ratby
Glenfield
Tamworth
Orton on the Hill
Market Bosworth
LEICESTER
Keame
Houghton on the Hill
Hilton
Brown Hills
Weeford
Bourne R.
Aldridge
Canwell Hall
Pipehall
Wilnecote
Ratcliffe Culey
Upton
Desford
Tooley Pa.
Newal Park
Oadby
Wednesfield
Walsall
Hill
Coldfield Wast
Sutton Park
Sutton
Atherstone
Fenny Drayton
Higham
Barwell
Hinckley
Potters Marston
Whetstone
Glen Magna
Burton Overy
Wednesbury
West Bromwich Heath
Great Bar
Cliffe
Horley
Oldbury
Coventry Canal
Stoney Stanton
Sharnford
Willoughby Waterless
Kilworth Beauchamp
Foxton
Handsworth
Hurst Gr.
Curdsworth
Minworth
Soho
Birmingham Can.
Maxtoke Castle
Arley
Burton Hastings
High Cross
Sherford
Frowleworth
Gilmorton
Mowsley
Market Harborough
BIRMINGHAM
Edgbaston
Stichford
Coleshill Hall
Park
Coleshill
Corley 521
Newland
Longford
Wolvershill
Wolvey
Over Claybrook
Withabrook
Bitteswell
Lutterworth
Theddingworth
Husbands Bosworth
Hales Owen
Leasowes
Selly Oak
Norton Wood
Northfield
Elmdon
Bickenhill
Pickford Gate
Radford
Hops Ford
Cottesbach
Shawell
Kilworth
Welford
Solihull
Hampton in Arden
Mereden
Estern Gr.
COVENTRY
Berkswell
Brinklow
Cosford
Swinford
Stanford Hall
Naseby
West Heath
Stratford on Avon Canal
Knoll
Redren
Millburn
Brettford Bridge
Wolston
Lawford
Lilbourn
Bromsgrove Lickey
Lickey Hill
Arrow Head
Alvechurch
Packwood
Kenilworth
Stoneley
Dunsmore Heath
Rugby
Clay Coaten
Warren Ho.
Wood End
Beoley Park
Bromesgrove
Hewel Grange
Tanworth
Lapworth
Wapenbury
Bourton
Dunchurch
Crick
Hollowell
Red Ditch
Hanbury Common
Liveridge Hill
Wedgnock Park
Wooton Leek
Blacklow Hill
Leamington
Marton
Birdingbury
Watford
Holdenby
Braunston
Studley
Henley in Arden
WARWICK
Radford
Warwick & Napton Can.
Woodwell
Wolfhampcote
Welton
Daventree
Brockhall
Hanbury Hall
Feckenham
Aswood
Wotton Wawen
Wolverton
Cornavy
Southam
Flecknoe
Dodford
Harpole
Feckenham Forest
Inkbarrow
Sherburn
Barford
Tachbrook Bishops
Napton on the Hill
Ladbroke
Weedon Beck
Heyford
Alcester
Black Hill
Ashorne
Harbury
Rodburne
Bugbrook
Grafton Flyford
Wixford
Red Hill
Stratford on Avon
Alveston
Wellesburne Hastings
Morton Morrell
Charwelton
Byfield
Woodford
Parthingstone
Cold Higham
Rodford
Kineton
Fenny Compton
Aston in the Walls
Maidford
Tiffield
Throckmorton
Sheriffs Lench
Bitford
Bickmarsh
Radway
Claydon
Edgcot
Wood End
Field Burcot
Towcester
Woferidge
Vale of Evesham
Pebworth
Halford
Upton
Ratley
Hornton
Culworth
Moreton Pinckney
Evesham
Ilmingdon
Ox Hill
Shotswell
Horley
Bengeworth
Tredington
Compton Winyate
Edge Hill
Hanwell
Gretworth
Aswell
Whittlebury
Wickhamford
Campden
Shipston
Epwell 836
Banbury
Warkworth
Radstone
Whittlebury Forest
Wakefield Lo.
Hinton
Weston
Lillingstone Lovell

WALES AND CENTRAL ENGLAND.

i.	*ii.*	*iii.*
14. CLUNCH CLAY & SHALE.	26. BLUE LIAS.	29. REDLAND LIMESTONE.
15. KELLOWAYS STONE.	27. WHITE LIAS.	30. COAL MEASURES.
16. CORNBRASH.*	28. RED MARL.	31. MOUNTAIN LIMESTONE.
17. SAND & SANDSTONE.		32. RED RHAB & DUNSTONE.
18. FOREST MARBLE.*		33. KILLAS.
19. CLAY OVER THE UPPER OOLITE.*		34. GRANITE, SIENITE & GNEISS.
20. UPPER OOLITE.*		
21. FULLER'S EARTH & ROCK.*		
22. UNDER OOLITE.		
23. SAND.		
24. MARLSTONE.		
25. BLUE MARL.		

COMMON REGIONAL FOSSILS.

* INDICATES ORGANIZED FOSSILS OF PREDOMINANT STRATA DISPLAYED IN THIS CHAPTER.

Natica.	*Ancilla.*	*Avicula costata.*	*Terebratula reticulata.*	*Madrepora turbinata.*	*Ammonites modiolaris.*

Bar chart of strata indicates the comparative area covered by each stratum in this region.

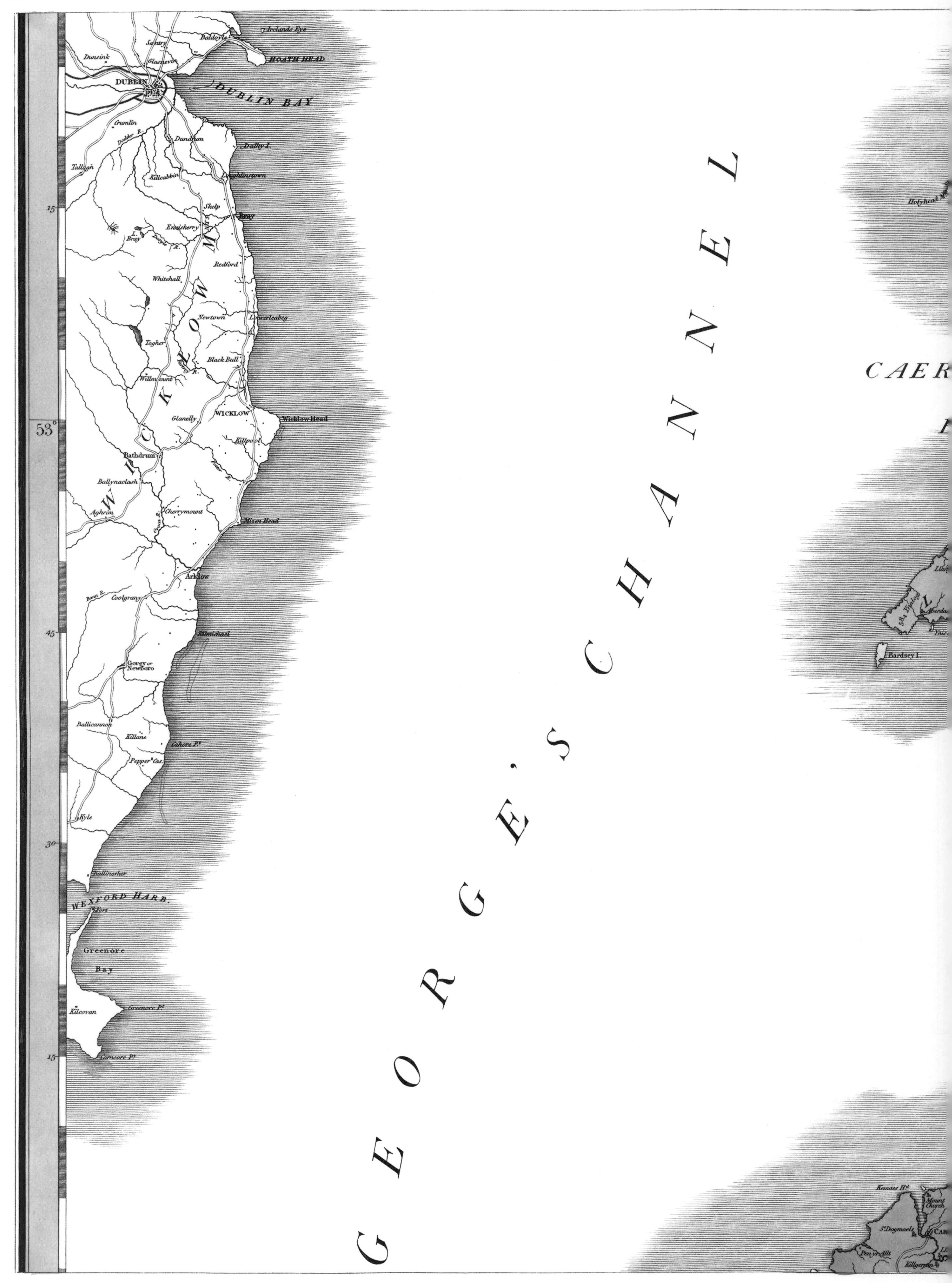
Irelands Eye
Baldoyle
Santry
HOATH HEAD
Dunsink
Glasnevin
DUBLIN
DUBLIN BAY
Crumlin
Dodder R.
Dundrum
Dalkey I.
Tallagh
Killcabbin
Loughlinstown
Skelp
Bray
Enniskerry
L. Bray
Redford
Whitehall
Newtown
Togher
Black Bull
Willmount
WICKLOW MTS.
Glanelly
WICKLOW
Wicklow Head
Killpool
Bathdrum
Ballynaclash
Aghrim
Cherrymount
Mizen Head
Arklow
Coolgrany
Kilmichael
Gorey or Newboro
Ballicannon
Killane
Cahore Pt
Pepper Cas.
Kyle
WEXFORD HARB.
Fort
Greenore
Bay
Kilcovan
Greenore Pt
Carnsore Pt
15
53°
45
30
15
GEORGE'S CHANNEL
CAER
Bardsey I.
Mount Church
St. Dogmaels
Killgerran

(see pages 88–89)

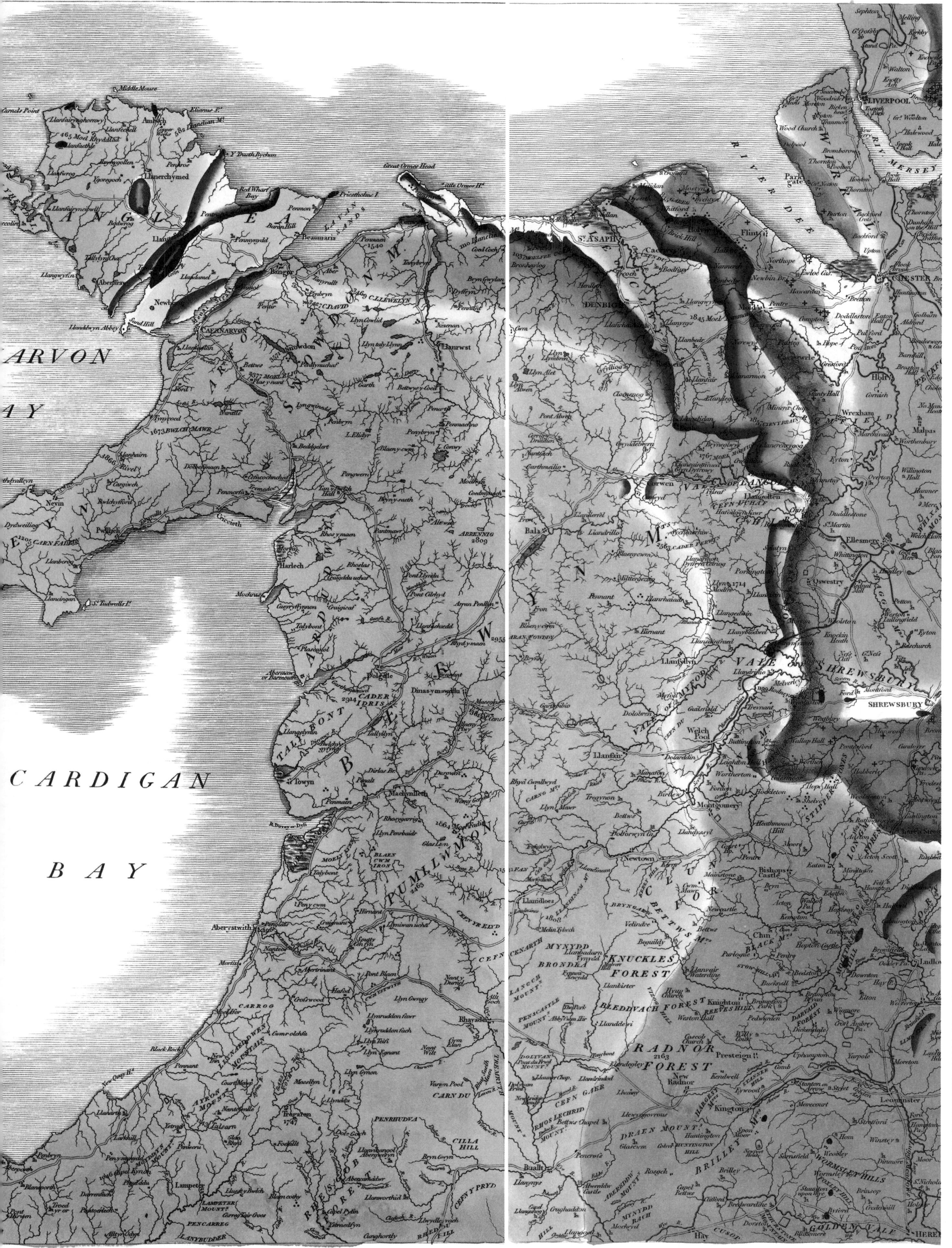

(see pages 86–87)

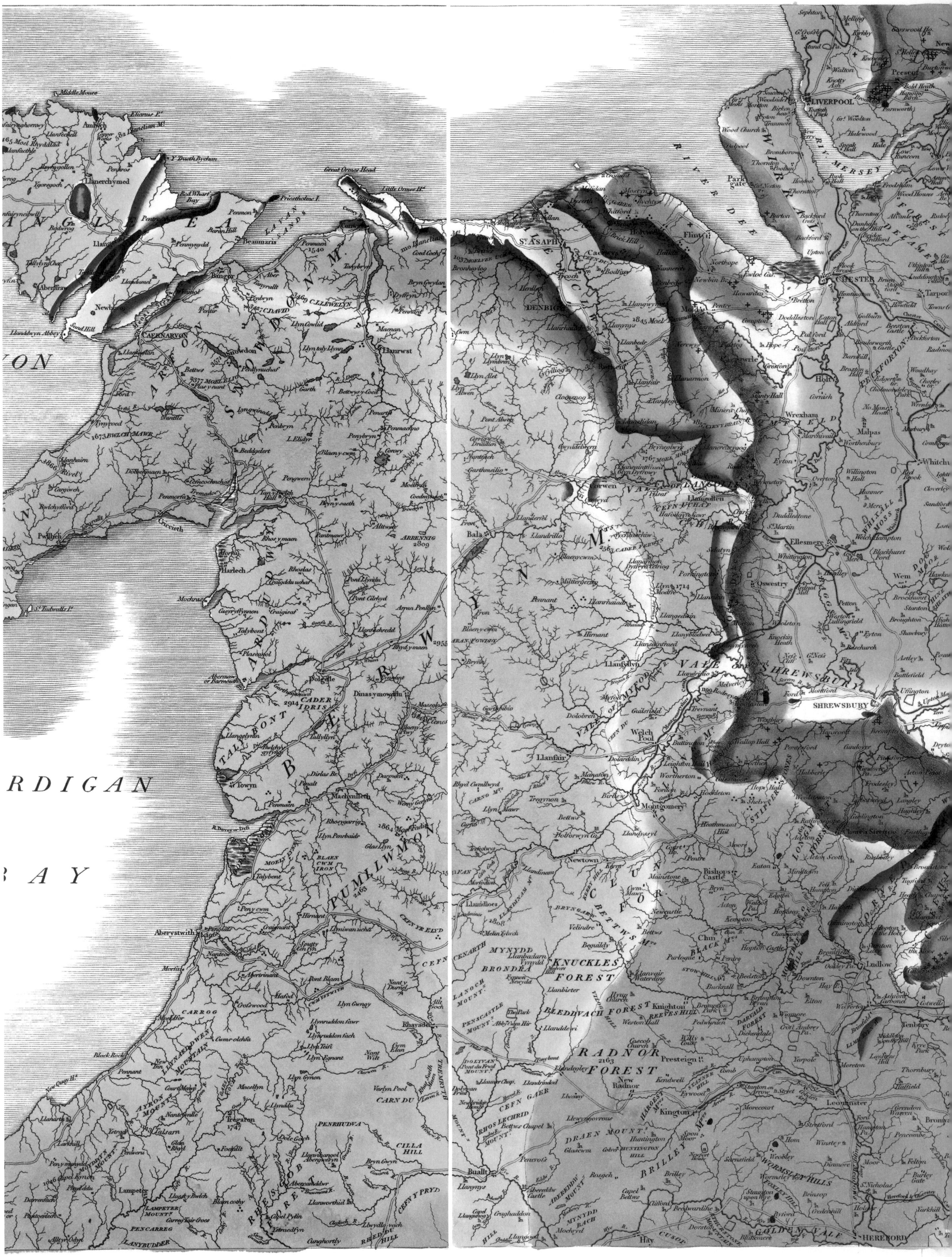

MANCHESTER
FEATHERBED MOSS
MIDHOPE DALE
CHAT MOSS
Stockport
Altringham
Knutsford
Macclesfield
SHEFFIELD
Rotherham
Bawtry
Gainsborough
Worksop
East Retford
Tuxford
Congleton
Sandbach
Middlewich
Leek
Chesterfield
Mansfield
Southwell
Wirksworth
Ashborn
Belper
Cheadle
Uttoxeter
DERBY
NOTTINGHAM
Bingham
STAFFORD
Newport
Stone
Eccleshall
Burton
Loughborough
CHARNWOOD FOREST
Ashby de la Zouch
LEICESTER
LITCHFIELD
Tamworth
Walsall
Wolverhampton
Wednesbury
BIRMINGHAM
Coleshill
COVENTRY
Kenilworth
WARWICK
Hinckley
Lutterworth
Rugby
Daventree
NORTHAMPTON
Kidderminster
Bewdley
Bromsgrove
Droitwich
WORCESTER
Stratford on Avon
Banbury
Towcester
WHITTLEBURY FOREST
Stony Stratford
NORTON WOOD
VALE OF BELVOIR
VALE or CATMOS
VII

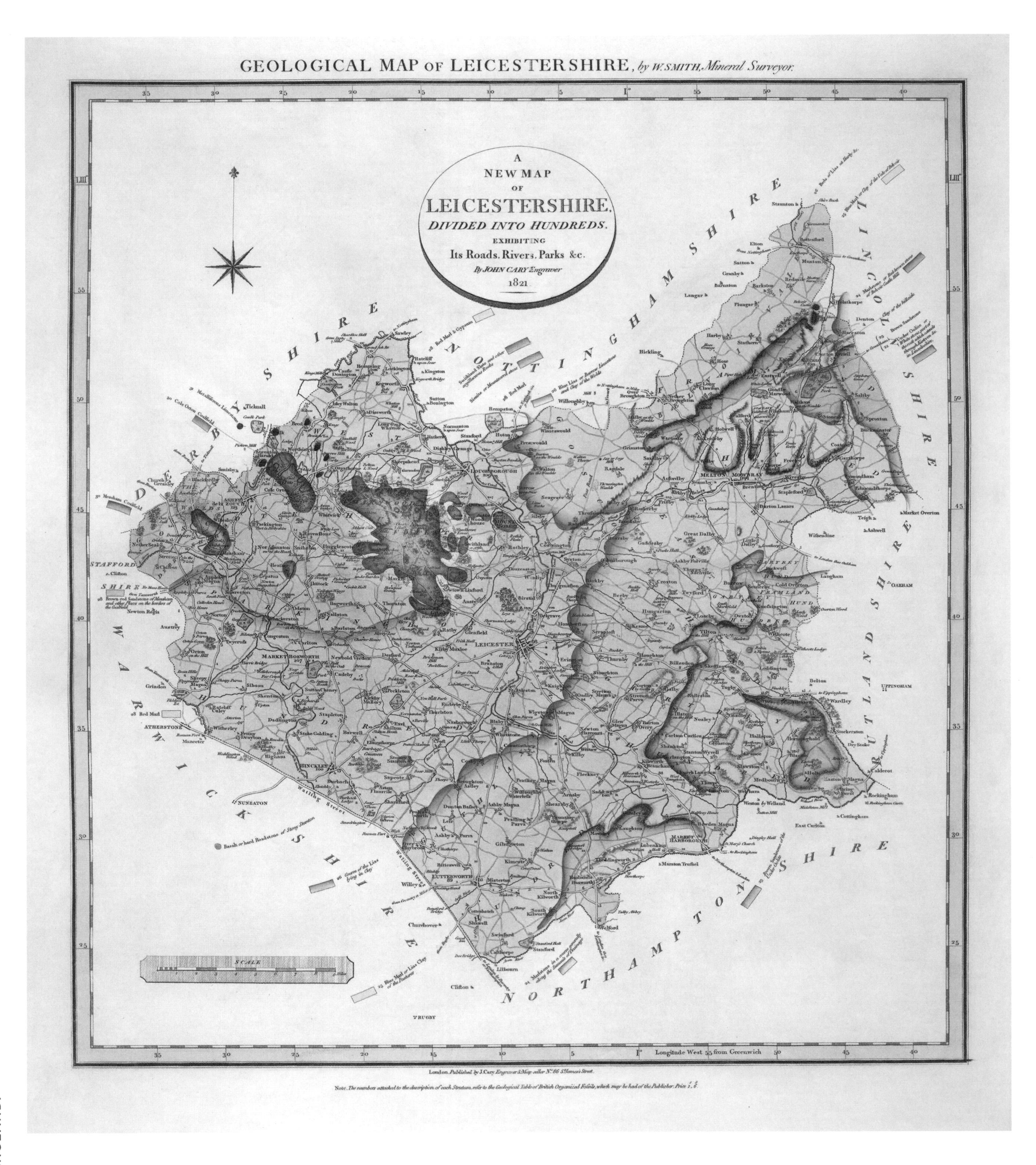

GEOLOGICAL MAP OF LEICESTERSHIRE, 1822.

First published in Part V of *A New Geological Atlas of England and Wales* with Nottinghamshire, Huntingdonshire and Rutlandshire. The uppermost stratum is the Under Oolite (22) which is to the east of the county. Further west the Lias and Red Marlstones overlie the Coal Measures. Igneous rocks, which Smith showed as sienite (34) (actually granodiorite) can be seen in at Mount Sorrel and although not given a stratigraphic position he also shows the famous Swithland Slate of the Charnwood Forest.

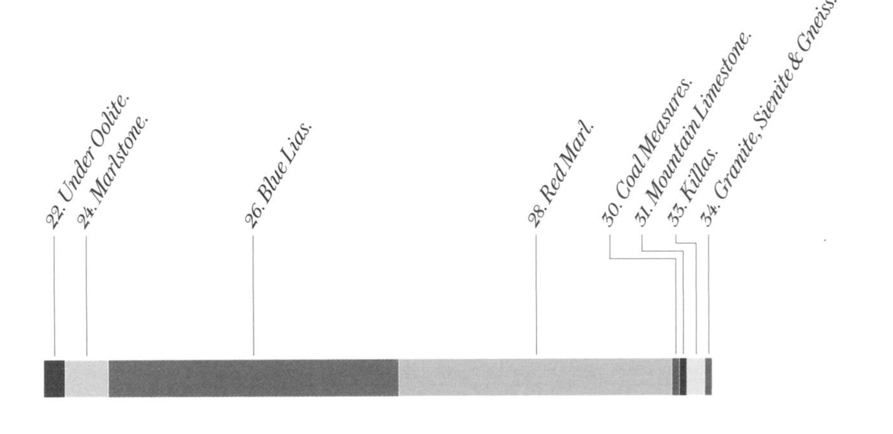

GEOLOGICAL MAP OF NOTTINGHAMSHIRE, 1822.

First published in Part V of *A New Geological Atlas of England and Wales* with Leicestershire, Huntingdonshire and Rutlandshire. The upper stratum is Blue Lias (26) above Red Marl and Sandstone (28) and the Redland Limestone (29) which is underlain by the Coal Measures (30).

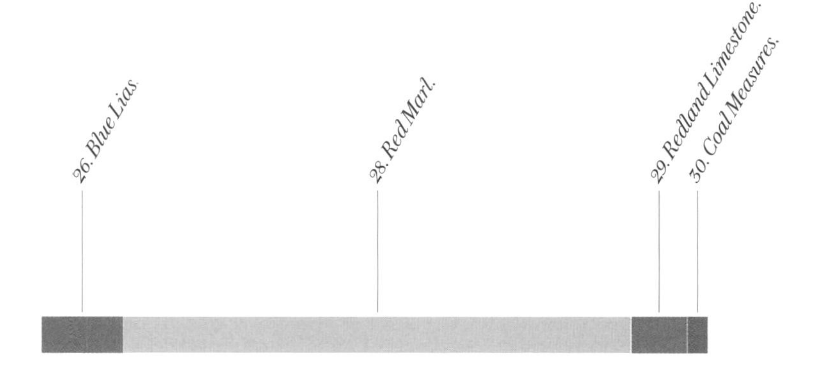

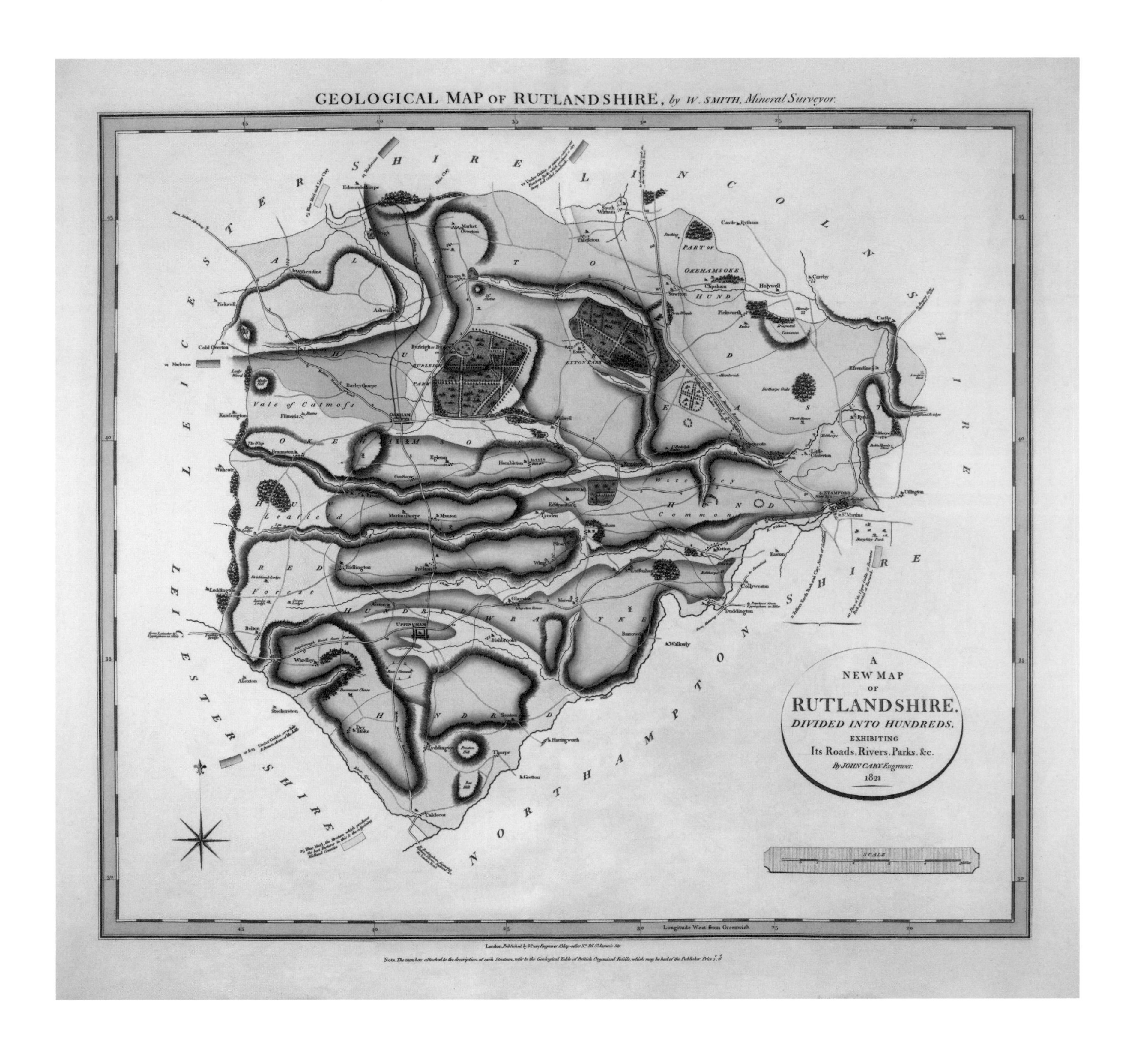

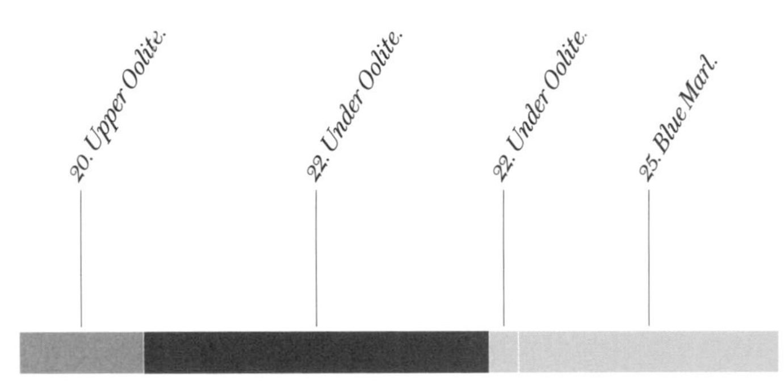

GEOLOGICAL MAP OF RUTLANDSHIRE, 1822.

First published in Part V of *A New Geological Atlas of England and Wales* with Nottinghamshire, Leicestershire and Huntingdonshire. The upper stratum is Upper Oolite (20) and the lower stratum is Blue Marl (25).

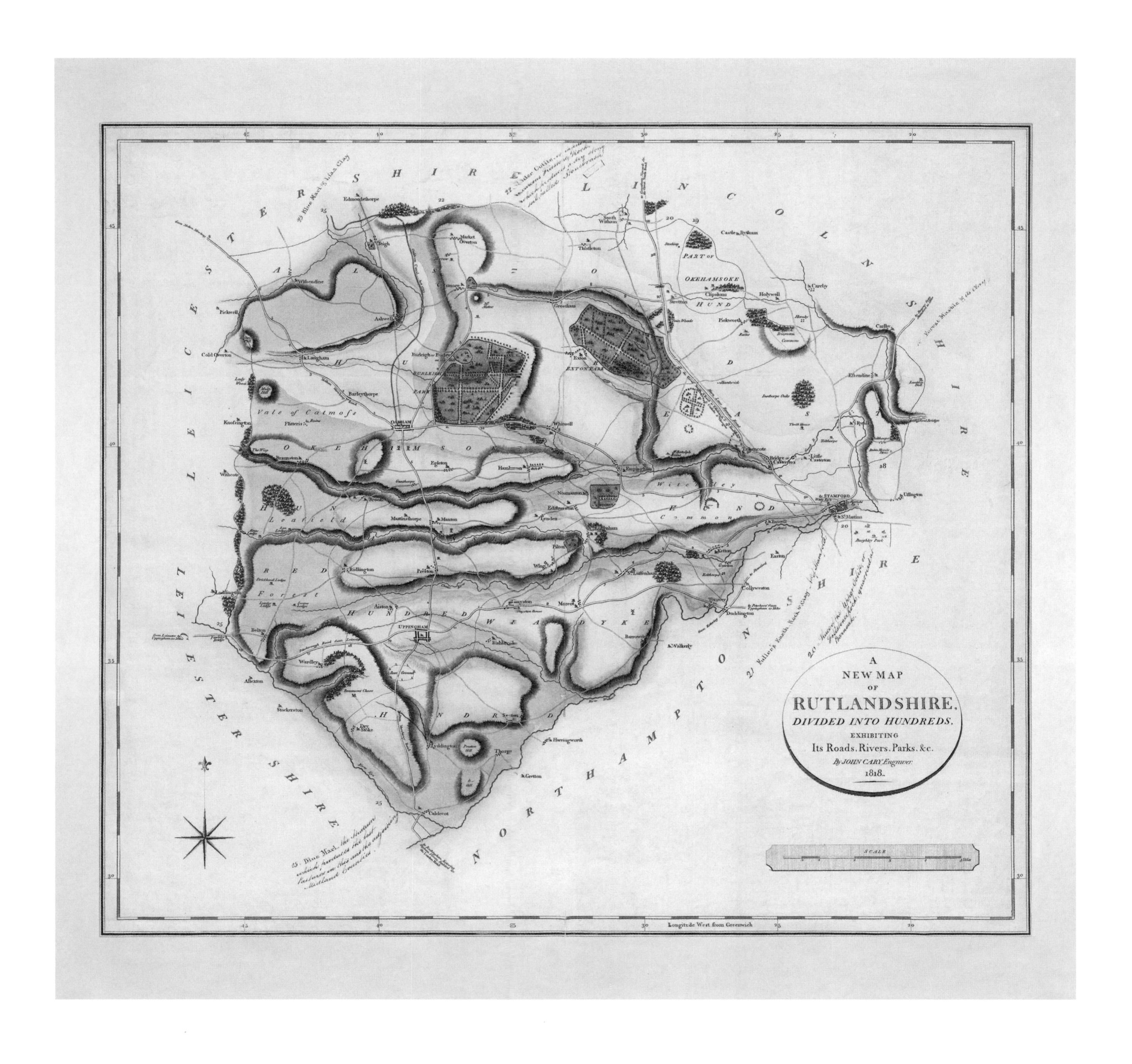

UNPUBLISHED GEOLOGICAL MAP OF RUTLANDSHIRE.

The base map is engraved by John Cary, and dated 1818. Smith appears to have annotated strata by hand including the Forest Marble (18), Upper Oolite (20), Fuller's Earth (21), Under Oolite (22–23), Marlstone (24) and Blue Marl (25). He has colour-washed the base map, to show the Under Oolite and Blue Marl.

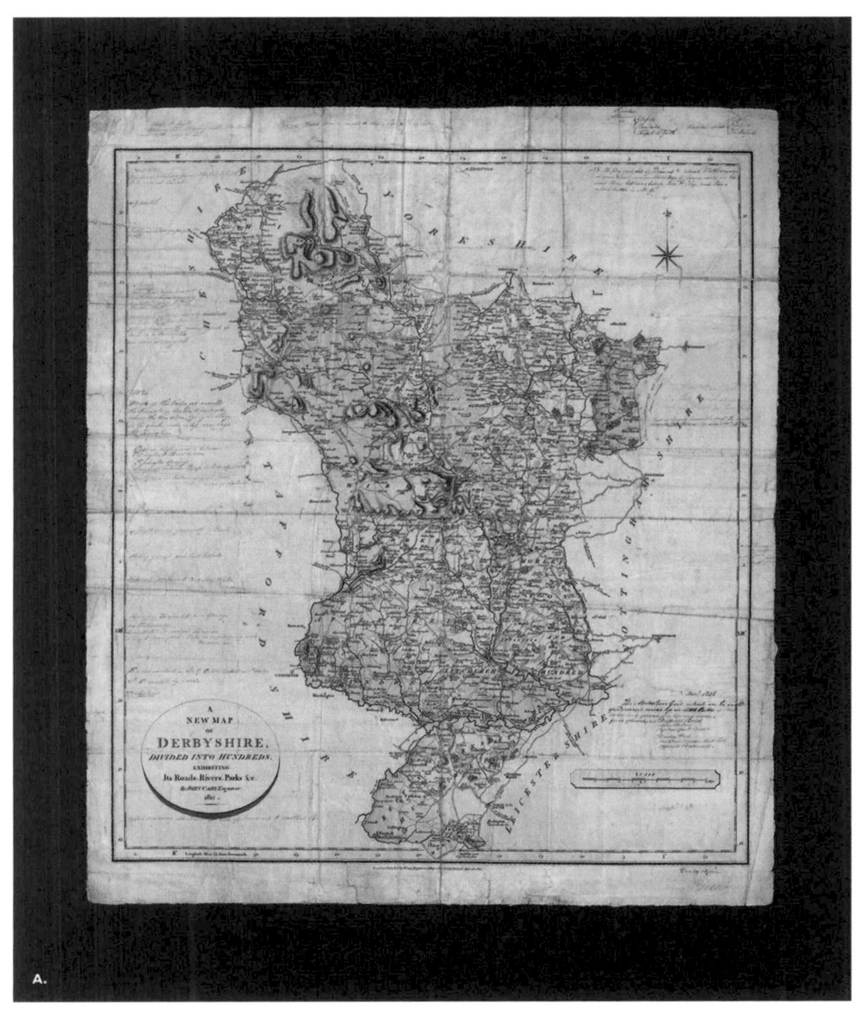

A.

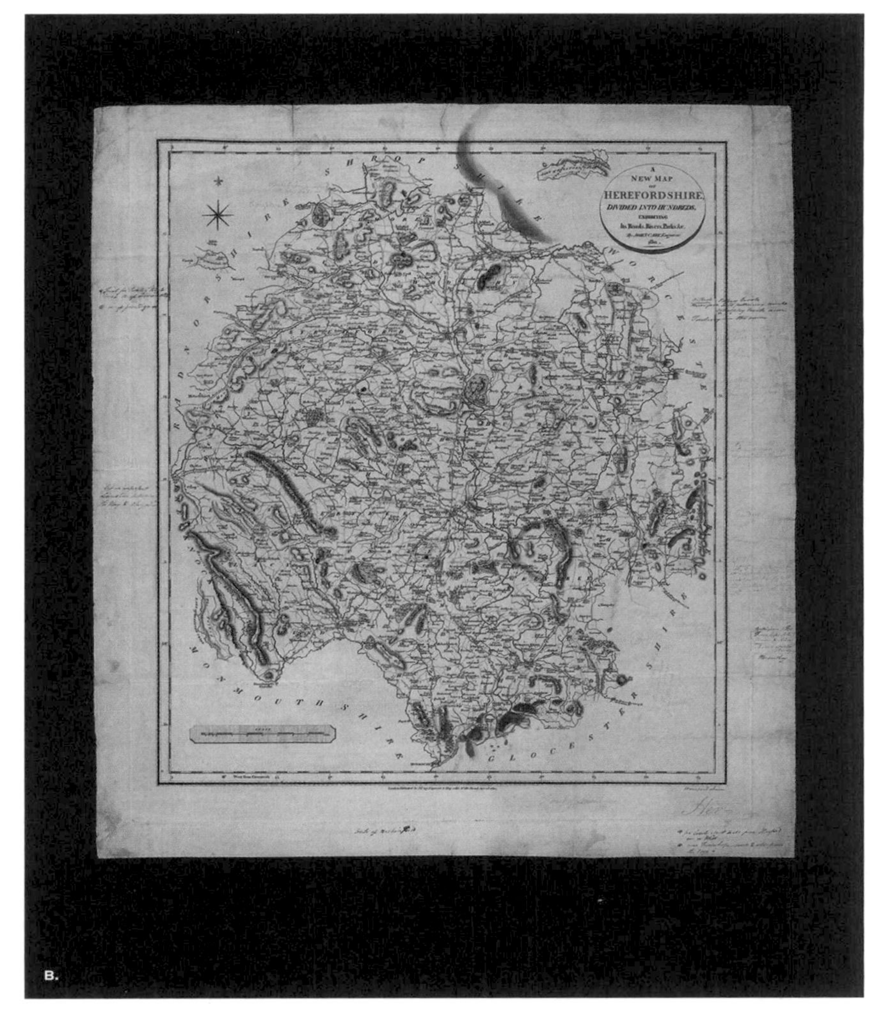

B.

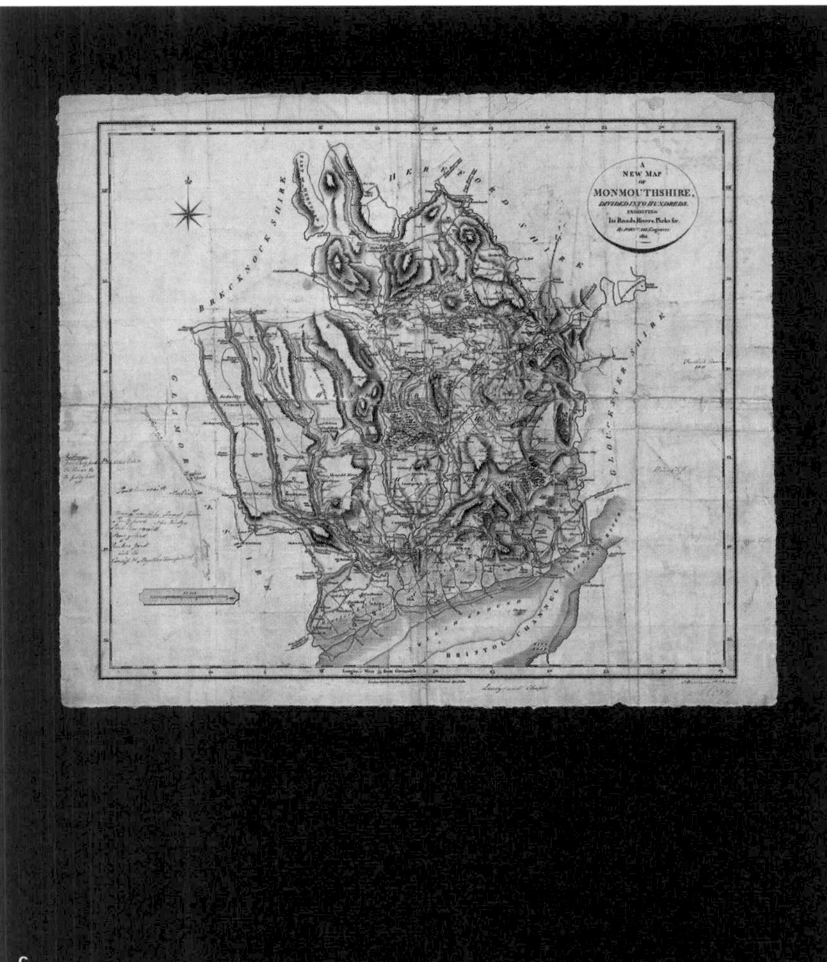

C.

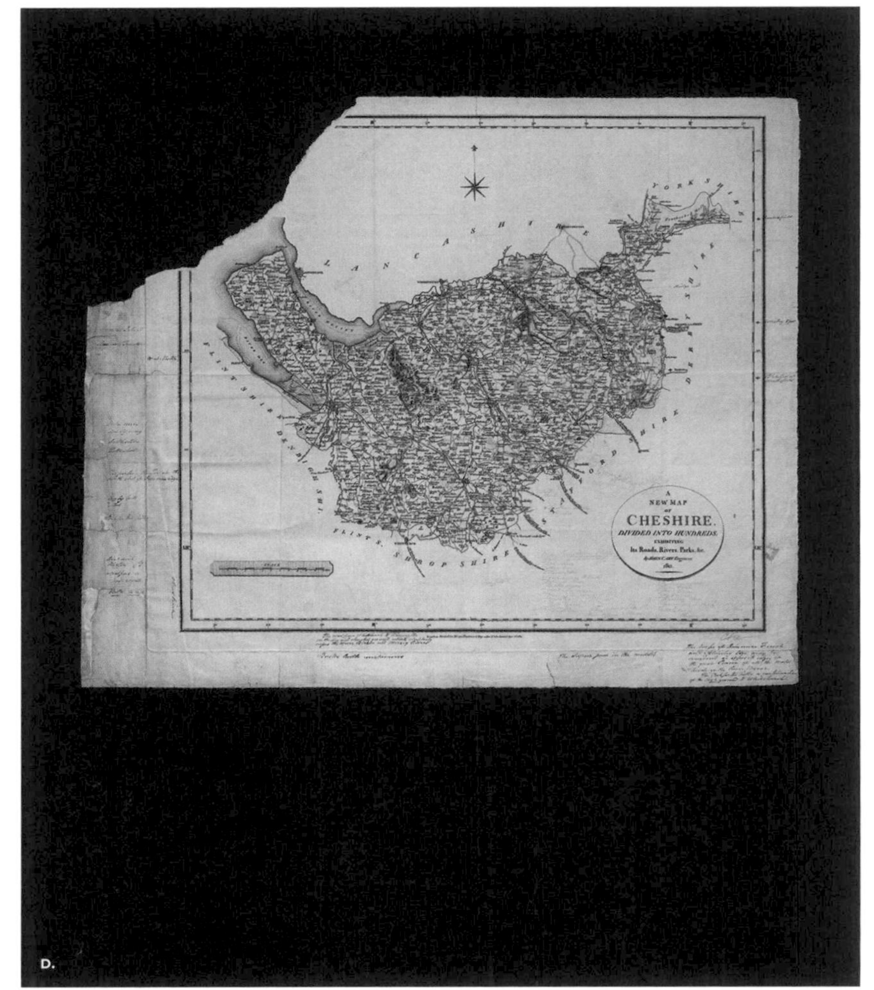

D.

A. UNPUBLISHED GEOLOGICAL MAP OF DERBYSHIRE.

There is geological colouring for Triassic, Magnesium Limestone and Mountain Limestone, an indication of collieries and mention of Millstone Grit dated 1838. Smith never published a completed map of Derbyshire.

B. UNPUBLISHED GEOLOGICAL MAP OF HEREFORDSHIRE.

There is a small amount of geological colouring referring to Silurian Limestone and reference in red to the Golden Vale. Smith never published a completed map of Herefordshire.

C. UNPUBLISHED GEOLOGICAL MAP OF MONMOUTHSHIRE.

There is a small amount of geological colouring and handwritten annotation indicating Blue Marl, Under Oolite and Fuller's Earth. Smith never published a completed map of Monmouthshire.

D. UNPUBLISHED GEOLOGICAL MAP OF CHESHIRE.

There is red-coloured Triassic in some of the river valleys and blue-coloured Marl south of Astbury. Smith never published a completed map of Cheshire.

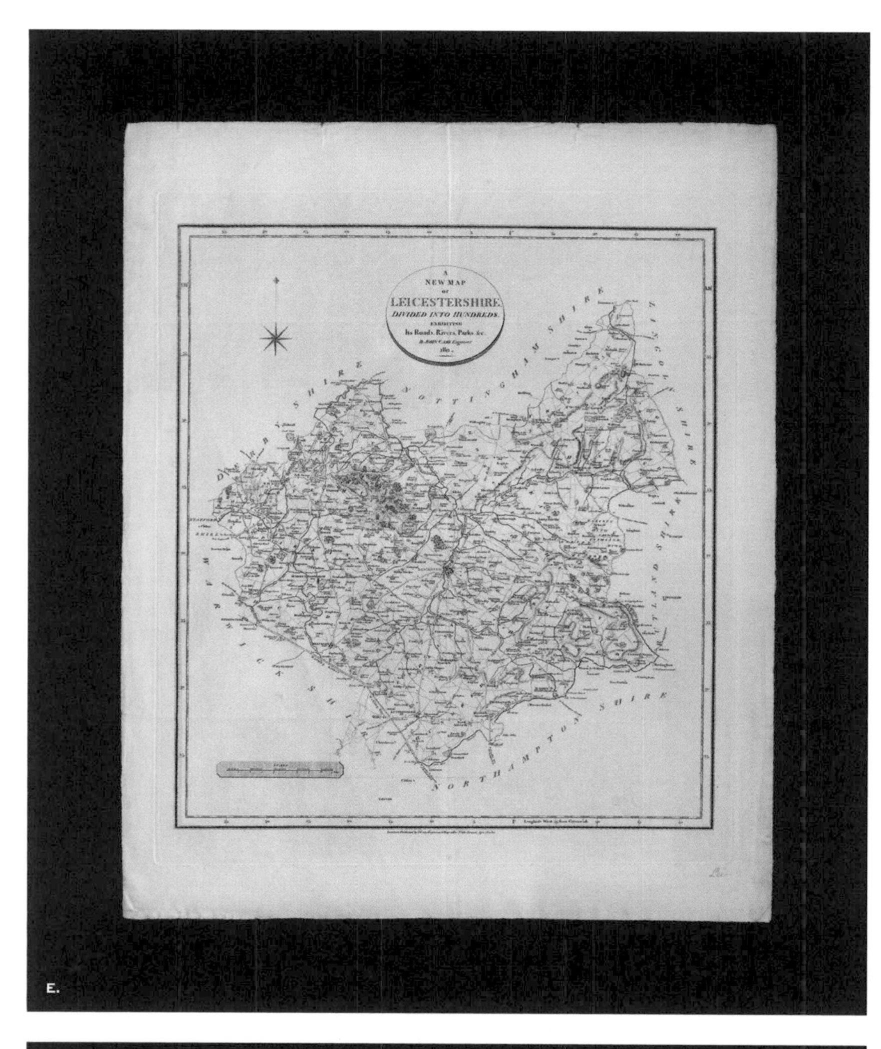

E.

F.

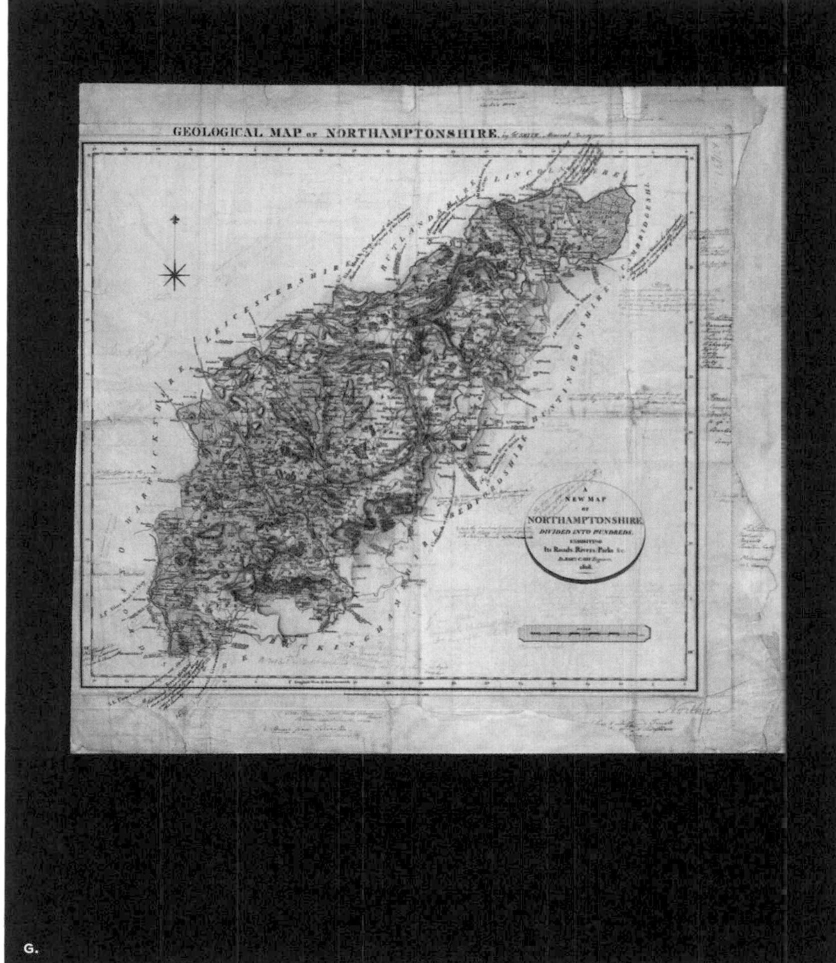

G.

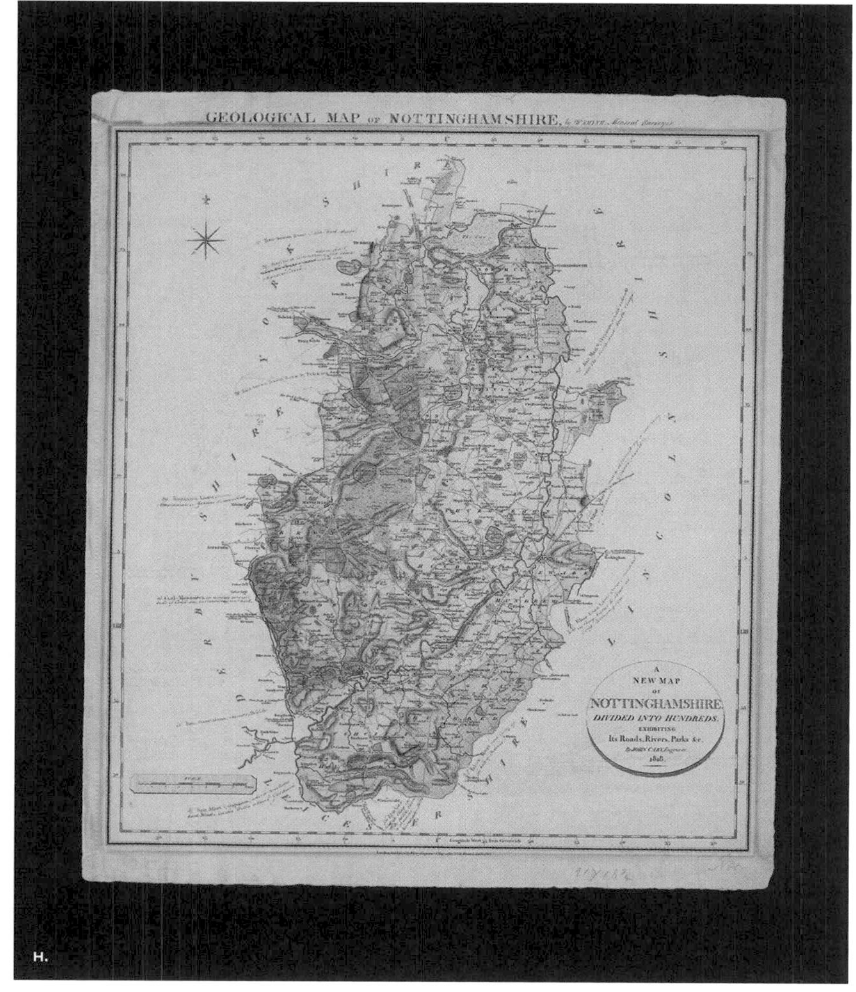

H.

E. UNPUBLISHED GEOLOGICAL MAP OF LEICESTERSHIRE.

There is no geology visible on this map.

F. UNPUBLISHED GEOLOGICAL MAP OF STAFFORDSHIRE.

There is some geological colouring relating to Trias and Carboniferous Limestone. Smith never published a completed map of Staffordshire.

G. UNPUBLISHED GEOLOGICAL MAP OF NORTHAMPTONSHIRE.

The geological colouring is mostly complete and there are hand-drawn legend tables and extensive notes. Smith never published a completed map of Northamptonshire.

H. UNPUBLISHED GEOLOGICAL MAP OF NOTTINGHAMSHIRE.

The geological colouring is mostly complete and there are hand-drawn legend tables and annotation.

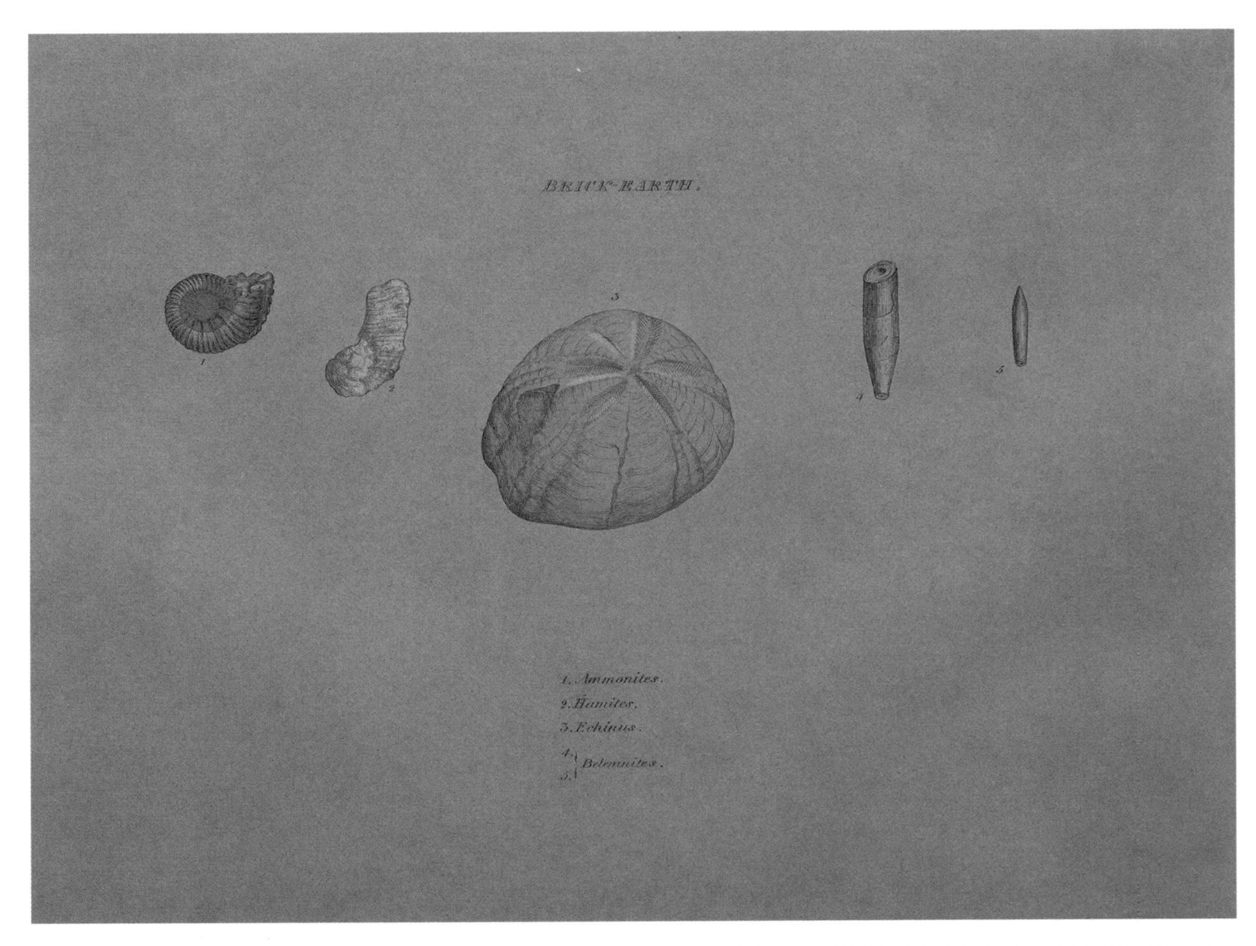

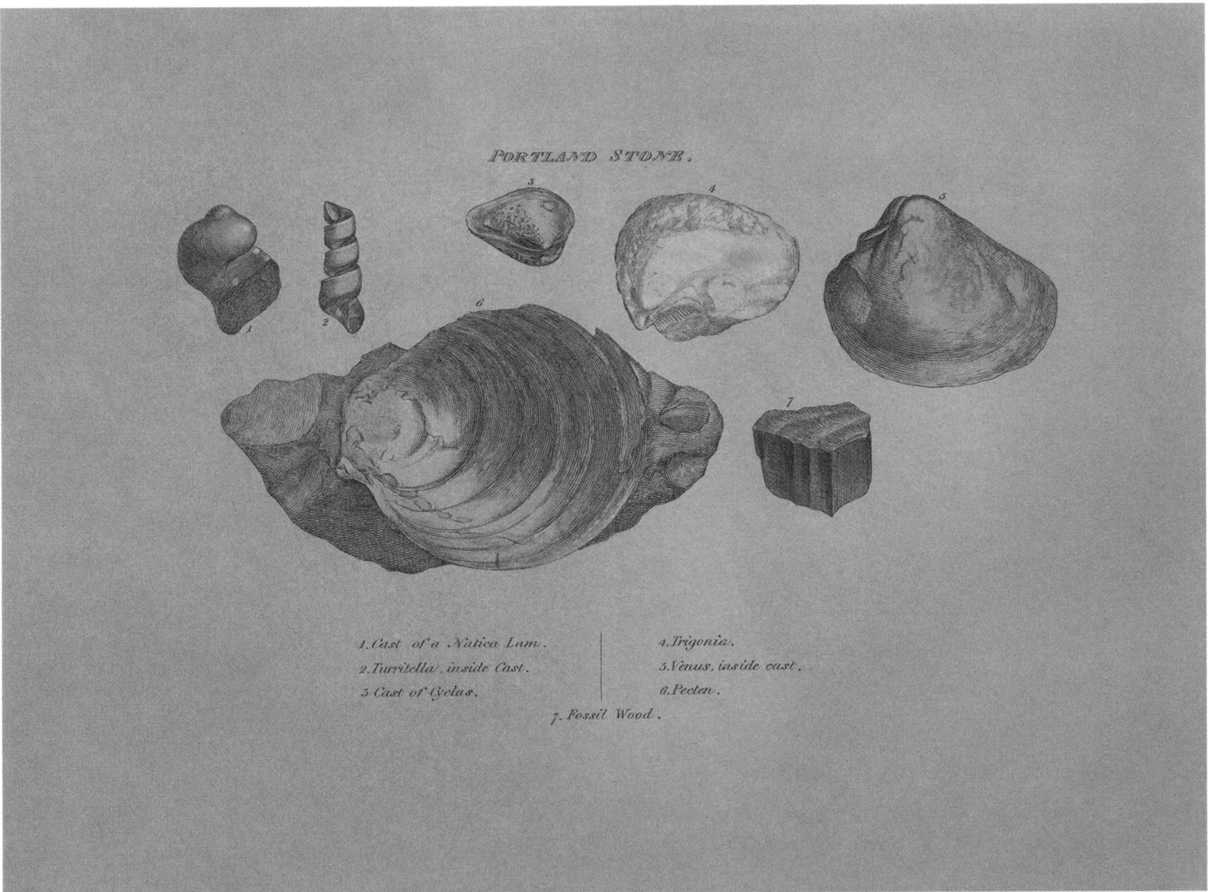

FOSSILS FOUND IN BRICKEARTH [ABOVE].

In Smith's hand-annotated copy of his catalogue he later crossed out Brickearth and replaced it with 'Golt' (now spelt Gault). As in other clays, many of the specimens from the Gault Clay are preserved in pyrite which can decay rapidly when exposed to the atmosphere and the original of his Figure 1 is missing, probably as a result of this problem.

FOSSILS FOUND IN PORTLAND STONE [BELOW].

The Portland Stone fossils in Smith's collection are mostly internal moulds which are not always useful in determining the species; it is the entire assemblage that Smith recognized as being important. The only fossil illustrated from the Isle of Portland is a gastropod, familiarly known as the 'Portland Screw'. The rest were collected in Swindon where he was working.

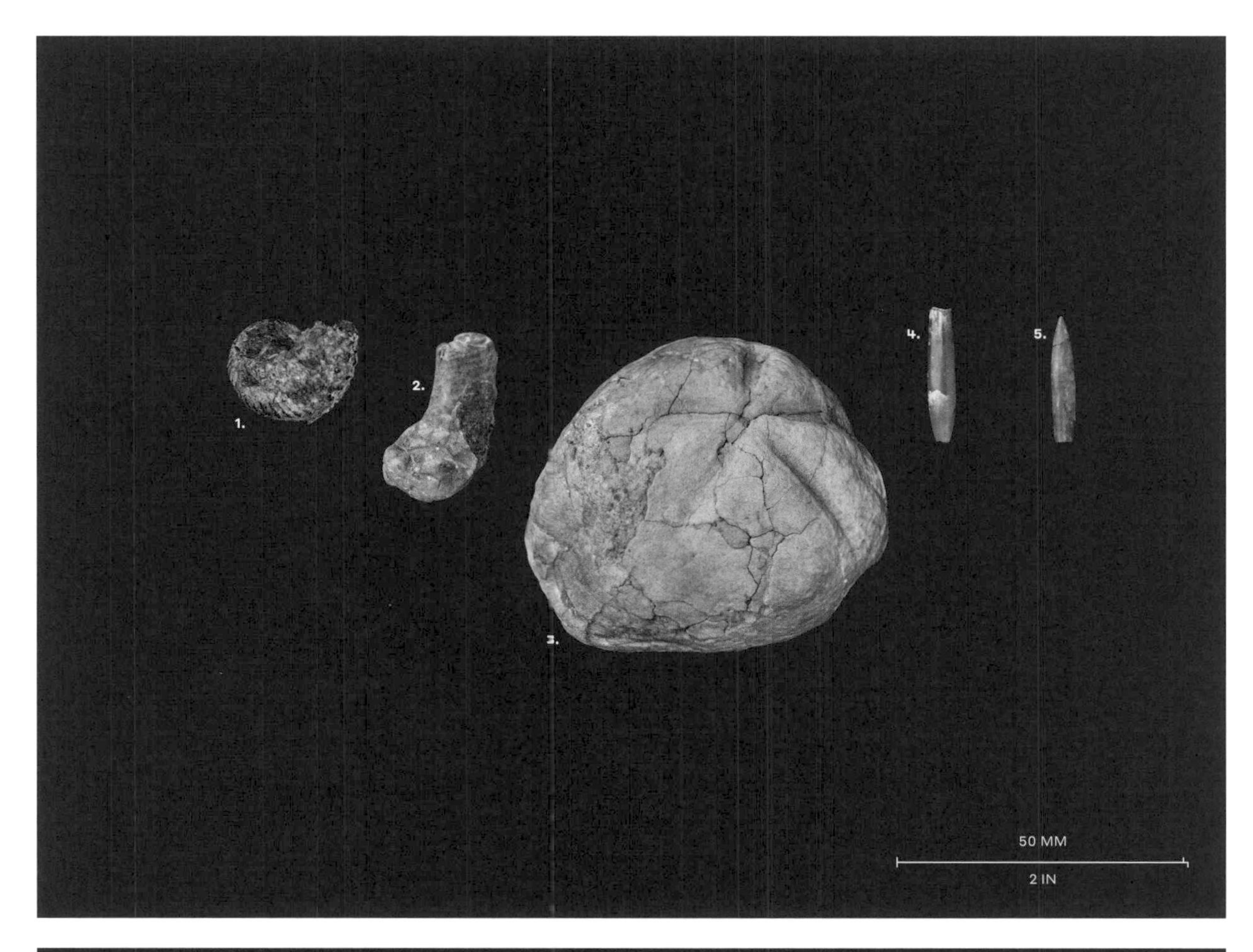

FOSSILS FOUND IN GAULT CLAY [ABOVE].

1. *Hoplites dentatus* C627 Steppingley Park (substitute) **2.** *Idiohamites* sp. C628 near Grimstone; **3.** *Pliotoxaster* sp. E489 near. Devizes; **4.** *Neohibilites minimus* C626, near Grimstone (subsitute); **5.** *Neohibilites minimus* C625 Prisley Farm Bedfordshire.

FOSSILS FOUND IN PORTLAND STONE [BELOW].

1. *Neritoma sinuosa* G1587 Swindon; **2.** *Aptyxiella portlandica* G1585 Portland; **3.** *Eomiodon* sp. L1486 Swindon; **4.** *Myophorella incurva* L1484 Swindon (substitute); **5.** *Protocardia dissimilis* L1487 Swindon; **6.** *Camptonectes lamellosus* L1492 Swindon; **7.** section of conifer V475 Woburn (?substitute).

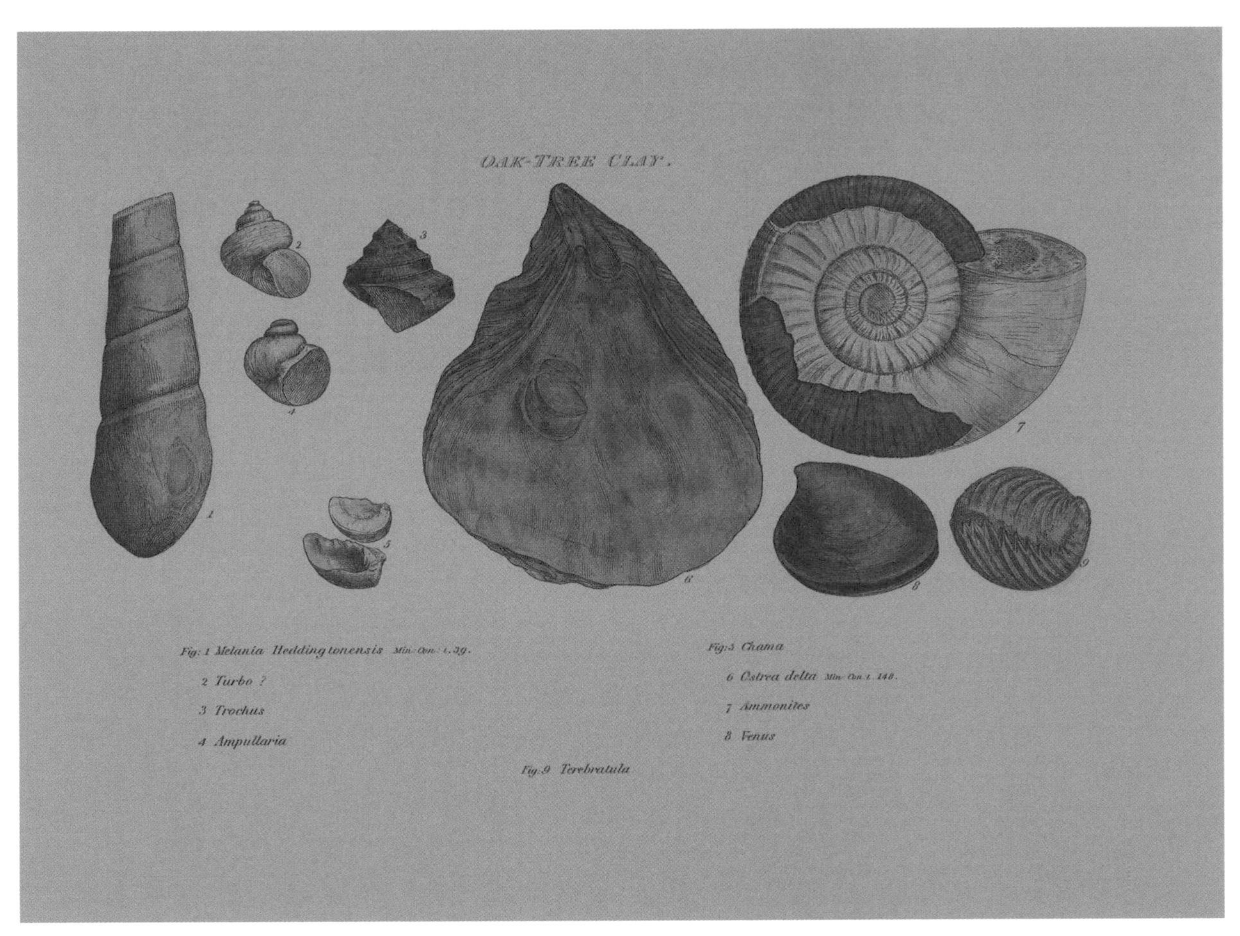

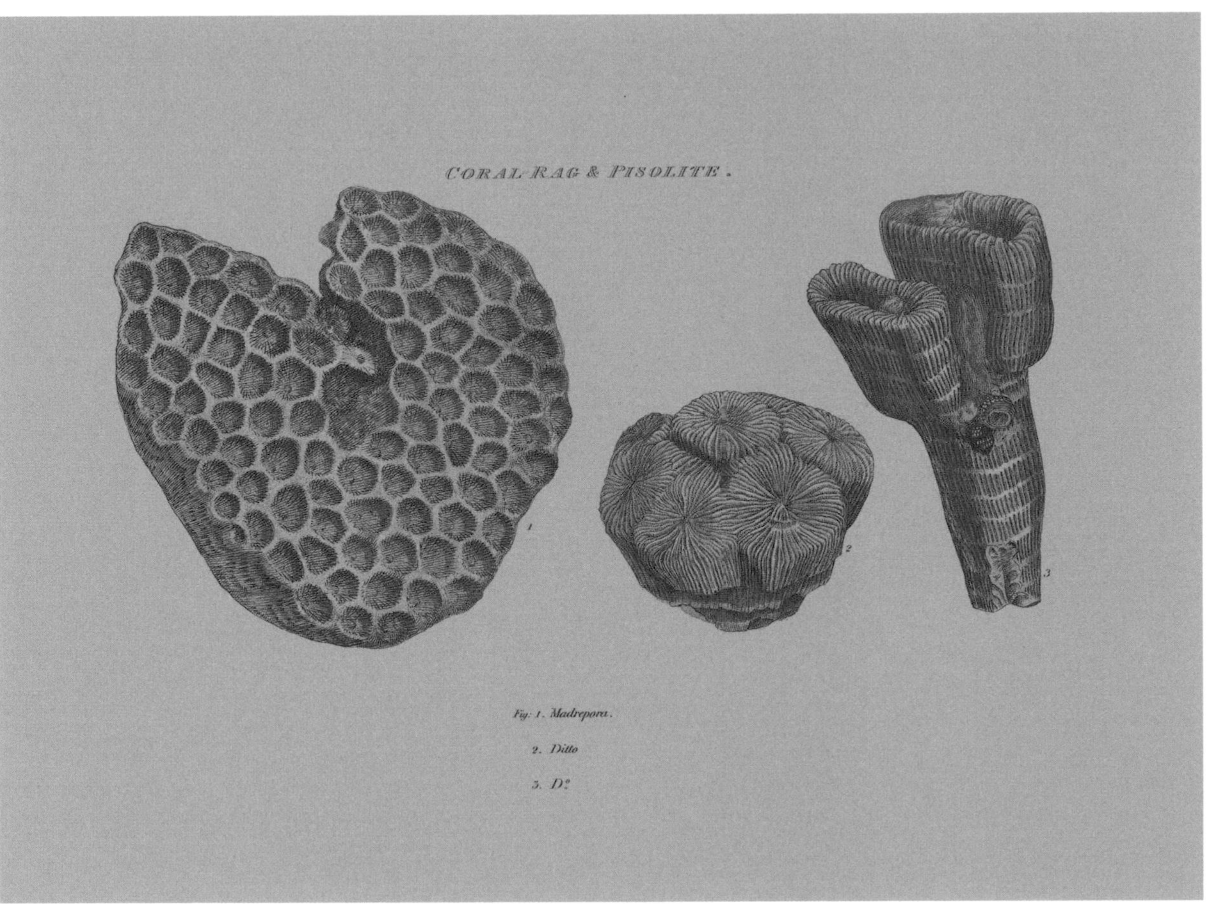

FOSSILS FOUND IN OAKTREE CLAY [ABOVE].

Smith's name, 'Oak Tree Clay' is unfamiliar to us today and was probably a quarrymen's term. It can best be seen in Kimmeridge Bay, where it was later described and acquired the formal name of Kimmeridge Clay. Smith encountered it while working on the canals, particularly the North Wilts Canal.

FOSSILS FOUND IN CORAL RAG & PISOLITE [BELOW].

As the name implies, this stratum contains a wide variety of some very handsome coral specimens. Smith chose three of the most common ones to illustrate this plate. Smith says that fossils turn up frequently in the plough but the sharper and better specimens come from quarries.

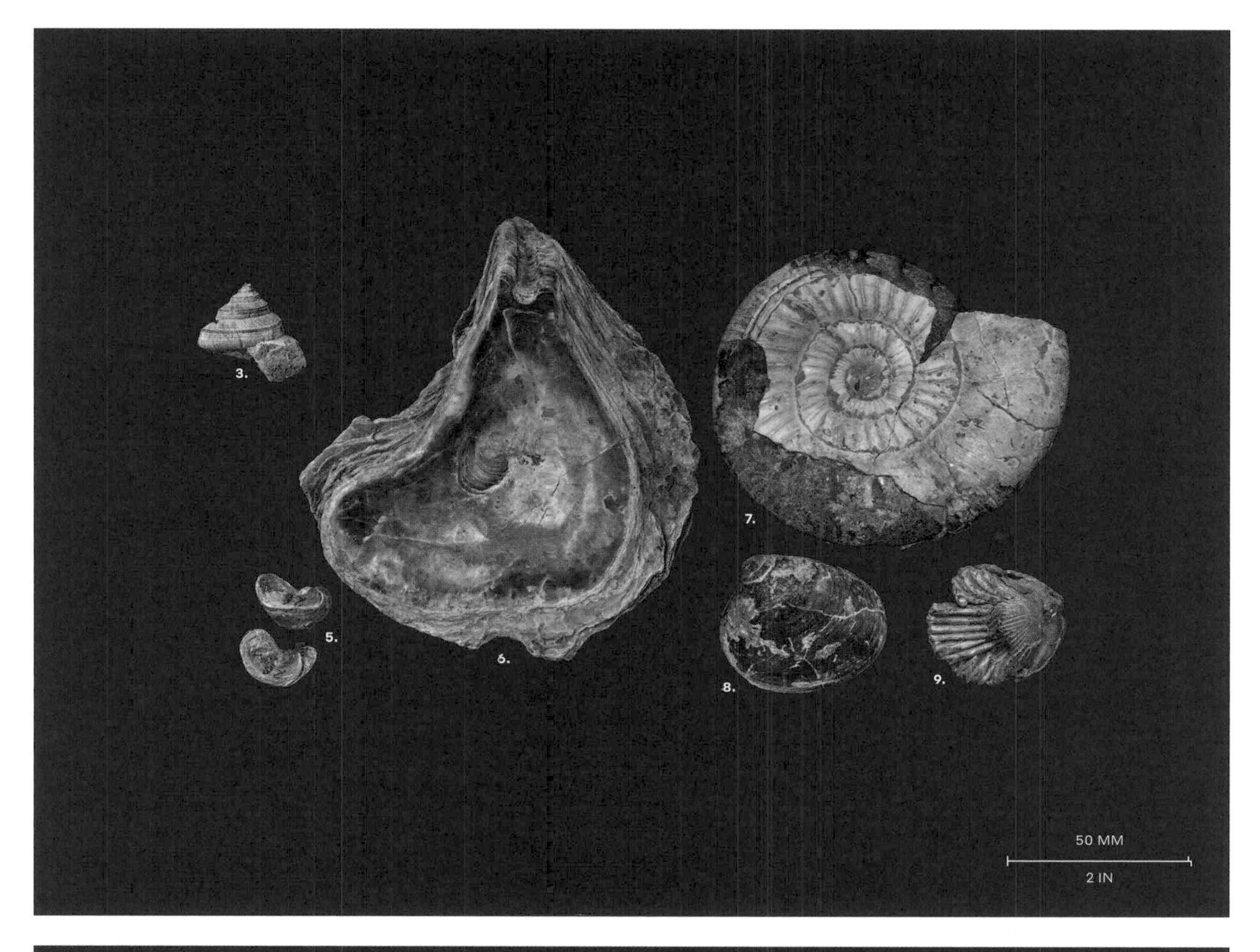

■ **FOSSILS FOUND IN KIMMERIDGE CLAY [ABOVE].**

3. *Bathrotomaria reticulata* 24817 North Wilts Canal; **5.** *Nannogyra nana* L53452-3 North Wilts Canal; **6.** *Deltoideum delta* L1495 North Wilts Canal; **7.** *Pictonia baylei* 37847 Wootton Bassett; **8.** *Neocrassina ovata* L256 (no location); **9.** *Torquirhynchia inconstans* B1409 Bagley Wood Pit (substitute).

■ **FOSSILS FOUND IN CORALLINE OOLITE [BELOW].**

1. *Isastrea explanata* R1076 Stanton near Highworth; **2.** *Complexastrea depressa* R1079 Steeple Ashton; **3.** *Thecosmilia annularis* 56336 Steeple Ashton.

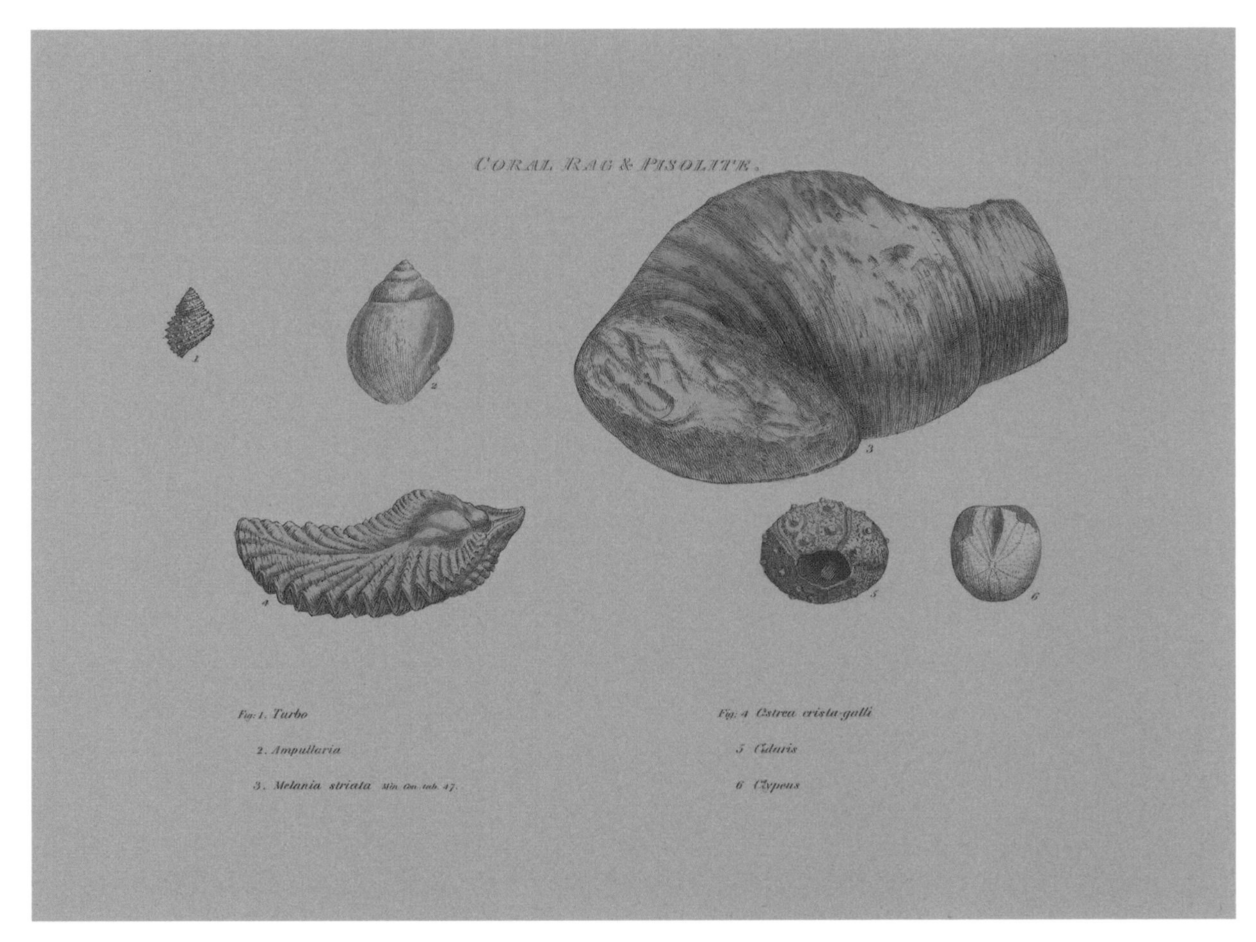

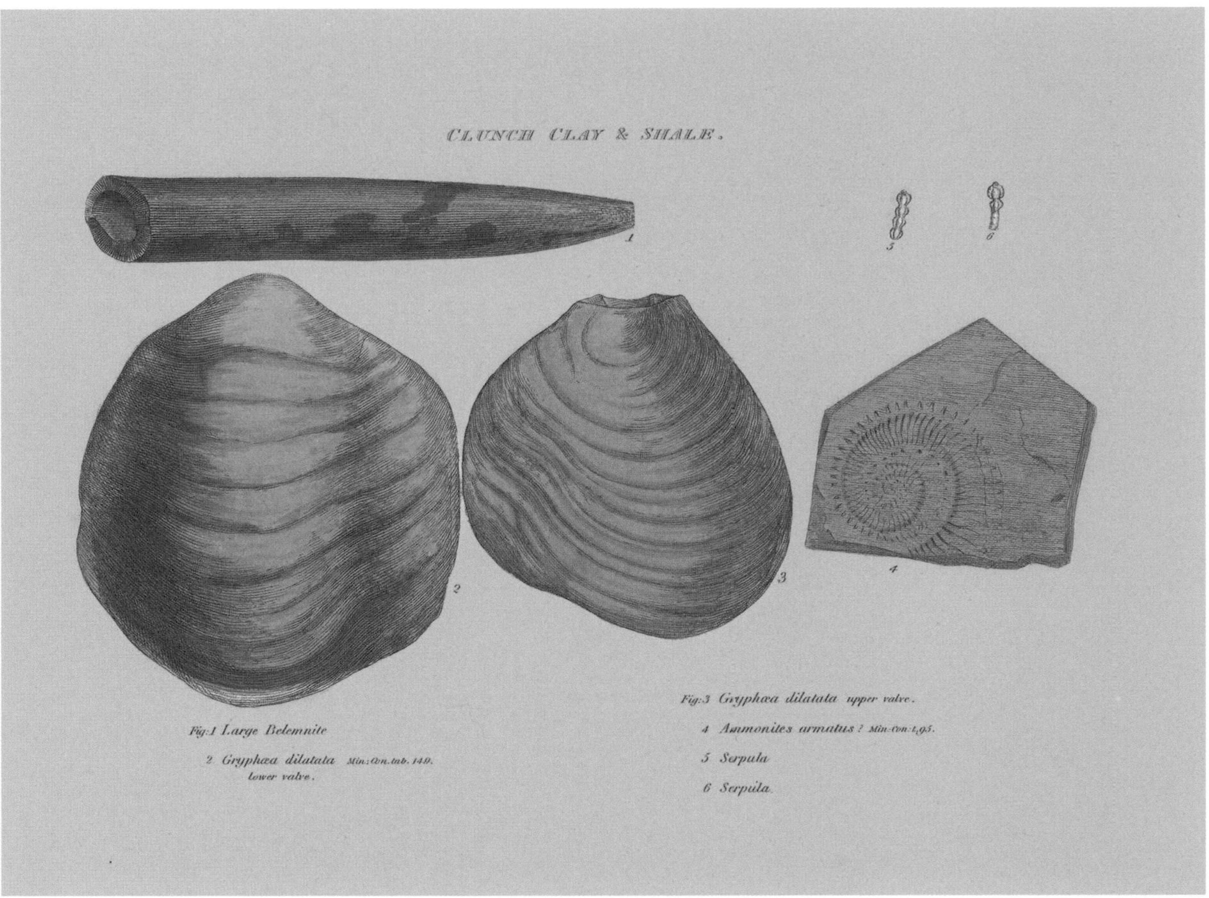

FOSSILS FOUND IN CORAL RAG & PISOLITE [ABOVE].

In 1816 Smith's advice was sought on the supply of water for the Wilts & Berks Canal and the specimens collected from the well near Swindon were almost certainly from that time (see Appendix). The well section allowed Smith to add the Coral Rag to his map, as well as contributing to his understanding the succession of clays in the Upper Jurassic.

FOSSILS FOUND IN CLUNCH CLAY & SHALE [BELOW].

The ammonite, **4.**, is from the Thames & Severn Canal, and more specifically from Siddington, near Cirencester, Gloucestershire. This is close to the canal tunnel that Smith visited in 1793 when researching details for the proposed Somersetshire Coal Canal. It is likely to have been an early acquisition in his collection.

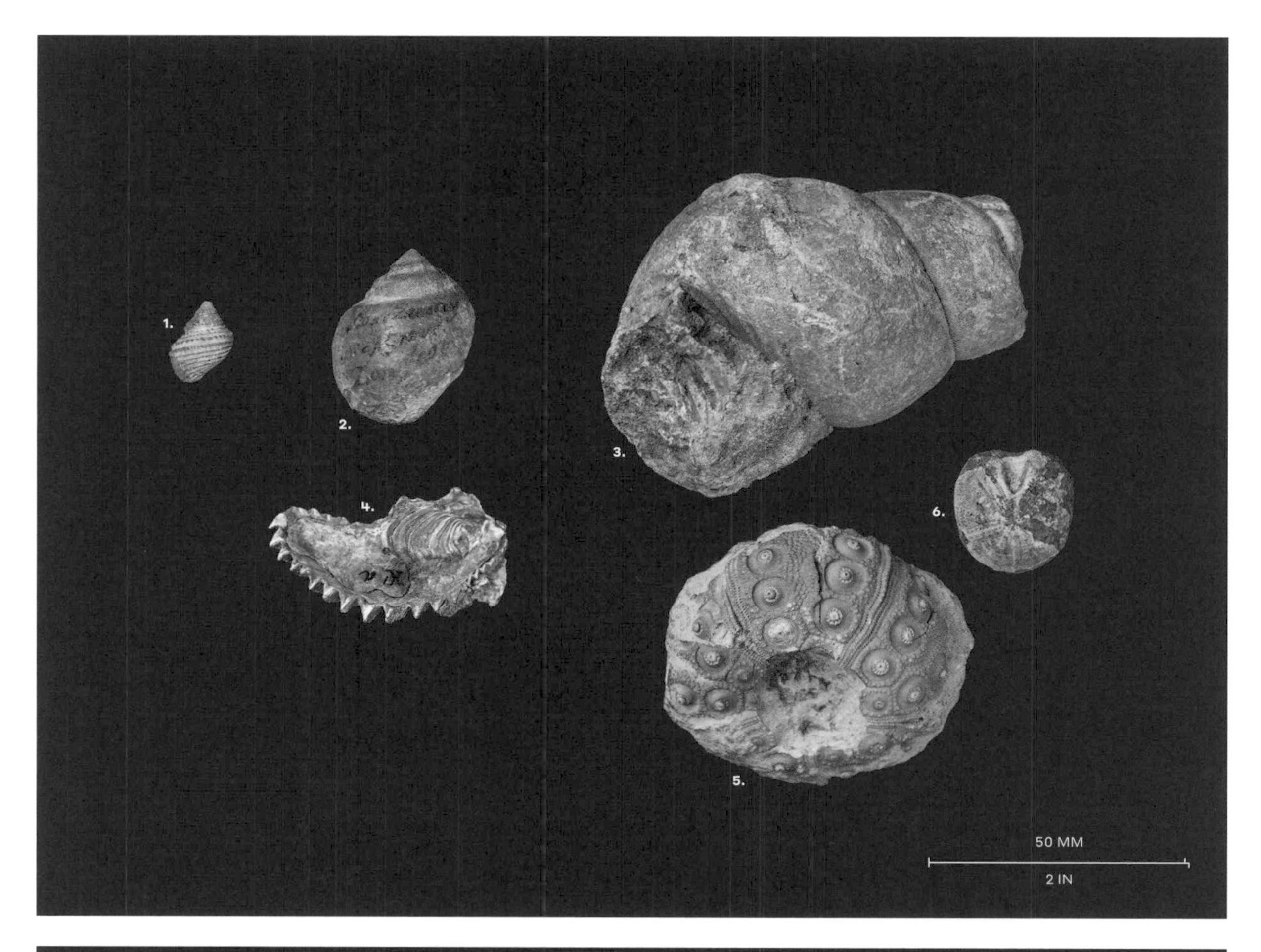

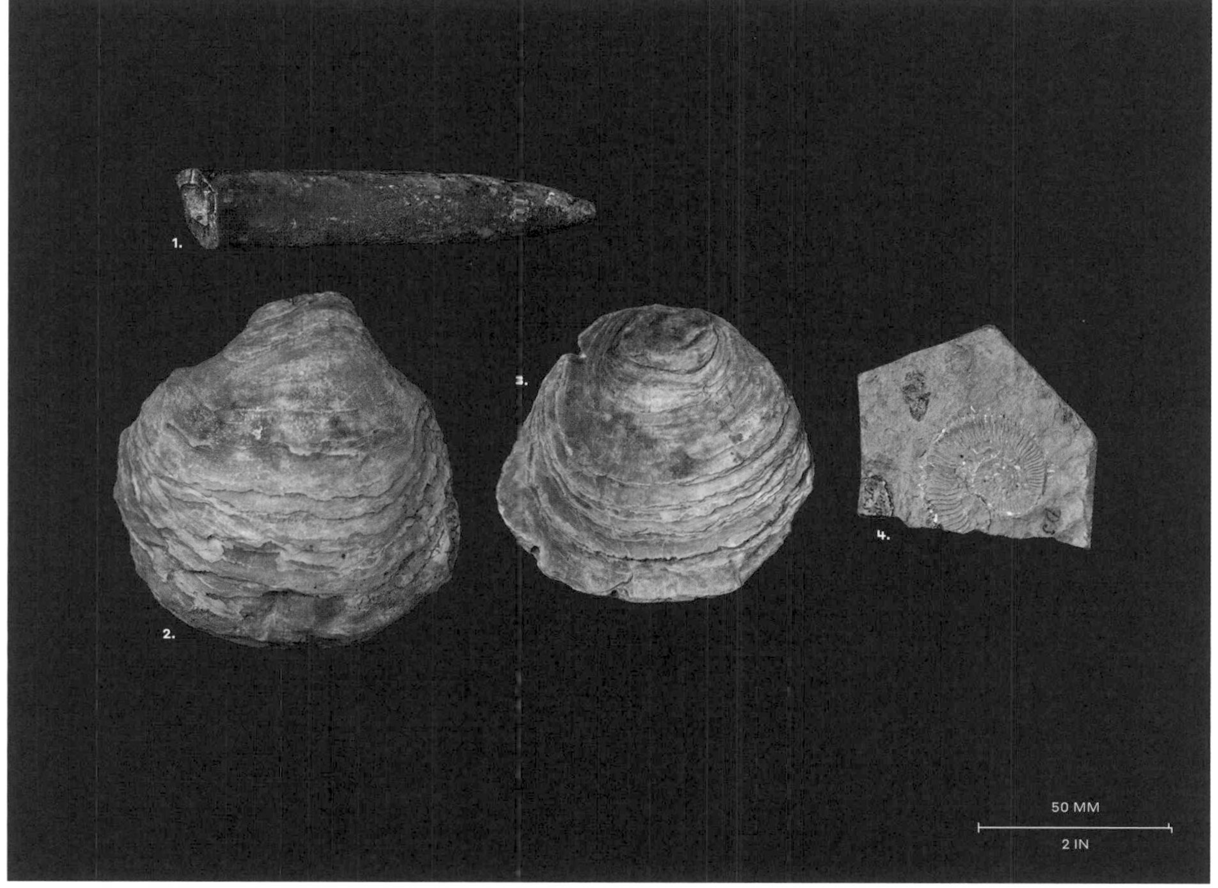

FOSSILS FOUND IN CORALLINE OOLITE [ABOVE].

1. *Ooliticia muricata* G1594 Derry Hill; **2.** *Ampullospira* sp. G1599 Longleat Park; **3.** *Bourguetia saemanni* G1646 Caisson, Wilts & Berks Canal (probable substitute); **4.** *Actinostreon gregarium* L1533 Wilts (substitute); **5.** *Paracidaris smithii* E492 Hilmarton (substitute); **6.** *Nucleolites clunicularis* E495 Meggot's Mill, Coleshill.

FOSSILS FOUND IN OXFORD CLAY [BELOW].

1. *Cylindroteuthis puzosiana* C640A Dudgrove Farm; **2–3.** *Gryphaea dilatata* L1518 Derry Hill (substitute); **4.** *Kosmoceras spinosum* C 636 Thames & Severn Canal.

II. MINERAL PROSPECTOR.

Observations on Smith's survey and cross section of Mearns Colliery; the surprising and discouraging results of a plethora of coal trials; the battle of the steam engines; a dialogue between our hero and Richard Trevithick; plotting coal and economic minerals on the 1815 geological map of England and Wales.

Coal is abundant in the British Isles and powered the Industrial Revolution; it had been worked since Roman times but came to the forefront in the eighteenth century, when it was used for smelting iron and later to power steam engines.

Initially, William Smith (1769–1839) had little knowledge of coal seams and their associated strata, but he learnt quickly. His introduction to coal mines came at the Mearns Colliery, in the small Somerset coalfield.

THE MEARNS COLLIERY.

In 1792, as part of his estate surveying work for Lady Elizabeth Jones (1741–1800), Smith produced *A Sketch of the Land in H. Littleton about the Coalworks*, showing shaft locations, levels and inclines, as well as local landowners. Another map, entitled *Original Sketch and Observations of my first Subterranean Survey of Mearns Colliery*, was a more detailed survey of one particular shaft and incline. Smith also recorded the depths and thicknesses of the various coal veins encountered in five other pits. The geology of the Somerset coalfields is much more complicated than in many other parts of Britain – the coal seams are often steeply inclined and are frequently faulted. Smith owned a copy of a cross section from John Strachey's *Observations on the Different Strata of Earths and Minerals, more particularly of such as are found in the Coal-Mines of Great Britain* (1727), which showed the coal seams. Strachey (1671–1743) had made some very astute observations on the sequences and attitudes of coal beds and associated strata in coal mines at Stowey and Littleton, in the area where Smith was working. These provided Smith with useful information about stratigraphic order and the attitudes of strata, particularly in relation to their dip and strike (Fuller, 1994), which he annotated on his copy of the section.

Smith's own first cross section of Mearns Colliery was rather cruder, with much of the detail obscured by textured shading. The main incline followed the dip of the Great Vein, and Smith showed the depth of the pit (300 ft or 90 m), with steep inclines (9 in. to 1 ft in a yard). He indicated standings (where it was possible to stand), guggs (roads), a twinway (a double branch road) and two additional shafts to the Little Vein at the base of the incline. Smith's description of the workings provides a flavour of just how difficult conditions must have been. Coal bushels were winched up the steeper guggs, but along the twinway the tunnels were no more than 1.2 m (4 ft) high, so the miners and boys had to negotiate them on hands and knees, dragging bushel-carts of coal chained to their waists. From the foot of the shaft the coal bushels were winched to the surface by a horse-drawn windlass. Smith was a well-built man, so clambering about in the dark along wet guggs and twinways must have been a challenge.

Smith hinted at the survey methods he had learnt in a drawing he made of surveys of the

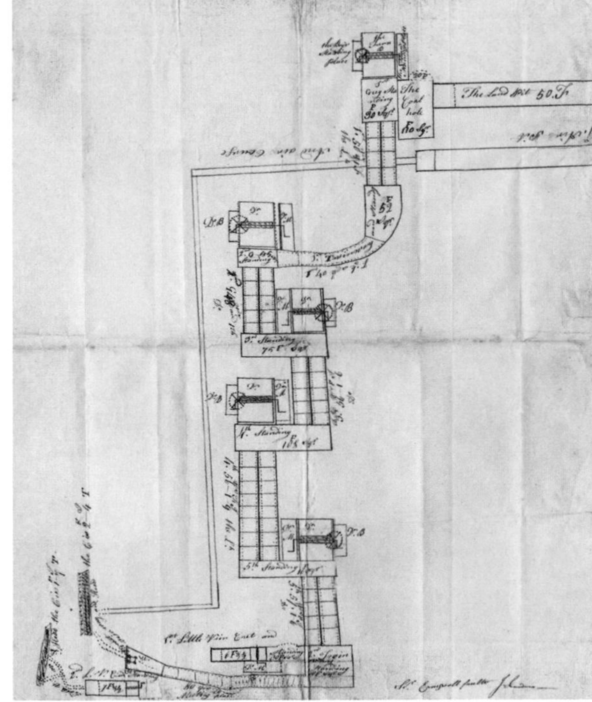

FIG. 1.

Fig. 1.
A plan of the Mearns Colliery made by J. Landsdown, a collier at the mine, in 1792. The plan clearly shows the cramped guggs, twinways and standing areas, as well as underground windlasses to assist in hauling the coal.

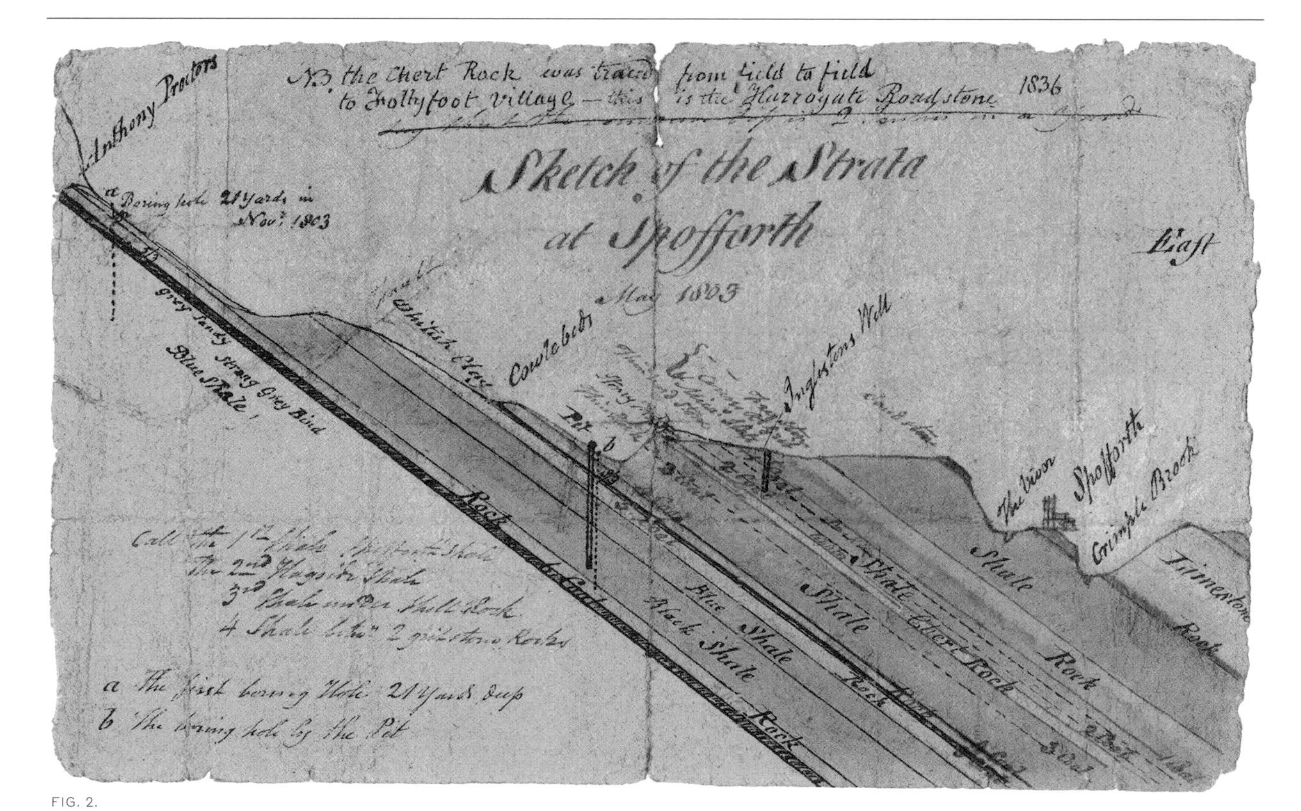

FIG. 2.

Fig. 2.
A sketch section of the strata below the Magnesian Limestone at Spofforth in Yorkshire by William Smith in 1803. The section also shows the locations of trial borings. There is further annotation on the section, possibly by John Phillips, made in 1836.

mine in the form of dial plates. He showed two dials marked not in degrees but as clock faces. The first dial is divided into 24 parts, with 12 hours of the day and 12 hours of the night. The second dial is similar, but uses a slightly different numbering system, though again, in time rather than degrees (Fuller, 1992). Marked on the dial charts are the alignment of the main fault and alignments of coals in various mines, such as Grove, one o'clock morning (compass bearing N30°E). The Somerset miners seem to have preferred hours rather than degrees to describe the strike and dip of the coal veins.

Building on Strachey's sections, Smith noted the sequence of the major coal seams encountered at the Water Pitt at Mearns, but did not differentiate their lithology – their physical characteristics – preferring to use the miner's dialect term 'shill' – a shale – often used to describe Liassic strata. (Smith recorded the presence of the Whore Vein at the top of the coal sequence, but this does not appear in later descriptions, possibly for reasons of propriety.) Later, Smith was to recognize sandstones (Greys), shales (Clift) and Pan (seat-earth or fireclay). Today we would interpret this sequence as a coal cycle or cyclothem.

John Phillips (1800–1874) noted in his *Memoir* that Smith had proposed to make a model of the mine where 'each stratum is composed, arranged in the same order as nature has placed them, and divided into sections, that may be taken apart to explain the method of mining for coal' and that it was his intention 'to make the model *true* and *proportionate* in vertical and horizontal measure' (Phillips, 1844, p. 7).

Smith's experience at the Mearns Colliery gave him an understanding of strata, their attitude – dip and strike – and the importance of faulting, as well as a practical knowledge of mines, shafts, inclines, cross-cuts, headways and other features. These would serve him well as he went on to consult on numerous coal mines, building on his knowledge of geology with each new commission.

IN SEARCH OF COAL.

Although coal was abundant in Britain, there were plenty of wealthy landowners who were keen to find more of this lucrative substance. Smith, who began to style himself as a 'Mineral Surveyor', was soon in demand. In early 1803 George Wyndham, 3rd Earl of Egremont (1751–1837), contacted Smith as he wanted to discuss some business via his agent John Claridge; evidently the Earl was not keen for the nature of it to be widely known. It emerged that Egremont's intention was for Smith to conduct a trial for coal on his land near Spofforth in Yorkshire. Smith duly made a survey and produced a report. Two boreholes were sunk, even though the locations were not in the Coal Measures but instead within the older Millstone Grit. Despite initial disappointment, Smith was optimistic that coal would be found. Thin coals seams were

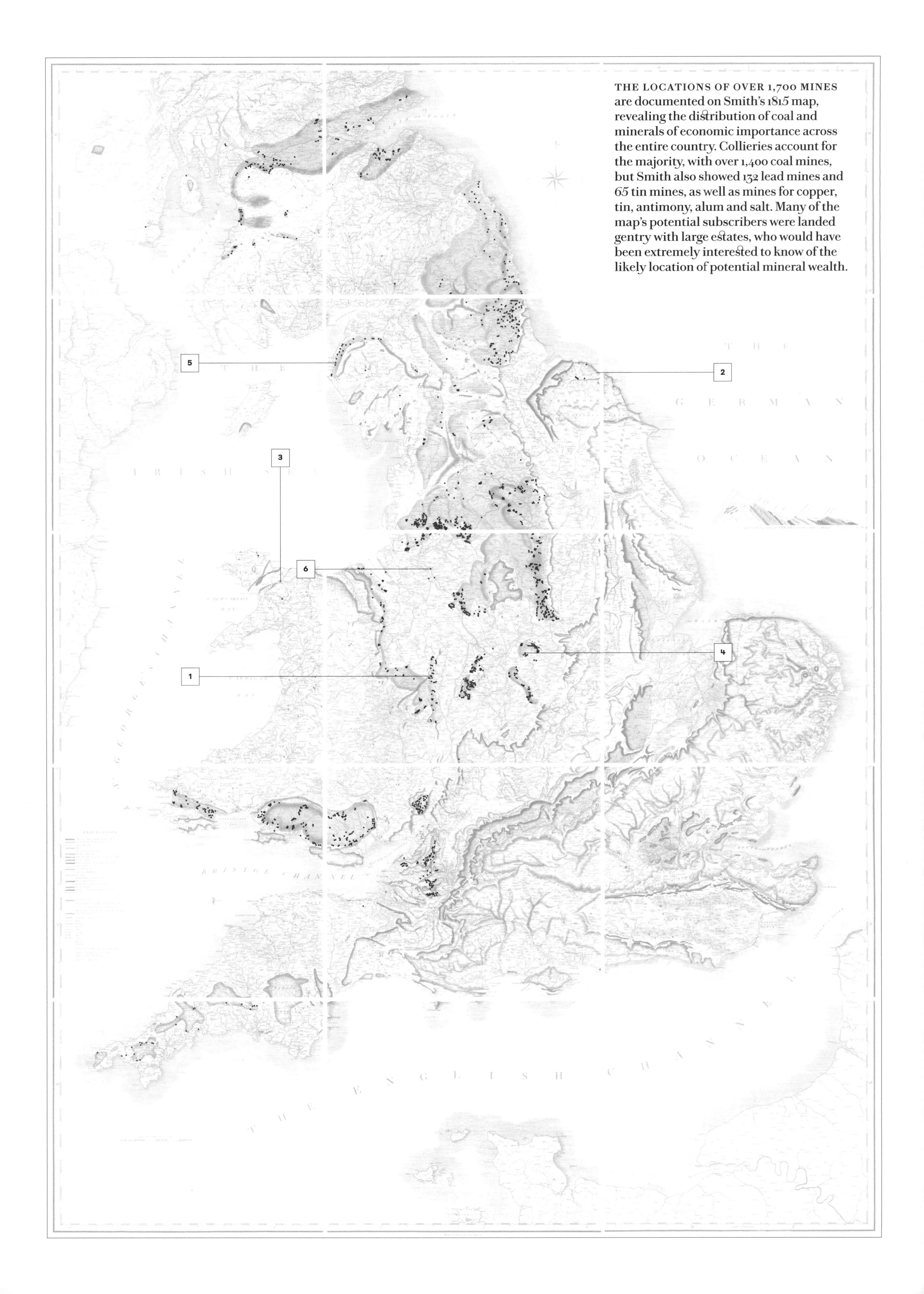

THE LOCATIONS OF OVER 1,700 MINES are documented on Smith's 1815 map, revealing the distribution of coal and minerals of economic importance across the entire country. Collieries account for the majority, with over 1,400 coal mines, but Smith also showed 132 lead mines and 65 tin mines, as well as mines for copper, tin, antimony, alum and salt. Many of the map's potential subscribers were landed gentry with large estates, who would have been extremely interested to know of the likely location of potential mineral wealth.

1.

IRON WORKS
Coalbrookdale, Shropshire. 1805.

The Iron Works at Colebrookdale [sic.], an engraving by William Pickett after the painting of Philippe J. de Loutherbourg, 1805.

2.

ALUM WORKS.
North Yorkshire. 1814.

Robert Havell's *Alum Works, North Yorkshire*, 1814, shows burnt shale being taken to steeping pits, where rainwater was used leach the alum salts.

3.

SLATE QUARRY.
Bangor, Wales. 1807.

John Nixon's watercolour of Lord Penrhyn's quarry near Bangor, 1807. At the end of the nineteenth century it was the world's largest slate quarry.

4.

COAL MINE, LIMEWORKS.
Moira, Ashby Would. 1804.

Elisabeth Watson's watercolour shows a steam-powered winding engine and two lime kilns, where Limestone would have been burned using local coal.

5.

MILL PIT COLLIERY.
Harrington, Cumbria. C. 1800.

This sketch shows a steam winding engine on the right and an older horse-drawn whim winder on the left.

6.

SALT MINE.
Marston, Northwich, Cheshire. C. 1800.

This illustration shows a large cavern ('The Cathedral') with pillars of salt left by the miners to support the roof.

MINES & QUARRIES IN EXISTENCE IN 1815.

ALUM.

12 QUARRIES.
LOCATED PRIMARILY IN NORTH ENGLAND.

On his 1815 map Smith indicated a number of alum works ringing the North Yorks Moors. Among his papers was also a north–south section drawn by the artist and naturalist John Bird (1768–1829), which showed the alum shales and overlying sediments.

LEAD.

132 MINES.
LOCATED ACROSS GREAT BRITAIN.

Smith located a large number of lead mines on his 1815 map but only specifically named a few, such as Leadhills in South Lanarkshire and Glendinning in Dumfries and Galloway, and the Old Gang Mines in the Yorkshire Dales.

ANTIMONY.

1 MINE.
LOCATED IN WEST SCOTLAND.

Smith shows one antimony mine on the map at Glendinning in Dumfries and Galloway. The Glendinning Mine, (Louisa vein) yielded some lead and over 200 tons of antimony.

SALT.

18 MINES.
LOCATED PRIMARILY IN CHESHIRE.

Salt production in Cheshire dates back at least to Roman times when salt was extracted from brine springs connected to outcropping strata containing a high proportion of salt. Rock salt or halite occurs widely in Triassic strata (Smith's Red Ground) of the Cheshire basin.

COAL.

1,446 MINES.
LOCATED ACROSS GREAT BRITAIN.

In Scotland coal is found in the Midland Valley. In England it is found in Cumberland, Northumberland, Durham, Yorkshire, Derbyshire the Midlands, Somerset and the Forest of Dean. The South Wales coalfield was the largest and perhaps most significant in the country.

SLATE.

3 QUARRIES.
LOCATED PRIMARILY IN NORTH WALES.

Smith did not specifically categorize slate quarries on his map legend, but he did label several slate quarries north of Moffat in Scotland. However, most slate in Britain came from quarries in North Wales, the largest of which was the Penrhyn Quarry near Bethesda.

COPPER.

24 MINES.
LOCATED PRIMARILY IN CORNWALL.

While copper and tin ores are commonly associated with Cornish mines, the largest copper mine was in fact at Parys Mountain in Anglesey in Wales.

TIN.

65 MINES.
LOCATED PRIMARILY IN CORNWALL.

Almost all the tin mines on Smith's map are in Cornwall. No individual mines are labelled, but one on the coast of west Cornwall is a close match with the Botallack Mine. Two near Pool are almost certainly Dolcoath and Cook's Kitchen; others are more uncertain.

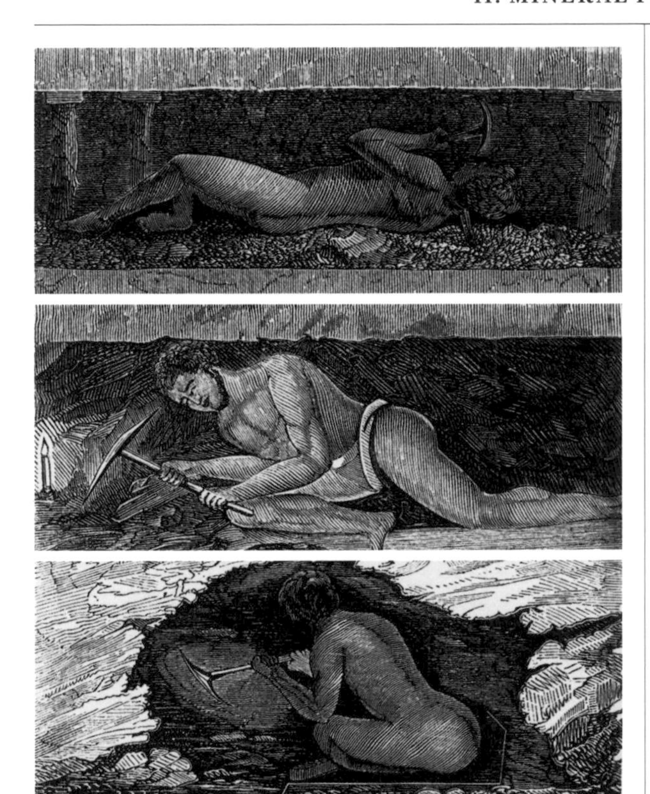

FIG. 3.

Fig. 3. Engravings of young men picking coal in stifling, narrow tunnels, taken from an extract of the 1842 report by the commissioners appointed to enquire into 'The Condition and Treatment of the Children Employed in the Mines and Collieries of the United Kingdom'.

later encountered and indeed Smith marked the Spofforth Colliery on his 1815 map. However, it is likely that little if any coal was ever produced commercially, in spite of the enormous expense Egremont is likely to have incurred.

STEAM ENGINES.

A more successful venture was undertaken by Smith for the Earl of Sefton in the development of the Torbock (Tarbock) Colliery to the east of Liverpool. William Philip Molyneux, 2nd Earl of Sefton (1772–1838), was a colourful character – a sportsman, gambler and friend of the Prince Regent. A need for steam engines for the coal trials at both Torbock and Spofforth led Smith to make contact with Richard Trevithick (1771–1833), the inventor and mining engineer. The two men had much in common: both came from the working classes and not the gentry and in appearance they were both muscular and strong. They also shared similar personalities – practical men with a 'can do' attitude. Unfortunately, brilliant as they both were, they were also unlucky in business and both suffered financial ruin.

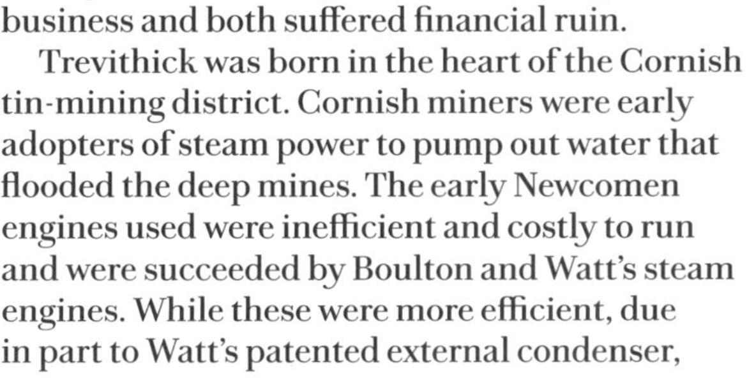

Trevithick was born in the heart of the Cornish tin-mining district. Cornish miners were early adopters of steam power to pump out water that flooded the deep mines. The early Newcomen engines used were inefficient and costly to run and were succeeded by Boulton and Watt's steam engines. While these were more efficient, due in part to Watt's patented external condenser, Boulton and Watt charged an excessive royalty for use of the patent, which was deeply resented by Cornish mine owners. The company was also extremely litigious in the enforcement of their patent right, a thorn in the side of both owners and engineers seeking to develop new engines. After the patent expired in 1800 the young Trevithick pioneered the use of high pressure or 'strong steam' in small, lightweight and extremely efficient engines.

In March 1804 Smith wrote to Trevithick congratulating him on the success of his 'Improved Steam Engine' and asking for quotes for the steam engines required for his coal trials. He went on to say, in a rather grandiose fashion, that he had contacts all over the country and implied that he would be able to pass considerable business Trevithick's way. There was no direct reply from Trevithick, but Samuel Homfray (1762–1822), who was manufacturing Trevithick's engines, responded stating that the price of the engines would be around half that of Boulton and Watt's, with lower maintenance costs and efficiency savings.

In August the same year Smith placed an order with Trevithick on behalf of Sefton for an engine for the Torbock Colliery. Later that month, in an effusive letter, Smith told Trevithick he would recommend his engine to the owners of the Batheaston Company for use in their coal trial, before going on to enquire about progress with the engine for Sefton. Trevithick's response assured Smith that all was going well with the engine and then enquired about a possible opportunity for him to acquire a coal lease on the banks of the Severn. The pair continued their correspondence on other matters, but not all the news was good. The Batheaston Company chose a Boulton and Watt engine instead of Trevithick's, and problems were foreseen with the capabilities of the engine at Torbock. There was talk of 'secret

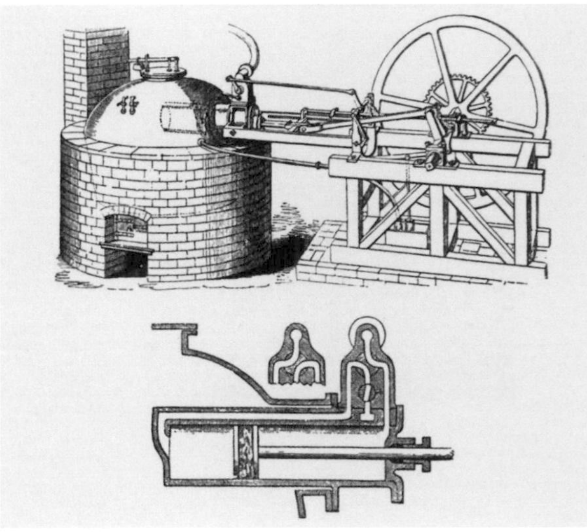

FIG. 4.

Fig. 4. An illustration of Trevithick's high pressure steam engine in 1803, taken from John Farey Jn.'s unpublished *Treatise on the Steam Engine*, Vol. II.

enemies' trying to discredit Trevithick's engine, while the fatal explosion of one of his engines at Greenwich the previous year had given the Boulton and Watt Company a golden opportunity to raise health and safety issues related to the use of high-pressure steam. In 1806, the problems with the Torbock engine, which were primarily the result of over-expectations on the part of the client, were resolved and the matter successfully concluded.

William Smith's involvement in the search for coal at Batheaston, to the east of Bath, seems to have been initially brought about – without his knowledge – by Mrs Mary Lane Browne (1741–1838) and a local entrepreneur, Thomas Walters (1757–1847). In several letters between Smith and Thomas Cruttwell, the Batheaston Company solicitor, Smith expressed surprise that he has been named in an advertisement as being involved with the 'Mining Scheme' and that a share in the company had been reserved for him. Cruttwell seems to have been equally surprised as he had assumed that Mrs Browne had informed him. Nevertheless, Smith did become involved with the venture and in a letter in late 1804, Walters provided Smith with a detailed account of the strata encountered in the pit down to 60 m (200 ft).

The choice of location for sinking a shaft was constrained by the requirement that it had to be on land owned by Walters and also close to a stream to provide water power. Smith knew that beneath the Lias, albeit at some depth, the Red Ground (Triassic) would be found, and beneath that there was a good chance that Coal Measures would be encountered. Unfortunately, instead of Coal Measures older sandstones were met with. Walters was aware of the unhappy news of what was beneath the Red Ground before March 1808, yet as late as March 1810 the company was still issuing new shares.

During the course of the operation there was a constant problem with water entering the shaft, hence necessitating the use of the Boulton and Watt steam engine to pump it out. An additional unanticipated side effect of sinking the shaft may have been the disruption to the water supply in the nearby thermal springs at Bath, much to the consternation of the local inhabitants. Smith was approached to help and opened up the hot springs and restored supplies, which fortuitously coincided with his successful efforts to stem the water flowing into the mine.

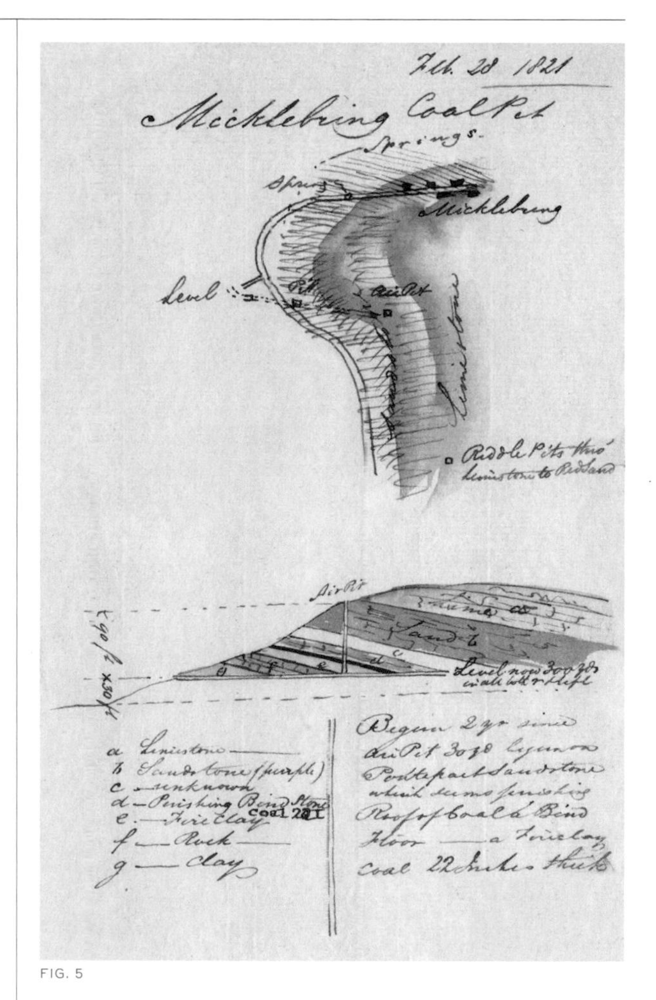

FIG. 5

Fig. 5.
A map and section of the Micklebring Coal Pit, south of Doncaster, made by Smith in 1821 during a tour of the area he made with John Phillips. The section below the Redland (Magnesian) Limestone shows the Coal Measure sequence in some detail. Note that Smith did not show this coal pit on his 1815 map.

COAL TRIALS.

Smith's approach to mining and mineral exploration was both meticulous and scientific. His understanding of the order of strata and the distinctive fossils they contained gave him a considerable advantage when asked for an opinion on a trial for coal. During the first decade of the nineteenth century an ill-fated attempt was made to find coal at Cook's Farm near Bruton in Somerset. Coal was known some 13 km (8 miles) to the north, and there was a related proposal to build a canal, the Dorset and Somerset canal, which would have connected Bristol to Poole on the English Channel. Investors in the coal trial were convinced that if they found coal then the canal would be a convenient way to transport it. The prospectus declared that the site was 'scientifically acknowledged to be the most eligible for such a trial' and a qualified engineer had apparently been consulted.

In 1803, a shaft was sunk and a waterwheel was later installed to assist in pumping operations. Glowing progress reports appeared in local newspapers and optimism ran high. Smith visited Cook's Farm in March 1805 and his examination of spoil from the bottom of the pit revealed specimens of a fossil now called *Gryphaea dilobotes*. Smith recognized it as characteristic of the 'Kelloways Stone' (Kellaways Rock) which was at the base of his Clunch (Oxford) Clay. From this Smith immediately knew that the trial was far too high in the geological succession to find coal.

A.

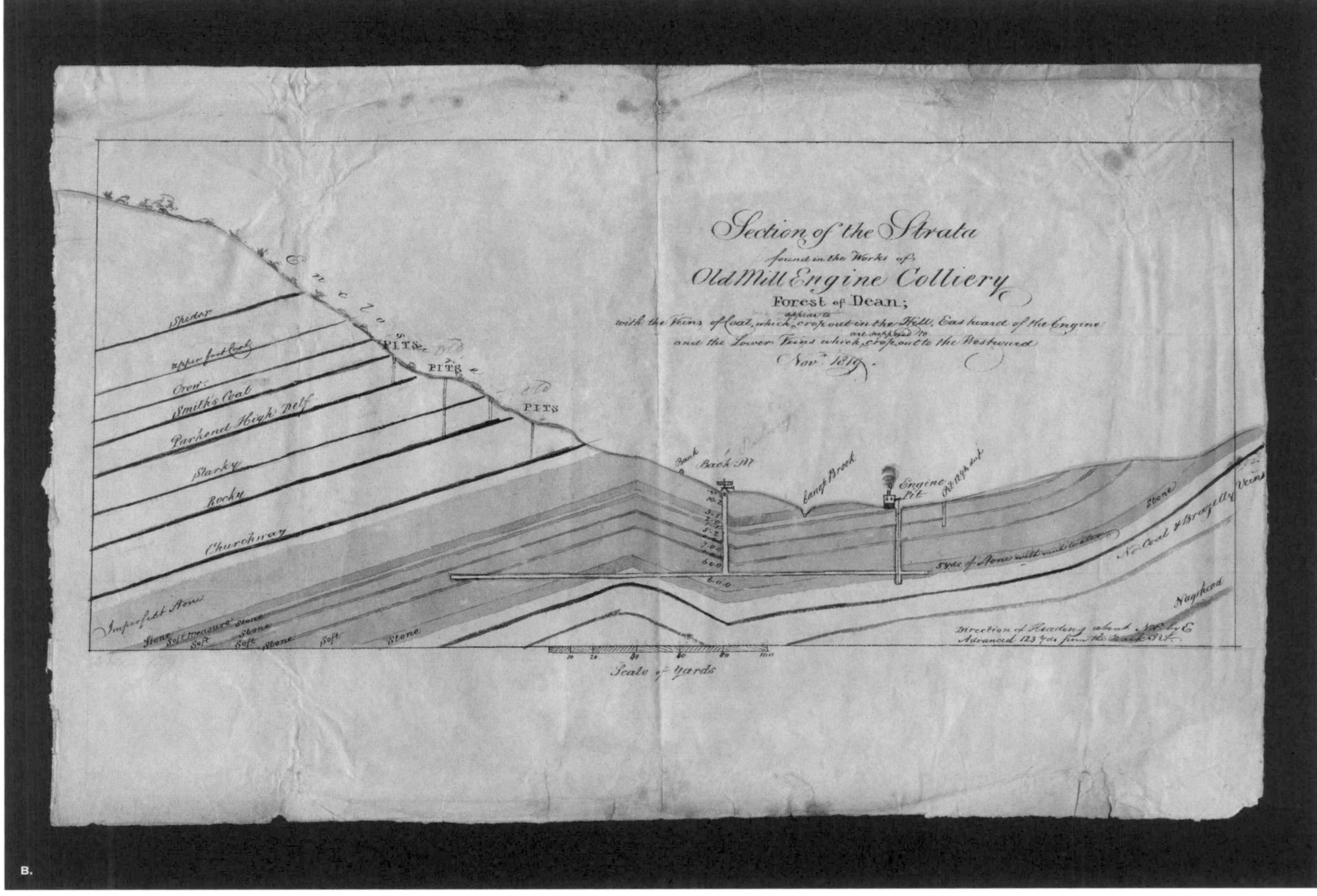

B.

A. MONMOUTHSHIRE BETWEEN THE COALFIELD OF THE FOREST OF DEAN AND SOUTH WALES, 1812.

This beautiful section is typical of Smith's panoramic style and shows the structure and sequence of strata beneath the Coal Measures from east to west across Monmouthshire. Demand for coal during the nineteenth century revolutionized the Welsh economy.

B. SECTION OF THE STRATA FOUND IN THE WORKS OF OLD MILL ENGINE COLLIERY, FOREST OF DEAN, 1819.

The section shows a named series of outcropping coal seams on the hill to the east,with two shafts, Back Pit and Engine Pit, further west. Back Pit has a whim winder (horse gin) whereas a steam engine driven pump had been installed at the Engine Pit.

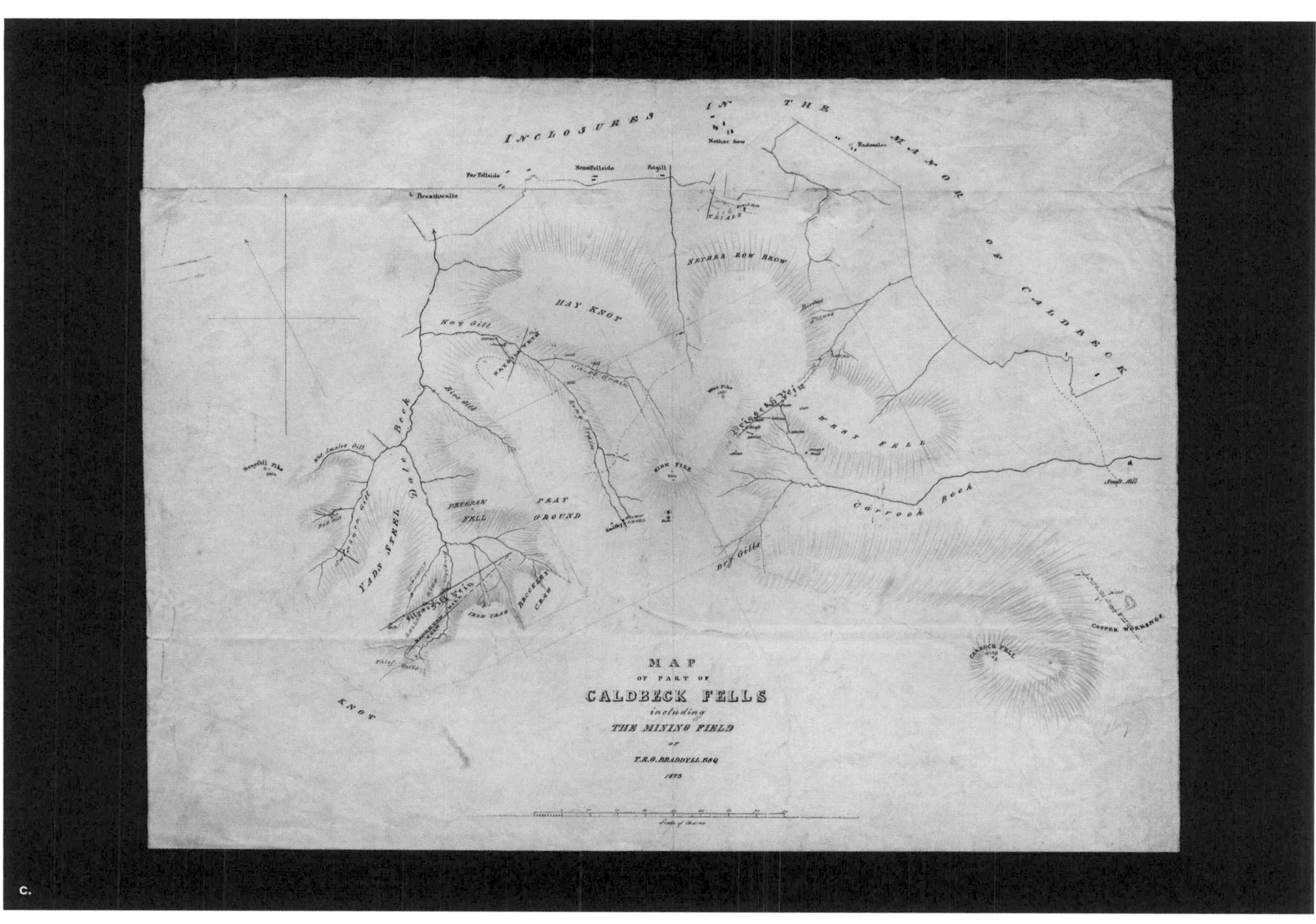

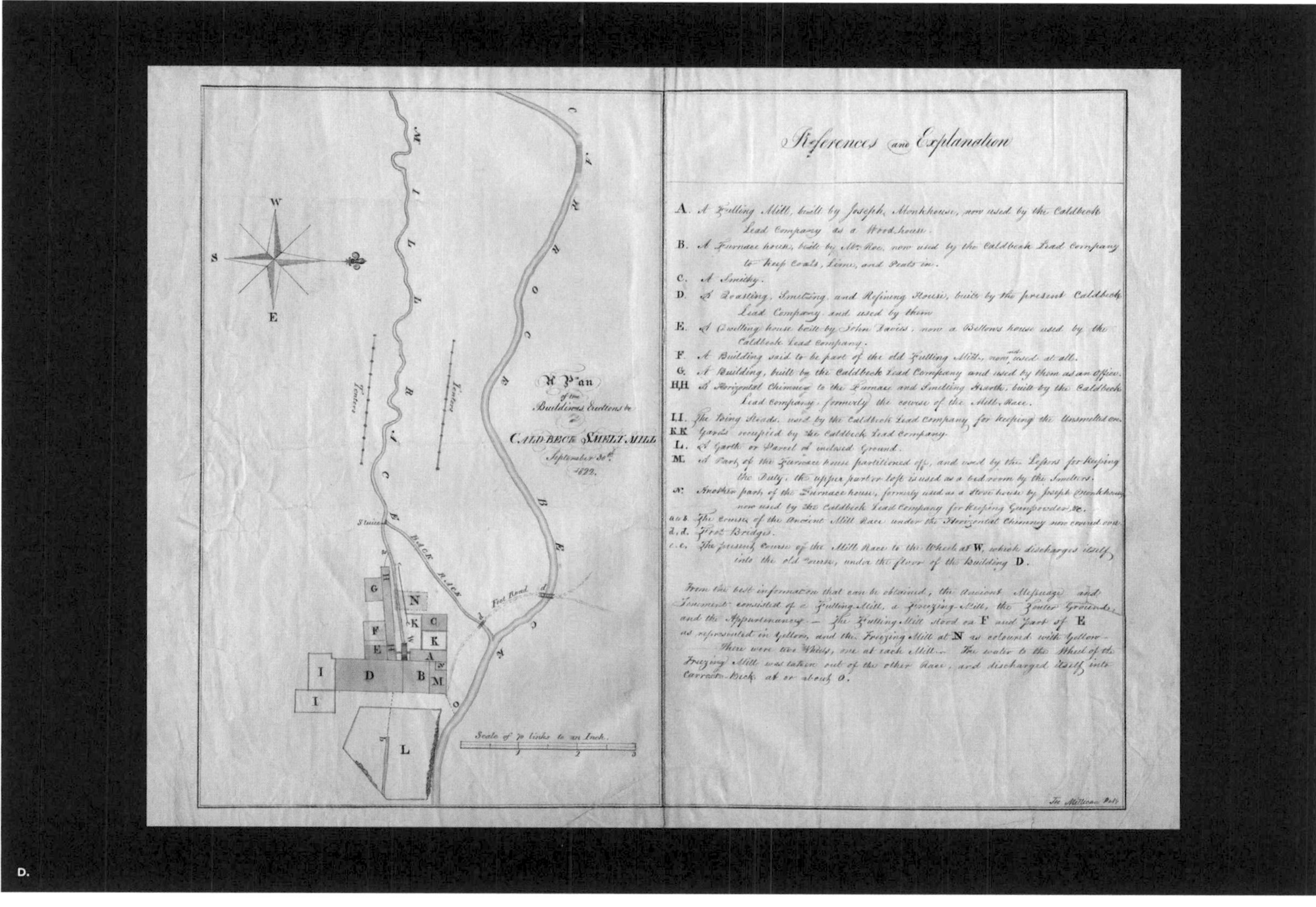

C. MAP OF PART OF CALDBECK FELLS INCLUDING THE MINING FIELD OF T. R. G. BRADDYLL ESQ., 1823.

A variety of minerals, principally copper and lead, had been mined here since the fourteenth century. The map shows several important mineral veins including the Silvergill, Roughton, Driggeth and Haygill veins. Smith and Phillips visited many mines in this area during the winter of 1822–23.

D. A PLAN OF THE BUILDINGS, ERECTIONS ETC. AT CALDBECK SMELT MILL, 1822.

A detailed plan of a complex of buildings used for the refining and smelting of lead, including furnaces, mills and dwellings owned by the Caldbeck Lead Company. The Caldbeck Fells in the Lake District are rich in a variety of minerals, in particular copper, lead and barytes, and there has been mining there since the thirteenth century.

FIG. 6.

Fig. 6.
Lead mining in Leadhills, Scotland. Once extracted from a mine the lead ore had to be sorted from ordinary rocks. This was often done by children (above). Smelting the ore in a furnace then separated the lead from any impurities before it was cast into an ingot (below). Working conditions were dangerous due to noxious fumes produced from melting lead.

As Phillips recorded, 'in spite of remonstrances from Mr Smith and his intelligent friends, the speculators proceeded at a ruinous expense' (Phillips, 1844, p. 66). Work on the shaft continued until Christmas 1807 when, at a depth of almost 200 m (652 ft), water from the Great Oolite aquifer flooded the shaft, effectively ending mining operations and bankrupting some of the speculators.

Canals and coal were often connected. In 1811, Smith and Edward Martin (1763–1818) were engaged by the Bristol and Taunton Canal Company to report on the productivity of coalfields in the Nailsea area in order to evaluate the viability of building a spur off the canal to carry the coal. The pair looked at Grace's, White's and Backwell collieries and their report gave an optimistic – probably over optimistic – assessment of the amount of coal that might be produced both from deepening existing mines and opening new ones. In the event the canal spur was never built, but the report provides an illuminating insight into the communication between mineral surveyors, such as Smith, and potential investors in canals during the early nineteenth century.

On a visit to north-east Yorkshire in 1813 Smith saw coal in a very different stratigraphic situation for the first time. Here the coal was not within the usual Coal Measures, but instead within the barren sandstones of the North Yorks Moors. The stratigraphic position of these sandstones was initially problematic for Smith. On early versions of his 1815 geological map he placed these strata relatively high in the succession beneath his OakTree Clay, but later discovered fossil evidence (in the Hackness Rock) which made him realize that the sandstones were below the Coral Rag. These sandstones are in fact the northern equivalents of the Oolitic limestone to the south. As with the Carboniferous Coal Measures, the coal-bearing sandstones were deposited in deltaic floodplains subject to periodic marine incursions.

Smith made two reports on these Yorkshire coals. The first concerned Gnipe Howe (Nape How), on the coast near Whitby, where a thin seam of coal was mined via an adit (a horizontal passage) into the cliff. The tunnel entrance was almost at sea level and only accessible by means of a dangerous pathway at low tide. Smith advised that the tunnel be driven deeper into the cliff and, if the seam proved to be thicker, a shaft might then be sunk into from the cliff top. In his second report he wrote an account of a similar coal seam at Borrowby (Boroby), 18 km (11 miles) north-west of Gnipe Howe. Here he made a correlation between the succession in a boring made near Danby with the cliff section at Gnipe Howe. He concluded that coal could be mined profitably for a local market, provided the method of extraction remained simple.

LEAD AND ANTIMONY.

Smith located a large number of lead mines on his map but only specifically named a few, including Leadhills in South Lanarkshire and Glendinning in Dumfries and Galloway and the Old Gang Mines in the Yorkshire Dales. Leadhills Mine in South Lanarkshire was a prolific source of lead with a number of veins, including the rich Susanna vein, running mostly north–south, related to Carboniferous mineralization of Ordovician sandstones. The primary ores are galena and sphalerite. The other claim to fame of Leadhills is the Leadhills Miners' Library. Founded in 1741 to provide miners with access to books for their self-improvement, it is the oldest subscription library in the British Isles.

The Glendinning Mine, in Dumfries and Galloway, with its Louisa vein, yielded over 200 tonnes of antimony in addition to some lead. Antimony was first discovered there around 1760 on land belonging to Sir William Pulteney (1729–1805), who in his day was reputed to be the richest man in the kingdom. It was worked from 1793 to 1798 and then again later in the nineteenth century. The principal ore was stibnite in association with galena and arsenopyrite, which is found within mineralized Silurian strata.

It is not known if Smith ever visited Leadhills or Glendinning, but he did have personal experience of the Old Gang Mines in Yorkshire.

FIG. 7.

Fig. 7.
The Mona and Parys copper mines on Parys Mountain. These two mines in Angelsey, Wales, were at one time the most productive in the world. The Great Copper Lode was worked by both underground mining and opencast mining. The mine also used precipitation pits into which scrap iron from London was dumped and used to precipitate copper.

Smith's connection with these lead mines began with a meeting at the office of the grandly named solicitors Desse, Dendy & Morphett on 18 May 1819. A letter on the same day instructed Smith to undertake an inspection of the mines in order to determine whether Mr Frederick Hall, manager of the mines from 1814 to 1818 on behalf of Messrs. Alderson, had 'for want of proper skill, talent and ability incurred a great expense in projecting works which have either been abandoned or are entirely useless'. The letter goes on to itemize two inclined planes inadvisedly constructed, with one later abandoned, a steam engine erected but never used, and other equipment also never used and now decaying. In the mines there were unfinished tunnels (rises and cross-cuts) and several shafts which had fallen into disrepair. Presumably, as a result of this litany of misdeeds, Hall was fired.

Smith visited the mines on 23 May 1819 and drew up a twenty-two-page report, in essence confirming all the allegations, which he signed and dated 1 June 1819. Litigation did indeed follow, but it was by Hall claiming Breach of Covenant, rather than his employers the Aldersons. Smith was summoned as a witness to appear in front of Sir Robert Dallas (1756–1824), Chief Justice of the Bench and well known as a defence council in the impeachment of Warren Hastings (1732–1818), governor-general of India. The summons was originally dated 24 May 1819, but was changed to 1 June. On that day Smith prevaricated, saying he needed more information in order to complete his report. The case was then rescheduled for two days later. Smith's diary entry for 2 June noted that he sent the report to the solicitors and spent the following day preparing his evidence. On 4 June he attended the trial court, but does not record the outcome, and the next day made out his account to Sir George Alderson. There is an illegible pencil diary entry relating to Sir George and the solicitors on 10 June, but no further entries after 11 June until September. In June 1819 Smith was prosecuted for debt, and until September that year, the unhappy William was confined in the King's Bench Prison in Southwark.

COPPER.

Copper in Cornwall is usually associated with tin, although there are some exceptions. Beyond Cornwall, Smith labelled Parys Mountain – the copper works near Amlwch on Anglesey. For a time in the eighteenth century this was the largest copper mine in the world, its dominance severely damaging the Cornish copper industry. Parys Mountain and the adjacent Mona Mine were controlled by Thomas Williams (1737–1802), known as the 'Copper King', who also owned copper mines in Cornwall.

Smith was also interested in the copper (and lead) mines in High Peak, Derbyshire and Caldbeck Fells, Cumberland (now Cumbria). With his nephew John Phillips he visited the mines and old workings there during the winter of 1822–23 (Phillips, 1844, p. 105). The Hay Gill copper mine was worked by Joseph Scott from 1785 until at least 1792 and Smith had in his

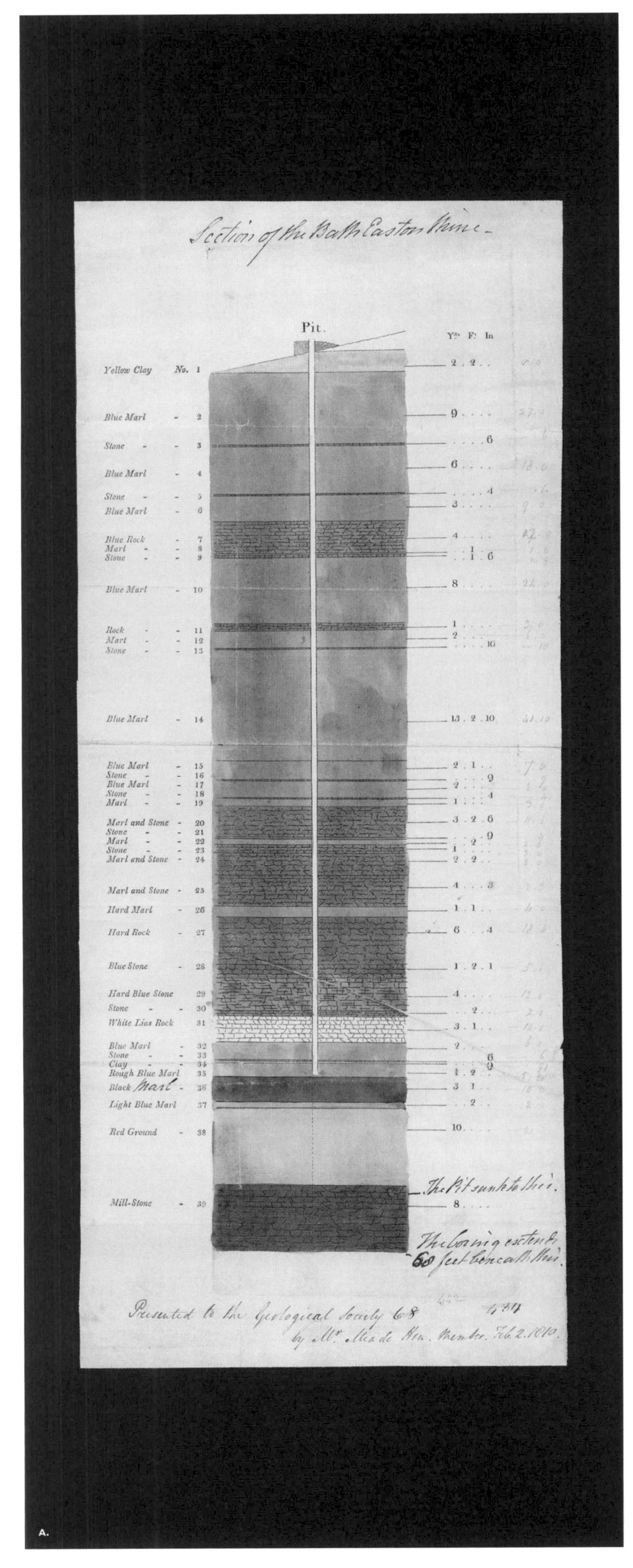

A.

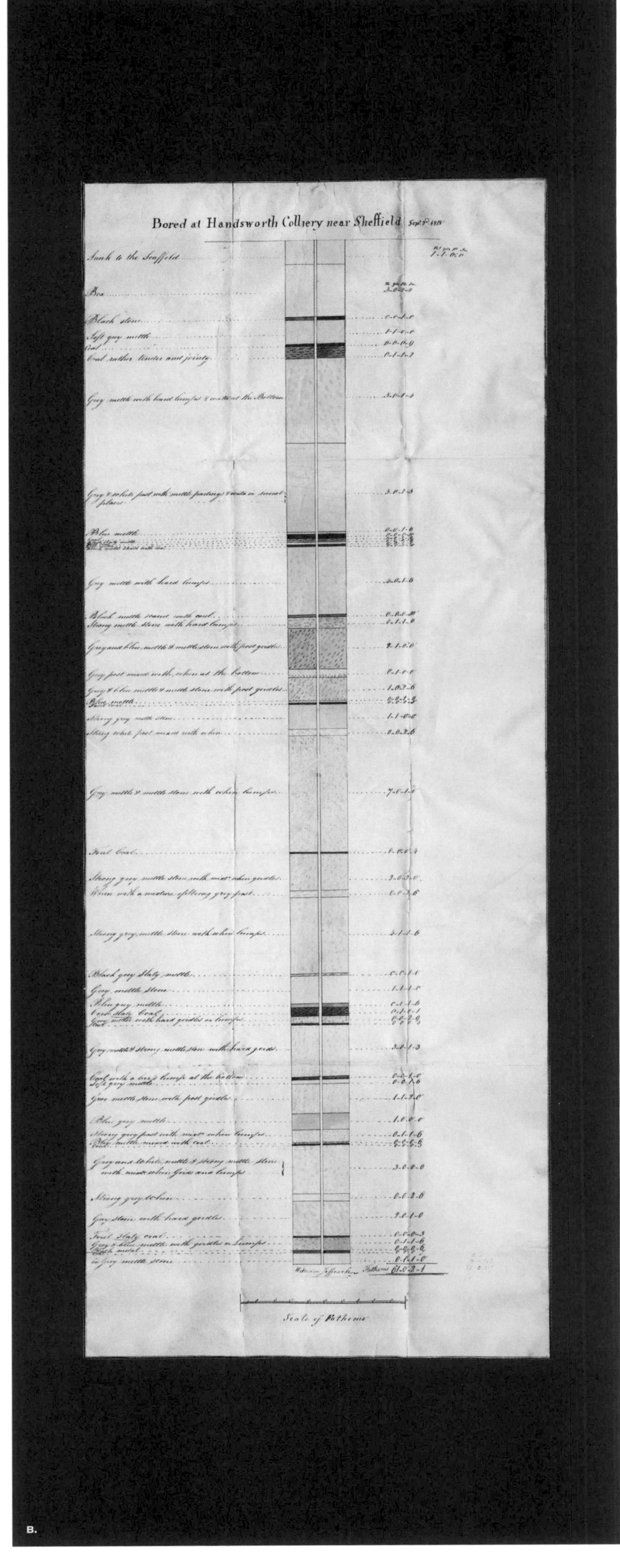

B.

A. SECTION OF THE BATHEASTON COAL MINE, AVON, 1808.
This strata section was prepared for the mine owner Thomas Walters. Walters sent Smith a detailed written log in 1804 which matches the depths and thicknesses shown on the section. The section clearly shows the fateful boring through the Red Ground and the absence of the Coal Measures.

B. BORED AT HANDSWORTH COLLIERY NEAR SHEFFIELD, 1818.
This section appears to have been prepared by a young William Jeffcock (1800–71), who became a coal-master of the company Handsworth belonged to: Jeffcock and Dunn, later known as the Sheffield Coal Company. The section shows a series of Sandstone, Shale, Limestone and Coal which he described using local miners' dialect terms (grey metal, post girdle).

Section of the Strata found in Sinking Lower Bilson Pits

Particular thickness of the different Strata found in Sinking Lower Bilson Pits, as under.—

No.	Strata	Thickness of the different Strata Yds	ft	In	Distance between each Vein of Coal Yds	ft	In	Depth from the Surface to each Vein of Coal Yds	ft	In
1	Wall & Wood above the Surface	3	1	..						
2	Soil Gravel &c.	9	1	6						
3	Hard Cliff Quar	5	1	..						
4	Soft Cliff	..	..	6						
5	Hard rock with a great quantity of Water	4	..	..						
6	Hard Cliff	3	1	..						
7	Hard Cliff Quar	6	2	6						
8	Cliff	5	2	..	38	1	6	38	1	6
9	Spider Delf	..	..	6						
10	Under Cliff	1	2	..						
11	Hard Cliff	2	1	..						
12	Soft Cliff	2	1	..						
13	Hard Cliff Quar	6	..	..						
14	Cliff	2	..	..						
15	Cliff Quar	1	..	..						
16	Harder Cliff Quar	9	1	..						
17	Soft Cliff	1	..	..	25	2	6	64	1	..
18	Upper Foot Coal	..	1	..						
19	Under Cliff	1	..	..						
20	Hard Cliff	2	..	..						
21	Hard Rock with a great quantity of Water	4	1	..						
22	Hard Cliff	2	1	..	10	..	..	74	1	..
23	Crow Delf	..	1	6						
24	Under Cliff	1	1	6						
25	Hard Cliff	4	1	..	6	1	..	80	2	..
26	Little Coal above the Twenty Inches	..	..	8						
27	Soft Clod between	..	1	4	2/4					
28	Twenty Inches Delf	..	2	3						
29	Under Cliff	1	..	..						
30	Hard Cliff	2	..	..	4	1	3	85	..	3
31	Lower Foot Coal	..	1	1						
32	Under Cliff	..	1	2						
33	Hard Cliff	8	..	..	8	2	3	93	2	6
34	Lowry Delf	1	..	..						
35	Under Cliff	..	1	6						
36	Hard Cliff	2	1	6	4	..	..	97	2	6
37	Little Coal	..	..	5						
38	Soft Clod between	..	1	2						
39	Little Vein of Coal	..	..	5						
40	Soft Cliff	3	2	..						
41	Hard Rock with a great quantity of Water	6	..	..						
42	Soft Cliff	1	..	..	11	1	..	109	..	6
43	Sterkey Delf	..	1	6						
44	Soft Cliff	1	2	6	2	1	..	111	1	6
45	Sharpnel Delf	..	..	6						
46	Soft Clod between	..	1	..						
47	Little Coal	..	..	1						
48	Hard Cliff	3	..	..						
49	Rock	..	1	6						
50	Hard Cliff	7	1	2	11	1	3	122	2	9
51	Rockey Delf	..	2	6						
52	Cliff Quar	2	1	..						
53	Hard Rock	2	2	..						
54	Soft Cliff	..	1	..	6	..	6	129	..	3
55	Breadless Delf	..	..	10						
56	Under Cliff	..	2	6						
57	Hard Cliff	3	1	2						
58	High Delf Stone with Water	8	2	..						
59	Cliff	..	1	6	13	2	..	142	2	3
60	High Delf Top-Coal	..	..	9						
61	Thin parting	..	..	..						
62	High Delf Middle Coal	..	2	3						
63	Stony Clod between	..	2	..						
64	High Delf Bottom Coal	..	1	1	2	..	1	144	2	4

Deep Pit

Land Pit

Spider Delf

Drift

Crow Delf

Lowry Delf

Drift

Rocky

Breadless

High Delf

c.

C. SECTION OF THE STRATA FOUND IN SINKING LOWER BILSON PITS, FOREST OF DEAN, DATE UNKNOWN.

This section includes a detailed listing of the various coal seams, showing their thicknesses and depths correlated between the two shafts. Pencil annotation appears to refer to some veins recorded elsewhere, for example the Spider Delf and Crow seams.

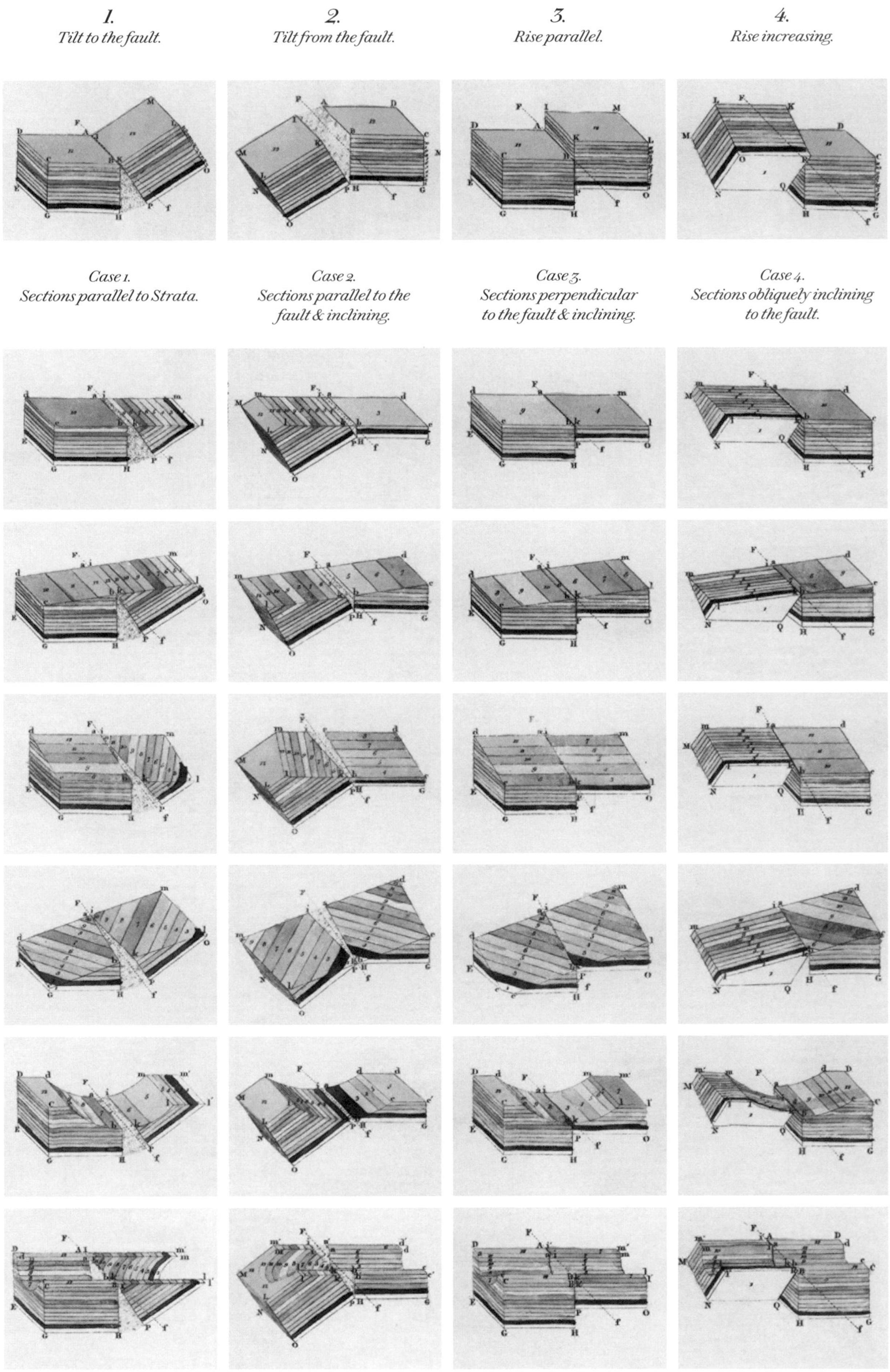

II. MINERAL PROSPECTOR.

1. Tilt to the fault.	2. Tilt from the fault.	3. Rise parallel.	4. Rise increasing.

Case 1. Sections parallel to Strata.	Case 2. Sections parallel to the fault & inclining.	Case 3. Sections perpendicular to the fault & inclining.	Case 4. Sections obliquely inclining to the fault.

FIG. 8.

Fig. 8. Illustrations explaining the nature of faults, as they would appear at the surface after denudation or excavation, taken from Chapter 1, Section 4 of John Farey's *General View of the Agriculture and Minerals of Derbyshire, with Observations on the Means of their Improvement* (1811). Farey was a friend and supporter of Smith and learned much from him. He was also a polymath and excellent geologist in his own right, as this diagram of faulting clearly shows. Indeed, his understanding of the nature of faulting was probably greater than Smith's.

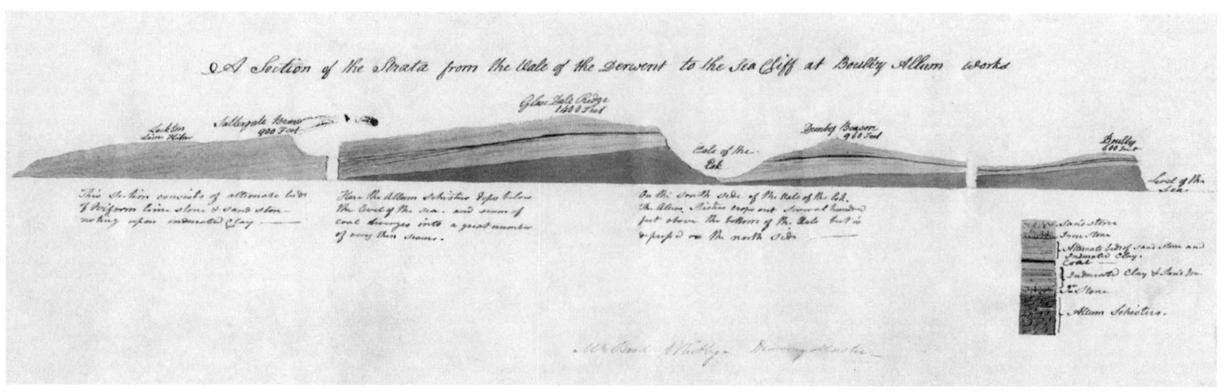

FIG. 9.

Fig. 9.
A section of the strata from the Vale of the Derwent to the sea cliffs at the Boulby Allum works, North Yorkshire. This north to south section was found amongst Smith's papers. It is drawn by John Bird (1768–1829), a Welsh artist and naturalist and one of the founders of the Whitby Museum. He had worked with the Rev. George Young (1777–1848) on *A Geological Survey of the Yorkshire Coast* in 1822.

possession the mine's 'Bargain Book', which recorded piecework contracts based on miners' tasks (for instance, driving a level) executed at an agreed price based on the ore extracted.

ALUM.

From the early seventeenth to the mid-nineteenth century, alum was a thriving extractive industry centred in north-east Yorkshire. Chemically, alum is a double sulphate of aluminium, with either potassium or ammonium. Its major use in pre-industrial Britain was in the dyeing of cloth. In order to fix a dye – make it colour fast – cloth was dipped in an alum solution which acted as a mordant. Until the mid-sixteenth century most alum was imported from Italy and Spain, where the mineral was mined and was virtually a papal monopoly. But Henry VIII's conflicts with the Pope during his reign (1509–47) resulted in severely restricted alum imports, and the hunt was on for a home-grown source.

What followed was a period of trial and error. Sir Thomas Chaloner (1559–1615) was responsible for developing the industry, initially on his land around Gisborough. The Lias shales in north-east Yorkshire were an ideal source, for they contained both aluminium silicates (in clay minerals) and sulphate-rich iron pyrites. The shale was quarried in vast quantities particularly along the coast. Huge pyramids of the rock were piled up, covered with wood and then burnt (calcined) for many months, the combustion aided by the high organic content of the shale. The residues were washed with water and the liquor drained off into a series of steeping cisterns. Stale urine and burnt seaweed were added, and eventually the liquor evaporated and the alum crystallized.

SALT.

Smith showed the presence of salt works at Northwich, Winsford and Middlewich in Cheshire, but for some reason not at Nantwich – the ending '-wich' indicating salt locally. Salt production has a long history in Cheshire, dating back at least to Roman times when salt was extracted from brine springs connected to outcropping strata containing a high proportion of salt. Rock salt or halite occurs widely in Triassic strata (Smith's Red Ground) of the Cheshire basin. With increasing demand, salt began to be mined in the seventeenth century and in the eighteenth century it was extracted from brines pumped out of boreholes. Excessive extraction by pumping resulted widespread subsidence in the area.

MAPPING THE MINES.

How did Smith gather such a wealth of detailed information? In some specific areas, such as the Somerset coalfields, he would have had first-hand experience, but elsewhere his personal knowledge may have been more limited. One source may have been a series of county surveys of agriculture commissioned by the Board of Agriculture and published in the late eighteenth and early nineteenth centuries, which contained information on mines and minerals. In particular, John Farey's monumental work, *General View of the Agriculture and Minerals of Derbyshire* (1811), listed some 500 collieries and around 250 lead mines in that county. Similarly, John Williams's *Natural History of the Mineral Kingdom* (1789) provided information on collieries and mines.

John Cary (1754–1835), the engraver and publisher of Smith's map, may also have assisted as he had produced topographic maps and surveys covering a large part of the country and must have had access to relevant material. A possible source for tin and copper mines on Smith's map may have been Richard Thomas's *Report on a Survey of the Mining District of Cornwall from Chasewater to Camborne*, which includes a detailed map and cross sections. Although Cary published the report in 1819, four years after Smith's map, he would have had sight of the original earlier manuscript. The cross sections from Thomas's work were sometimes bound in with the atlas of Smith's geological sections.

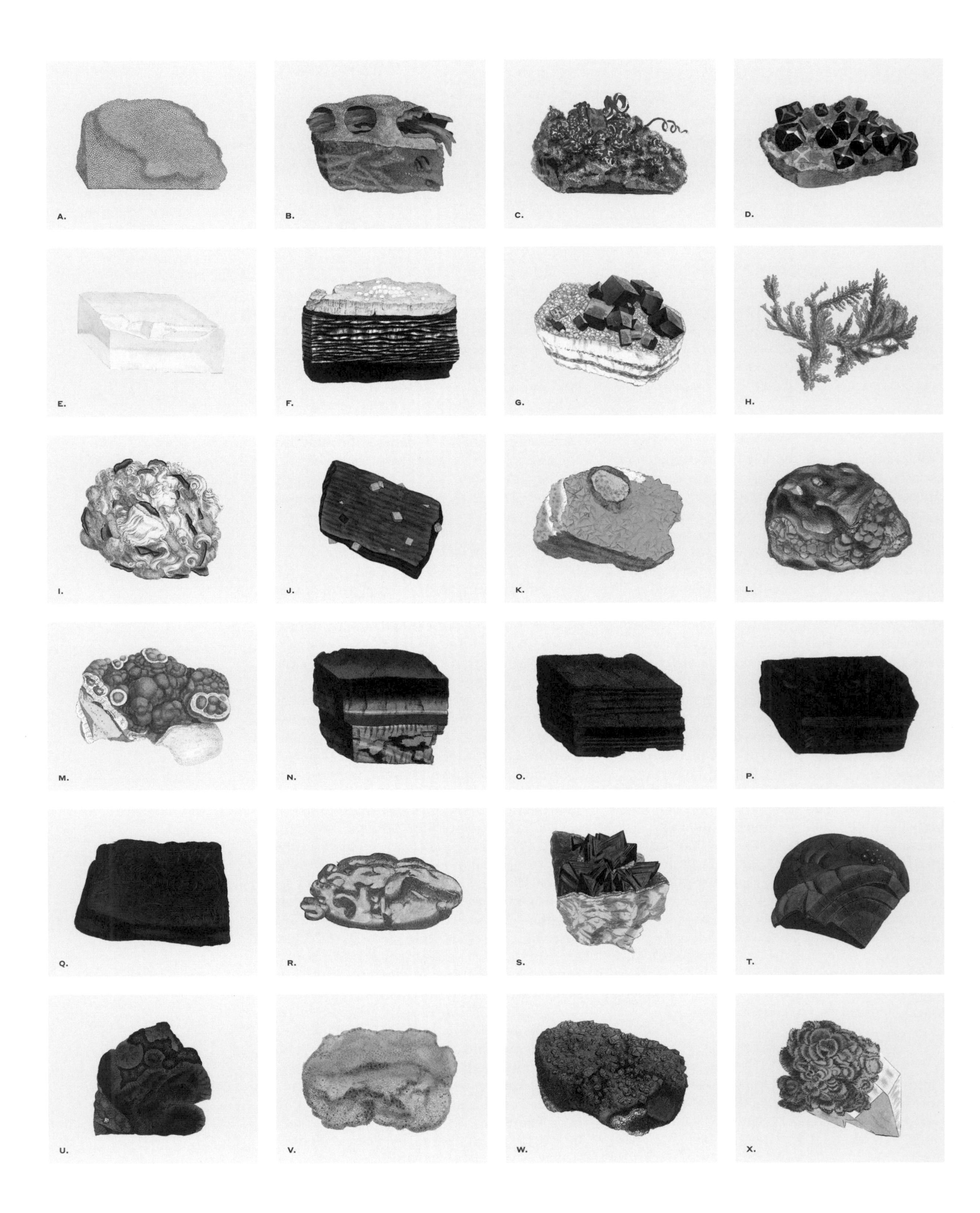

A. LIMESTONE.
B. AZURE IRON ORE.
C. CAPILLARY SILVER.
D. OXYGENIZED TIN.
E. COMMON SALT.
F. IRON SULPHATE.
G. SULPHURE OF LEAD.
H. NATIVE COPPER.
I. SULPHATE OF IRON.
J. SULPHURET OF IRON.
K. ARSENIATE OF COPPER.
L. ARSENIATE OF COPPER.
M. ARSENIATE OF COPPER.
N. OXYGENIZED CARBON.
O. PIT-COAL.
P. PIT-COAL.
Q. JET.
R. SULPHATE OF IRON.
S. ZINC.
T. RADIATED OXIDE OF IRON.
U. CRYSTALLIZED OXIDE OF IRON.
V. LEAD SULPHATE.
W. CARBONATE OF LEAD.
X. CARBONATE OF COPPER.

PLATES FROM JAMES SOWERBY'S, *BRITISH MINERALOGY OR, COLOURED FIGURES INTENDED TO ELUCIDATE THE MINERALOGY OF GREAT BRITAIN* (1802–17).

James Sowerby (1757–1822) was perhaps the most talented engraver and naturalist of his generation. During his lifetime he produced several landmark publications, including the thirty-six-volume *English Botany* (1791–1814). *British Mineralogy* was the first comprehensive illustrated work on mineralogy. It was issued in parts over fifteen years and ultimately contained 550 plates. Sowerby also collaborated with William Smith on *Strata Identified by Organized Fossils* (1816–19).

'THE COLLIER' ENGRAVED BY ROBERT HAVELL, FROM GEORGE WALKER, *THE COSTUME OF YORKSHIRE*, 1814.

The locomotive shown in the background was the first commercially successful steam locomotive, Matthew Murray and John Blenkinsop's *Salamanca*, which ran on the edged railed Middleton Railway between Leeds and Middleton in 1812, eleven years before the famous Stockton and Darlington Railway, the first public railway to use steam locomotives. Unfortunately *Salamanca* was destroyed six years later when its boiler exploded after its driver tampered with the safety valve.

'THE RUDDLE PIT' ENGRAVED BY ROBERT HAVELL, FROM GEORGE WALKER, *THE COSTUME OF YORKSHIRE*, 1814.

Ruddle is a type of red ochre used in coarse paints and marking sheep. The location of this pit may have been Micklebring, South Yorkshire, where ruddle was mined and processed. Mining the ruddle involved sinking a shaft around 7 metres (23 ft) in depth and 1.5 metres (5 ft) in diameter. The miner could then excavate the ruddle using a small axe. It was then sent to a mill where it was ground to a powder and mixed with water.

A.

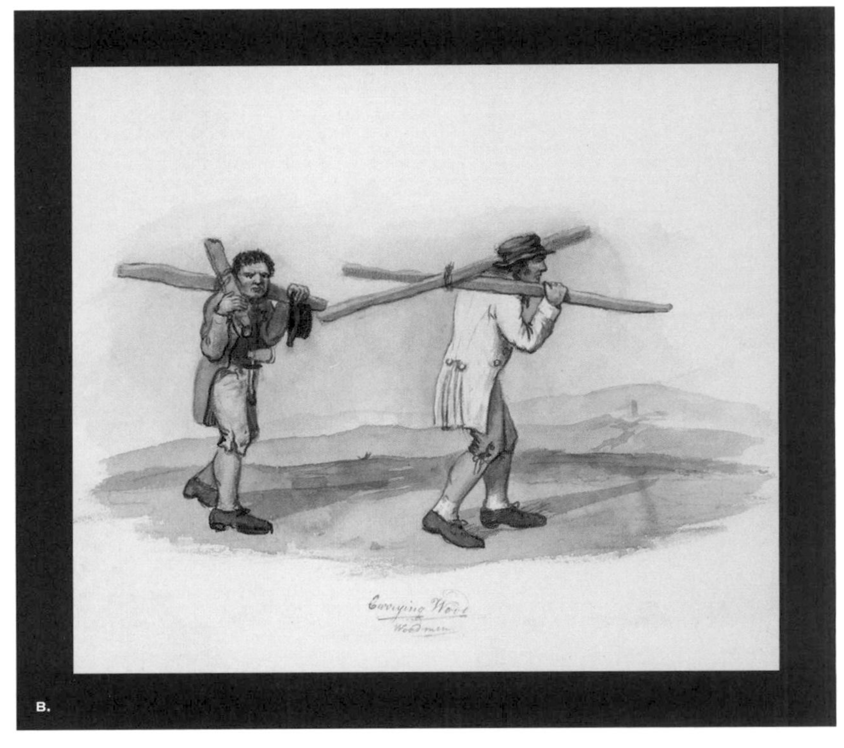

B.

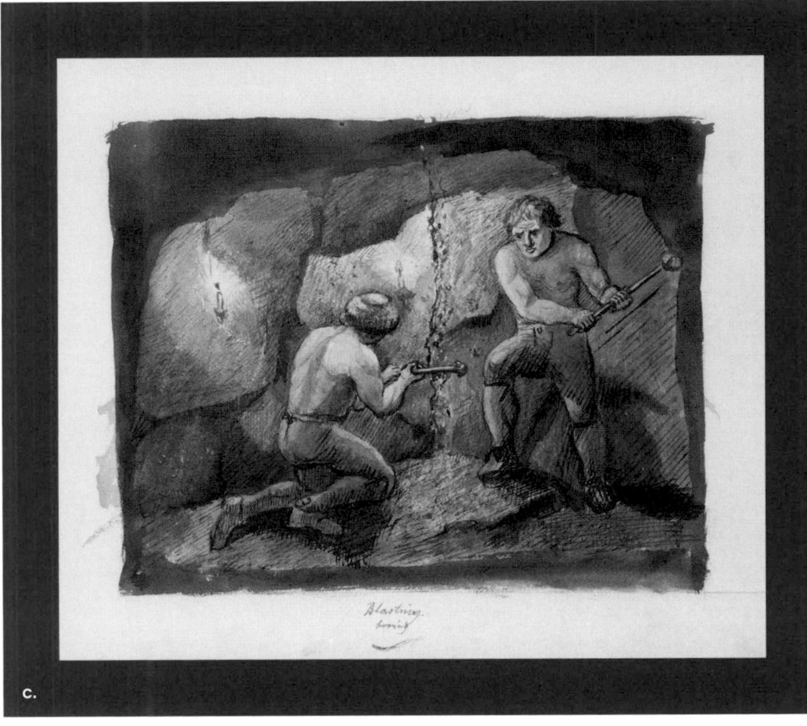

C.

D.

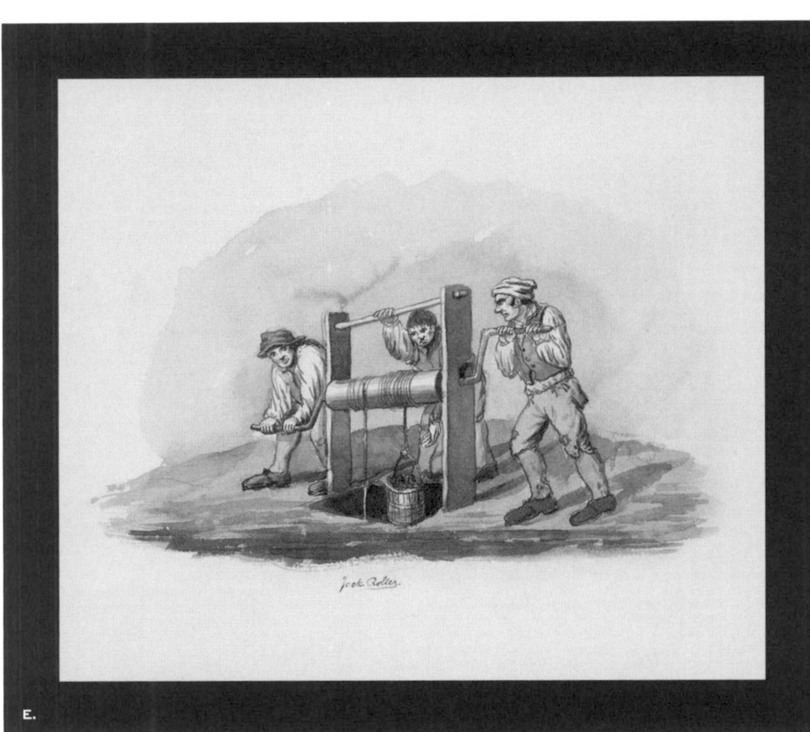

E.

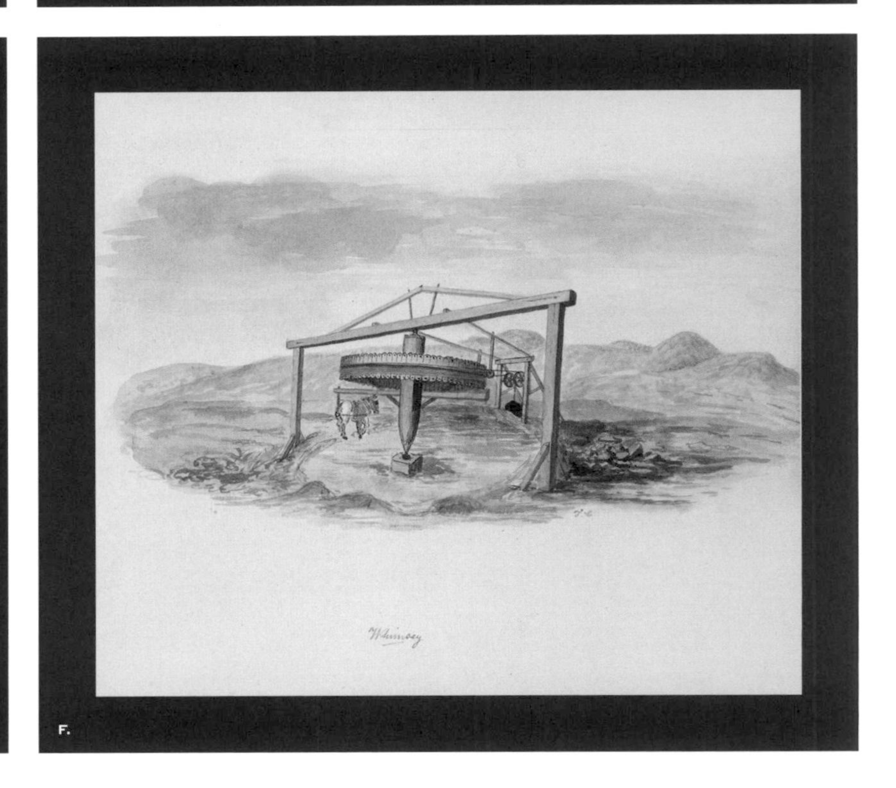

F.

LEAD MINING IN NORTHUMBERLAND, POSSIBLY ALSTON MOOR OR ALLENDALE, C. 1805–20.

Illustrations taken from a set of 67 watercolours and sketches showing aspects of lead mining above and below ground, the processing of ore, the tools used and the life of the miners. The artist is unknown, however it may be Joseph Crawhall, a talented local artist and the son of a lead mining agent for W.B. Lead Co. The watercolours appeared in two stitched volumes, and were accompanied by a manuscript dissertation 'On the Washing of Lead Ore', and a related diagram of a proposed vessel.

G.

H.

I.

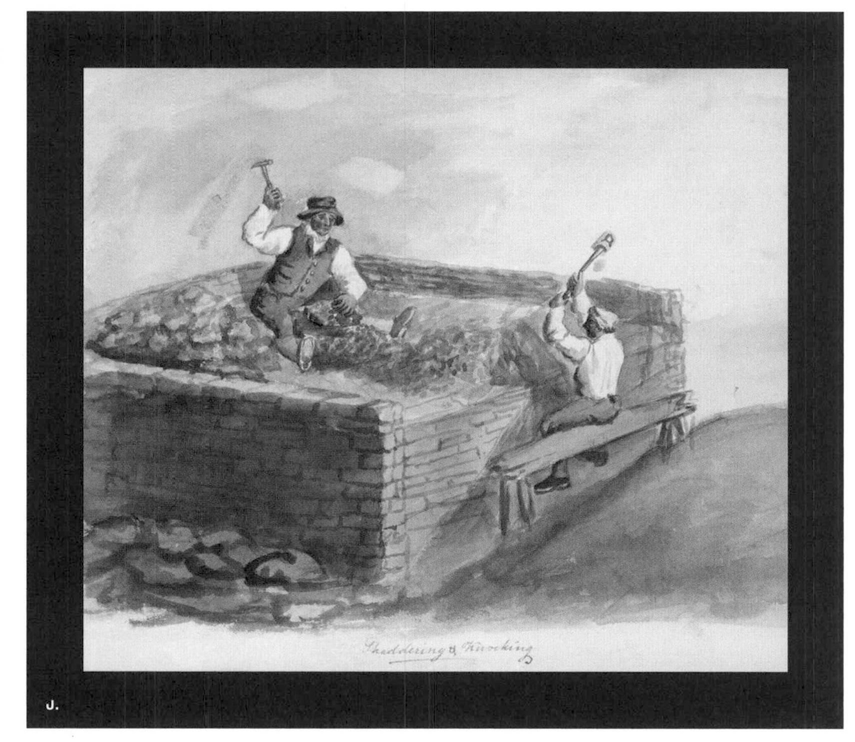

J.

K.

L.

A. MONDAY MORNING. Miners walking into work after the weekend.

B. CARRYING WOOD, WOODMEN. Wooden props supported mining tunnels.

C. BLASTING, BORING. Despite the title, the technique used here appears to be wedge and feather.

D. PICK. The pick is a key tool used for mining minerals.

E. JACK ROLLER. A jack roller is the type of winch used here.

F. WHIMSEY. 'Whimsey' refers to a whim winder, used to haul materials from a mine to the surface.

G. WAGON. Wagons were used to transport the lead ore through the mine.

H. WEDGE. A man drives a wedge into a stone in order to split it.

I. FILLING BUDDLE BARROW. Ore is carried to a buddle – a shallow vat – where it is washed.

J. SHUDDERING & KNOCKING. Hammers are used to break up the ore.

K. GRATING. The ore is passed over a grate to separate out the small from the large chunks. Large chunks were sent back to be further broken down.

L. FILLING & WEIGHING ORE CARRIERS. Carrier bags are filled and weighed.

Hope
Scale Knowle
Ravensworth
ARKENGARTH FOREST
Lead Mines
Arkendale
Richmond
Gilling
Aske
Middleton Tyas
Sockburn
Enter Common
Gt. Smeaton
East Cowton
Ailey Hill
Welbury
Scorton
Love some Hill
Whitwell
Hudswell
Hipswell
Downholme
Stanton
Catterick
Feetum Holme
Gragpot
Summer Lodge
COAL PIT MOOR
Redmire
Leyburn
Hawkswell
Hackforth
Constable Burton
Patrick Brompton
Fencoat
Yafforth
Leases Hall
Bedale
Leeming
Hutton Hang
Thornton Steward
Rockwith
Sunnywit
Middleham
Nappa
Yore R.
Swale R.
Coverham
Est. Witton
WALDEN
COVERDALE
Thornton Wattas
Snape Water
Longwith
Pickhill
Thornton le Moor
High Ellington
MASHAMSHIRE
Masham
CHACE
Leighton Hall
Swinton & Hall
Well
Coal Pits
Woodale
Sand Hutton
Catton
Topcliffe
Birdforth
Wath
Norton Hall
Ripon
Newby Pa.
Ditchford
MOORS
1263 G. Whernside
Middlesmoor
Ketsmoor
Kettlewell
NETHERDALE
FOREST
KITTLEDALE
High Grantley
Norton in the Clay
Easingwold
Boroughbridge
Aldborough
Alne
Coniston
Cliffe Hou.
Pateley Bridge
Greenhow Hill
WHARFDALE
NETHER DALE
Gt. Ouseburn
Thornbrough Park
Ripley Hall
Ripley
Knaresborough
Longhill House
Rilstone
Winterburn
Fellisciffe
Bolton Park
Harrowgate
High Harrowgate
Thornville
Moor Monckton
KNARESBORO
GAWKHALL RIDGE
Brackenthwaite
Oakton
Spofforth
Ribston
Walshford
Knapton
Rufforth
Gargrave
Skipton
FOREST
Rigston
Swinton Hall
Wetherby
Thorp Arch
Wighill Park
Wighill
AINSTY
ROMBALDS MOOR
1308
Wood Side
Otley
Harewood
Harewood Ho.
Clifford
Tadcaster
Eastburn
Hawkeworth
Keighley
Bingley
Cookridge
BLACK MOOR
Moor Allerton
Horsforth
Shadwell
WINN MOOR
Abberford
Church Fenton
Hainworth
Okeworth
Cullingworth
Howorth
Eccleshill
LEEDS
Moor Garforth
Micklefield
Sherburn
Temple Newsham
Thornton
Clayton
Grimworth Dean
SAWOOD HILL
Grimsworth Hall
Illingworth Chap.
Wadsworth
Wisket Hill
Wibsey Chap.
Lightcliff
Chitwell
ROTHWELL HAIGH
Rothwell
Howley Hall
Methley Pa.
Castleford
Hambleton Haugh
Ferry Bri.
Halifax
Rochdale
Canal
Calder R.
Studley Pike
Greetland
Mirfield
Alverthorp
OUTWOOD
WAKEFIELD
Wakefield
Pontefract
Greave Hall
Fixby Hall
Heath
Nostal Hall
Ackworth Park
Stapleton Park
Huddersfield
Moor Top
Hemsworth
BLACK HAMBLETON
1246
Coxwold
Rivaulx
Hawnby
Leeds & Liverpool Can.
Air R.
Nidd R.
Ouse R.

'It is to the establishment of these great works [canals and railways] and the minerals which they have distributed, that England owes half her present consequence in the scale of nations; for these works are the result of those energies of the mind which have called forth the labours of the industrious, and furnished humanity with every thing that is great and good; and it is hoped that many more such works may yet be established, with the aid of a better knowledge of the rich contents of our sub-strata.'

SMITH'S *MEMOIR*, 1815.

←
DETAIL FROM SHEET V, SHOWING COAL MEASURES IN WEST YORKSHIRE.
The dark grey-coloured area broadly describes the extent of the Leeds–Bradford coalfield, part of the larger Yorkshire coalfield. Smith marked working collieries on the map with crosses.

III. FIELD WORK.

Considering the agricultural revolution and the usefulness of sheep shearings in the calendar; Smith's excellent practice of draining and flooding boggy lands; the intransigence of gentleman farmers; hints concerning the correct education of farmers' sons; admiration for agricultural reformer and patron Sir Thomas Coke.

The late eighteenth and early nineteenth centuries saw great strides in the improvement of land and increased crop production in England. Charles 'Turnip' Townsend (1674–1738) had pioneered a system of crop rotation based on roots, barley, clover and wheat, and Robert Bakewell (1725–1795) scientifically bred livestock to improve desirable characteristics. But perhaps the major contributory factor to the agricultural revolution in the country was the series of Enclosure Acts, especially between 1750 and the General Enclosure Act of 1801, which allowed large open fields of what was once common land to be enclosed. These consolidated lands were now owned by the rich and influential, to the great detriment of the general population.

However iniquitous they may have seemed, the Enclosure Acts offered an economy of scale, facilitating large-scale crop rotation and mechanization using equipment such as the Dutch plough, seed drills and threshing machines that used horse rather than human power.

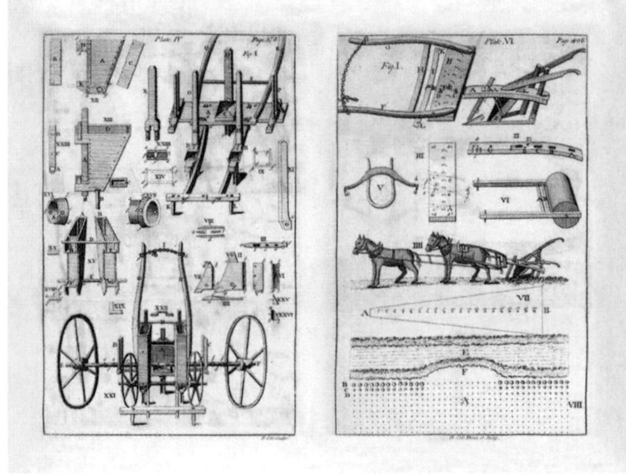

FIG. 1.

Fig. 1. Illustrations of a drill (left) and plough (right) taken from Jethro Tull's *Horse-hoeing husbandry* (1762). Tull's development of the horse-drawn seed drill and horse-drawn hoe was a major influence on the British agricultural revolution.

LAND IMPROVEMENT.

William Smith (1769–1839) played a key part in aspects of land improvement, itself an important element of the agricultural revolution. After his work with the Somersetshire Coal Canal Company came to an end in 1799, Smith was largely employed in improving agricultural land by draining and irrigation, usually for landed gentry. He used his knowledge of strata and skills learnt on the canal to work as a consulting engineer.

The late eighteenth century was particularly wet, and John Phillips (1800–1874) in *Memoirs of William Smith* noted the need for improved drainage particularly on the heavy clay soils where much of the country's wheat was grown. He goes on to say: 'Mr. Stephens of Camerton, the chairman of the canal company, and Mr. T. Crook of Tytherton, one of the best farmers of the Bath district, set the example of encouraging Mr. Smith in his new occupation; and from this time forward, for several years, he was almost daily occupied in various parts of the country, first in *draining* land, and, as a second improvement *irrigating* it when drained' (Phillips, 1844, p. 34).

Thomas Crook seems to have been a model farmer. Arthur Young (1741–1820), Secretary of the Board of Agriculture, visited his farm in 1798 and wrote an account in the *Annals of Agriculture*. He was full of praise for Crook, particularly his stock, describing his half-French and half-Devon crossed cattle and some Prussian cattle: 'strange animals! —Immense ears! and legs as big as mill posts! — All of a dun colour. Such bones are not often seen!' In 1800, the drainage and irrigation improvements made by Smith at Crook's farm at Tytherton Kellaways were inspected by Thomas William Coke (1754–1842), of Holkham Hall in Norfolk, who would employ him in a great variety of works and recommended him to others, in particular Francis Russell, 5th Duke of Bedford (1765–1802). It was at Tytherton Kellaways that Smith found his 'Kelloways Stone' (Kellaways Rock), at the base of the Clunch Clay, which was later to establish the Callovian as an important stage in the Jurassic period.

Smith's life must have been busy at a time when nothing moved faster than a horse. While travelling, he had to keep to a schedule, maintain his accounts and remember the stage each project had reached on his previous visit. The cover of his diaries was often used as a year-planner and itinerary, summarizing his activities and forming a record for preparing invoices for his many clients (see p. 14). Inside, he kept details of time spent on individual projects, including draining and irrigating, sea defence work, prospecting for coal and other minerals, often working under pressure for aristocratic land owners. Smith must have had a remarkable memory, and been very well organized in those days when everything had to be written down by hand.

'PRISLEY BOG' AND WATER MEADOWS.

One of Smith's first major projects was on the Duke of Bedford's estate at Flitwick, Bedfordshire, then called 'Prisley Bog', now Priestly Farm (Lewis, 2018). It was in this context that Smith first met John Farey (1766–1826), agent to the Duke, who would later become a great supporter of Smith. Drainage improvements had been attempted previously at Prisley Bog by the famous 'drainer' of agricultural land, Joseph Elkington (1739–1806), who was awarded £1,000, plus a gold ring for his work by Parliament. Yet when Smith began work in 1802, he found the land in a poor state, so he set about draining and irrigating it again. To do this he built ridges at intervals across the meadow and created channels, or water 'feeders', along the tops of those ridges, and 'drains' along the hollows between them to take the water out. The slopes were carefully controlled to achieve as even a flow of water as possible over the whole field. Flow into the feeders was managed by opening and closing three 'hatches', similar to those used in controlling water in canal locks. The result was that 'Mr. Smith, after thoroughly depriving the bog of stagnant water, converted it to valuable meadows' (Phillips, 1844, p. 49). Some of the remains of Smith's work, revealing the kind of hatches used to control water flow, can still be found in Oxfordshire (Walton, 2016).

Smith then wrote his first book: *Observations on the Utility, Form and Management of Water Meadows and the Draining and Irrigating of Peat Bogs, with an account of Prisley Bog and other extraordinary improvements*. Dedicated to Thomas Coke, it was published in 1806 to coincide with the June Sheep Shearing at Holkham Hall. Precursors of today's county shows, annual Sheep Shearings, as at Woburn

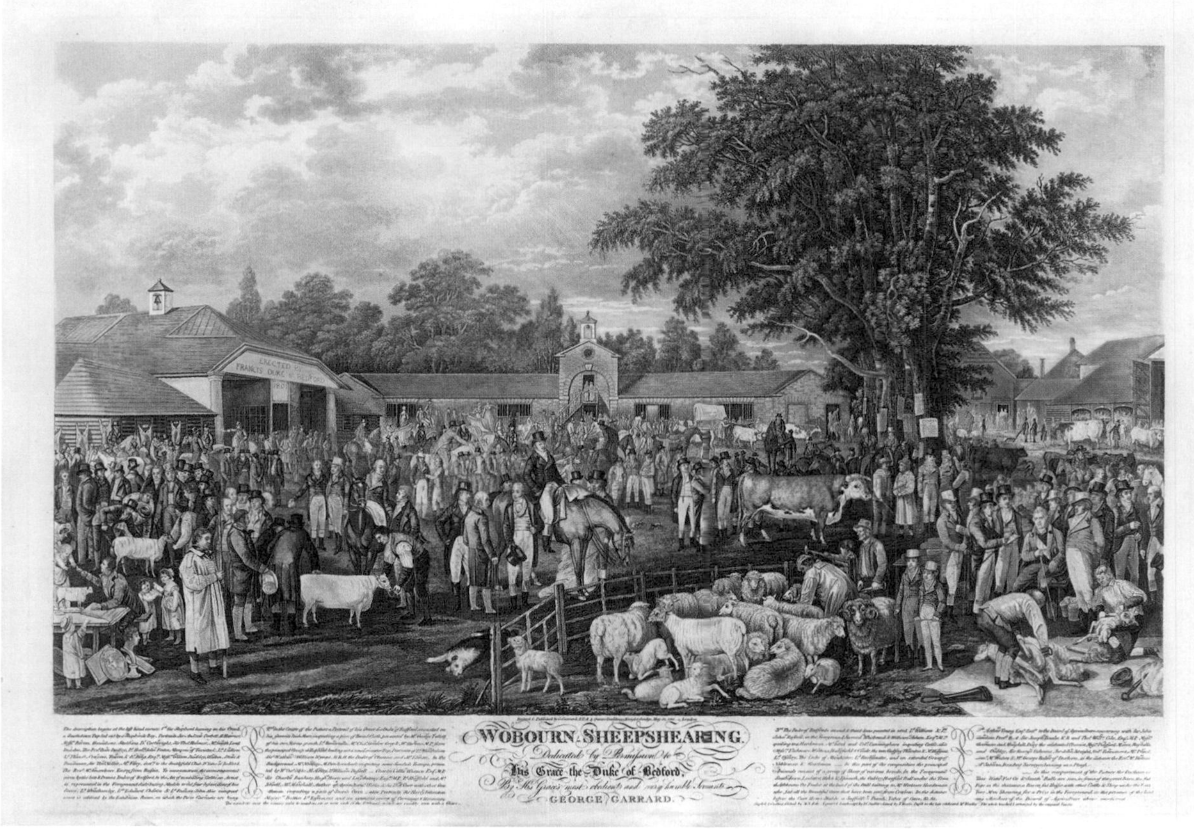

FIG. 2.

Fig. 2.
Engraving after George Garrard's 1811 painting of the 1801 Woburn Sheep Shearing, a 'Who's Who' of early nineteenth-century celebrity, with 921 attendees. Royalty is represented by the Duke of Clarence, later William IV, and MPs, mostly Whigs, and scientists including Humphry Davy and Sir Joseph Banks, mingled with tenant farmers and herdsmen. Presiding over all is Francis Russell, 5th Duke of Bedford, seated on his favourite Irish mare. In the background is the back view of 'the Drainer', William Smith.

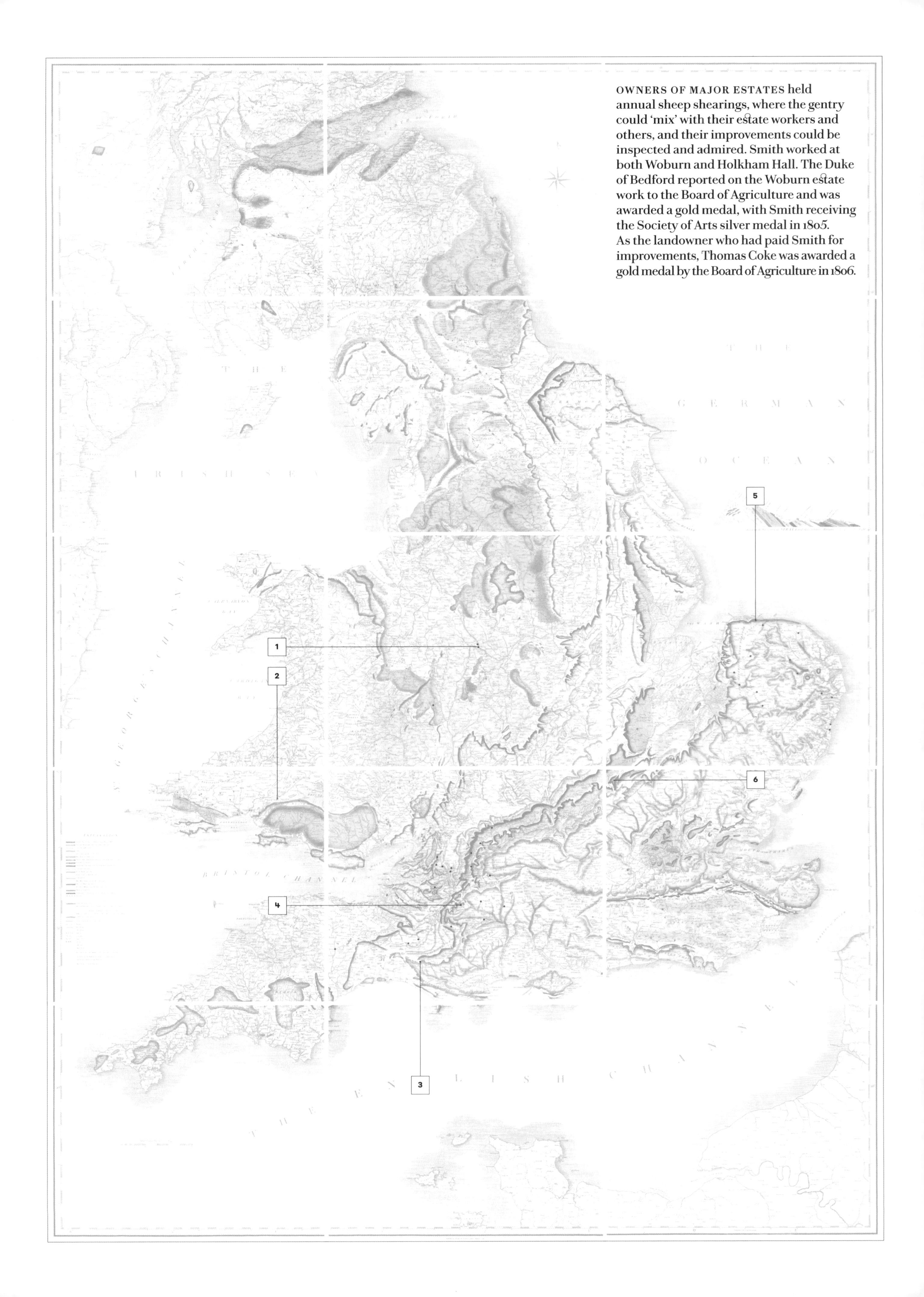

OWNERS OF MAJOR ESTATES held annual sheep shearings, where the gentry could 'mix' with their estate workers and others, and their improvements could be inspected and admired. Smith worked at both Woburn and Holkham Hall. The Duke of Bedford reported on the Woburn estate work to the Board of Agriculture and was awarded a gold medal, with Smith receiving the Society of Arts silver medal in 1805. As the landowner who had paid Smith for improvements, Thomas Coke was awarded a gold medal by the Board of Agriculture in 1806.

1.

SHUGBOROUGH HALL.
Staffordshire. 1693.

Smith was employed to drain the estate by Lord Anson. The Hall had been acquired by the Anson family in 1624, and remained theirs until 1966.

2.

GOLDEN GROVE.
Carmarthenshire. 1754.

The first mansion on the Golden Grove estate was built in 1560. However, it was rebuilt in 1754 and again in 1827.

3.

MELBURY.
Dorset. 1546.

Originally a medieval manor house, Melbury was rebuilt in 1546. Smith was employed to drain the estate by the second Earl of Ilchester in 1801.

4.

LONGLEAT.
Wiltshire. 1580.

Longleat was built in the sixteenth century as the seat of the Thynn family. Smith was employed to drain the estate in the early nineteenth century.

5.

HOLKHAM HALL.
Norfolk. 1764.

The house was built in the eighteenth century for the first Earl of Leicester, Thomas Coke, who employed Smith to create the water meadows.

6.

WOBURN ABBEY.
Bedfordshire. 1744.

The Abbey has been the seat of the Dukes of Bedford since the sixteenth century. This was Smith's most important drainage and irrigation commission.

ESTATES SMITH WORKED ON.

CENTRAL ENGLAND.

INGESTRY.	Earl Talbot.	*Staffordshire.*
SHUGBOROUGH.	Lord Anson.	*Staffordshire.*
KINLET HALL.	William Childe.	*Shropshire.*

WALES.

GOLDEN GROVE.	Lord Cawdor.	*Carmarthenshire.*

THE WEST.

PUCKLECHURCH.	Lord Whitmore.	*Gloucestershire.*
TRACY PARK.	Mr Bush Esq.	*Gloucestershire.*
TIVERTON.	Mr Hay.	*Devon.*
SILTON.	Mr Samuel Davis.	*Dorset.*
MELBURY.	Earl of Ilchester.	*Dorset.*
CAMERTON.	James Stephens.	*Somerset.*
BATHFORD.	Mr Eleazer Pickwick.	*Somerset.*
LILIPUT.	Charles Gordon Grey.	*Somerset.*
MUDFORD.	John Old Goodford.	*Somerset.*
MONTACUTE.	The Rev. William Phelips.	*Somerset.*
DAUNTSEY.	Earl of Peterborough.	*Wiltshire.*
TYTHERTON KELLAWAYS.	Thomas Crook.	*Wiltshire.*
BOWOOD.	Marquess of Lansdown.	*Wiltshire.*
ALL CANNINGS.	Mr J. Grant.	*Wiltshire.*
BOX.	William Northey.	*Wiltshire.*
LONGLEAT.	Marquess of Bath.	*Wiltshire.*
CHITTERN.	Paul Methuen.	*Wiltshire.*
AMESBURY.	Duke of Queensbury.	*Wiltshire.*
DINTON PARK.	William Wyndham.	*Wiltshire.*

EAST ANGLIA.

HOLKHAM HALL.	Thomas William Coke.	*Norfolk.*
WIGHTON.	Thomas William Coke.	*Norfolk.*
HANWORTH HALL.	Robert Doughty.	*Norfolk.*
LEXHAM.	Thomas William Coke.	*Norfolk.*
TAVERNHAM HALL.	Miles S. Branthwayte.	*Norfolk.*
KESWICK HALL.	Richard Gurney.	*Norfolk.*
BEECHAMWELL.	John Motteux.	*Norfolk.*
LYNFORD HALL.	George Robert Eyres.	*Norfolk.*
WRETHAM.	William Colhoun.	*Norfolk.*
RIDDLESWORTH.	Silvanus Bevan.	*Norfolk.*
ELLINGHAM.	The Rev. William Johnson.	*Norfolk.*
WORLINGHAM HALL.	Robert Sparrow.	*Suffolk.*
DUNWICH.	Barne Barne.	*Suffolk.*
MINSMERE.	Minsmere Level Drainage.	*Suffolk.*

THE SOUTH.

KIMBOLTON.	Duke of Manchester.	*Huntingdonshire.*
WOBURN PARK.	Duke of Bedford.	*Bedfordshire.*
PRISLEY BOG.	Duke of Bedford.	*Bedfordshire.*
ASHFORD.	Earl Thanet.	*Kent.*

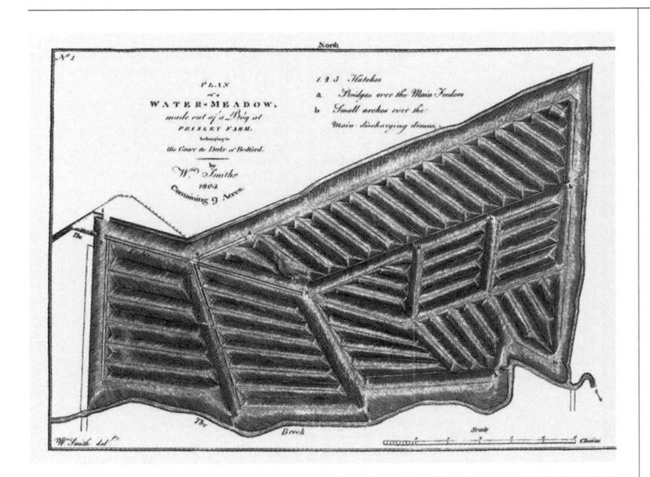

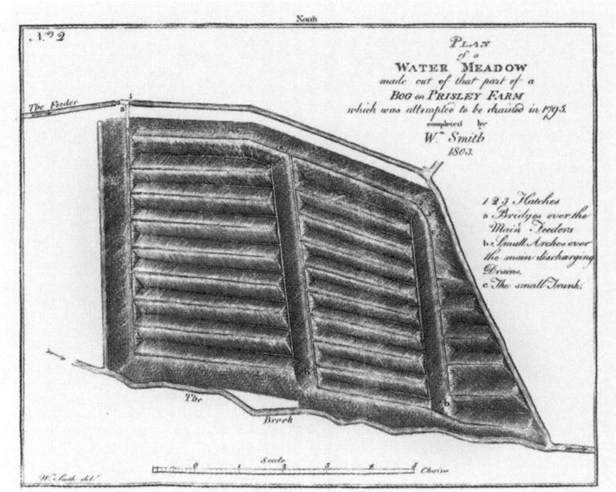

FIG. 3.

Fig. 3.
Irrigation plans for the two fields that Smith drained at Prisley Bog between 1802 and 1803. Smith turned the fields into water meadows. This means feeding water into the meadow through controlled hatches, allowing it to drain through the meadow with excess water feeding into ditches that discharge into a river or brook. This process keeps the ground damp, ideal for larger crop yields and higher quality grass. In both these cases water enters at top left and drains into the brook at bottom right.

and Holkham, had the serious purpose of encouraging agricultural reform but were also great occasions for networking. Smith, who had been a member of one of the oldest agricultural societies, the Bath and West, for several years, used these events to seek work and gain support for his geological mapping project, carrying drafts of his maps to show prospective subscribers. His great patrons Francis Russell, Thomas Coke and Sir Joseph Banks (1743–1820), the eminent botanist and long-time President of the Royal Society, were regular attendees.

Despite Smith's efforts the book was not a financial success: 2,000 copies were printed, but only a handful were bought at $\frac{1}{2}$ a guinea, so Smith sold off 1,000 to the publisher at 2 shillings (Cox, 1942, p. 22). In his Preface, he confesses (Smith, 1806, p. vii): 'If I could have felt the same confidence in writing that I have in Draining and Floating, this Essay might have made its appearance sooner; but I find it less difficulty in directing the labours of the spade, than those of the pen.'

Smith was a practical man and he laid out his distinct views about the need and type of education best suited to farmers' sons (Smith, 1806, p. viii): 'there can be no doubt but it would be much better for society, and much more conducive to improvements in agriculture, if farmers' sons were well instructed in practical geometry and the use of mathematical instruments, with the principles of machines intimately connected with their profession, instead of spending their time learning latin, or pursuing other studies.'

Productive water meadows at the base of valleys were very fertile places for hay making and stock-grazing, as Smith was aware (Smith, 1806, p. 20): 'A water meadow is a piece of ground so formed by nature or art, that water may flow quickly over its surface, for the purpose of promoting an early and increased vegetation of grass. There are but few natural water meadows, and Wiltshire has to boast of some which probably gave the first idea of making them artificially.' The main objective in creating water meadows was to convert poorly drained or boggy land into productive pasture. By irrigating or 'flooding' the land in winter and then alternately flooding and draining it in spring, sheep could be put out to graze earlier. After their 'first bite', the water could be reintroduced to encourage an early crop of hay. More flooding and draining could then produce a good second crop of hay. This was a managed system, which involved the daily moving of sheep and careful control of water on to and off the meadows.

Smith showed his knowledge of grass growing and stock management, advocating floating (flooding) during most of the winter to bring warmth and moisture to the soil. He recommends holding stock in small areas using hurdles, to ensure controlled grazing (now known as 'mob grazing') and moving them off the pasture each night until well after dawn. Managing all this was a skilled task (Smith 1806, pp. 108–9):

> *Much of the perfection of a water meadow likewise depends upon the care and pride which the floater takes in doing his work well; it would therefore be very advisable not to change these men when it can be avoided.... Water meadows will never be brought to perfection in any country till the proprietors and managers of them shall take pride in conducting them properly, and strive to rival each other in excellence. Land owners and agricultural societies should therefore offer premiums for the greatest produce which can be obtained from a given quantity of a water meadow, and a smaller premium to the floater.*

For the 3.6-ha (9-acre) Prisley water meadows Smith listed the value of the produce, hay and grazing for sheep and bullocks in 1803 as more than £165, giving details of the timetable and numbers of stock grazed and amounts of hay cut.

LAND IMPROVEMENT AT HOLKHAM.

Soon after this Thomas Coke invited Smith to work on his large estates around Holkham Hall, a grand eighteenth-century house near the north Norfolk coast. Smith started on a plot at Lexham, about 32 km (20 miles) from Holkham, farmed by Mr Beck, from whom he requested a detailed timetable for watering and amounts of materials grown. Smith monitored activities here from 1804 to 1806, and remarks that in the second year of controlled irrigation, the grass was much improved, and that 'cows fed on it in the winter, and it was observed to produce good milk'; in all the value of the meadows was nearly £11 an acre per year, at a time when boggy land was worth less than £1 per acre.

These fields were viewed by John Russell, 6th Duke of Bedford (1766–1839), Sir Joseph Banks and 'other great personages and eminent agriculturists' at the Holkham Sheep Shearing in 1805. Remnants of the irrigation channels can be seen today between the branches of the Nare at Lexham. Smith was based in Norwich for several years, working on land improvement for Coke and sea defences on the Norfolk coast, and is commemorated in a relief at Holkham Hall.

DRAINING MINSMERE MARSH.

The Ipswich Journal of 26 May 1810 carried an announcement stating that an Act of Parliament had been passed for embanking and draining of marsh and fen land known as the Minsmere Levels in Suffolk. John Phillips recorded that Smith was appointed engineer to drain the Minsmere marshes and his friend Henry Jermyn, an antiquarian of Sibton Abbey (1767–1820), was one of the commissioners (Phillips, 1844, pp. 73–74). Another trustee and marsh owner was Barne Barne MP (1754–1828), of Sotterley Hall. His name appeared on Smith's 1808 client list, having written to him in 1806 telling him, on behalf of a Mr Robinson, of storm damage to a pre-existing sluice, and enquiring about a plan and costs for draining and irrigating his land.

This was work on a considerably grander scale, as Smith's report on Minsmere (1812–14) and his costings for the work (£5,656 8s) show. Phillips (1844, p. 74) noted that the ingenious plan

> *was accomplished (in 1814) by bringing to one point close to the beach the great discharging drains, and uniting their currents in a hexagonal channel or well, from which a large cast-iron pipe was laid right through and deeply buried in the pebbles and sand, which were the natural barrier against the sea. This pipe was stopped by the sea rising at every tide, and often buried at its mouth by the accumulated load of "shingle" and sand thrown up by the waves. But on the turn of the tide, the pent-up inland water gradually made its appearance by the opening of a pair of doors in the upper part of the tube; then a second, and afterwards a third pair of valves was opened by the force of the water, which soon left the marsh ditches and swept away the accumulated pebbles from the aperture of the tube. In process of time such a tube may be expected to become in parts converted to plumbago, but the iron removed as oxide may cement the surrounding pebbles into a mass more durable than the original tube.*

The New Iron Sluice is shown on Smith's 1819 geological map of Suffolk (p. 143). Smith was successful in draining the marshes, but in 1940 the land was deliberately flooded to act as a barrier in the event of an enemy invasion.

THE WATERMILL AT MUDFORD.

Between 1800 and 1805 Smith was employed by Maria Goodford (1757–1848), a widow managing an estate of around 810 ha (2,000 acres) near Yeovil in

FIG. 4.

Fig. 4.
Part of the frieze decorating Thomas Coke's memorial on the Holkham estate depicts Smith and his right-hand man, Jonathan Crook, shown holding a spade. Crook was Smith's foreman for many years, working with him on the Somersetshire Coal Canal, at Woburn Abbey and at Holkham Hall. Crook was eventually made Engineer for the Holkham estate.

Somerset. Hinton Mill, Mudford, was an ancient watermill in the valley of the River Yeo, serving the whole estate, but suffered badly from flooding. The work included building a new weir, with hatches to control water flow, and relocating the mill machinery into a huge mid-eighteenth-century tithe barn on higher land. Water was brought from the river more than 400 m (437 yd) along an underground tunnel, more than 6 m (20 ft) deep, in the Lower Lias to a waterwheel. A fossil belemnite from Mudford, probably from this tunnel, is in the Smith fossil collection at the Natural History Museum, London.

A wooden underpass wheel was installed below ground in the barn. The machinery was on the three floors above, with three grindstones as well as a threshing mechanism in the barn. The waterwheel was replaced by a turbine in the late nineteenth century, which generated power until very recently, and the mill race still runs today through the tunnel under the barn (now a farm shop). All this work was done without disturbing the main structure, with the tunnel being dug as in a mine. The overall cost of the work to relocate Hinton Mill must have been considerable, and it is noteworthy that Smith was apparently chosen for this large project, outside the scope of his previous experience, and by a female landowner.

In his autobiographical notes written much later (probably 1830s), Smith recorded a visit to this work: 'Year 1805 commenced with a journey to Bath with a hope that I should be able to close my long unsettled affairs with the coal company but the same perverseness I had before experienced still subsisted. During my stay in the west I revisited some of my former works in Drainage and Machinery which I established in a barn at Mudford near Yeovil for thrashing and grinding corn.'

THE RELATION OF SOIL TO ROCK STRATA.

Smith's 1801 *Prospectus* shows that he believed that his order of strata was intrinsic to the nature of the soil, thus making his maps of practical use. As he put it: 'for what can be of greater importance in human science, than a Complete Theory of the Soil, which man is under a divine injunction to cultivate and replenish, that he may derive from that labour his daily subsistence.' In his *A Memoir to the Map and Delineation of Strata of England and Wales, with part of Scotland* (1815) one section was headed the 'Characteristic distinctions of soil and surface in the courses of the respective strata'. It is clear that Smith used the various types of vegetation and soils he observed as indicators of the strata beneath, and knew which soils were best for agriculture, as can be seen in his descriptions of three adjacent rocks. Chalk, he wrote can 'render the thin soil unfit for cultivation ... soil turned up by the plough is white ... The whole course of this stratum is destitute of timber.' Greensand, on

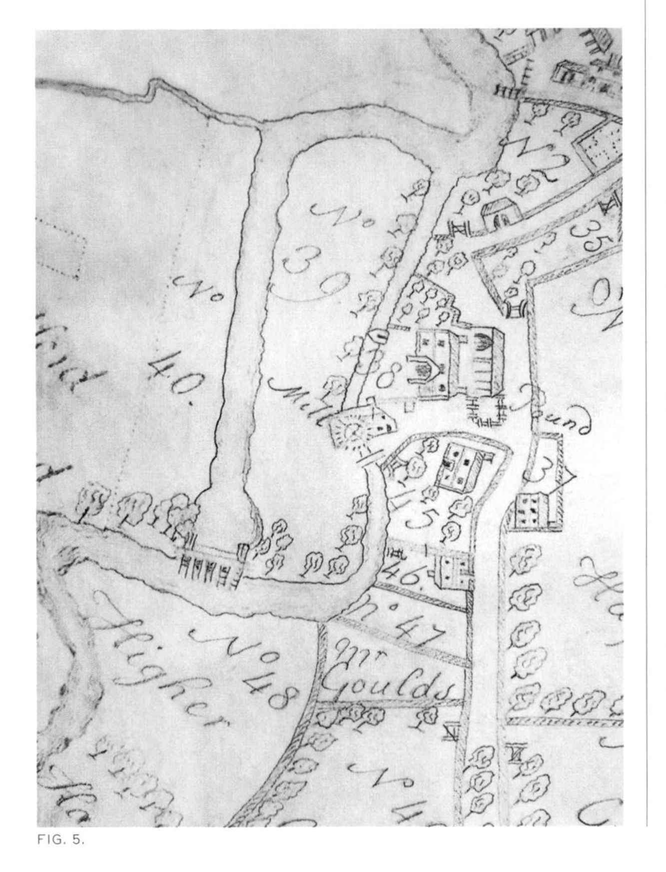

FIG. 5.

Fig. 5. A tithe map from the 1760s showing the original location of the Hinton watermill on the east branch of the River Yeo. Smith's underground tunnel carried water from here around 400 metres (437 yds) to the north-east, where it reached the new watermill, just off this map.

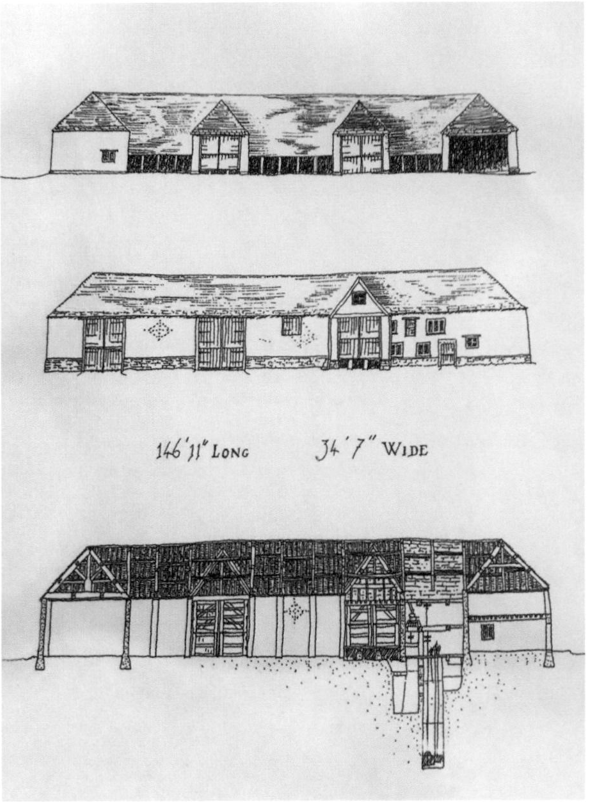

FIG. 6.

Fig. 6. The magnificent eighteenth-century tithe barn that Smith used to house the new Hinton watermill. A wooden waterwheel was installed beneath the ground, and the milling machinery placed on the floors above.

FIG. 7.

Fig. 7.
An engraving by William Walker after George Zobel's *Distinguished Men of Science in 1807–8*, made in 1862. At the time of the 1801 Woburn sheep shearing, William Smith – 'the Drainer' – was named but his back turned. How different was his position now: Smith stands prominently, taking his place as an established member of the fifty scientific elite of his generation, behind Hon. Henry Cavendish, the seated figure in profile to the left of centre.

the other hand, is 'fine, dry, mellow, deep, loamy soil, which is some of the best arable land and remarkable for the growth of wheat.' And 'Blue marl or oak-tree soil is distinguished by the tenacity of its soil, and some of the finest oaks in the kingdom, in parks and woods along the various parts of its course' (Smith, 1815, pp. 42–43).

The majority of Smith's agricultural commissions were on land underlain by strata of Jurassic or later age, so beds of limestone, sandstone or clay rocks. Clays predominate in the less well-drained land in the valleys, which provided much of Smith's work.

Smith planned another book, *Description of Norfolk, its Soil and Substrata*, probably written in 1807. Only one printed copy of the first 56 pages exists, now in the Oxford University Museum of Natural History. It provides fascinating glimpses into Smith's views of agricultural matters, and many other subjects. The first eighteen pages form a synopsis of the general topography and scenery of Norfolk, which did not impress him after the hills and vales of Somerset, describing it as: 'the extremes of dreariness ... the inhabitants of this district have all the comforts of the bogs of Ireland and the deserts of Arabia united. One side blowing sands only are seen, and on the other nothing but shaking bogs. The sand hills abound with vipers, and the marshes with frogs.'

In his Preface he sets out the case for scientific investigations of substrata, drainage and irrigation, and criticizes the farming community for their slow uptake of new ideas and crops, such as potatoes and turnips, which have made 'slow progress toward universal adoption'. Similarly, drilling and thrashing machines and 'improved breeds of sheep and all other stock have, amongst the farmers more enemies than friends'.

To Smith, the contrast with industry was very stark: 'The most ingenious men are employed by the manufacturer to make new discoveries and are liberally rewarded by him and the legislature; but in agricultural concerns, such men are looked upon as chimerical beings, and most dangerous advisers.' In his opinion part of the problem was that 'many practical farmers are not accustomed to travel far from home, and too many of them seldom read much'. But he chided landowners too: 'Much of the indolence and mis-management of farmers ... too often arises from the supineness of the landlords.' His patron, however, is a notable exception, noting 'the spirited desire for improvement which has ever been manifested by Mr. Coke'.

FROM 'DRAINER' TO 'SCIENTIST'.

Smith was employed all over the country in more than fifty projects related to water: draining, irrigating, sea defences, water supply, as well as coal and mineral prospecting. The geological work for which he is best known today, his geological maps and sections, and his work on fossils and strata, still lay in the future. In the painting of the 1801 Woburn Sheep Shearing, Smith was shown as the anonymous 'Drainer', hidden at the back of crowds. In 1862, when George Zobel drew the *Distinguished Men of Science in 1807–8*, Smith was shown in a new light. Now he was a leading member of the scientific elite, prominent among his fellows, including 'scientists' who are better remembered today for their engineering works, such as Watt, Boulton, Crompton and Trevithick. As a practical man all his life, Smith would have been very content to be shown in such distinguished company.

ROBERT HAVELL, *RAPE THRESHING*, 1813.

A group of agricultural labourers in Yorkshire are shown threshing the rape that they have harvested. Threshing is the process of removing the edible grain from the chaff, and until the process was mechanized at the end of the nineteenth century this had to be completed by hand. The men are shown using jointed wooden tools called flails to beat the rape, and the women at the front hold a sieve. It was laborious work, and around one quarter of agricultural labour was devoted to it.

ROBERT HAVELL, *PEAT CART*, 1814.

Here a horse-drawn cart is shown waiting to be loaded with the peat being dug from the Yorkshire hillside. In waterlogged, acidic conditions dead plants will not fully decompose and instead form peat. Once dug, the peat would have been cut into blocks and then dried, after which it could be used as a fuel.

PAUL SANDBY MUNN, *A CORNFIELD*, 1823.

Here corn is shown being cut, tied into sheafs and then gathered into stooks: several sheafs leaning against each other to allow drying air to flow through. Harvesting crops was back-breaking work. However, it was followed by the harvest celebration, one of the most important events in the eighteenth-century agricultural calendar.

JOHN NIXON, *EASTBOURNE FROM LORD G. CAVENDISH'S SEAT IN THE PARK*, 1787.

During the eighteenth century Eastbourne, situated on England's south coast, was the site of several small, rural villages. Here it is viewed from the park of Compton Place, a seat of the Cavendish family since 1759. William Cavendish, seventh Duke of Devonshire, would develop Eastbourne into an elegant seaside resort during in the nineteenth century. Beyond Eastbourne is Beachy Head, Great Britain's highest chalk sea cliff.

FRANCIS JUKES, *A VIEW OF MR METCALF'S MILL NEAR BROMLEY*, 1785.

The windmill depicted here is found on Three Mills island on the River Lea, London. The two major mills on the site were watermills. There have been mills on this site since the time of the *Domesday Book*. By the sixteenth century they were producing flour for bread, and briefly gunpowder. From the eighteenth century, however, they ground grain to be used for distilling gin.

WILLIAM HINCKS, *THIS VIEW TAKEN NEAR SCARVA IN THE COUNTY OF DOWNE, REPRESENTING PLOUGHING, SOWING THE FLAX SEED AND HARROWING*, 1791.

In this scene the man in the foreground is ploughing the field, while behind him another man is sowing flax seeds and another drives cattle pulling a harrow, an implement used to smooth the surface of the soil. In the foreground a woman is seated with her children, and beckons her husband to join them for lunch.

Brancaster Bay
Holkham Bay
High Cape
Blakeney Meals
Brancaster
Hunstanton
Ringstead
Burnham Market
Holkham Park
Wells
Morston
Cley
Glantord
Bayfield
FELBRIG
HEATH
Wighton
Binham
Holt
Docking
Stanhoe
South Creak
New Walsingham
Beaconsthorpe
Hexham Harb.
Sedgeford
Great Bircham
Syderstone
Thornage
Edgefield
Gunthorpe Hall
Melton Park
Saxthorpe
Wolterton
Shernborne
Houghton
Fakenham
Anmer
Snettisham Harb.
Shereford
Tattersford
Wensum R.
Norton
Wood Dalling
Blickling
Bentinck Hole
North Sand
Aylsham
Hungry Hill
Hillington
Rising
Grimston
E. Rainham
Foulsham
CAWSTON & HORSFORD
Reepham
Wellingham
Weasenham
Elmham Pa.
Elmham
Foxley
Whitwell
Clenchwarton
Lynn Regis
Billingford
Mileham
Litcham
Attlebridge
Castle Acre
Lexham
Longham
Morley
Hoe
Weston Hou.
Sparrow Gr.
East Dereham
Honingham Hall
Taverham
Gt. Dunham
Wiggenhall St. Mary
Narford
Wendling
Scarning
Well Hill
Tuddenham
Swaffham
Necton
Costesey Park
SWAFFHAM HEATH
Shipdam
Garvestone
Yare R.
Marham
Wallington
Stradset
Beechamwell
Ashill
Woodrising
Kimberley
Hetherset
Downham Market
Marshland Cut
Cockley Cley
Denver
Oxburgh
Scoulton
Hingham
Hethel Park
Meer
Watton
Wymondham
Rockland St. Peters
Merton
Ashwell Thorpe
Langford
Stoke R.
Stow Beden
Attleborough
Methwold
Southery
Moundford
Stanford
Buckenham
Lynford
Hockham Magna
Hargham
Bunwell
Tofts
Wretham
Wacton Mag. & Parva
Feltwell
Larling Ford
Quiddenham Hall
New Buckenham
Foul Mere
HEATHS
Hockwold
Littleport
Santon Downham
Croxton
East Harling
Pulham M.
Gissing
West Harling
Winfarthing
Wangford
Thetford
Dickleborough
Riddlesworth
Rushford
BRANDON HEATH
High Lodge
Elvedon Hall
Euston
Lopham Ford Ga.
Fen Street
Diss
Euston Pa.
Coney Weston
Wortham
Holywell Row
Isleham
Mildenhall
Hinderclay
Bottesdale
Soham
Barton Mills
Hepworth
Hoxne
Brome
Wordwell
SUFFOLK SANDS
Ampton Park
Eye
Cavenham
Culford
Ixworth
Rush Green
Timworth
Thorndon
Badwell Ash
Snailwell
Kentford
Conyard Green
Wyverston
Thwaite
Southolt
High Street
WOOD LANDS OR HIGH
Elmswell
Mendlesham
Dalham Hall
BURY St. EDMUNDS
Bedfield
NEW MARKET HEATH
New Market
Gt. Saxham
Woolpit
Debenham
Horningsheath
Rushbrooke & Hall
Wood Ditton
Ickworth
Tot Hill
Stretchworth
Owsden Hall
Drinkstone
Stow Market
Fen Street

'I am also the more desirous to bring the practice of draining and irrigation into general use, because the late scarcity and the pressure of the times, has called aloud for every improvement that can be made in landed property. Trade has already gained too great an ascendancy over the landed interest ... we are compelled to send foreigners such immense sums of money as thirty million sterling, for grain to feed our manufacturers ... and while there is such a large proportion of wet and cold late vegetating land in this island, as to lose much of its produce by a few rainy summers, ought we not to exert our utmost abilities to remedy these defects in our soil, by draining and irrigation?'

SMITH'S 'OBSERVATIONS ON THE UTILITY, FORM & MANAGEMENT OF WATER MEADOWS', 1806.

DETAIL FROM SHEET IX, SHOWING NORTH AND WEST NORFOLK.
The Holkham Hall and Lexham estates owned by Thomas Coke can both be seen in this segment of the 1815 map. Regular flooding and irrigation channels were important in improving the quality of the pastures on both estates.

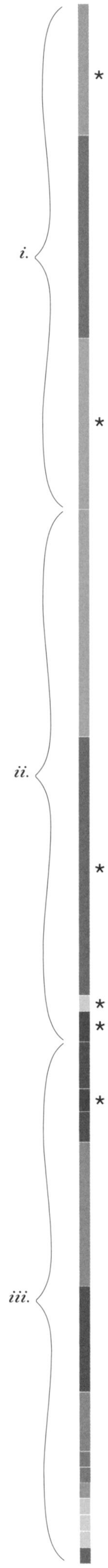

EAST ANGLIA AND THE SOUTH EAST.

i.	ii.	iii.
1. LONDON CLAY.*	4. SAND & LIGHT LOAM.	8. SAND.
2. SAND.	5. CHALK.*	9. PORTLAND ROCK.*
3. CRAG.*	6. GREEN SAND.*	10. SAND.
	7. BRICKEARTH.*	11. OAKTREE CLAY.
		13. SAND.
		14. CLUNCH CLAY & SHALE.
		16. CORNBRASH.
		18. FOREST MARBLE.
		20. UPPER OOLITE.
		21. FULLER'S EARTH & ROCK.
		24. MARLSTONE.
		25. BLUE MARL.
		26. BLUE LIAS.

COMMON REGIONAL FOSSILS.

* INDICATES ORGANIZED FOSSILS OF PREDOMINANT STRATA DISPLAYED IN THIS CHAPTER.

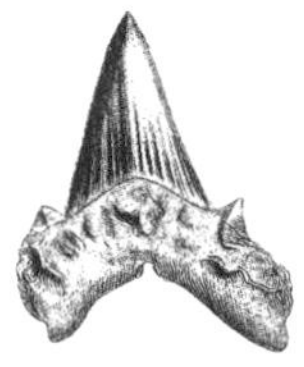

Large sharks tooth.	*Murex contrarius.*	*Echinus*	*Ostrea concentrica.*	*Belemnites.*	*Turritella.*

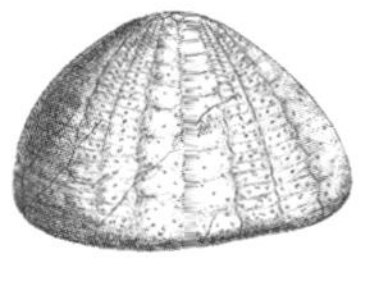

Bar chart of strata indicates the comparative area covered by each stratum in thi. region.

(see pages 88–89)

SHEET VIII.

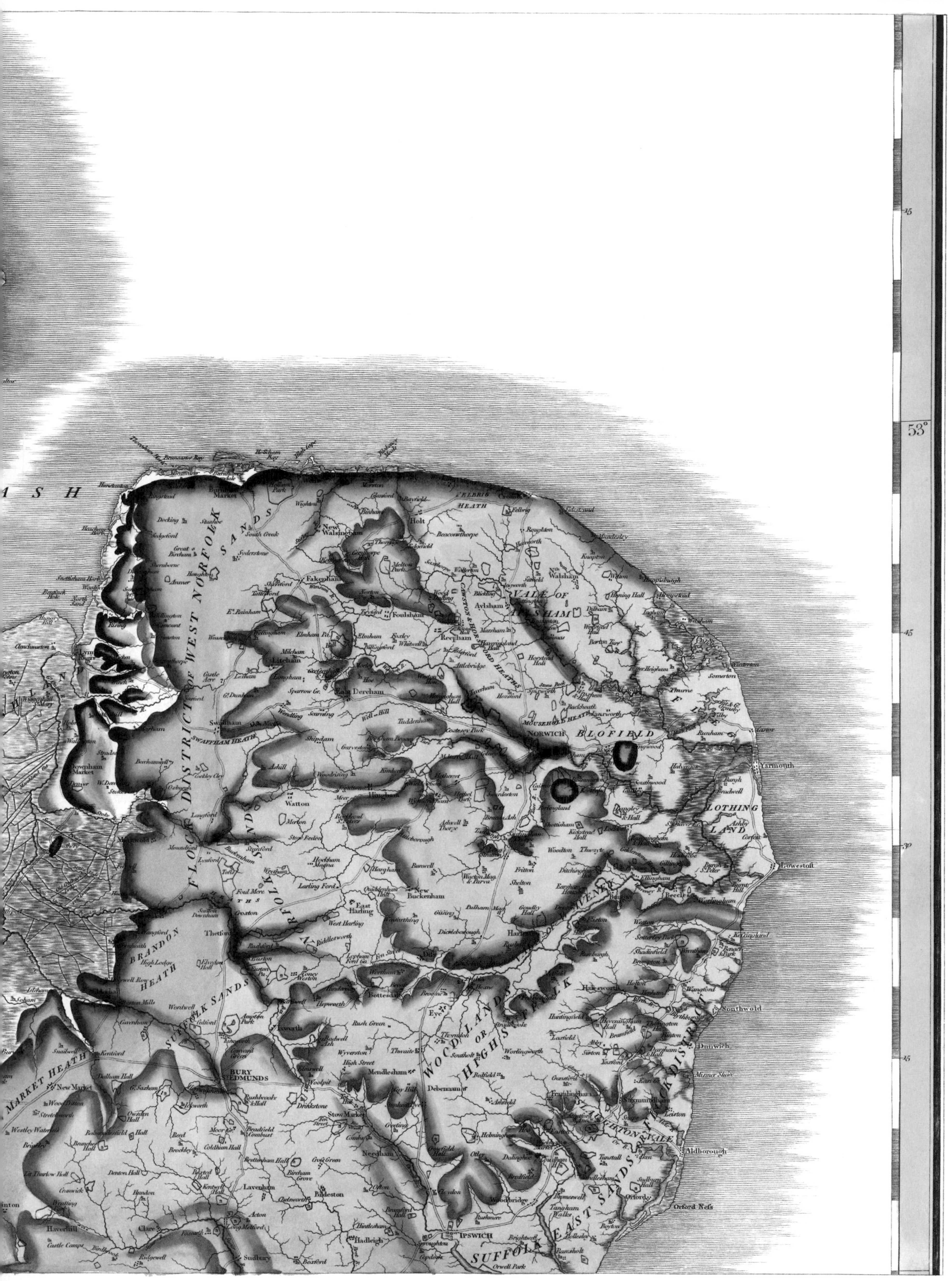

(see pages 204–205)

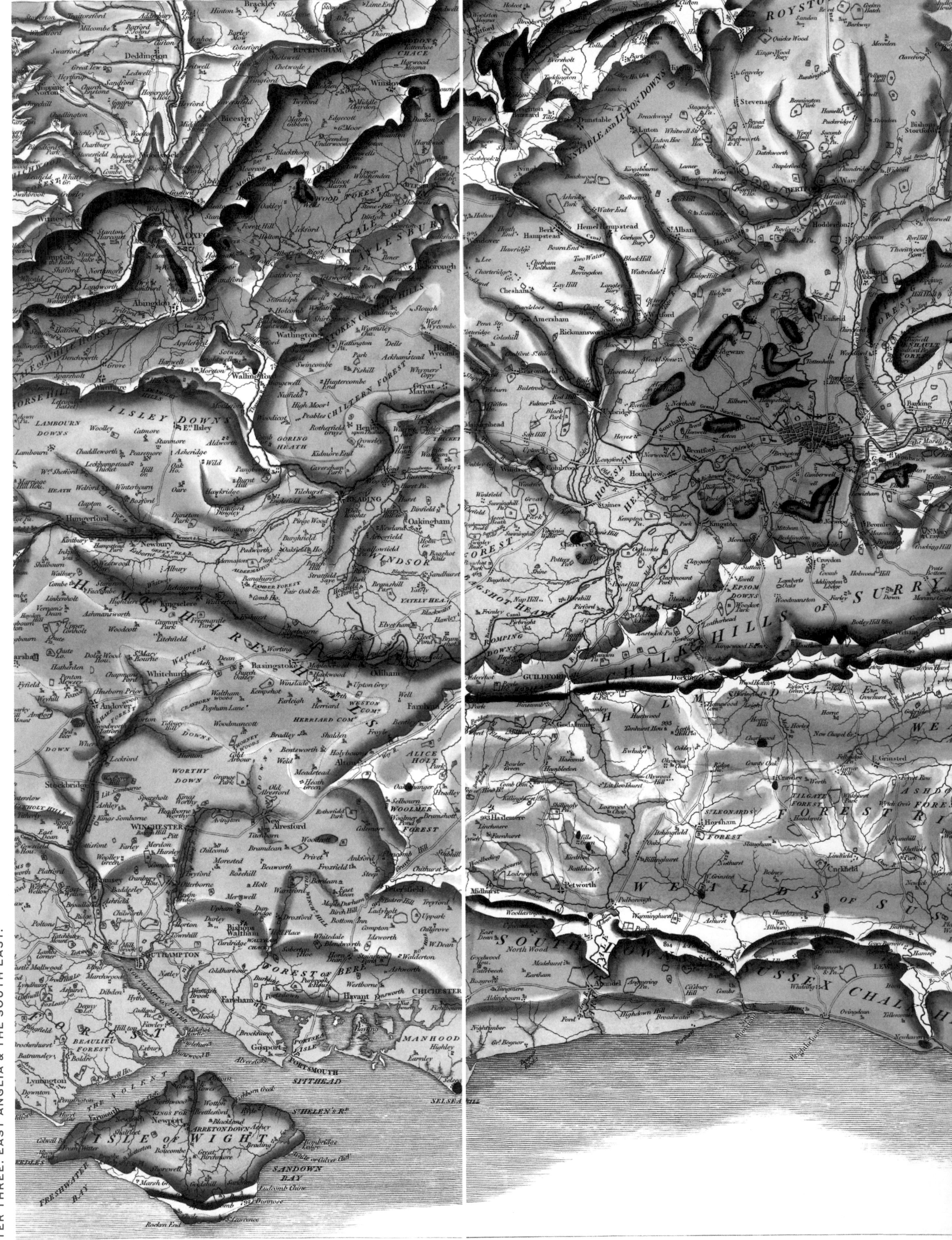

52°
Harwich
Colchester
Neyland
Halstead
Braintree
Witham
Coggeshall
CHELMSFORD
Malden
BLACK WATER RIV.
HUNDREDS
ESSEX
Rochford
Foulness Isle
MOUTH OF THE THAMES
Sheerness
Margate
Queenboro
ROCHESTER
Chatham
Faversham
CANTERBURY
CHALK HILLS
KENT
Deal
THE DOWNS
Dover
Tenterden
Cranbrook
Appledore
ROMNEY MARSH
New Romney
Lidd
Rye
DENGENESS
Hastings
Battel
PEVENSEY LEVEL
BEACHY HEAD
STRAITS of DOVER
51°
CALAIS
Ambleteuse
BOULOGNE
Marquise
Samer

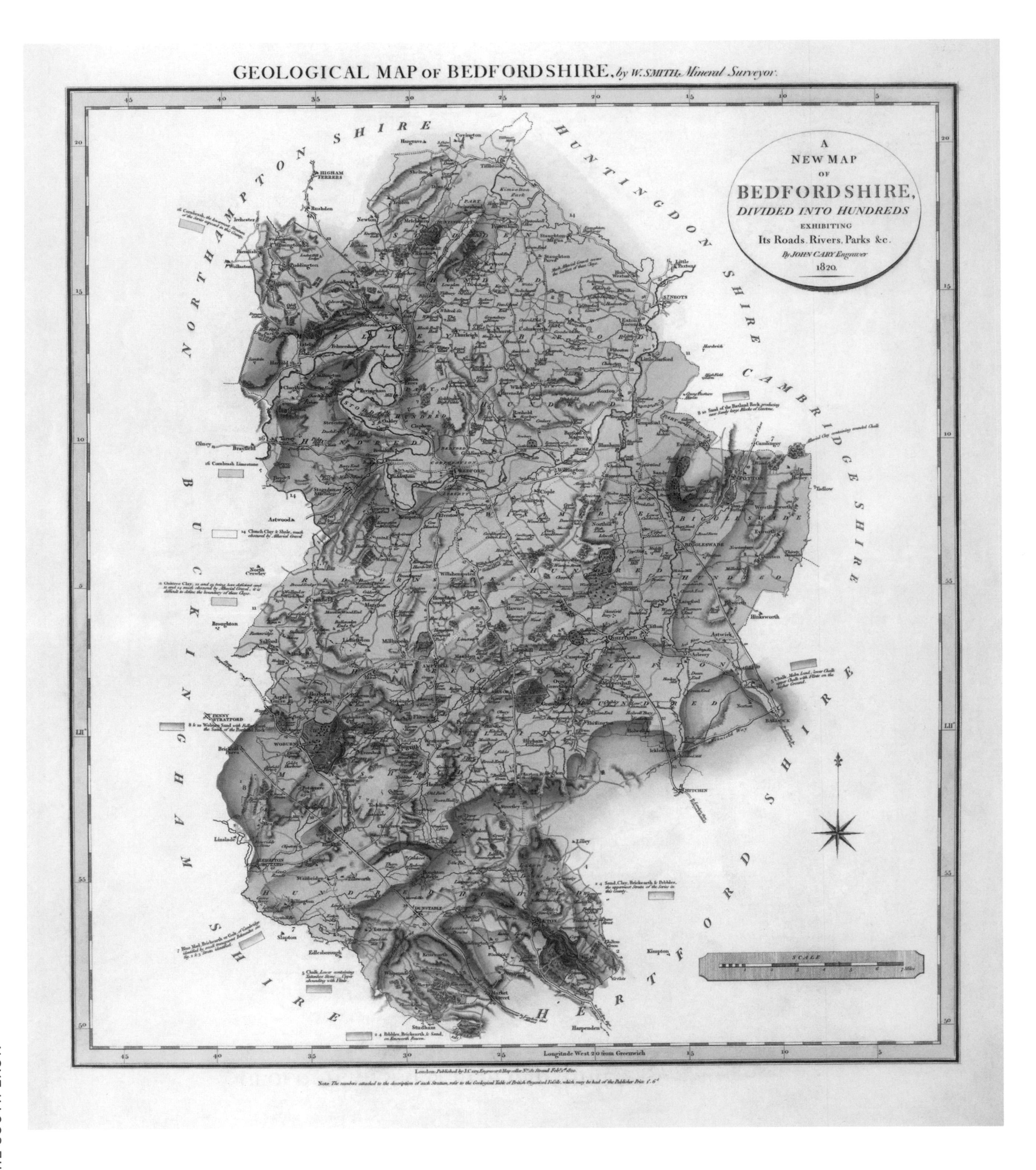

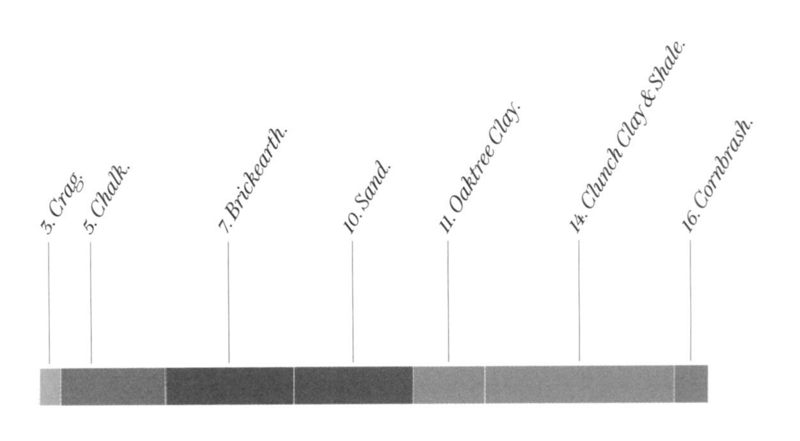

GEOLOGICAL MAP OF BEDFORDSHIRE, 1820.

First published in Part III of *A New Geological Atlas of England and Wales* with the counties Oxfordshire, Buckinghamshire and Essex. The upper stratum is composed of Brickearth (7), Sand (10) and small pebbles over chalk (5) which in turn overlies Blue Marl or Brickearth or Golt (7). The lowest stratum is the Cornbrash in the north of the county.

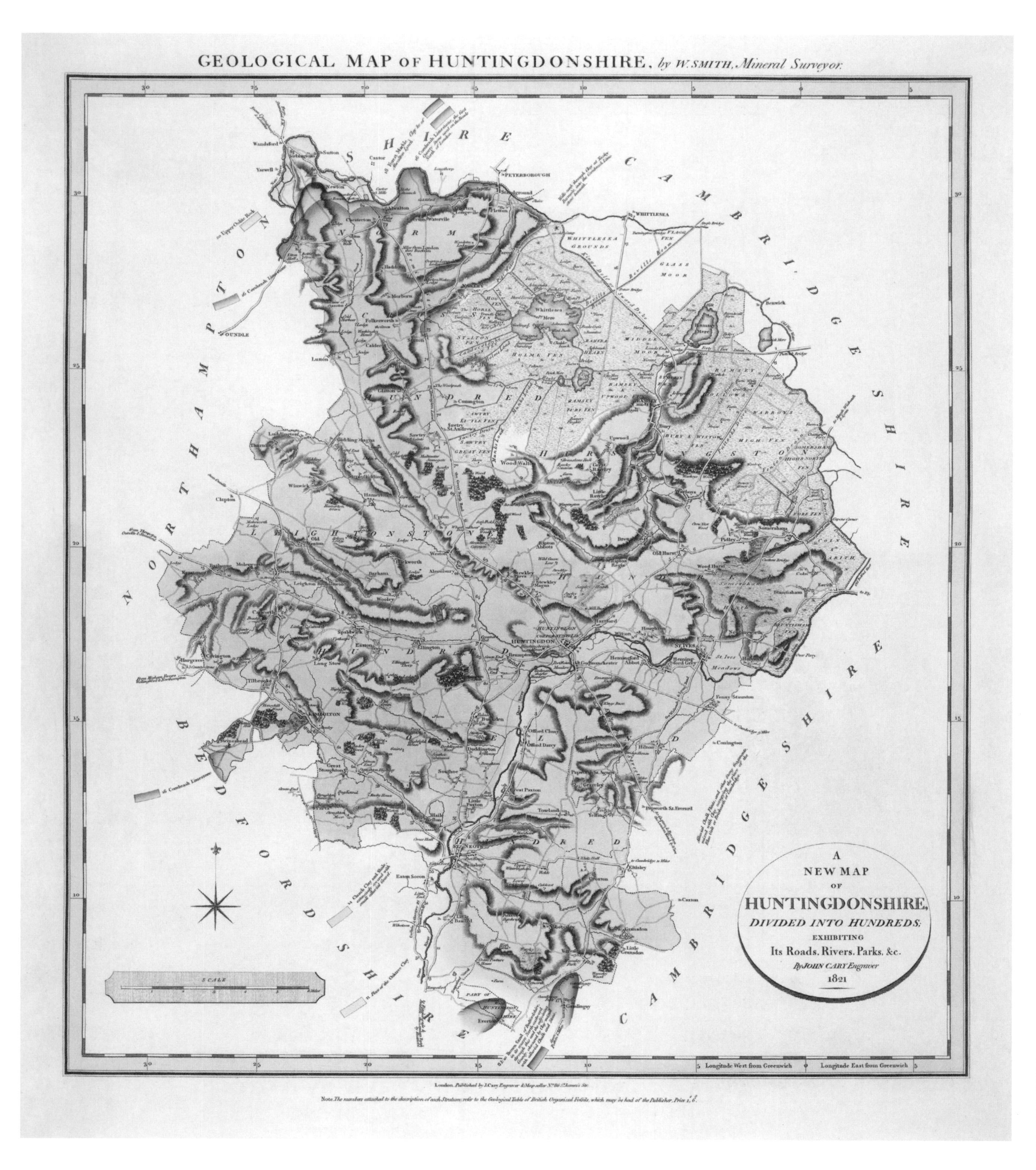

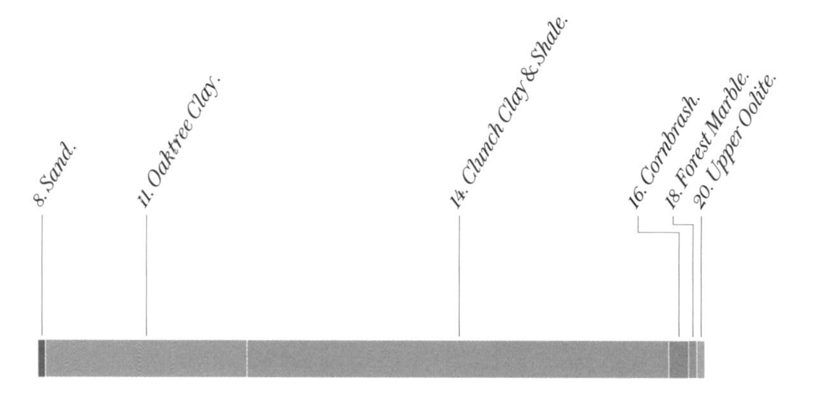

GEOLOGICAL MAP OF HUNTINGDONSHIRE, 1822.

First published in Part V of *A New Geological Atlas of England and Wales* with the counties Nottinghamshire, Leicestershire and Rutlandshire. To the northeast of the county Oaktree Clay (11) and Clunch (14) are overlain by fenland alluvial sediments. Further north Cornbrash (16), Forest Marl (18) and finally Upper Oolite (20) occur.

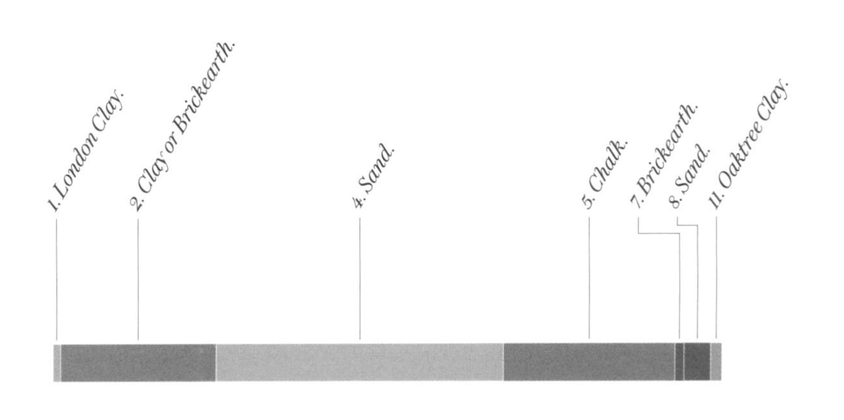

GEOLOGICAL MAP OF NORFOLK, 1819.

First published in Part I of *A New Geological Atlas of England and Wales* with the counties Kent, Wiltshire and Sussex. The highest stratum shown by Smith are two outcrops of London Clay (1) with Brickearth, Sand and Crag (2, 3 and 4) overlying Chalk (5) with flinty nodules in the upper part. The deepest stratum in the county is the Oaktree Clay (11) in the Kings Lynn and Downham area.

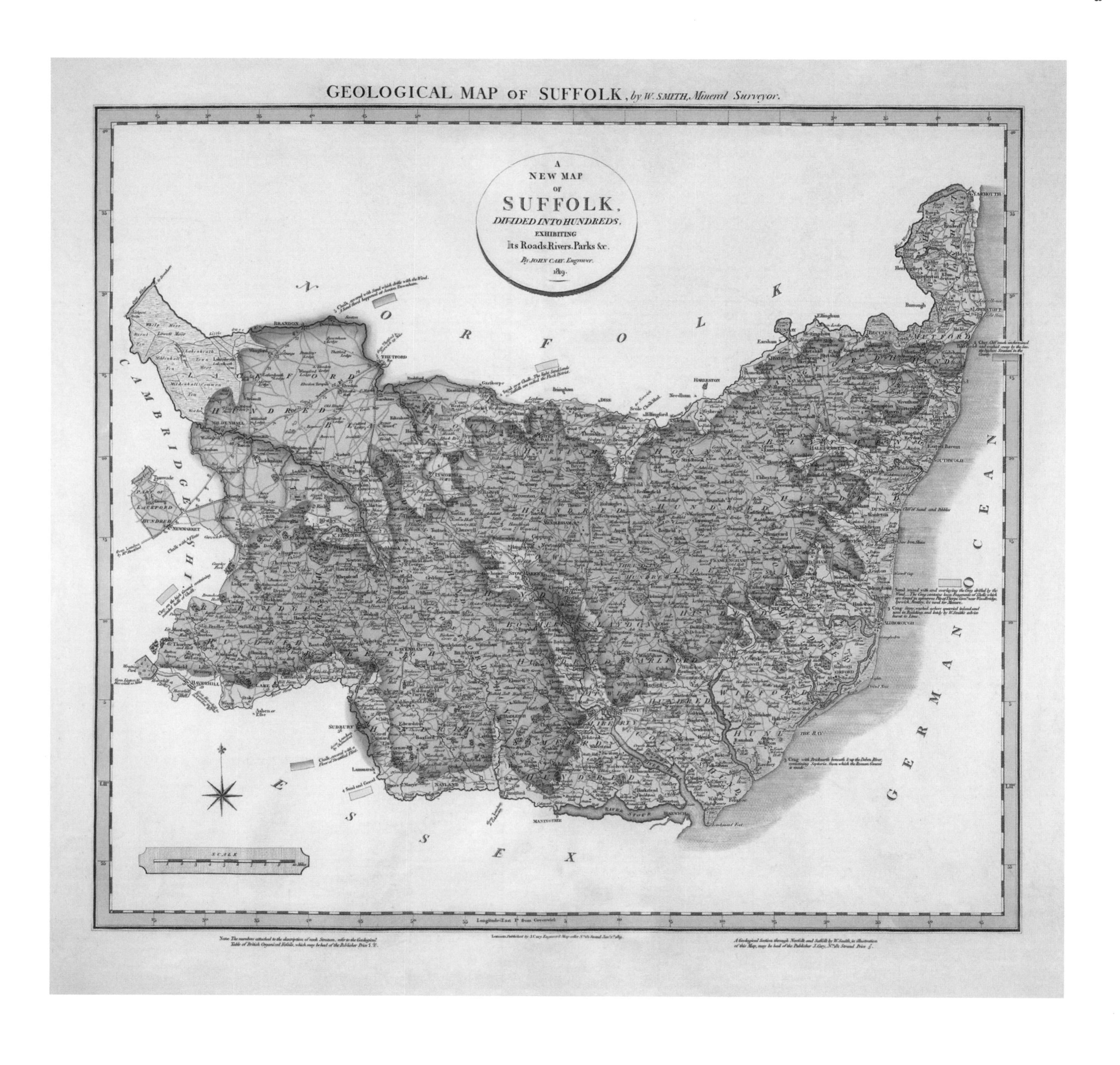

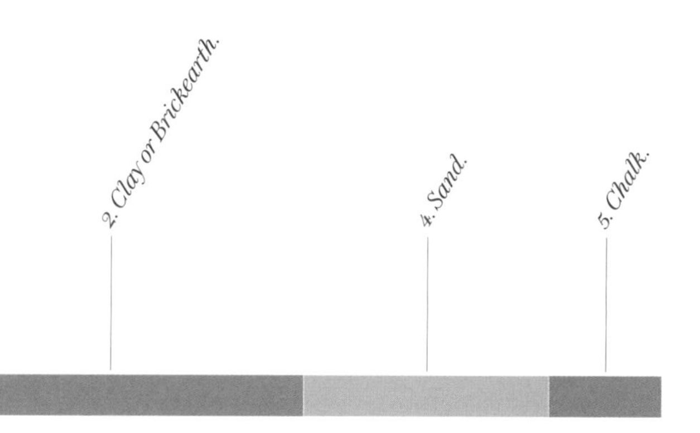

GEOLOGICAL MAP OF SUFFOLK, 1819.

First published in Part II of *A New Geological Atlas of England and Wales* with the counties Gloucester, Berkshire and Surry (Surrey). The highest strata are composed of Brickearth (2) and Sand (4) with pebbly gravel and Crag (3) which overlies Chalk (5), with stratified flint.

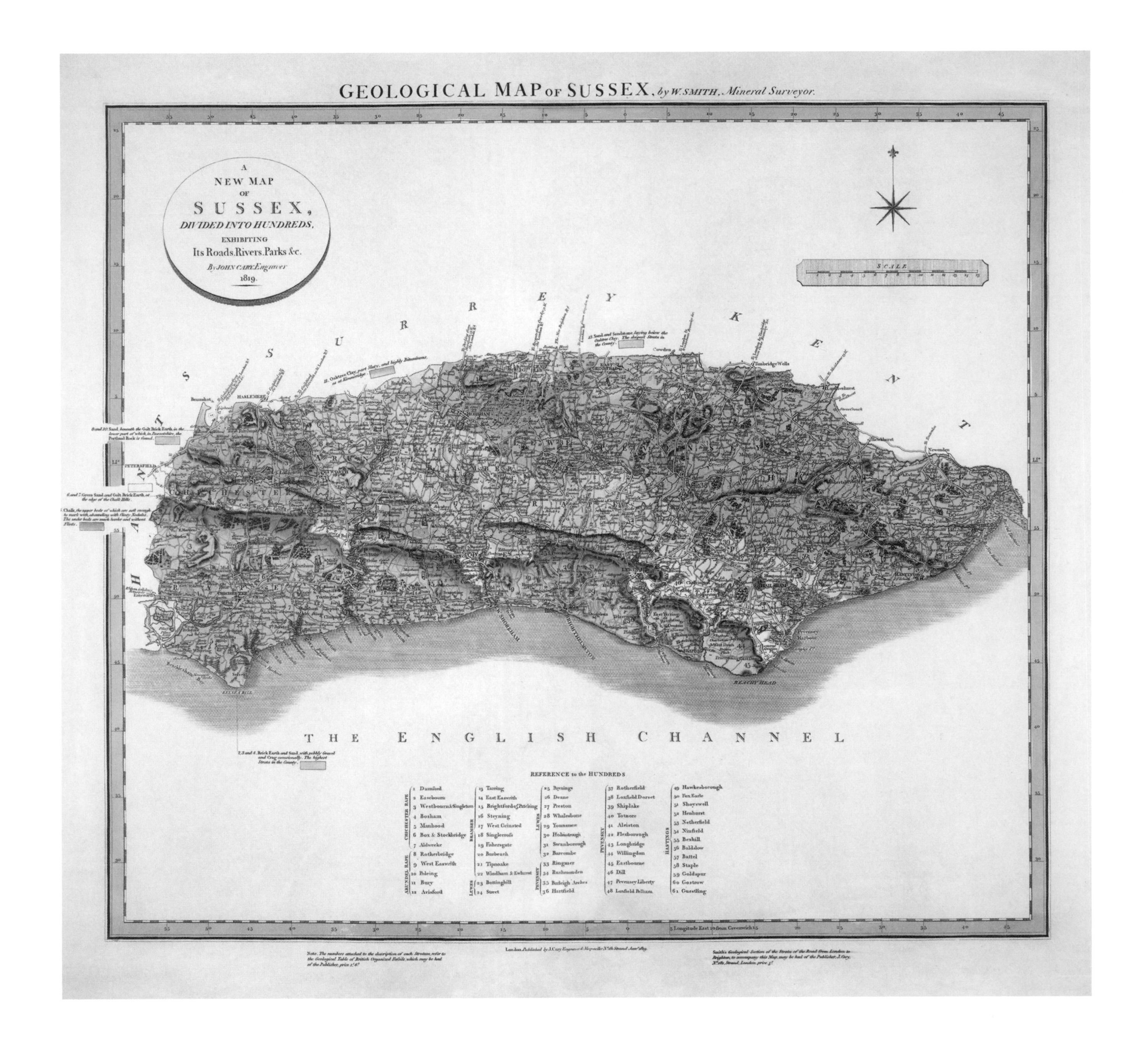

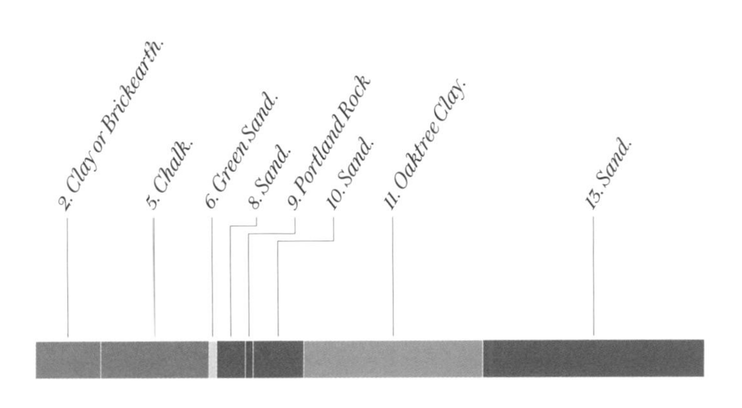

GEOLOGICAL MAP OF SUSSEX, 1819.

First published in Part I of *A New Geological Atlas of England and Wales* with the counties Norfolk, Kent and Wiltshire. Smith's uppermost strata consisted of Sands and Clays (2, 3, 4) within his Crag beneath which were the Chalk (5), Golt brickearth (6, 7). He also showed what he erroneously considered to be Oaktree Clay (11) and Sandstone (13).

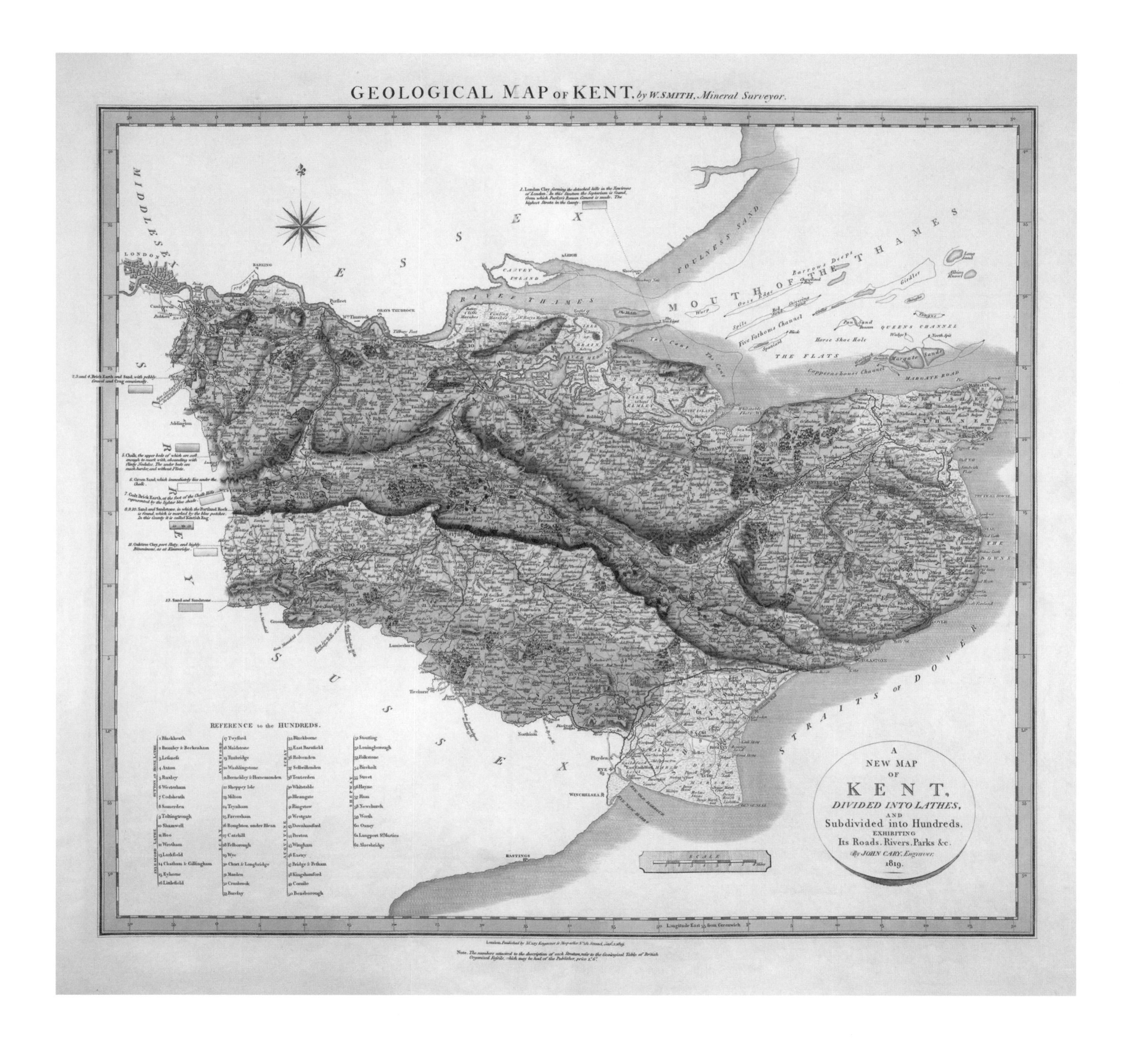

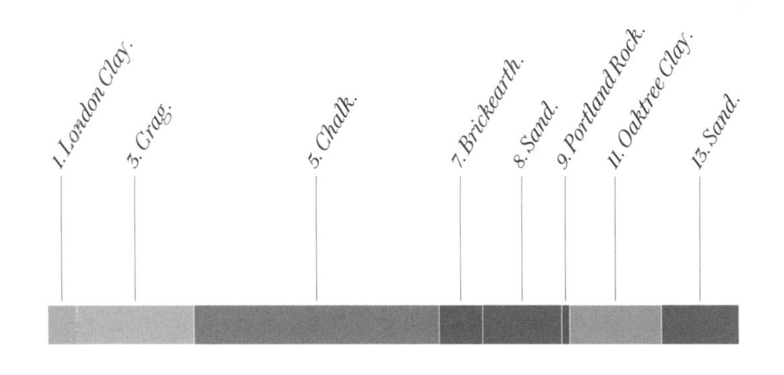

GEOLOGICAL MAP OF KENT, 1819.

First published in Part I of *A New Geological Atlas of England and Wales* with the counties Norfolk, Wiltshire and Sussex. The top stratum is composed of London Clay (1) and the bottom stratum Sand (13) which is overlain by high bituminous Oaktree Clay (11) similar to that seen at Kimmeridge.

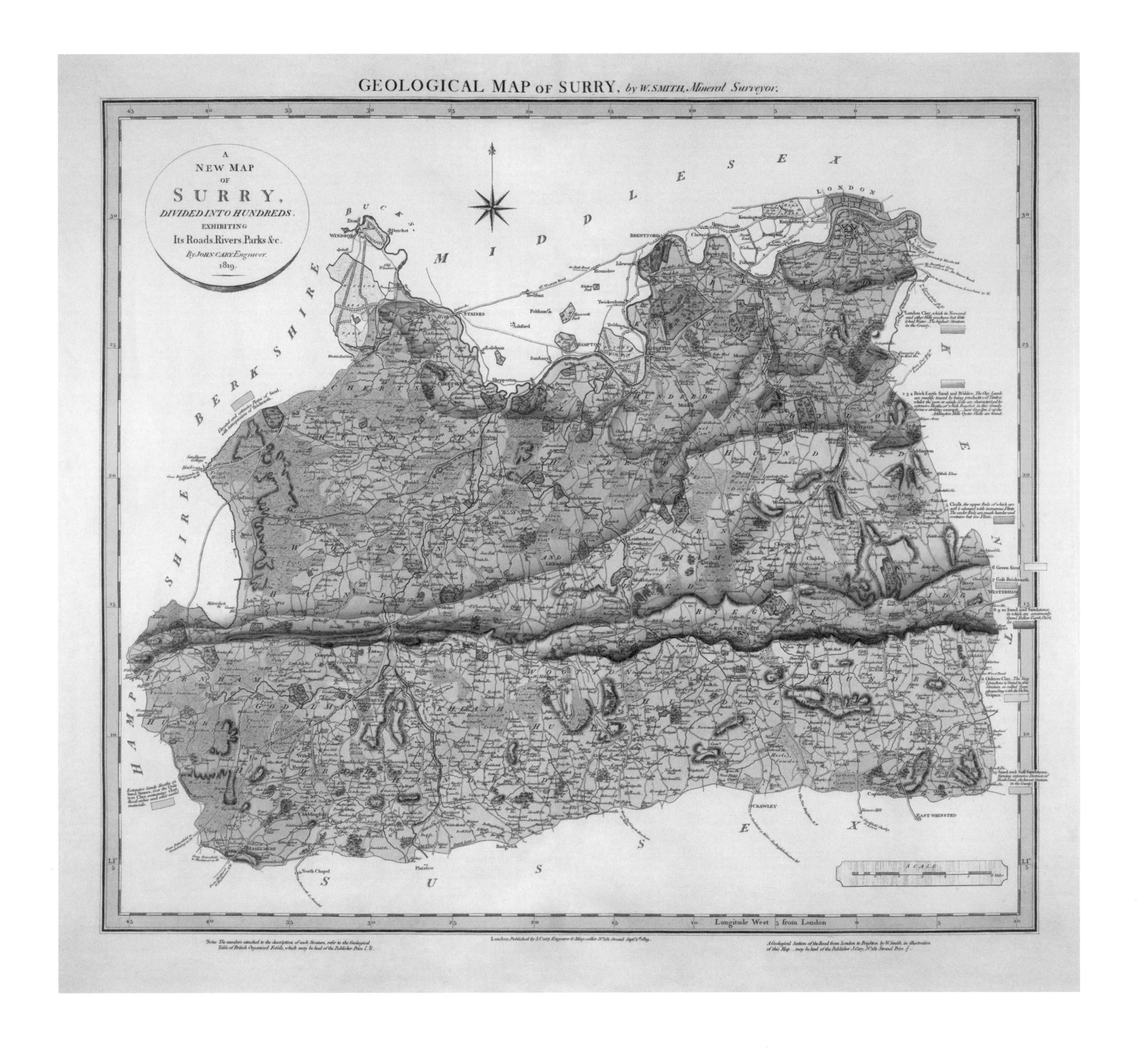

GEOLOGICAL MAP OF SURRY [SURREY], 1819.

First published in Part II of *A New Geological Atlas of England and Wales* with the counties Gloucestershire, Berkshire and Suffolk. The highest stratum is composed of London Clay (1) and the lowest stratum Sand and Sandstone (13) below the Oaktree Clay (11) containing the Sneg Limestone.

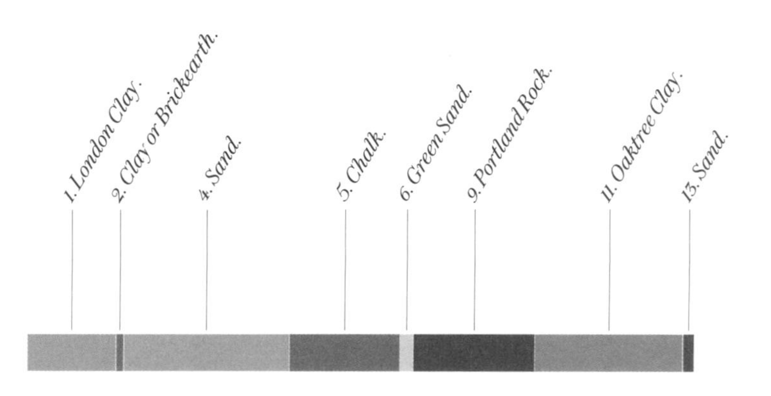

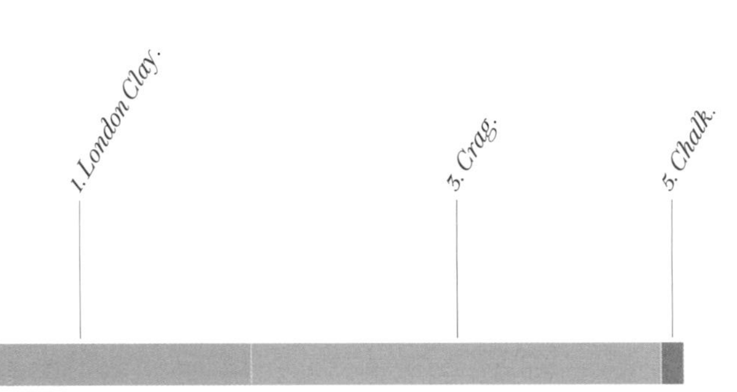

GEOLOGICAL MAP OF ESSEX, 1820.

First published in Part III of *A New Geological Atlas of England and Wales* with the counties Oxfordshire, Buckinghamshire and Bedfordshire. The highest stratum is composed of London Clay (1) and the lowest stratum Chalk (5).

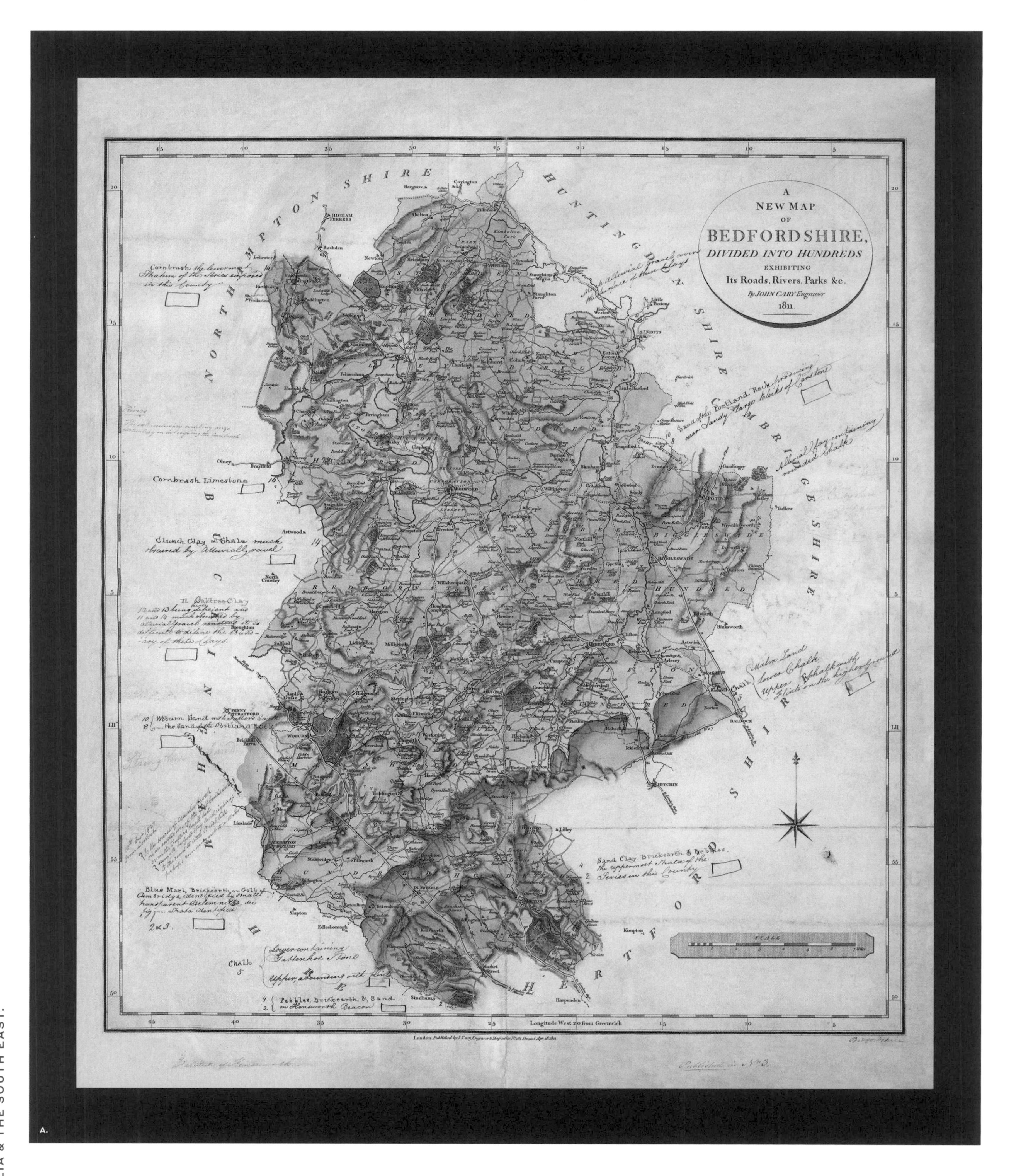

A.

A. UNPUBLISHED GEOLOGICAL MAP OF BEDFORDSHIRE.

A draft map, with Smith's handwritten notes and strata boxes sketched in. This is probably the map from which the published plate was engraved.

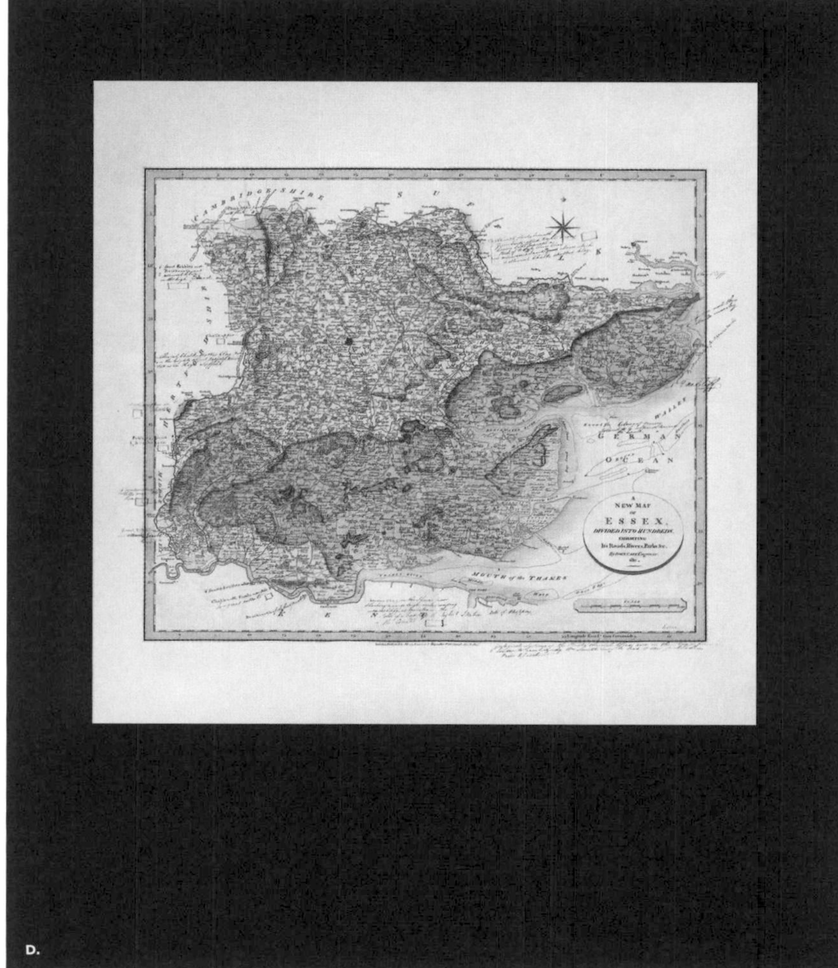

B. UNPUBLISHED GEOLOGICAL MAP OF LINCOLNSHIRE.

A draft map showing some pencil annotation by Smith, but with strata lines, key boxes and text all engraved.

C. UNPUBLISHED GEOLOGICAL MAP OF CAMBRIDGESHIRE.

A draft map extensively annotated by Smith and with some outline geological colouring. Smith never published a completed map of Cambridgeshire.

D. UNPUBLISHED GEOLOGICAL MAP OF ESSEX.

A draft map with some geological notes written in script. This is probably the map from which the published plate was engraved.

E. UNPUBLISHED GEOLOGICAL MAP OF HUNTINGDONSHIRE.

A draft map partly coloured with Smith's handwritten strata notes and with a manuscript title.

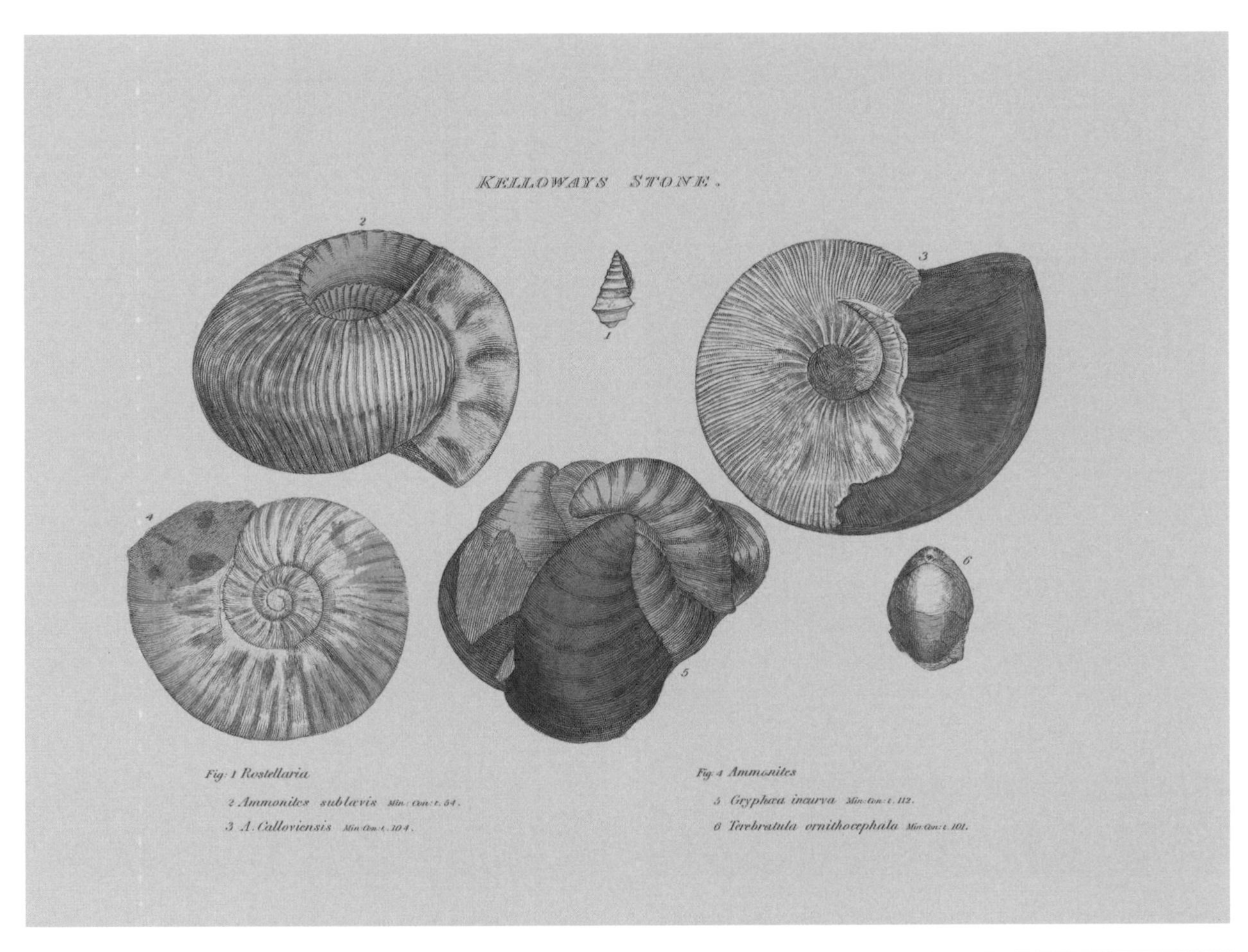

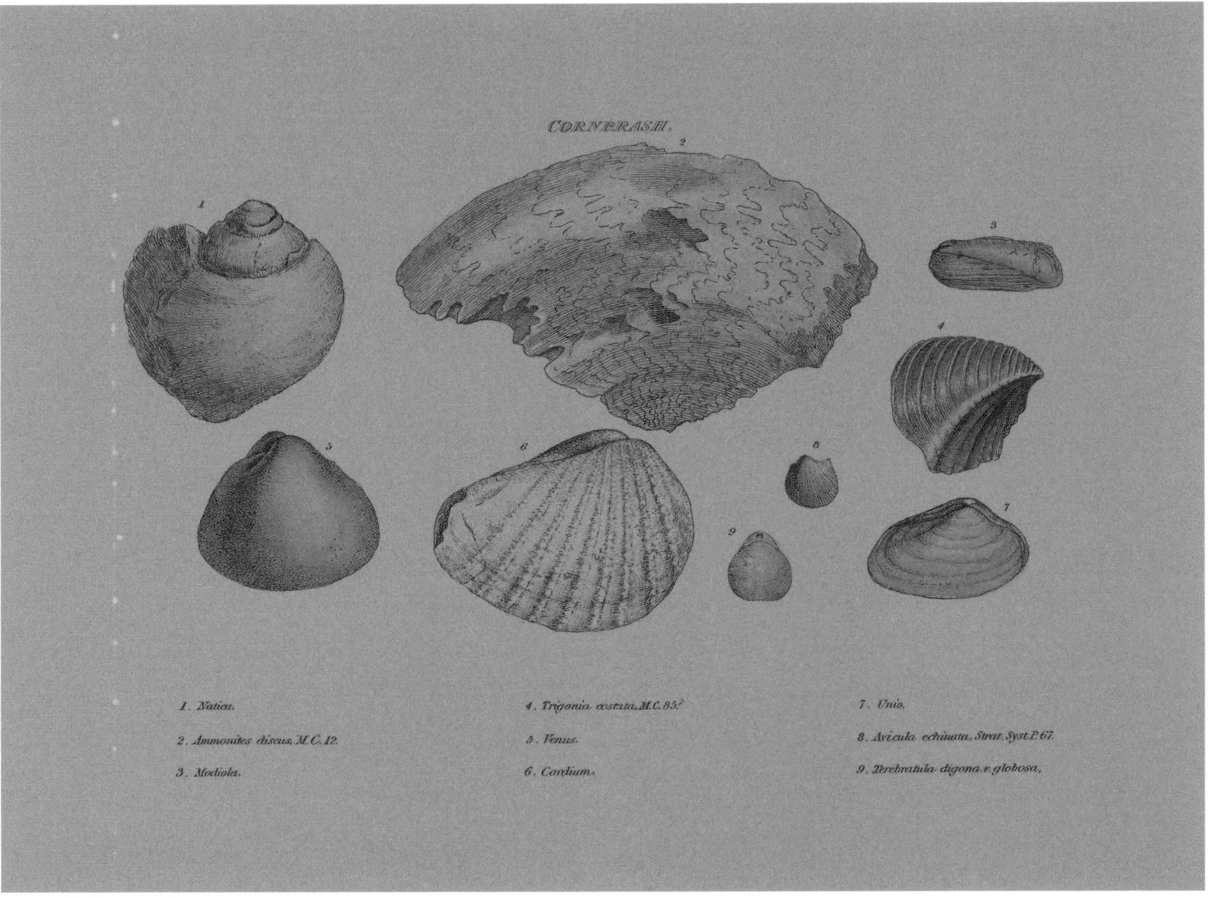

FOSSILS FOUND IN KELLOWAYS STONE [ABOVE].

Smith lists Kelloways as a location for all the fossils illustrated. It is now spelled Kellaways and is the name of a Formation, so Smith's name endures. He could also match his fossils with those found on the Wilts & Berks, Kennet & Avon, and Thames & Severn Canals.

FOSSILS FOUND IN CORNBRASH [BELOW].

The Cornbrash was named by Smith as the youngest member of the 'Stonebrash Hills' that form a band right across England from Dorset to Yorkshire. The named locations run the length of it from Sleaford near the Wash in the north-east to Chillington in Dorset in the south-west. Smith easily recognized it by its productive arable pasture. *Meleagrinella echinata*, **8.**, was named by Smith.

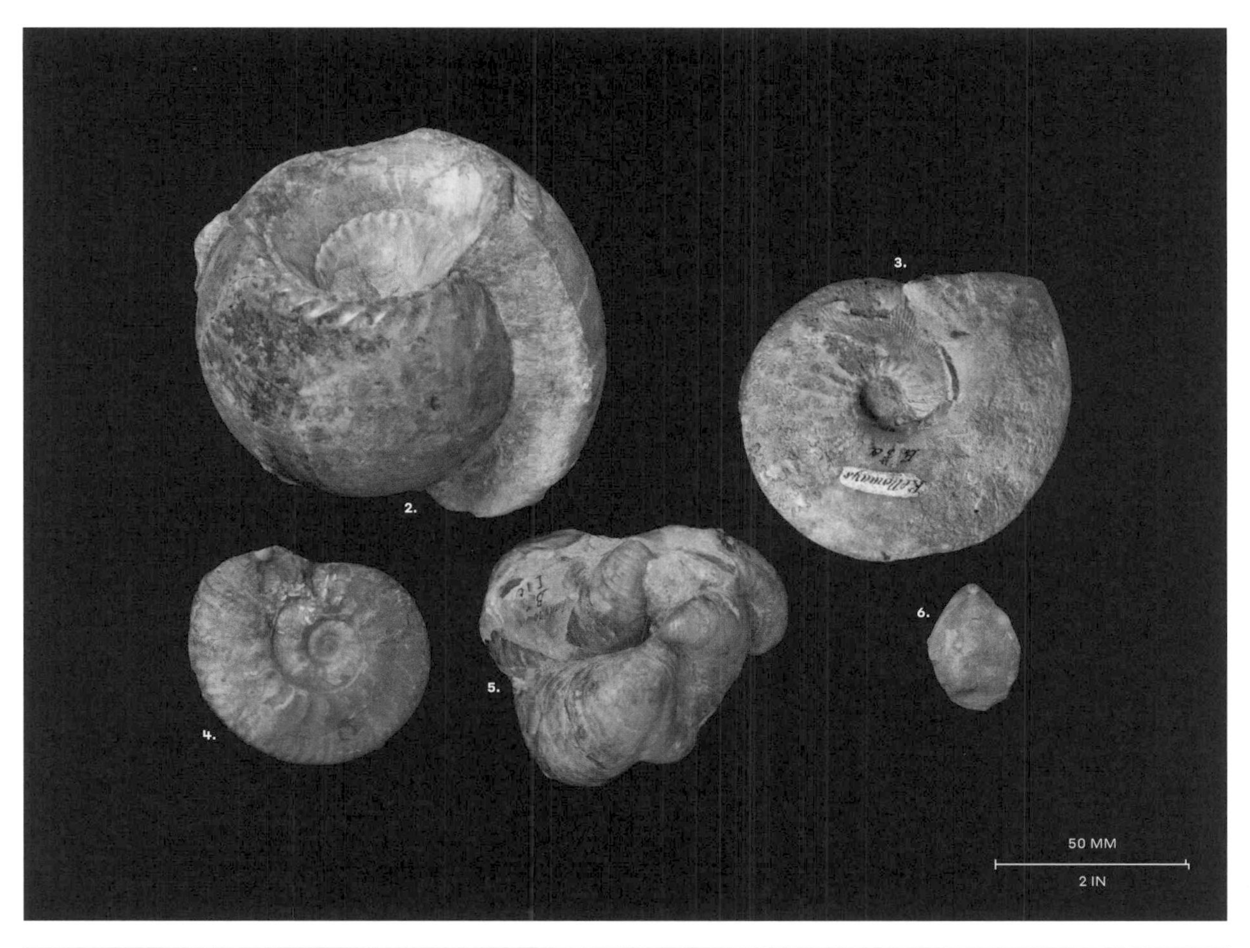

FOSSILS FOUND IN KELLAWAYS ROCK [ABOVE].

2. *Cadoceras sublaevis* C748 Kellaways; **3.** *Sigaloceras calloviense* C642b Kellaways (substitute); **4.** *Proplanulites koenigi* C643 Kellaways; **5.** *Gryphaea dilobotes* L1778 Ladydown; **6.** *Ornithella ornithocephala* B1417 Wilts & Berks Canal, near Chippenham.

FOSSILS FOUND IN CORNBRASH [BELOW].

1. cf. *Ampullospira* sp. G1605 Road; **2.** *Clydoniceras discus* C649 south-west of Wincanton; **3.** *Modiolus imbricatus* L1536 Wick Farm; **4.** *Trigonia crucis* L1551 north side of Wincanton; **5.** *Protocardia buckmani* L1556 Trowle; **6.** *Pholadomya deltoidea* L1552 Road; **7.** *Pleuromya uniformis* L1543 Road; **8.** *Meleagrinella echinata* L1579 Draycot; **9.** *Digonella siddingtonensis* B1423 Latton.

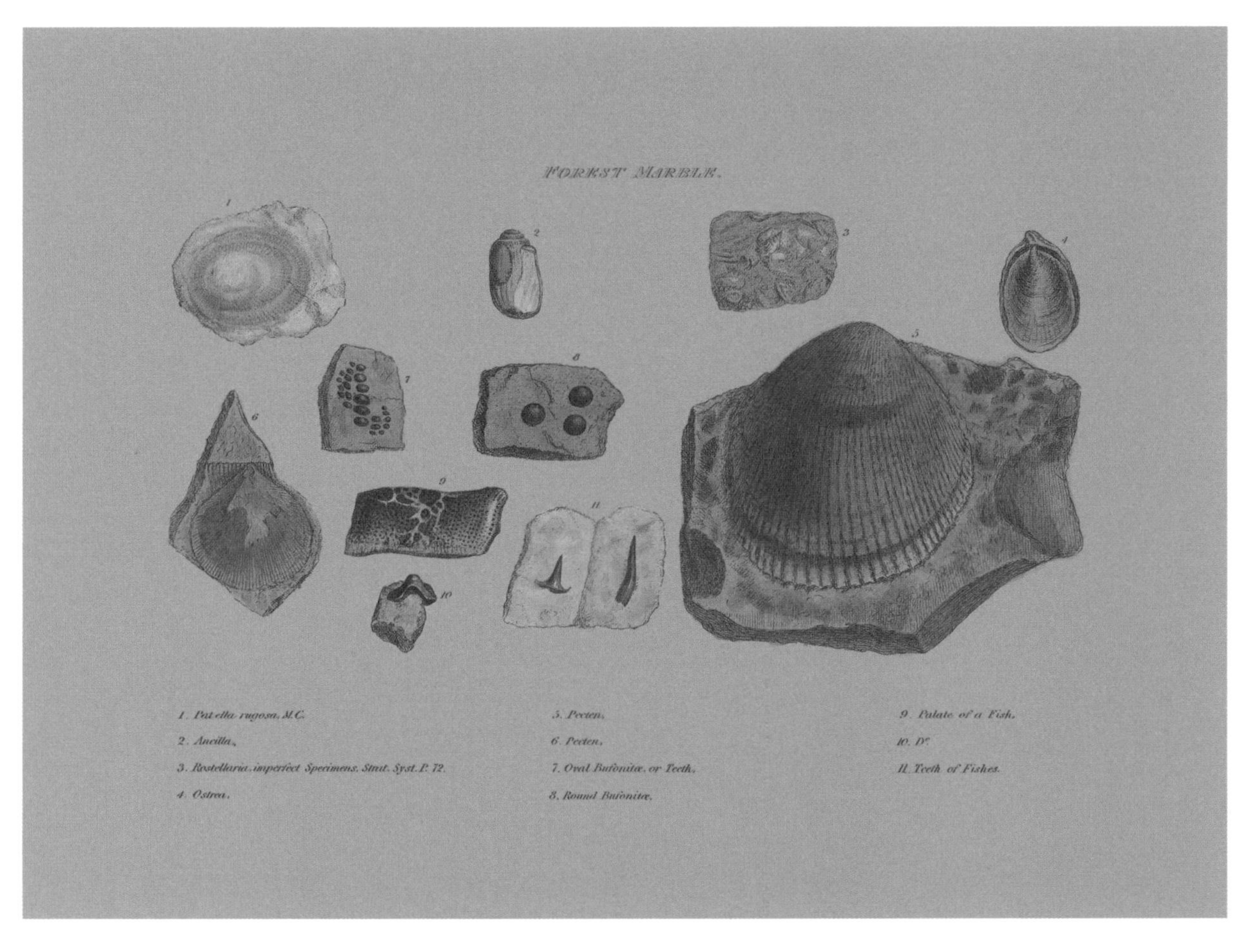

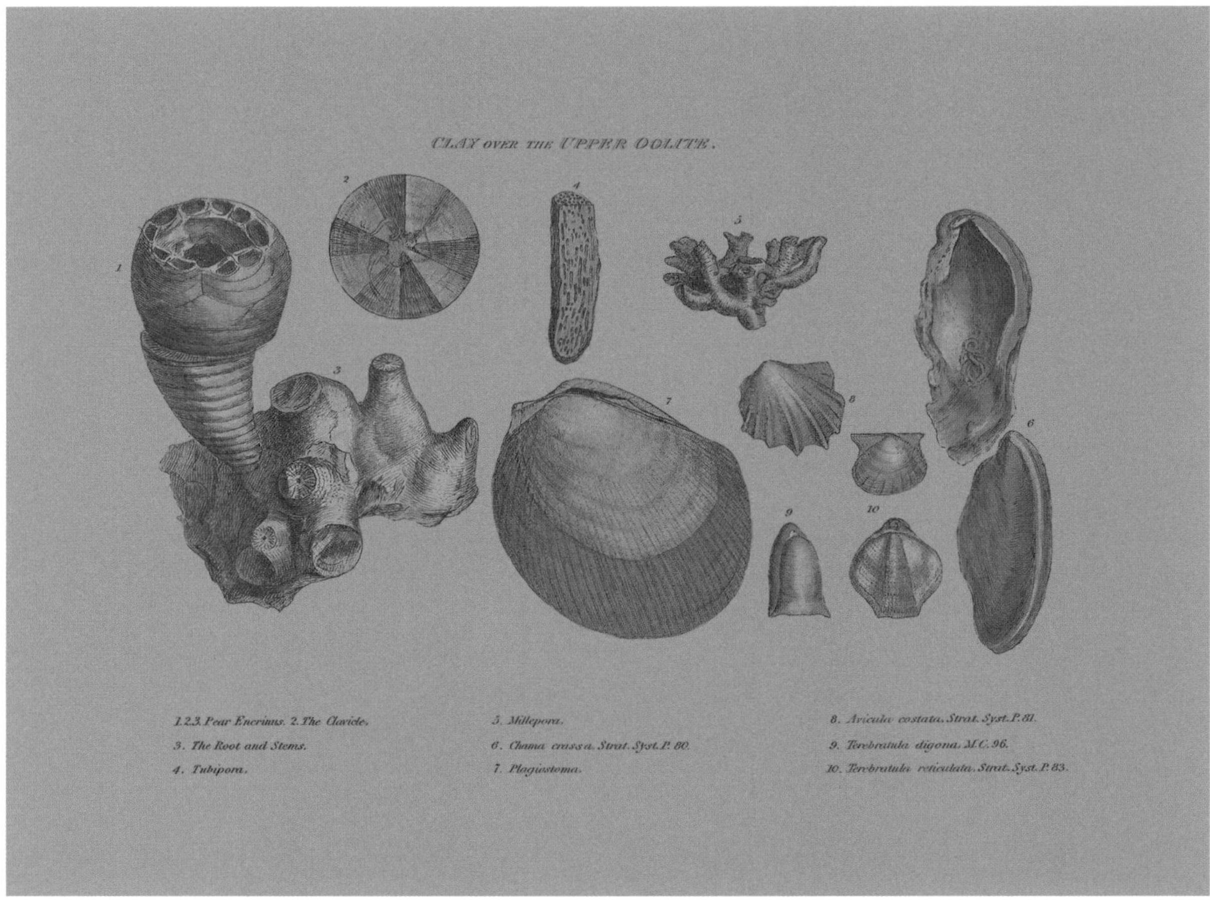

FOSSILS FOUND IN FOREST MARBLE [ABOVE].

As the name suggests, stone from the Forest Marble takes a fine polish and was used as an ornamental stone for such things as marble chimney pieces and as slabs. Smith notes that whole specimens are rare and extracted with great difficulty. Bones, teeth and wood are some of its most characteristic identifications. The quarrymen termed **7.** 'bird's eyes'.

FOSSILS FOUND IN CLAY OVER THE UPPER OOLITE [BELOW].

The clay over the Upper Oolite is now known as the Bradford Clay. Smith says that coming from the clay the fossils lie loose and are easily cleaned. Articulated crinoids are very rare and the fossils illustrated (**1–3.**) are fine examples. The bryozoan, **5.**, is the only example illustrated by Smith.

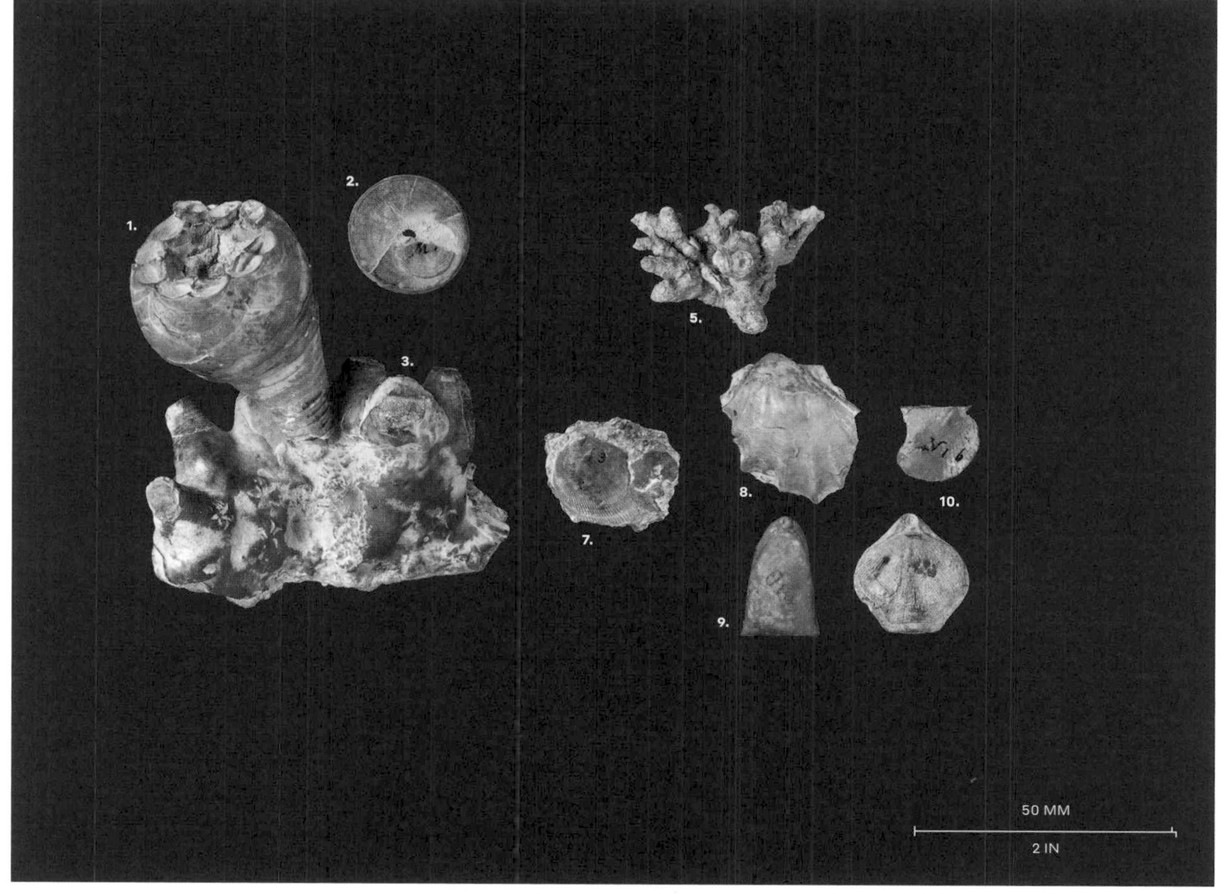

■ **FOSSILS FOUND IN FOREST MARBLE [ABOVE].**

2. *Cylindrites archiaci* G1608 Farley; **3.** Rostellaria in block L1596 (no location, possibly Poulton); **4.** *Catinula* sp. L1589 Wincanton; **5.** *Plagiostoma subcardiiformis* L1591 Siddington; **6.** *Camptonectes auritus* L1594 Farley; **7.** ?*Eomesodon* sp. P4826 Stonesfield; **9.** ?*Asteracanthus magnus* P4820 Pickwick; **10.** *Asteracanthus tenuis* P4823 Pickwick.

■ **FOSSILS FOUND IN BRADFORD CLAY [BELOW].**

1–3. *Apiocrinites elegans* E559 Bradford; **5.** *Terebellaria ramosissima* D34537 Hinton; **7.** *Plagiostoma cardiiformis* L1605 Combhay (substitute); **8.** *Oxytoma costata* a) 43258 Bradford b) L1611 Hinton; **9.** *Digonella digona* B1424 Farley; **10.** *Dictyothyris coarctata* B1430 Farley.

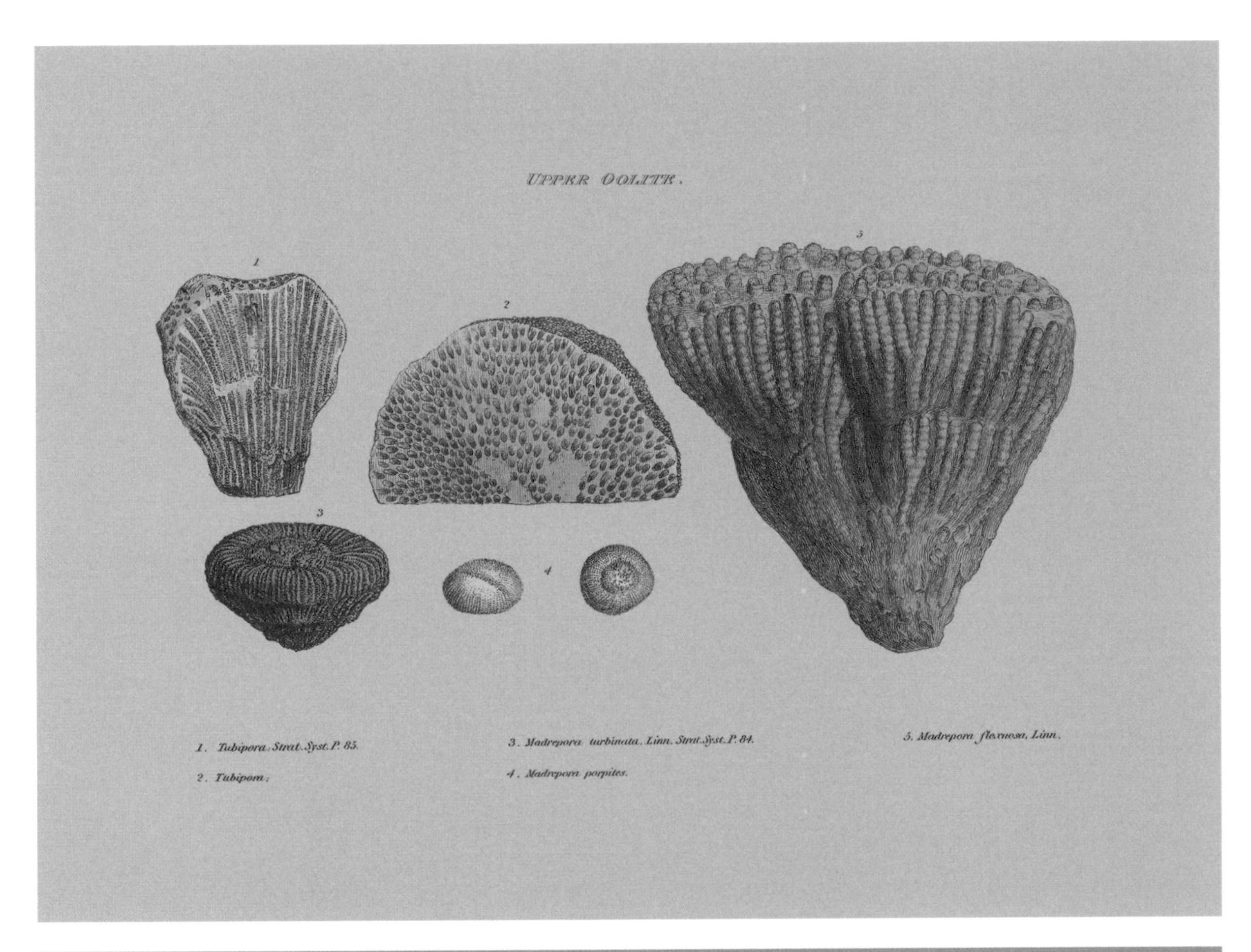

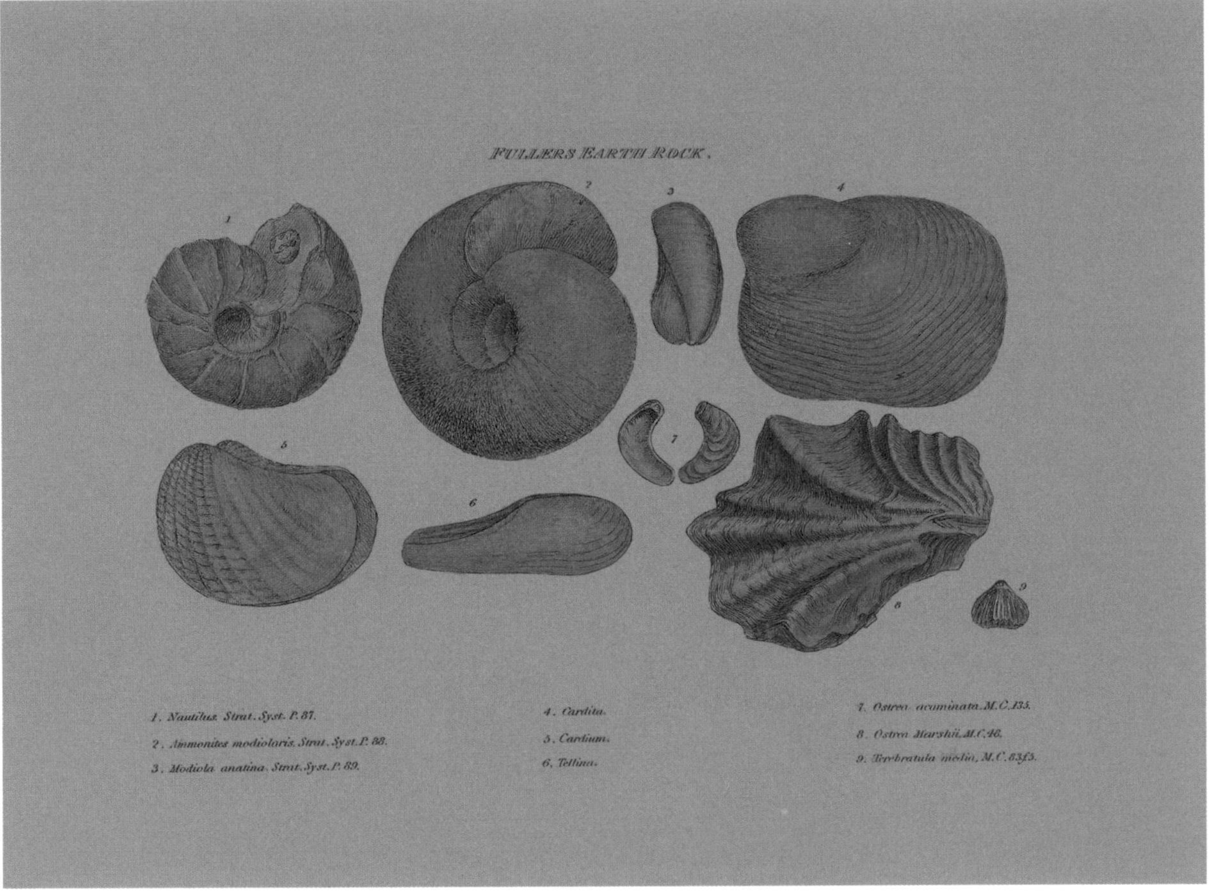

FOSSILS FOUND IN UPPER OOLITE [ABOVE].

Stone from the oolite will be familiar to many as the warm honey-coloured limestone of the Cotswolds where its use in building and walling is ubiquitous. It is mostly a freestone composed of round calcite ooids, precipitated from sea water without obvious whole fossils, but in certain horizons fossils are numerous.

FOSSILS FOUND IN FULLER'S EARTH ROCK [BELOW].

Fuller's Earth is the name given to a clay that expands with water. It was used in cleaning sheep (fulling) and is found throughout Smith's strata. The name still exists for this particular layer. Unfortunately, the distinctive oyster *Actinostreon marshii* from the Cotswold Hills is quite wide ranging; it is this assemblage that Smith recognized.

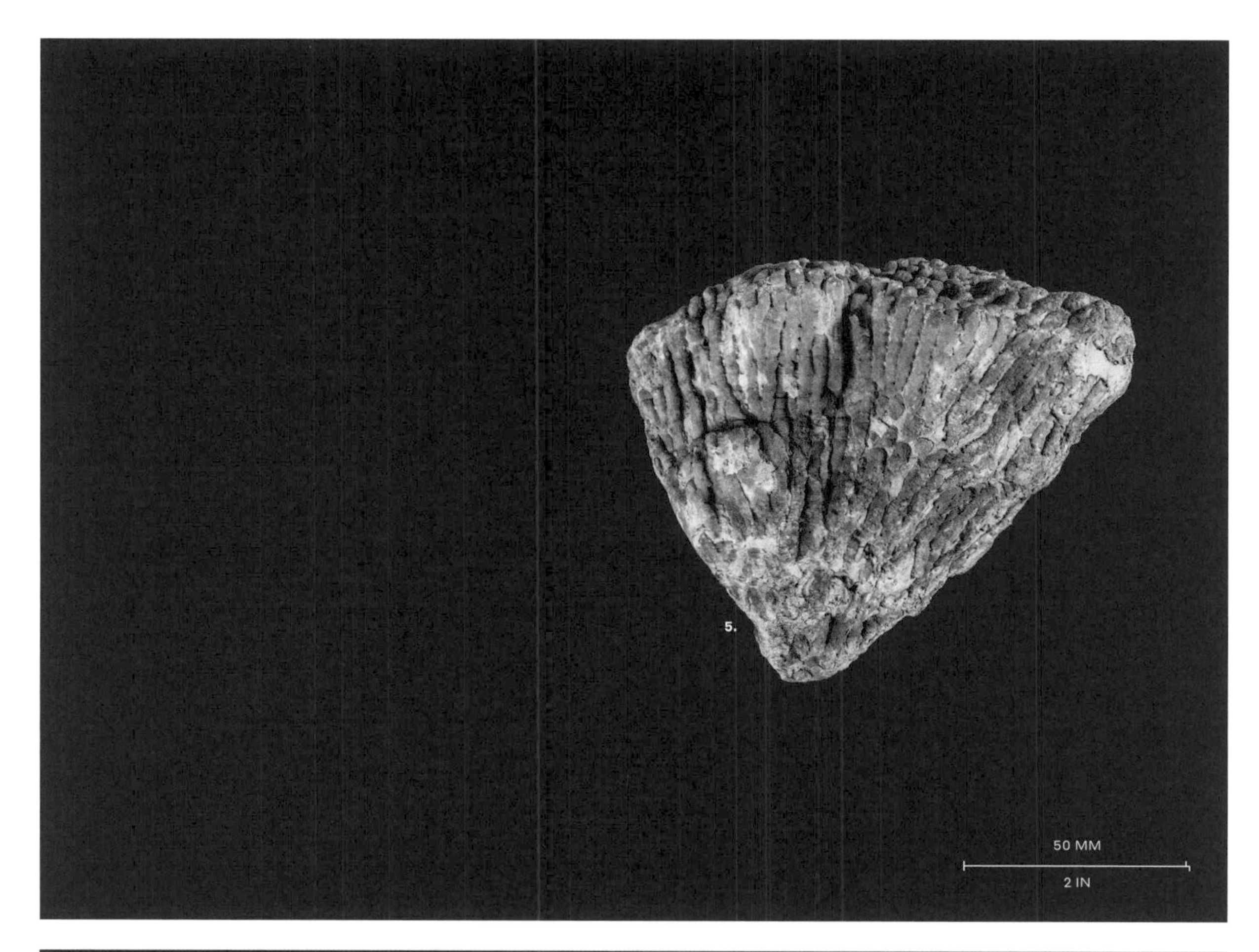

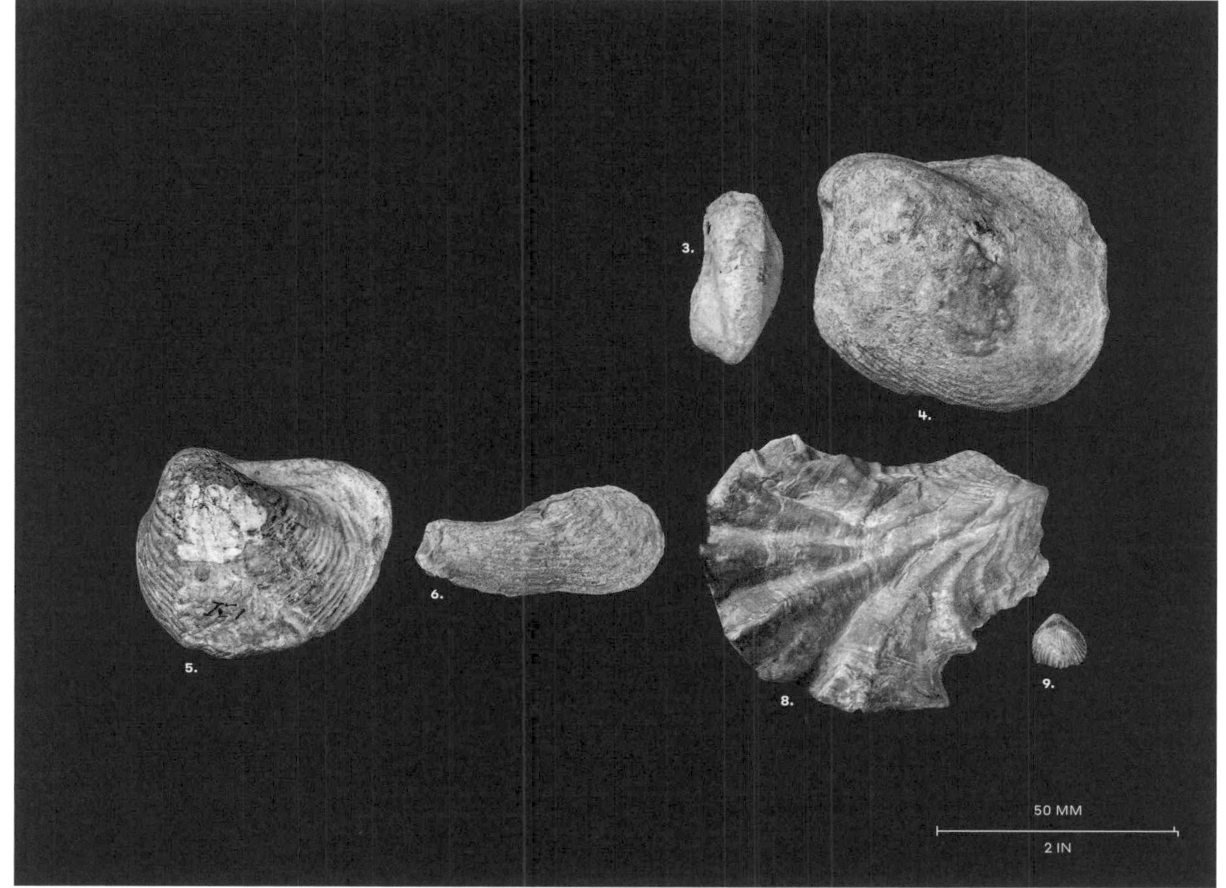

FOSSILS FOUND IN GREAT OOLITE [ABOVE].

5. *Cladophyllia babeana* B55 Castle Coombe.

FOSSILS FOUND IN FULLER'S EARTH [BELOW].

3. *Modiolus anatinus* 66930 Avoncliff; **4.** *Ceratomya* aff. *striata* L53451 Hardington; **5.** *Pholadomya* aff. *lirata* L1685 Charlton Horethorn; **6.** *Cercomya* aff. *pinguis* L1689 Avoncliff; **8.** *Actinostreon marshii* LL40856 Cotswold Hills; **9.** *Rhynchonelloidella media* B1488 no location (substitute).

IV. CARTOGRAPHER.

A short description of the protracted and meticulous preparation of Smith's geological map of England and Wales with assistance from the pre-eminent cartographer-publisher John Cary and the generous patronage of Sir Joseph Banks; a singular method of colouring; rivalry with George Bellas Greenough; pleasing geological cross sections; the great unfinished county map project.

On 22 May 1815, William Smith (1769–1839) received the long-awaited, first completed copy of his geological map of England and Wales. Comprising fifteen printed sheets, each measuring 635 by 542 mm (25 by 21 in.), on a scale of 5 miles to the inch, hand-coloured over the course of seven or eight days by a professional map colourist called Rogerson, supervised by Smith, and mounted on canvas and rollers, this was an impressively large map which measured, overall, some 2.5 by 1.7 m (8 ft 6 in. by 6 ft). It was the culmination of fifteen long years of work which had seen many delays and setbacks. The next day Smith exhibited the map to the Board of Agriculture and, a week later, to the Society of Arts at its premises in John [Adam] Street a few minutes' walk from Smith's home at 15 Buckingham Street off the Strand in central London. During the week he may also have taken the map to the Royal Institution on Albemarle Street off Piccadilly.

Smith had shown an earlier, probably incomplete, version of the map to the Society of Arts on 8 February 1815 and, supported by testimonials from his friends Benjamin Richardson (1758–1832) and Joseph Townsend (1739–1816), Smith claimed, and was awarded in June 1815, the prize of 50 guineas which the Society had been offering since 1802 for a mineralogical map of England and Wales on a scale of not less than 10 miles to the inch. In fact, Smith had written to the Society of Arts as early as July 1804 to ask if they would consider giving him the award then as a contribution towards his proposed publication.

The full title of Smith's 1815 map, *A Delineation of the Strata of England and Wales, with part of Scotland: exhibiting the collieries and mines, the marshes and fen lands originally overflowed by the sea, and the varieties of soil according to the variations in the substrata*, demonstrates what he saw as the great value of his map: it was a practical tool for mineral exploration, land drainage and agriculture (Sharpe, 2015).

In Bath in 1799, after his dismissal from the Somersetshire Coal Canal Company, Smith had been encouraged by Richardson and Townsend to publish what he had discovered about the regularity of the sequence of strata in England and his idea of the use of fossils to indicate stratigraphical position and correlation. On 11 June 1799 at Townsend's house at 27 Great Pulteney Street, Bath, Smith had dictated his *Table of Strata* to Townsend and Richardson and copies were soon in circulation, but his friends continued to push Smith to publish more formally. To this end, on 1 June 1801, Smith issued *Prospectus of a Work, entitled, Accurate Delineations and Descriptions of the Natural Order of the Various Strata that are Found in Different Parts of England and Wales:*

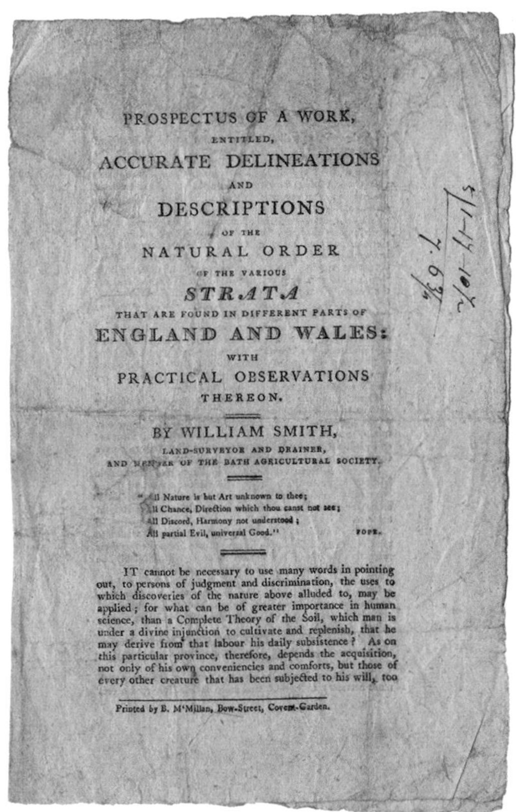

PROSPECTUS OF A WORK,
ENTITLED,
ACCURATE DELINEATIONS
AND
DESCRIPTIONS
OF THE
NATURAL ORDER
OF THE VARIOUS
STRATA
THAT ARE FOUND IN DIFFERENT PARTS OF
ENGLAND AND WALES:
WITH
PRACTICAL OBSERVATIONS
THEREON.

BY WILLIAM SMITH,
LAND-SURVEYOR AND DRAINER,
AND MEMBER OF THE BATH AGRICULTURAL SOCIETY.

"All Nature is but Art unknown to thee;
All Chance, Direction which thou canst not see;
All Discord, Harmony not understood;
All partial Evil, universal Good." POPE.

IT cannot be necessary to use many words in pointing out, to persons of judgment and discrimination, the uses to which discoveries of the nature above alluded to, may be applied; for what can be of greater importance in human science, than a Complete Theory of the Soil, which man is under a divine injunction to cultivate and replenish, that he may derive from that labour his daily subsistence? As on this particular province, therefore, depends the acquisition, not only of his own conveniencies and comforts, but those of every other creature that has been subjected to his will, too

Printed by B. M'Millan, Bow-Street, Covent-Garden.

FIG. 1.

Fig. 1. A four page octavo *Prospectus* issued by Smith and dated at 'Midford, near Bath, June 1st 1801'. It outlines his plan for a one-volume quarto work on *Accurate Delineations and Descriptions of the Natural Order of the Various Strata*, 'Price, to subscribers, Two Guineas ... which will be published in November next.' As a result of the bankruptcy of his publisher, John Debrett, the plan described in the prospectus was never realized.

with Practical Observations Thereon. By William Smith, Land-Surveyor and Drainer, and Member of the Bath Agricultural Society. This was a four-page pamphlet calling for subscriptions for a proposed one-volume quarto work accompanied by a map and section to be published 'in November next' at a cost of 2 guineas.

Immediately, however, the project foundered when his intended publisher, John Debrett (1753–1822), was declared bankrupt in 1801 (and again in 1804) (Torrens, 2016). Smith then changed his plans, and instead of publishing a book with a small map he would publish a large, separate map with a small explanatory book.

Over the course of the next ten years, Smith took every opportunity to promote and explain the practical utility of his ideas on strata and fossils and to exhibit drafts of his map at events such as agricultural fairs, earning himself in the process the sobriquet 'Strata' Smith. By this means, he garnered the support of the gentry, nobility, industrialists and agriculturalists, among them Thomas William Coke (1754–1842), the progressive agriculturist of Holkham Hall, Norfolk; the ironmaster Richard Crawshay (1739–1810) of Merthyr Tydfil in South Wales; and both Francis Russell (1765–1802), 5th Duke of Bedford, and John Russell (1766–1839), the 6th Duke, as well as Sir Joseph Banks (1743–1820), President of the Royal Society. At the Woburn Sheep Shearing hosted by the Duke of Bedford in June 1804, Banks contributed £50 towards Smith's map. He was to be Smith's most notable and influential supporter, and sponsor, and it is to Banks that the final, published map and *Memoir* is gratefully dedicated.

Through Banks, Smith's ideas even reached the continent of Europe. In addition to inviting visitors to dinner at the Royal Society, Banks hosted Thursday morning breakfasts and Sunday evening *conversaziones* at his home in Soho Square, where savants and literati could drop by and discuss the latest developments in art and science. During the brief Peace of Amiens (1802–3) during hostilities between France and Britain, the French mineralogist Alexandre Brongniart (1770–1847) and his father-in-law Charles Étienne Coquebert de Montbret (1755–1831), who had taught at the École des Mines in Paris and edited the *Annales des Mines*, visited England. Coquebert dined with Banks several times and although it is not clear whether Banks and Brongniart met, they were certainly in correspondence. At the time, Banks had recently learnt of Smith's discoveries of the value of fossils as stratigraphical identifiers, and it seems unlikely that he would have omitted to mention them to his French visitors. Soon after, in 1803 or 1804 and with war having resumed, Brongniart began work on mapping the geology of the Paris Basin in collaboration with Georges Cuvier (1769–1832), using fossils to distinguish the strata. Their map, *Carte géognostique des environs de Paris*, was published in 1811 (Eyles, 1985; Rudwick, 1996).

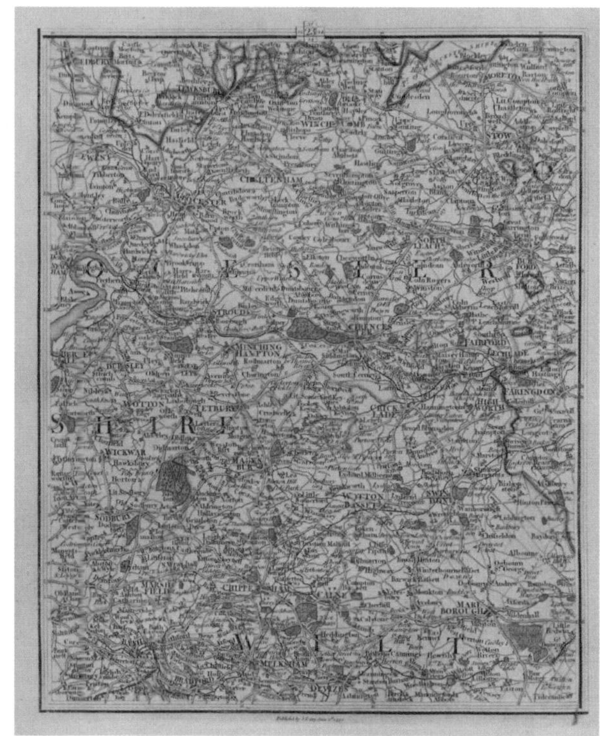

FIG. 2.

Fig. 2.
Cary's New Map of England and Wales, With Part of Scotland, sheet 23, 1794. John Cary's 1794 map provided the topographic base for Smith's 1815 geological map, but much of the detail of roads and settlements – as seen on this map of Smith's home ground of the Cotswolds between Bath and Stow-on-the-Wold – had to be removed to avoid obscuring the geological boundaries.

PUBLICATION AT LAST.

It was not until late in 1812 that Smith eventually found a publisher for his map – perhaps the leading British cartographer of the time, John Cary senior (1754–1835), whose premises were in the Strand, close to Smith's London house. Both Cary and Smith had worked together before on the publication of the plans for the Somersetshire Coal Canal. Cary brought to the project an ideal topographic base map on to which Smith's geological data could be superimposed. This was *Cary's New Map of England and Wales, with Part of Scotland* published in 1794 on a scale of 5 miles to the inch. It required some modification, however, and the map was redrawn as Smith and Cary removed much extraneous topographic detail so that the clarity of the geological boundaries would not be obscured. With an eye, no doubt, to potential purchasers, the map retained the seats and country estates of the gentry and aristocracy.

Work began on engraving the fifteen large copper plates for Smith's great map in January 1813, and although progress was rapid at first, with three plates topographically engraved within a month, the labour continued for two years.

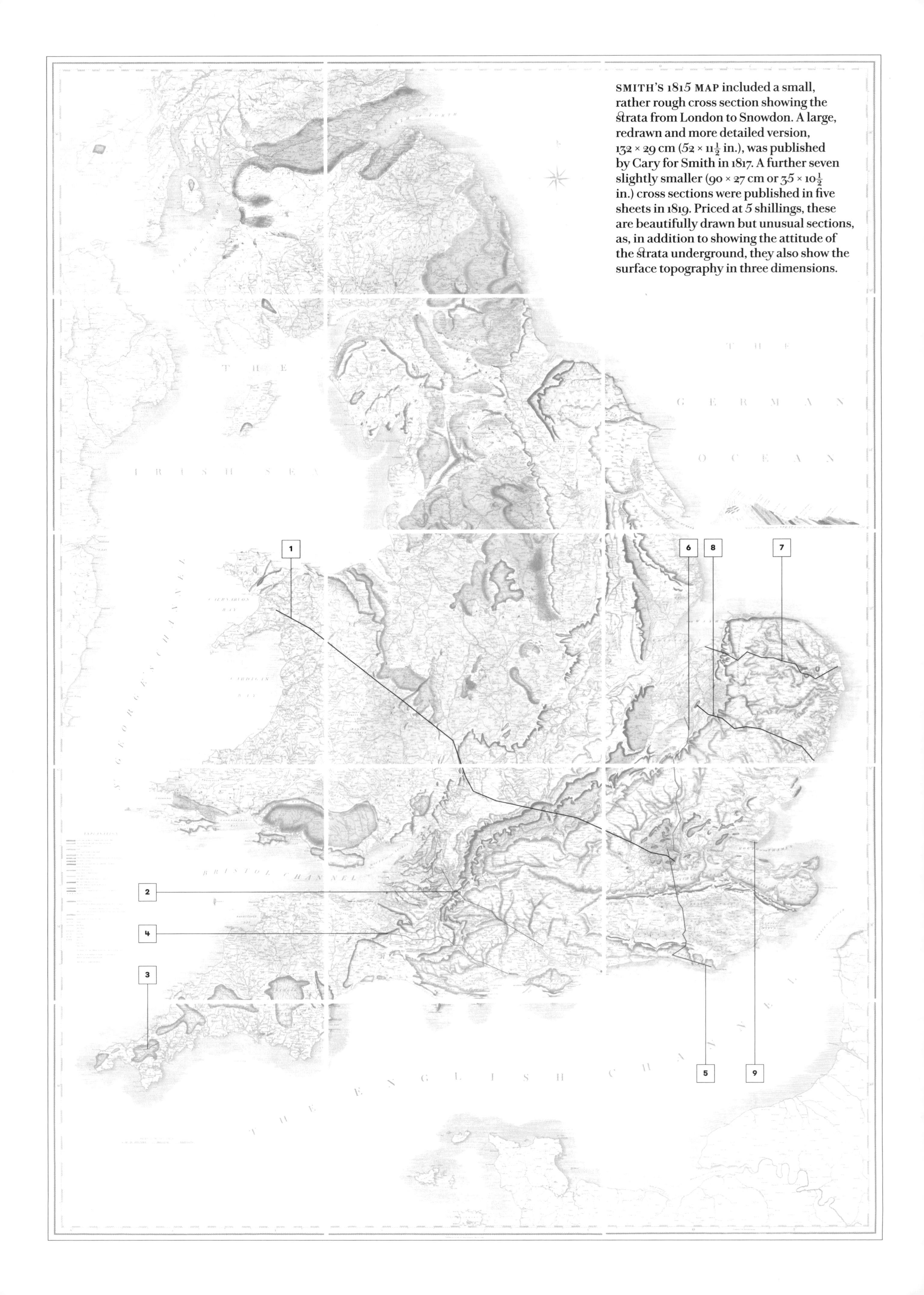

SMITH'S 1815 MAP included a small, rather rough cross section showing the strata from London to Snowdon. A large, redrawn and more detailed version, 132 × 29 cm (52 × 11½ in.), was published by Cary for Smith in 1817. A further seven slightly smaller (90 × 27 cm or 35 × 10½ in.) cross sections were published in five sheets in 1819. Priced at 5 shillings, these are beautifully drawn but unusual sections, as, in addition to showing the attitude of the strata underground, they also show the surface topography in three dimensions.

1. **LONDON TO SNOWDON.**
William Smith. 1817.

A detailed and enlarged version of the rough section on the 1815 map, showing the south-eastward dip of the strata away from the rocks of Snowdonia and mid-Wales, through the Cotswolds and Chalk escarpments to the London Basin.

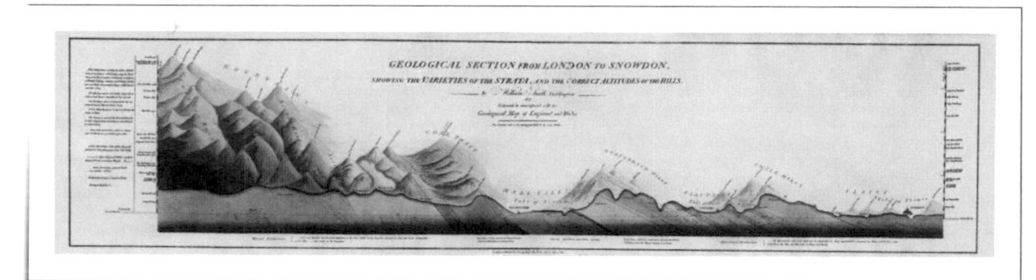

2. **BATH TO SOUTHAMPTON.**
William Smith. 1819.

A section stretching from the deep combes of Smith's home ground around Bath and Midford in Somerset through the Upper Oolite and Chalk escarpments to Salisbury, Wiltshire, and the overlying Brickearth around Southampton, Hampshire.

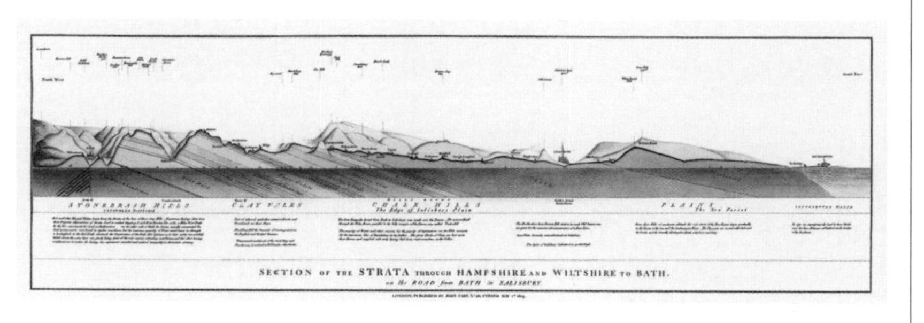

3. **CHASEWATER TO CAMBORNE.**
Richard Thomas. 1819.

A section through the mining area to the east of Camborne in Cornwall. The depths the mines are extended to are noted. The section was created by the mining engineer Richard Thomas.

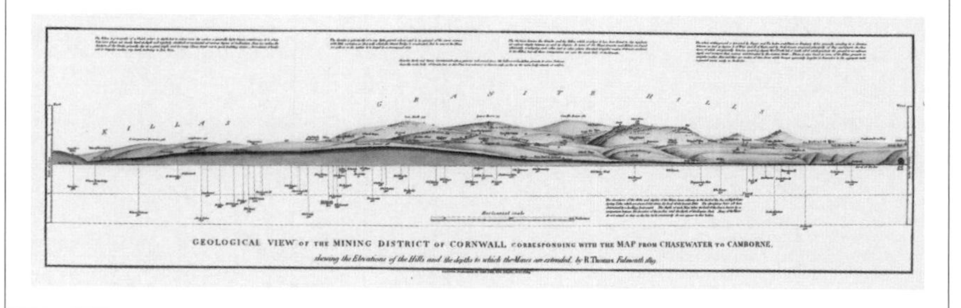

4. **TAUNTON TO CHRISTCHURCH.**
William Smith. 1819.

The south-easterly dip of the strata is clearly displayed in this section, from the Red Marl vales around Taunton, Somerset, to the Chalk hills of Wimborne Minster and the clay plains around Christchurch Bay, both in Dorset.

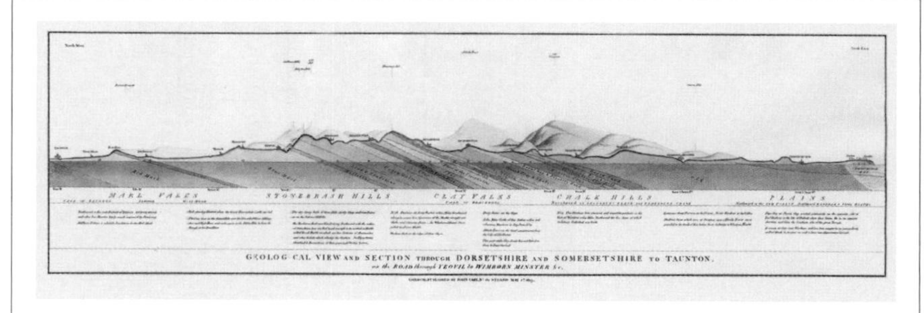

5. **LONDON TO BRIGHTON.**
William Smith. 1819.

A section south from the London Basin through the anticline of the Weald between the Chalk hills of the North Downs and the South Downs in Sussex.

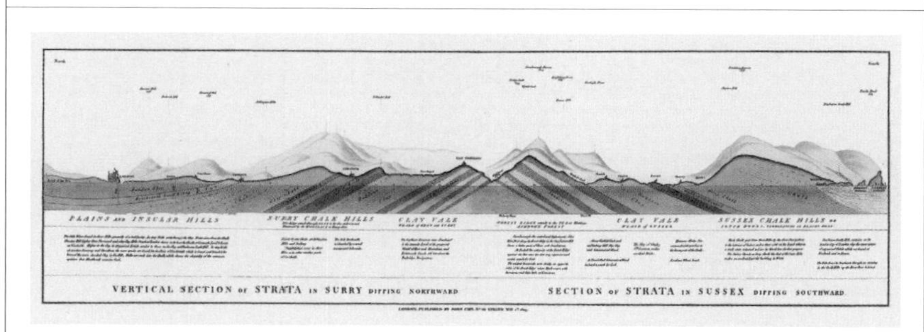

6. **CAMBRIDGE TO LONDON.**
William Smith. 1819.

A section from the Sand and Brickearth of Cambridge, Cambridgeshire, though the Chalk around Royston, Hertfordshire, to the Clay of the London Basin.

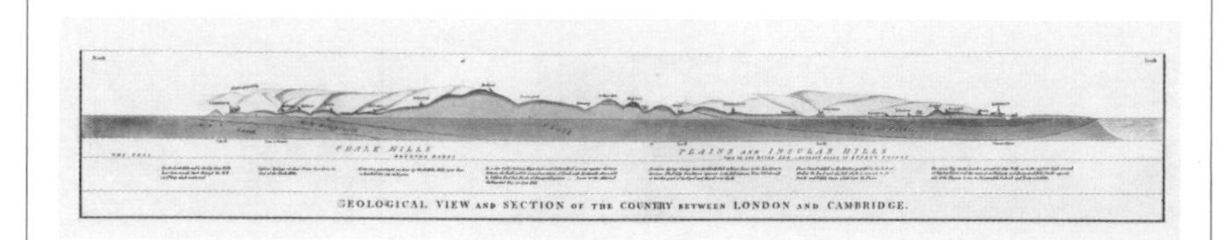

7. **LYNN TO YARMOUTH.**
William Smith. 1819.

A section east from the Clay vales of King's Lynn and The Wash through Chalk hills to the Sand and Crag on the coast at Great Yarmouth, all in Norfolk.

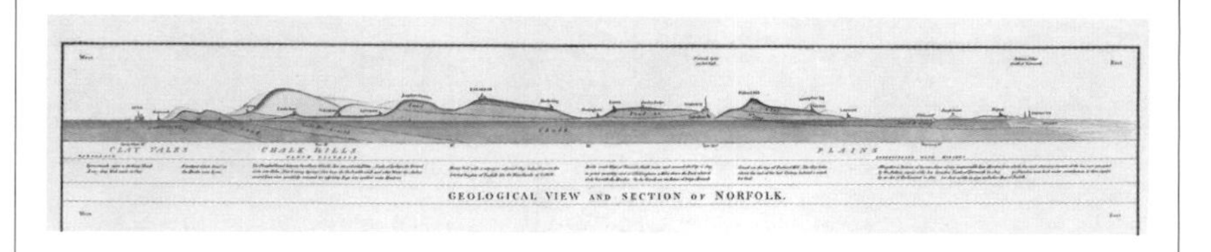

8. **ELY TO BAWDSEY.**
William Smith. 1819.

A section from the Oaktree Clay of Ely, Cambridgeshire, through the Chalk around Stowmarket, Suffolk, to the softer rocks on the Suffolk coast at Bawdsey.

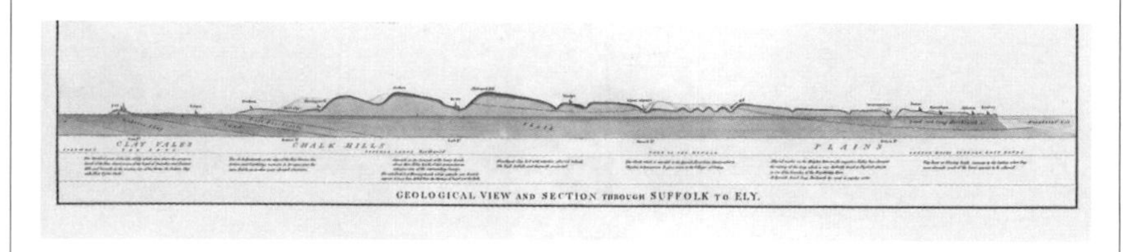

9. **ROYSTON TO SOUTHEND.**
William Smith. 1819.

A section from the Chalk of Royston, Hertfordshire, through the Sands and pebbles of Chelmsford, Essex to the London Clay at Southend, Essex.

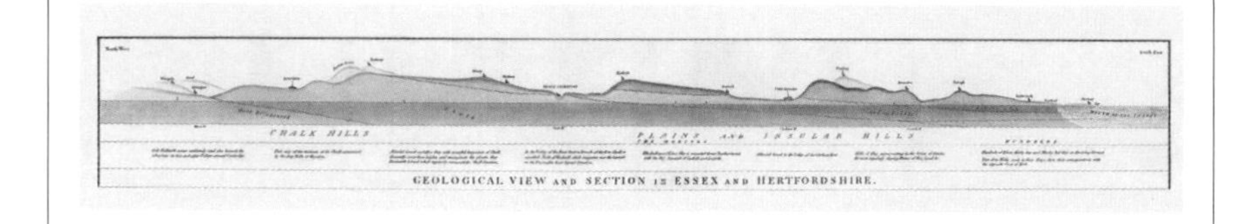

GENERAL MAP, C. 1801.

One of Smith's earliest maps of the geology of England and Wales, showing the outcrop across the country of two strata: the Red Marl indicated in brown and the overlying Blue Lias indicated in blue. They are drawn on Cary's *General Map* index to his 1794 atlas.

GENERAL MAP OF STRATA IN ENGLAND & WALES, 1801.

In another of his earliest maps, Smith again uses Cary's 1794 *General Map*, adding to its title in manuscript, 'of STRATA of England & Wales by Wm. Smith 1801'. Strata prominent on this map are the Red Marl, Blue Lias and Great Oolite, but younger strata are also shown in south-east England and Norfolk. Small black crosses mark collieries in the coalfields of Wales, Scotland and the north of England.

The topographic detail was engraved first and the geological boundary lines, as faint dots, later. Smith was unable to devote all his time to the production of the map as he also needed to earn a living through surveying and drainage work, which frequently took him away from London. This had been the situation throughout the ten years since the issuing of the 1801 *Prospectus*. Smith travelled widely, perhaps covering as much as 16,000 km (10,000 miles) a year, 'sometimes on foot, sometimes on horseback, riding on the tops of stage coaches, often making up by night-travelling the time he had lost by day, so as not to fail in his ordinary business engagements' (Smiles, 1859, p. 59). As he did so, he was able to gather additional information and collect specimens and add detail to his geological map.

COLOURING.

As a result, it was May 1815 before the first completed map was available for Smith to exhibit to the various institutions. It then took two months to colour the pattern sheets needed to guide Cary's team of colourists, Morse, Mitchell and Trout, when colouring the main production run. Smith chose to represent the twenty-three strata with a different colour, each selected to be close to the actual colour of the rock. The exception was the Chalk, 'which being colourless [white] seems best represented by green, as strong colours are necessary and there being no stratum of equal extent which required that colour', he explained (Smith, 1815b, p. 59). Smith's 1815 choice of colours, such as blue for limestone, green for Chalk, purple for slates, largely lives on today in the colours chosen to represent different rock units on the modern maps of the British Geological Survey.

During the preparation of his earliest geological maps, of the Bath area in 1799 and a small map of England and Wales in 1801, Smith had developed a novel and characteristic method of shading the strata to show their orientation and arrangement. Using a darker hue at the base of each stratum and fading to white towards the top allowed Smith to show clearly how the strata are tilted and dip beneath one another and hence which strata are the older and which are the younger, thus indicating the three-dimensionality of the strata. This is particularly clear in what we would now call the Mesozoic (Triassic, Jurassic and Cretaceous) rocks, from Smith's Marl to the Chalk, the outcrops of which, broadly, sweep across England from south-west to north-east and have a shallow dip towards the south-east (Rudwick, 2005).

With a key in the correct stratigraphical order, and the colouring rendering the relationship between different rock units apparent, the viewer can predict where strata are likely to continue underground away from their surface outcrop. This makes Smith's map a true geological map, in contrast to other contemporary or earlier 'mineralogical' maps which showed only the spatial distribution of rock types or minerals on the surface, and from which no inferences could be made of structure or the underground disposition of the rocks.

Unfortunately, Smith's method of colouring was difficult to execute well on the scale of the large map. As his nephew, John Phillips (1800–1874) noted in 1844, while it 'gave a picturesque effect to the map', it 'required more than usual skill and patience to be correctly executed, and occasioned great trouble in examining the copies', making the process slow and expensive (Phillips, 1844, p. 76). On today's geological maps, colours

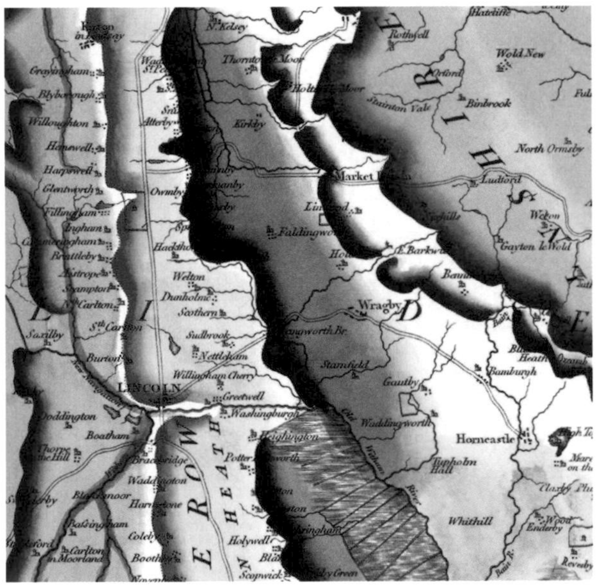

Fig. 3.
Smith employed a novel – and expensive – style of colouring to illustrate the dip of the strata, each stratum shown with a denser tone at its base and fading to a lighter shade upwards. This gave an immediate visual impression of the disposition of the beds, in this case with beds tilted towards the east.

Fig. 4.
Smith's 1815 map was accompanied by a quarto *Memoir* which included an explanation of the strata and their colours; a description of soils and strata by county, plus Wales and Scotland; and the characteristics of soils produced by the different strata.

FIG. 3.

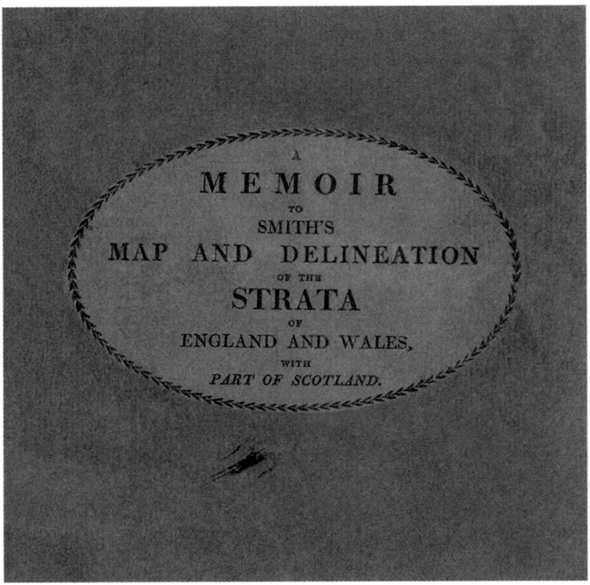

A
MEMOIR
TO
SMITH'S
MAP AND DELINEATION
OF THE
STRATA
OF
ENGLAND AND WALES,
WITH
PART OF SCOTLAND.

FIG. 4.

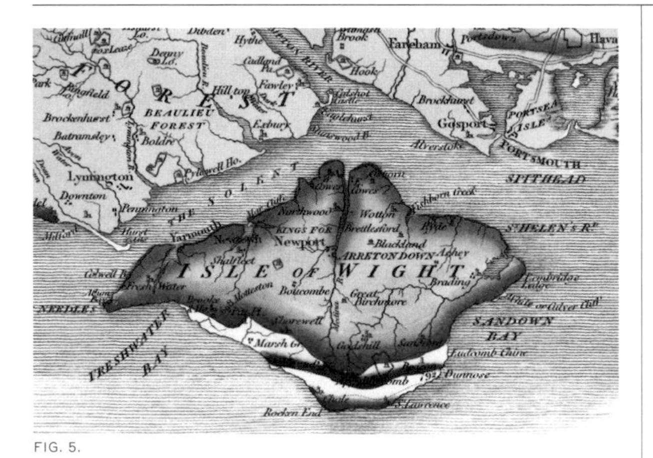

FIG. 5.

FIG. 6.

Figs. 5 & 6.
As new geological information came to Smith's attention, changes were made to the 1815 map as new copies were coloured and issued. This is most clearly shown in changes to the representation of the geology of the Isle of Wight, of which at least five variants are known. Figure 5 shows an early version while Figure 6 is later.

are still used to differentiate rock units, but the attitude of the strata – their dip and strike – is indicated by symbols.

Although the map carries a date of 1 August 1815, it was not published until possibly the first week of September, and at least twenty copies were produced and distributed by the end of October, including ones for Sir Joseph Banks, the Duke of Bedford and the Geological Society. On 2 November, Smith decided to begin numbering and signing every copy of the map, each of fifteen sheets, as batches arrived from Cary. Production rose through November and December 1815, then dropped in January and February 1816, but in those four months, Smith numbered 244 copies and signed all except 31. These unsigned maps were ones on which he considered the colouring to be imperfect, complaining that Cary was getting 'maps done at a cheap rate by new hands', no doubt trying to speed up production and sales to recover some of his costs (Sharpe, 2016).

An additional complication for Cary was that Smith insisted on making changes to his map during the course of its production as new information came his way. Sometimes these modifications involved additional engraving of lines on the plates, but largely they were accommodated by adjustments to the colouring. This is most readily apparent in the colouring of the Isle of Wight, of which at least six variants are known. One significant addition was the insertion of a new stratum, the Coral Rag, which first came to Smith's attention during work on a deep well near Swindon in April 1816 and appears, gradually, on various sheets from Wiltshire to Yorkshire (Eyles and Eyles, 1938; Sharpe, 2016). As a result of such changes, as well as the fact that the map is hand-coloured, no two Smith maps are identical.

Smith's map was available for purchase in a variety of formats at prices ranging from 5 guineas to £12. The fifteen coloured sheets with an uncoloured index map and the explanatory *Memoir* cost 5 guineas; as a bound atlas, it was £6 12s; for £7 the map was available folded on to three large canvas sheets in a travelling case or mounted on canvas and rollers; for an extra pound the rolled map would be varnished; for £10 the purchaser could have the map on spring rollers, and for £12, on canvas, varnished and on spring rollers.

To accompany the map, in the summer of 1815, Smith hurriedly prepared a quarto, paper-bound, volume, *A Memoir to the Map and Delineation of the Strata of England and Wales, with part of Scotland*. This summarizes the geology of forty English counties, along with a brief description of the geology of Wales and Scotland, and a description of the soils and surface produced by each of the strata on the map. Smith also explains the system of colours. It opens with the lines:

> *After twenty-four years of intense application to such an abstruse subject as the discovery and delineation of the British Strata, the reader may easily conceive the great satisfaction I feel in bringing it to its present state of perfection.*

The *Memoir* includes a list of 410 'subscribers'. This was, in reality, a list of those who had expressed an interest in purchasing the map. While a few of these had made an advance financial contribution, most notably Banks, many had not. Some on the list refused to take the map when it was published, and because it had taken so long to produce, at least ten of the 'subscribers' had died. September 1815 was not a good time to launch such an expensive product. Following the end of the Napoleonic Wars a few months earlier, the British economy had entered a severe recession. This particularly affected landowners and agriculture and must surely have impacted on Smith's map sales. Despite the list of 410 interested parties and additional advertising in newspapers in early 1816, it is likely that fewer than half of the 750 copies of the map that it had been Smith's ambition to print were in fact ever produced (Sharpe and Torrens 2015; Sharpe 2016).

Overleaf.
William Smith's Geological Table of British Organized Fossils. First published in 1817 in Smith's *Stratigraphical System of Organized Fossils*, this table was also available separately for 1 shilling and 6 pence in the form shown here, with additional information on which strata are found in each county. It was reprinted and sold until at least 1829. The table has been recreated on pp. 10–11, and the colouring and numbering system it establishes employed throughout this book.

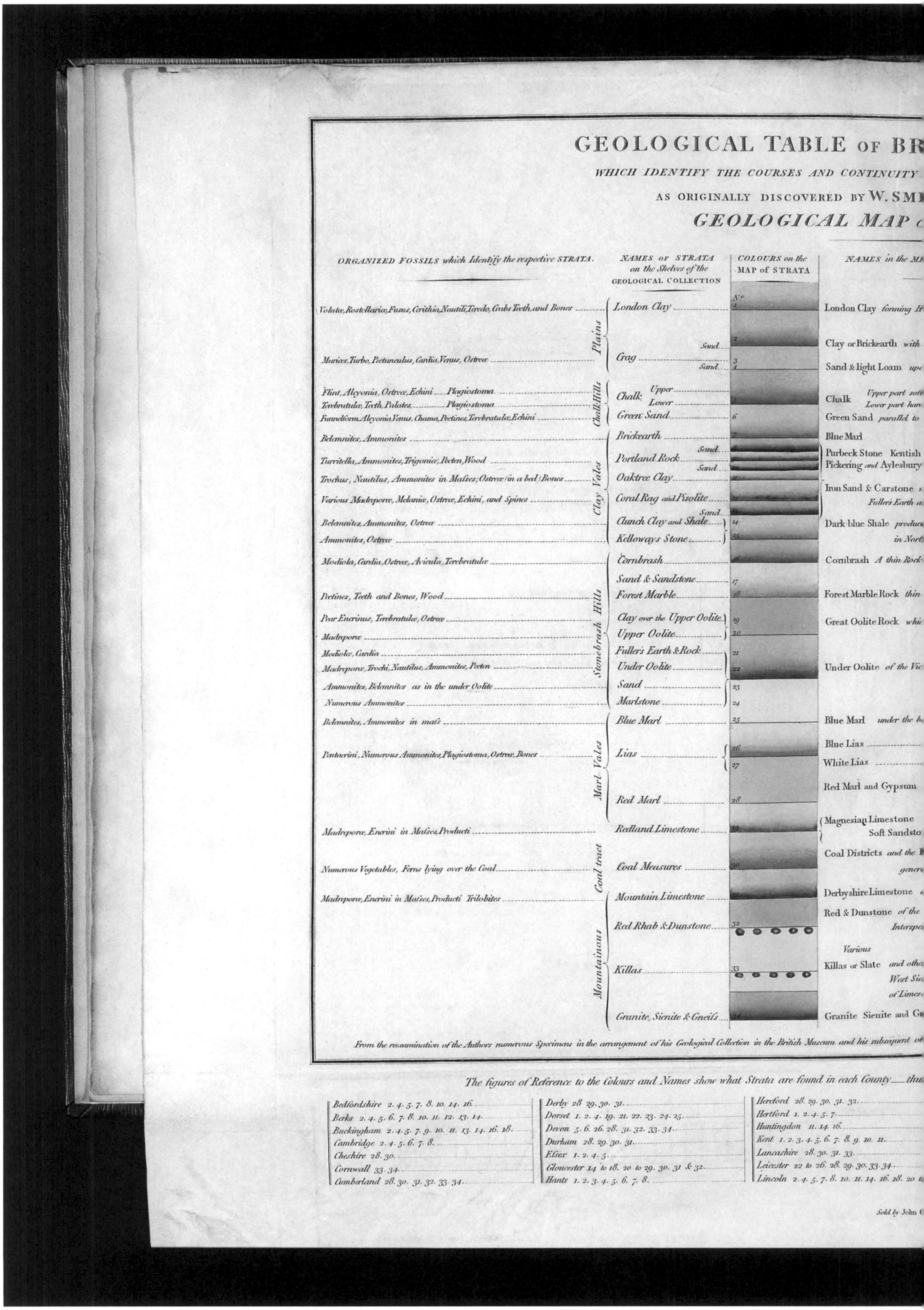

GEOLOGICAL TABLE OF BR

WHICH IDENTIFY THE COURSES AND CONTINUITY

AS ORIGINALLY DISCOVERED BY W. SM

GEOLOGICAL MAP

ORGANIZED FOSSILS which Identify the respective STRATA.		*NAMES OF STRATA on the Shelves of the* GEOLOGICAL COLLECTION	*COLOURS on the* MAP of STRATA	*NAMES in the M*
			Nº	
Volutæ, Rostellariæ, Fusus, Crithia, Nautili, Teredo, Crabs Teeth, and Bones	*Plains*	*London Clay*	1	London Clay *forming H*
		Sand	2	Clay or Brickearth *with*
Murices, Turbo, Pectunculus, Cardia, Venus, Ostreæ		*Crag*	3	
		Sand	4	Sand & light Loam *up*
Flint, Alcyonia, Ostreæ, Echini ... Plagiostoma	*Chalk Hills*	*Chalk: Upper*		Chalk *Upper part so*
Terebratulæ, Teeth, Palates, ... Plagiostoma		*Lower*	5	*Lower part har*
Funnelform Alcyonia Venus, Chama, Pectines, Terebratulæ, Echini		*Green Sand*	6	Green Sand *parallel to*
Belemnites, Ammonites	*Clay Vales*	*Brickearth*	7	Blue Marl
		Sand	8	Purbeck Stone Kentish
Turritella, Ammonites, Trigoniæ, Pecten, Wood		*Portland Rock*	9	Pickering *and* Aylesbury
		Sand	10	
Trochus, Nautilus, Ammonites in Maſses; Ostreæ (in a bed) Bones		*Oaktree Clay*	11	Iron Sand & Carstone
Various Madreporæ, Melaniæ, Ostreæ, Echini, and Spines		*Coral Rag and Pisolite*	12	*Fuller's Earth a*
		Sand	13	
Belemnites, Ammonites, Ostreæ		*Clunch Clay and Shale*	14	Dark-blue Shale *produc*
Ammonites, Ostreæ		*Kelloways Stone*	15	*in Nort*
Modiola, Cardia, Ostreæ, Avicula, Terebratulæ	*Stonebrash Hills*	*Cornbrash*	16	Cornbrash *A thin Rock*
		Sand & Sandstone	17	
Pectines, Teeth and Bones, Wood		*Forest Marble*	18	Forest Marble Rock *thin*
Pear Encrinus, Terebratulæ, Ostreæ		*Clay over the Upper Oolite*	19	Great Oolite Rock *whi*
Madreporæ		*Upper Oolite*	20	
Modiolæ, Cardia		*Fuller's Earth & Rock*	21	
Madreporæ, Trochi, Nautilus, Ammonites, Pecten		*Under Oolite*	22	Under Oolite *of the Vi*
Ammonites, Belemnites as in the under Oolite		*Sand*	23	
Numerous Ammonites		*Marlstone*	24	
Belemnites, Ammonites in maſs	*Marl Vales*	*Blue Marl*	25	Blue Marl *under the b*
Pentacrini, Numerous Ammonites, Plagiostoma, Ostreæ, Bones		*Lias*	26	Blue Lias
			27	White Lias
				Red Marl and Gypsum
		Red Marl	28	
Madreporæ, Encrini in Maſses, Producti		*Redland Limestone*	29	Magnesian Limestone Soft Sandsto
Numerous Vegetables, Ferns lying over the Coal	*Coal tract*	*Coal Measures*	30	Coal Districts *and the* *gener*
Madreporæ, Encrini in Maſses, Producti Trilobites		*Mountain Limestone*	31	Derbyshire Limestone
	Mountainous	*Red Rhab & Dunstone*	32	Red & Dunstone *of the* *Interspe*
		Killas	33	*Various* Killas or Slate *and othe* *West Si* *of Limes*
		Granite, Sienite & Gneiſs	34	Granite Sienite and G

From the reexamination of the Authors numerous Specimens in the arrangement of his Geological Collection in the British Museum and his subsequent o

The figures of Reference to the Colours and Names show what Strata are found in each County — thu

Bedfordshire 2. 4. 5. 7. 8. 10. 14. 16.	*Derby* 28 29. 30. 31.	*Hereford* 28. 29. 30. 31. 32.
Berks 2. 4. 5. 6. 7. 8. 10. 11. 12. 13. 14.	*Dorset* 1. 2. 4. 19. 21. 22. 23. 24. 25.	*Hertford* 1. 2. 4. 5. 7.
Buckingham 2. 4. 5. 7. 9. 10. 11. 13. 14. 16. 18.	*Devon* 5. 6. 26. 28. 31. 32. 33. 34.	*Huntingdon* 11. 14. 16.
Cambridge 2. 4. 5. 6. 7. 8.	*Durham* 28. 29. 30. 31.	*Kent* 1. 2. 3. 4. 5. 6. 7. 8. 9. 10. 11.
Cheshire 28. 30.	*Eſsex* 1. 2. 4. 5.	*Lancashire* 28. 30. 31. 33.
Cornwall 33. 34.	*Gloucester* 14 to 18. 20 to 29. 30. 31 & 32.	*Leicester* 22 to 26. 28. 29. 30. 33. 34.
Cumberland 28. 30. 31. 32. 33. 34.	*Hants* 1. 2. 3. 4. 5. 6. 7. 8.	*Lincoln* 2. 4. 5. 7. 8. 10. 11. 14. 16. 18. 20

Sold by John

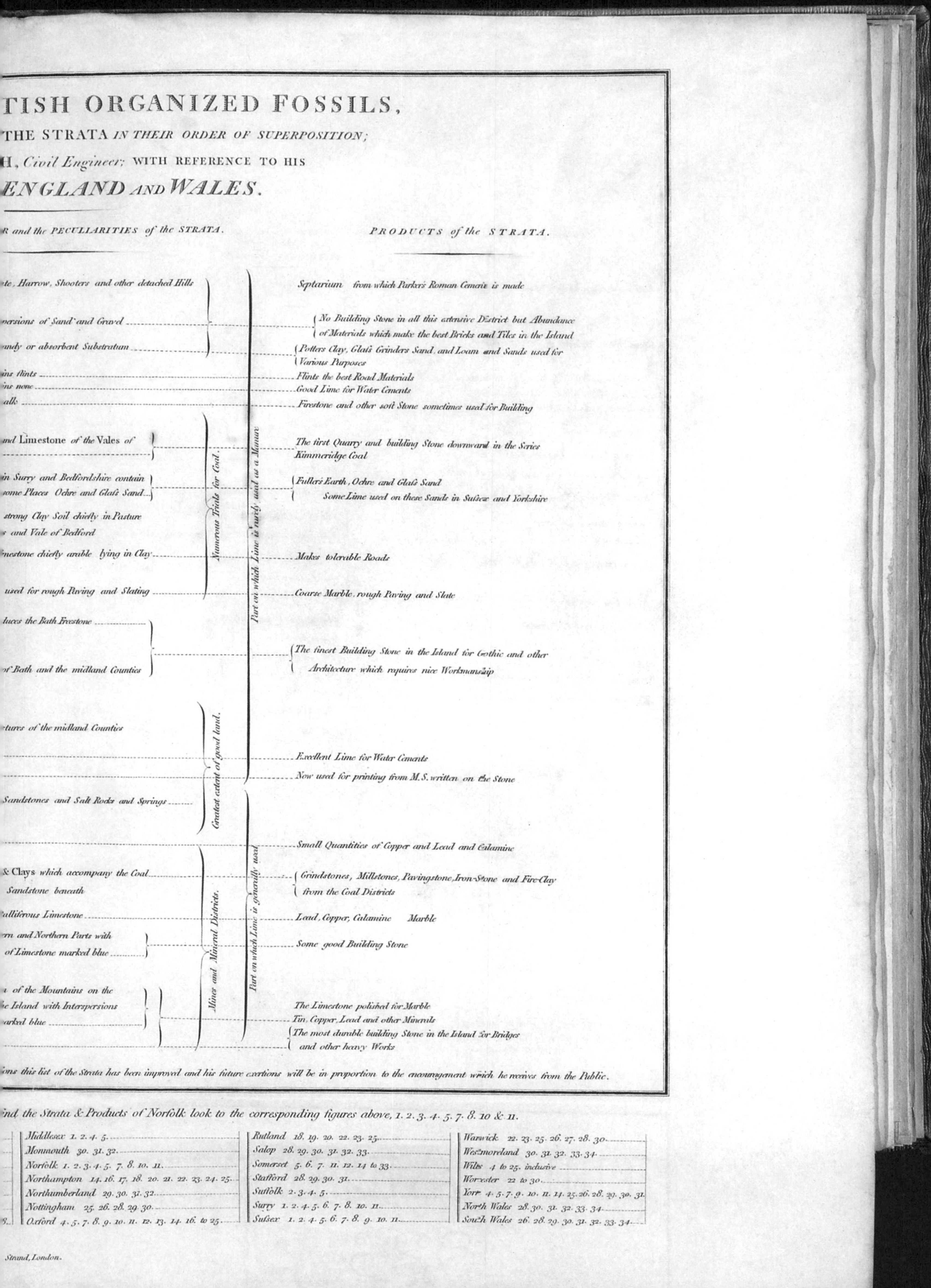

TISH ORGANIZED FOSSILS,

THE STRATA *IN THEIR ORDER OF SUPERPOSITION;*

H, *Civil Engineer;* WITH REFERENCE TO HIS

ENGLAND AND WALES.

R and the PECULIARITIES of the STRATA.	PRODUCTS of the STRATA.
te, Harrow, Shooters and other detached Hills	Septarium from which Parker's Roman Cement is made
persions of Sand and Gravel	No Building Stone in all this extensive District but Abundance of Materials which make the best Bricks and Tiles in the Island
andy or absorbent Substratum	Potters Clay, Glass Grinders Sand, and Loam and Sands used for Various Purposes
ins flints	Flints the best Road Materials
ins none	Good Lime for Water Cements
alk	Firestone and other soft Stone sometimes used for Building
and Limestone of the Vales of	The first Quarry and building Stone downward in the Series Kimmeridge Coal
in Surry and Bedfordshire contain some Places Ochre and Glass Sand	Fullers Earth, Ochre and Glass Sand Some Lime used on these Sands in Sussex and Yorkshire
strong Clay Soil chiefly in Pasture s and Vale of Bedford	
nestone chiefly arable lying in Clay	Makes tolerable Roads
used for rough Paving and Slating	Coarse Marble, rough Paving and Slate
duces the Bath Freestone of Bath and the midland Counties	The finest Building Stone in the Island for Gothic and other Architecture which requires nice Workmanship
tures of the midland Counties	
	Excellent Lime for Water Cements
	Now used for printing from M.S. written on the Stone
Sandstones and Salt Rocks and Springs	
	Small Quantities of Copper and Lead and Calamine
& Clays which accompany the Coal Sandstone beneath	Grindstones, Millstones, Pavingstone, Iron-Stone and Fire-Clay from the Coal Districts
allifrous Limestone	Lead, Copper, Calamine Marble
rn and Northern Parts with of Limestone marked blue	Some good Building Stone
of the Mountains on the e Island with Interspersions arked blue	The Limestone polished for Marble Tin, Copper, Lead and other Minerals The most durable building Stone in the Island for Bridges and other heavy Works

Vertical brace labels: *Numerous Trials for Coal.* — *Part on which Lime is rarely used as a Manure* — *Greatest extent of good land.* — *Mines and Mineral Districts.* — *Part on which Lime is generally used*

ons this list of the Strata has been improved and his future exertions will be in proportion to the encouragement which he receives from the Public.

nd the Strata & Products of Norfolk look to the corresponding figures above, 1. 2. 3. 4. 5. 7. 8. 10 & 11.

Middlesex 1. 2. 4. 5.	Rutland 18. 19. 20. 22. 23. 25.	Warwick 22. 23. 25. 26. 27. 28. 30.
Monmouth 30. 31. 32.	Salop 28. 29. 30. 31. 32. 33.	Westmoreland 30. 31. 32. 33. 34.
Norfolk 1. 2. 3. 4. 5. 7. 8. 10. 11.	Somerset 5. 6. 7. 11. 12. 14 to 33.	Wilts 4 to 25. inclusive
Northampton 14. 16. 17. 18. 20. 21. 22. 23. 24. 25.	Stafford 28. 29. 30. 31.	Worcester 22 to 30.
Northumberland 29. 30. 31. 32.	Suffolk 2. 3. 4. 5.	York 4. 5. 7. 9. 10. 11. 14. 25. 26. 28. 29. 30. 31.
Nottingham 25. 26. 28. 29. 30.	Surry 1. 2. 4. 5. 6. 7. 8. 10. 11.	North Wales 28. 30. 31. 32. 33. 34.
Oxford 4. 5. 7. 8. 9. 10. 11. 12. 13. 14. 16. to 25.	Sussex 1. 2. 4. 5. 6. 7. 8. 9. 10. 11.	South Wales 26. 28. 29. 30. 31. 32. 33. 34.

Strand, London.

A. ***STRATA DELINEATED…*, 1815.**
The earliest copies of Smith's map were issued unsigned and unnumbered. This is the Geological Society's subscription copy, dating from September 1815.

B. ***STRATA DELINEATED…*, 1815.**
From 2 November 1815, Smith began to number and sign copies of the map. This copy, now at Stanford University, was examined by Smith on 20 November 1815 and numbered 34 but he left it unsigned as he was displeased by the colouring.

C. ***STRATA DELINEATED…*, 1815.**
On 17 December 1815 Smith examined, signed and numbered this copy 100 before beginning a second series of numbering a.1 to a.100. This copy, now in the National Museum of Wales, belonged to Elisha de Hague of Norwich.

D. ***STRATA DELINEATED…*, C. 1820S.**
The geology shown on the maps continued to evolve after 1816. This map, in Nottingham University, may date from the 1820s.

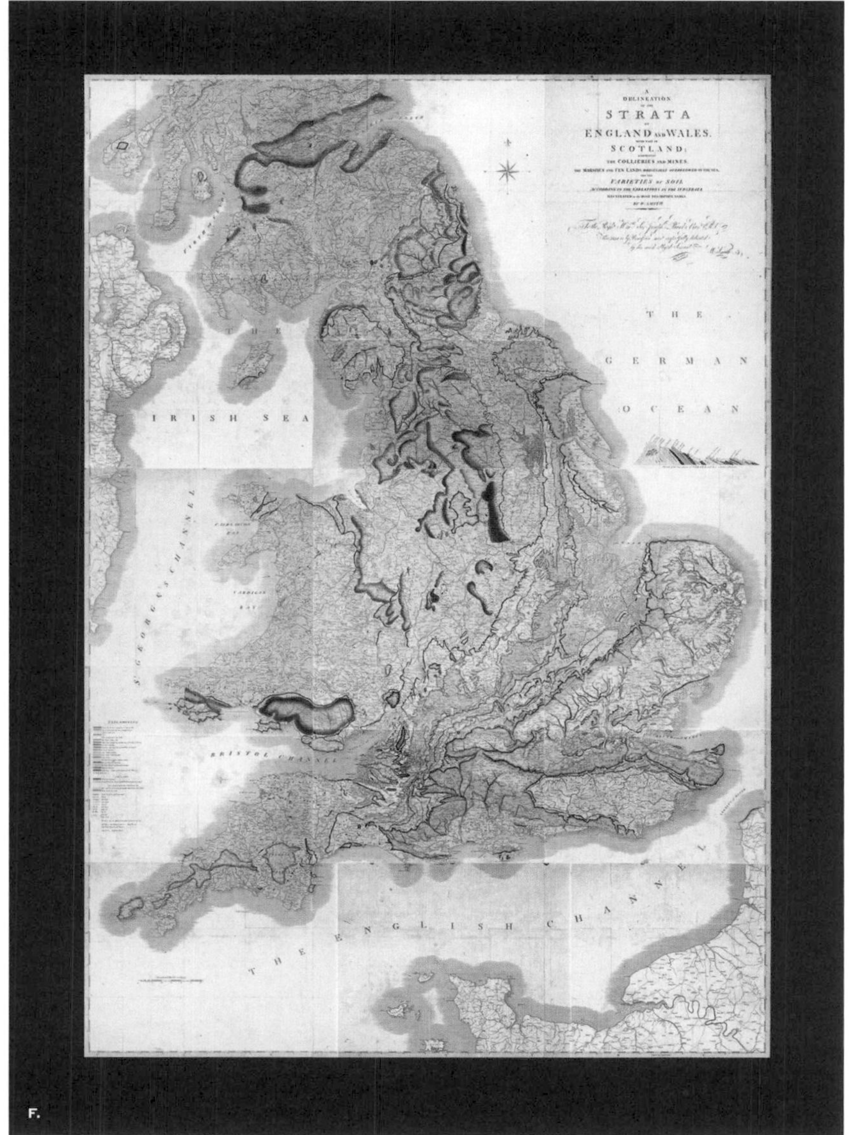

E. ***STRATA DELINEATED…*, C. 1824.**
Smith's map was issued in various formats. Some were issued on rollers for display, such as this late-issue map in the Yorkshire Museum, which exhibits the fading typical of most surviving rolled maps.

F. ***STRATA DELINEATED…*, 1834.**
In the 1830s Smith seems to have considered a second edition of his map which may be represented by this unsigned and unnumbered map in the National Museum of Wales, printed on paper dating from 1832.

G. ***A NEW GEOLOGICAL MAP OF ENGLAND AND WALES*, 1820.**
In 1820 Smith and Cary issued a small and more convenient geological map which included for the first time geological colouring of the rocks along the east coast of Ireland.

H. ***A NEW GEOLOGICAL MAP OF ENGLAND AND WALES*, 1828.**
The success of *A New Geological Map* map saw it go through at least four editions by 1828.

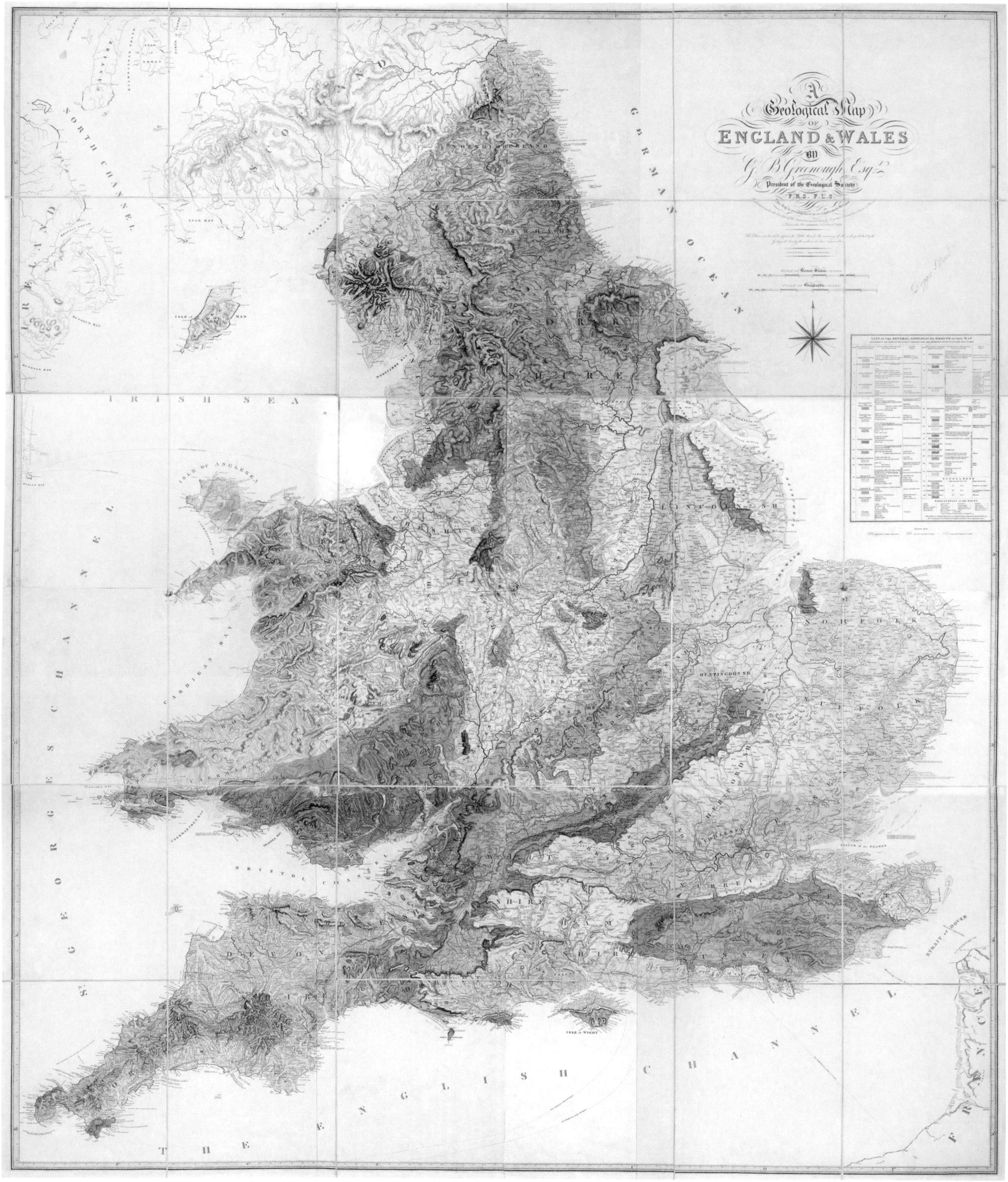

***A GEOLOGICAL MAP OF ENGLAND & WALES* BY G. B. GREENOUGH ESQUIRE., PRESIDENT OF THE GEOLOGICAL SOCIETY, F.R.S. F.L.S., 1819.**

On 1 May 1820, the Geological Society of London published its own version of a geological map of England and Wales. Unlike William Smith's map, this was a collaborative effort overseen by the Society's President George Bellas Greenough. Although dated 1 November 1819, the map was not issued until six months later.

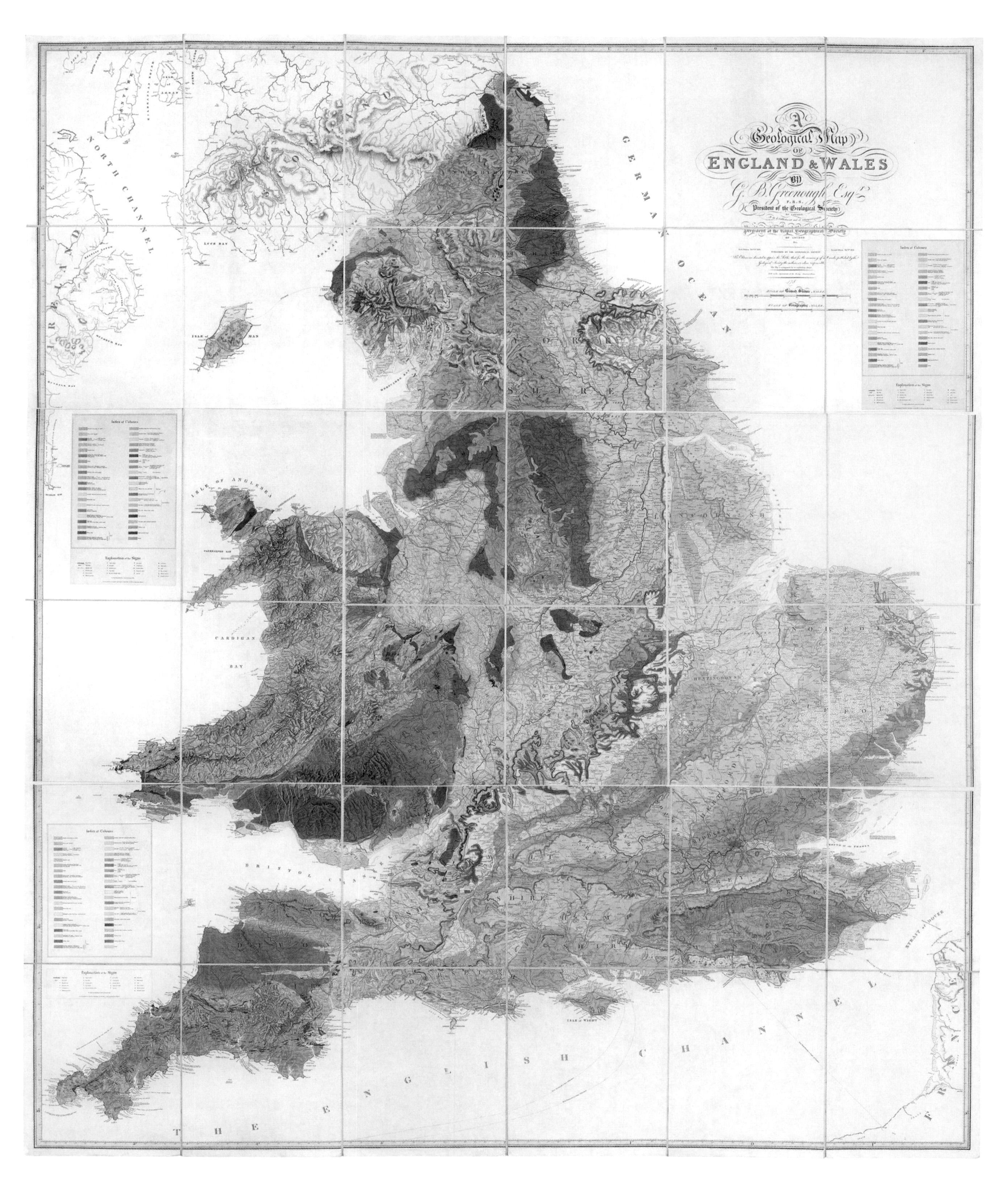

***A GEOLOGICAL MAP OF ENGLAND & WALES* BY G. B. GREENOUGH ESQUIRE., PRESIDENT OF THE GEOLOGICAL SOCIETY, F.R.S. F.L.S., 1839.**

The Geological Society map went through several editions between 1820 and 1865. Publication of this second edition, like the first, was delayed, so the map carries a date of 1 November 1839 but was not published until 12 September 1840. By the time of the 1865 edition, Greenough was dead and the map, for first time, acknowledged the work of William Smith.

COMPETITION.

In May 1820, just four and a half years after publication of Smith's masterpiece, a competitor appeared in the form of a map published by the Geological Society. Created in 1807 as a gentlemanly dining club, the Society soon set about gathering 'geological facts' – not just from books, specimens and maps but also local geological details – sent in by members. The Society's President, George Bellas Greenough (1778–1855), began to compile the information on topographic maps. In contrast to Smith's solitary endeavour, this was a collaborative effort, making use of many sources – including Smith's map.

From 1803, Smith had based himself in London, renting a house on Buckingham Street, off the Strand. Here he encouraged influential visitors – potential funders and sponsors – to view his fossil collection, which was arranged in stratigraphical order on sloping shelves, and to hear his ideas and see his draft maps. Among these visitors, in March 1808, was a delegation from the Geological Society, including Greenough. They had left unimpressed by Smith and his bare, sparsely furnished house and unconvinced of his notions on the use of fossils to identify strata. One thing they were convinced of, however, was that the Society could do a better job (Herries Davies, 2007). On his part, Smith felt that 'the theory of geology is in the possession of one class of men, the practice in another'.

As the Geological Society's map was a compilation of all available information on the geology of England and Wales, it was inevitable that use would be made of Smith's map. This led John Farey (1766–1826), a friend and pupil of Smith's, to accuse Greenough of plagiarism, a charge which the latter was forced to deny in the brief memoir that accompanied the Society's map (Greenough, 1820).

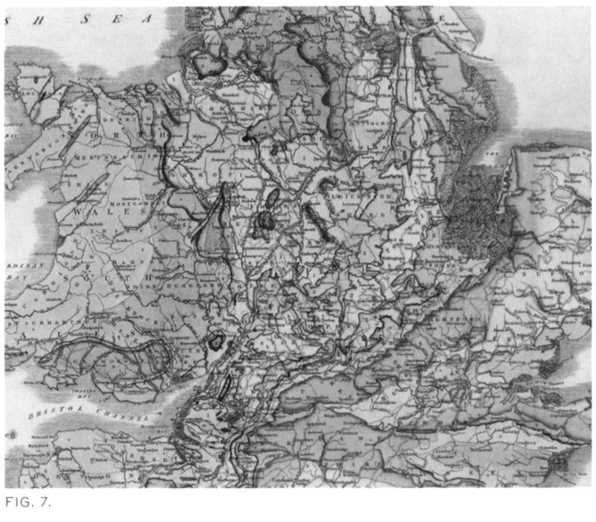

Fig. 7.
Detail of *A New Geological Map of England and Wales*, 1820. The small map published by Smith in the 1820s maintained the same style of colouring as the 1815 map, although it was easier to apply to the younger rocks of south-east England than the older and more complex rocks of mid and north Wales.

FIG. 7.

When the Geological Society map was published in May 1820, six months later than the 1 November 1819 date on the map, a copy was sent to Smith. By now disillusioned, debts having necessitated the sale of his fossil collection and the seizure of his property, books and papers, and having endured a spell in the King's Bench Prison, Smith was in no mood to look kindly on Greenough's work: 'The copy seemed like a ghost of my old map, intruding on business and retirement, and mocking me in the disappointments of a science with which I could scarcely be in temper. It was put out of sight.' (Cox, 1942.) It may have been some consolation to Smith, had he ever known it, that the Geological Society's map sold even fewer copies than his.

The Society's map is less eye-catching than Smith's, with much more subdued colours on a more detailed, cluttered topographic base map with relief shading. It does, however, display much greater detail, especially in regions of older rocks and more complex geology such as in Wales and Cornwall, and is extensively annotated around the coast.

RECOGNITION.

Despite poor sales of his own map, and disillusionment caused by that of the Geological Society, by 1834 Smith was considering a second edition of his map and *Memoir*. A few copies of what appears to be this second edition map seem to have been issued and are recognizable by their full colouring, without the uncoloured areas of the early maps, and with the inclusion of additional rocks such as the Cheviot granite and 'trap' (volcanic rocks). Annotated copies for this second edition of the *Memoir* are preserved in the Smith Archive at the Oxford University Museum of Natural History (Sharpe, 2016).

The stimulus for this new lease of life for his map and *Memoir* may have been Smith's final acceptance and commendation by the Geological Society in 1831 and his receipt of the Society's newly instituted Wollaston Prize 'in recognition', as the Society's President, Adam Sedgwick (1785–1873), declared, 'of his being a great and original discoverer in English Geology; and especially for his having been the first, in this country, to discover and to teach the identification of strata, and to determine their succession by means of their imbedded fossils' (Sedgwick, 1831, p. 271).

The copper plates used for printing Smith's map survived until 1877, by which time they had come into the possession of the map-seller Edward Stanford (1827–1904). A public appeal for funds to purchase them for the Geological Society was unsuccessful and the plates were melted down (Boulger, 1877; Sheppard, 1917).

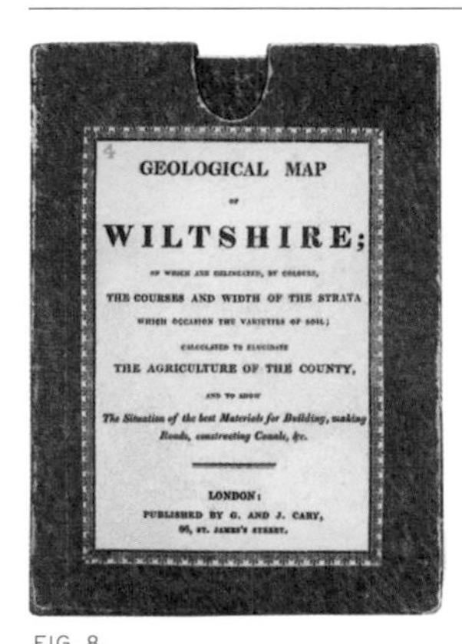

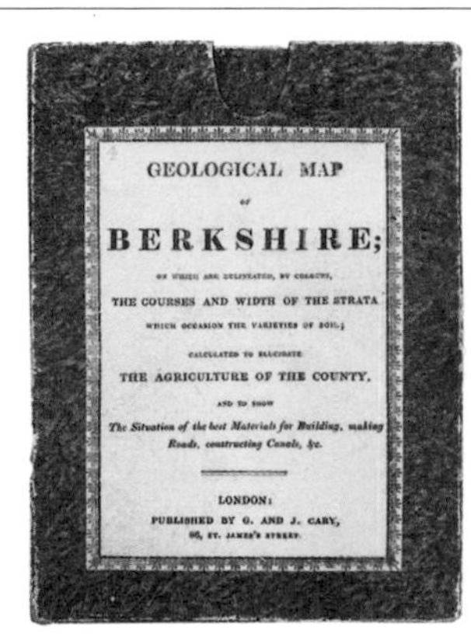

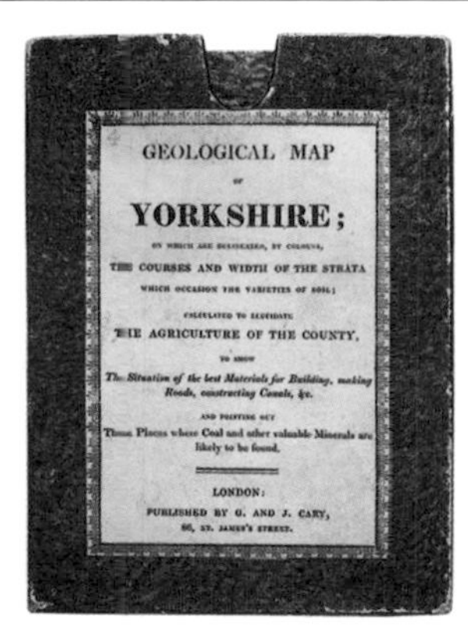

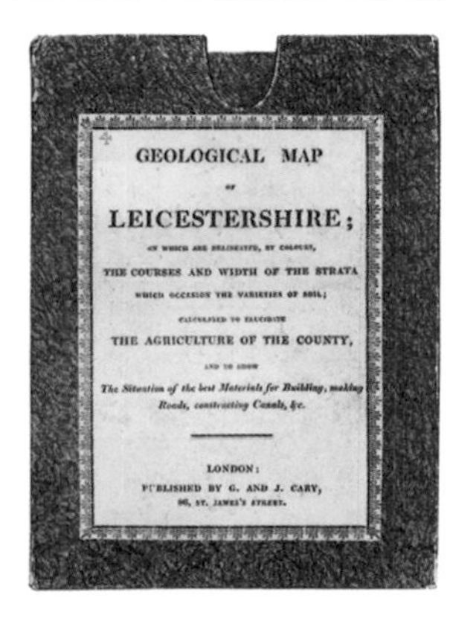

FIG. 8.

Fig. 8.
The county maps from Smith's *Geological Atlas* were available both flat and folded, the latter dissected, mounted on linen and inserted into a slip case as shown here. The Wiltshire and Berkshire maps were first issued in 1819, Yorkshire in 1821, and Leicestershire in 1822.

A NEW GEOLOGICAL MAP OF ENGLAND AND WALES.

In March 1820, Smith and Cary issued a smaller, one-sheet geological map, *A New Geological Map of England and Wales, with the Inland Navigations; Exhibiting the Districts of Coal and other Sites of Mineral Tonnage by W. Smith, Engineer*. On a scale of 15 miles to the inch, and using a redrawn version of Cary's 1796 map, it is fully coloured, with nineteen strata and, unlike the large 1815 map, includes geological colouring for the east coast of Ireland and the Channel coast of France. Two tables on the map list mineral railways, canals and rivers, and the mined materials that they transported. In January 1820, Smith took lodgings at 20 Suffolk Street, off Pall Mall, while examining and correcting proofs of *A New Geological Map*, so he was fully engaged with the additional geological and other information on the map.

That this more conveniently sized map, selling for 14 shillings, or 18 shillings in a travel case or on rollers, was a commercial success is suggested by the fact that it was reissued several times throughout the 1820s. Another likely factor helping sales was that it was marketed as a 'reduction from Smith's large map; exhibiting a general view of the stratification of the country; intended as an elementary map for those commencing the study of geology' – a field which was becoming a popular pursuit. Its success may also have led the map-seller James Gardner of Regent Street to issue a reduced version of the Geological Society map in 1826.

A NEW GEOLOGICAL ATLAS.

On 21 November 1815 Smith had been joined in London by his orphaned nephew, John Phillips (1800–1874), the son of Smith's sister, Elizabeth. Having been educated at various schools at his uncle's expense, Phillips, then almost fifteen years old, would become Smith's invaluable assistant, apprentice draughtsman and surveyor. He also became skilled in the new technique of lithographic printing, and produced several cross sections for Smith in 1817, including *Section of Strata North Wilts.* and *Strata south of London dipping northward* (Eyles, 1969; Twyman, 2016).

In 1819 Cary and Smith, with the assistance of Phillips, began publication of a series of geological county maps based on the much-praised county sheets of the 1818 issue of Cary's *New English Atlas*, which had first been published in 1809. The scale of these maps varies, but each is contained within an engraved border measuring 53 × 48 cm (21 × 19 in.). Six parts, covering twenty-one counties, were issued between 1819 and 1824 with the title *A New Geological Atlas of England and Wales, on which are delineated by colours, the courses and width of the strata, which occasion the varieties of soil; calculated to elucidate the agriculture of each county, and to show the situation of the best materials for building, making of roads, the construction of canals, and pointing out those places where coal and other valuable materials are likely to be found. By William Smith, Author of the Geological Map of England and Wales*. Each part contained four maps of counties as they then existed (many of which have since changed, as have some spellings), with the exception of Part IV, as all its four sheets were taken up by the large county of Yorkshire.

The strata are coloured in the same manner as the 1815 map, with a darker tone marking the base of each stratum and fading upwards to a lighter tone or even to white, and using similar colours. The identifying key, however, is different, with coloured boxes adjacent to each stratum around the margins of the outcrop. The geological colouring is shown only on the relevant county of the map, with bordering counties left uncoloured. The larger scale of the county maps meant that considerably more geological and local detail could be included than on any of Smith's published maps, with the exception of his 1832 geological map of the Hackness Hills in Yorkshire.

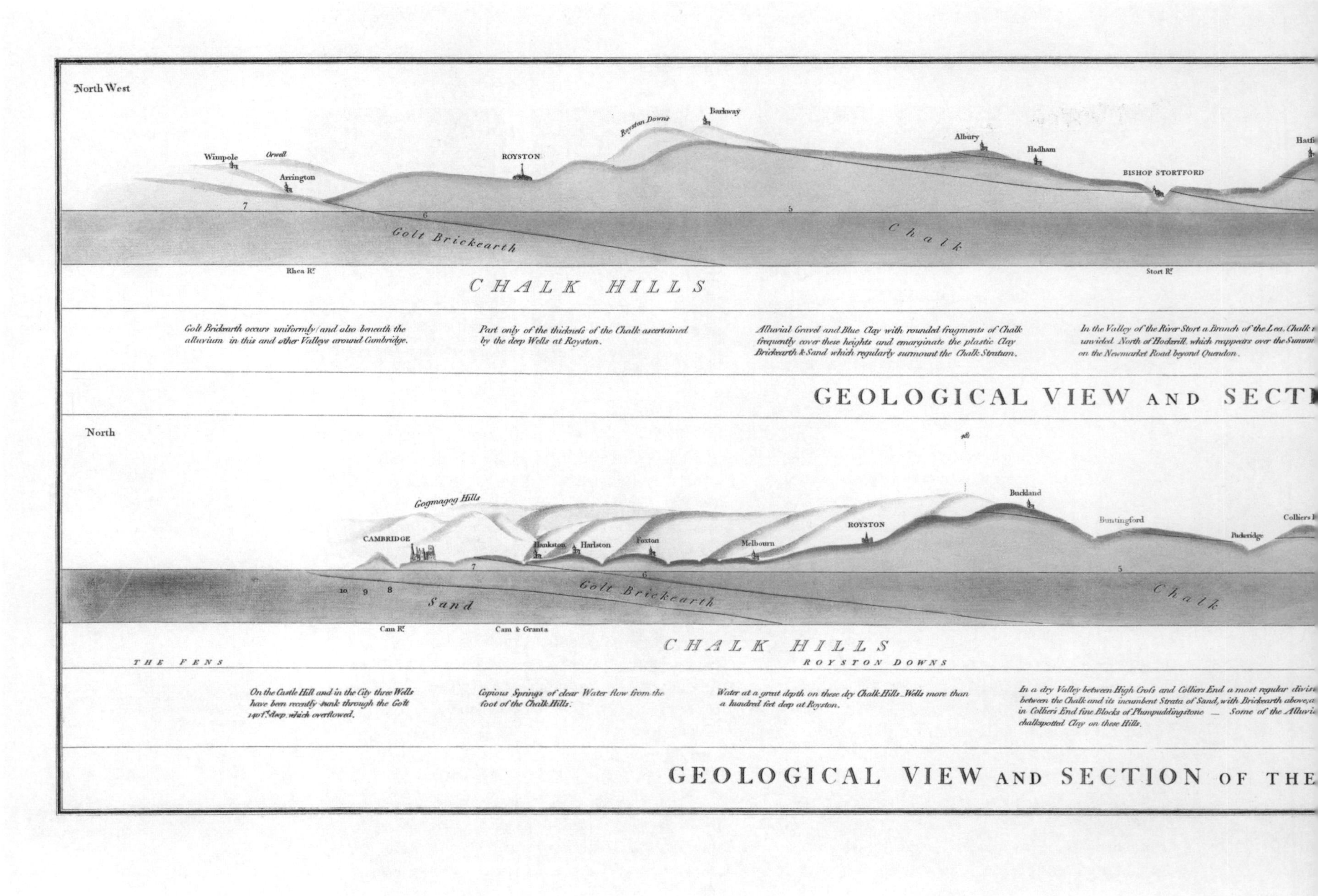

North West

Wimpole Orwell Arrington ROYSTON Royston Downs Barkway Albury Hadham BISHOP STORTFORD Hatfi

7 6 5

Golt Brickearth

Chalk

Rhea R.^t Stort R.^t

CHALK HILLS

Golt Brickearth occurs uniformly/ and also beneath the alluvium in this and other Valleys around Cambridge.

Part only of the thickness of the Chalk ascertained by the deep Wells at Royston.

Alluvial Gravel and Blue Clay with rounded fragments of Chalk frequently cover these heights and emarginate the plastic Clay Brickearth & Sand which regularly surmount the Chalk Stratum.

In the Valley of the River Stort a Branch of the Lea, Chalk i[...] unveiled. North of Hockerill, which reappears over the Summi[t] on the Newmarket Road beyond Quendon.

GEOLOGICAL VIEW AND SECTI

North

Gogmagog Hills CAMBRIDGE Haukston Harlston Foxton Melbourn ROYSTON Buckland Buntingford Puckeridge Colliers E

10 9 8 7 6 5

Sand Golt Brickearth Chalk

Cam R.^t Cam & Granta

THE FENS

CHALK HILLS

ROYSTON DOWNS

On the Castle Hill and in the City three Wells have been recently sunk through the Golt 140 F.t deep which overflowed.

Copious Springs of clear Water flow from the foot of the Chalk Hills.

Water at a great depth on these dry Chalk Hills. Wells more than a hundred feet deep at Royston.

In a dry Valley between High Cross and Colliers End a most regular divis[ion] between the Chalk and its incumbent Strata of Sand, with Brickearth above,[...] in Colliers End fine Blocks of Plumpuddingstone — Some of the Alluvi[...] chalkspotted Clay on these Hills.

GEOLOGICAL VIEW AND SECTION OF THE

A.

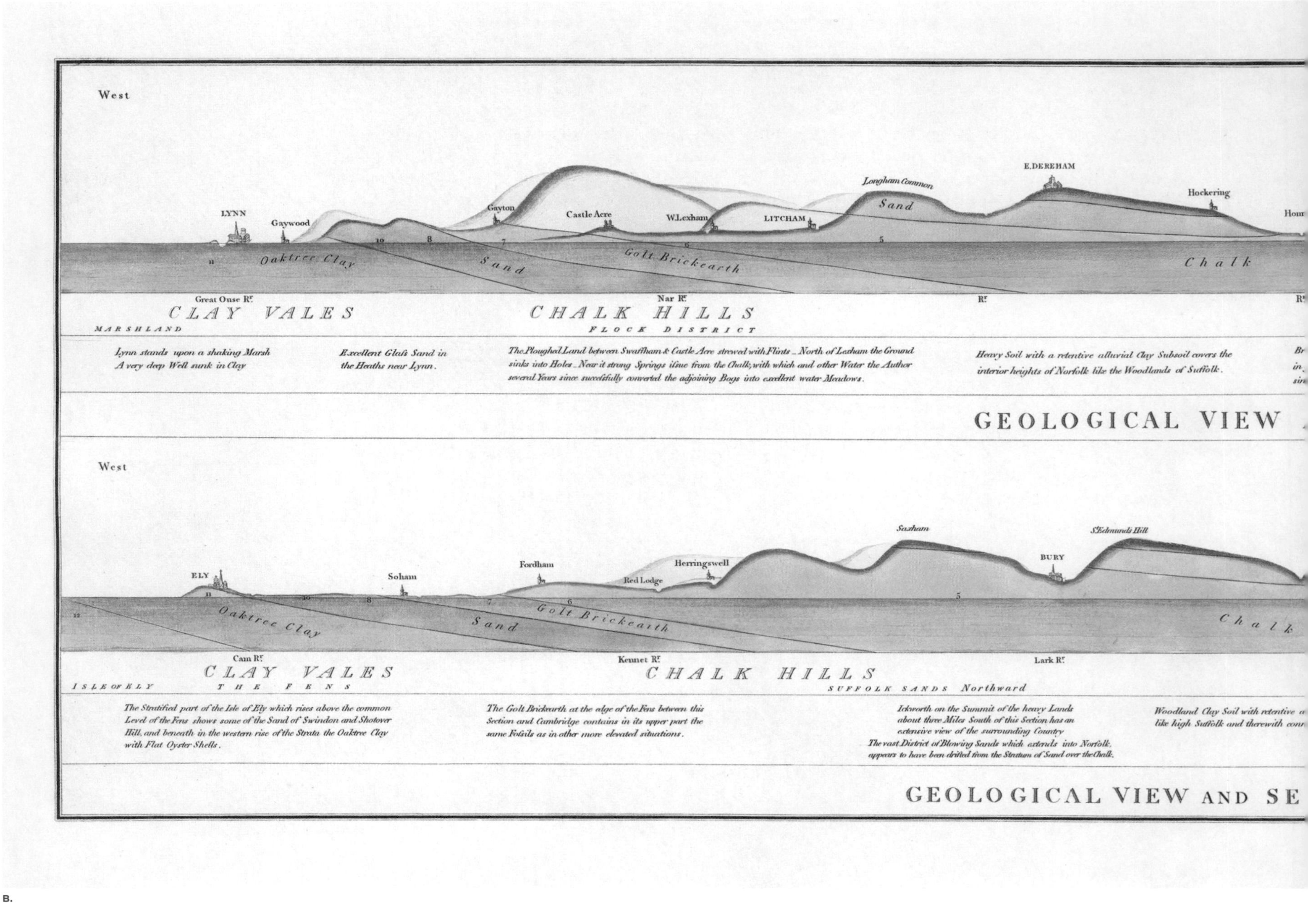

West

LYNN Gaywood Gayton Castle Acre W.Lexham LITCHAM Longham Common E.DEREHAM Hockering Hon[...]

11 10 8 7 6 5

Oaktree Clay Sand Golt Brickearth Sand Chalk

Great Ouse R.^t Nar R.^t R.^t R.^t

CLAY VALES

MARSHLAND

CHALK HILLS

FLOCK DISTRICT

Lynn stands upon a shaking Marsh A very deep Well sunk in Clay

Excellent Glass Sand in the Heaths near Lynn.

The Ploughed Land between Swaffham & Castle Acre strewed with Flints. North of Lexham the Ground sinks into Holes. Near it strong Springs issue from the Chalk; with which and other Water the Author several Years since successfully converted the adjoining Bogs into excellent water Meadows.

Heavy Soil with a retentive alluvial Clay Subsoil covers the interior heights of Norfolk like the Woodlands of Suffolk.

GEOLOGICAL VIEW

West

ELY Soham Fordham Red Lodge Herringswell Saxham BURY S.t Edmunds Hill

11 10 8 7 6 5

Oaktree Clay Sand Golt Brickearth Chalk

Cam R.^t Kennet R.^t Lark R.^t

ISLE OF ELY

CLAY VALES

THE FENS

CHALK HILLS

SUFFOLK SANDS Northward

The Stratified part of the Isle of Ely which rises above the common Level of the Fens shows some of the Sand of Swindon and Shotover Hill, and beneath in the western rise of the Strata the Oaktree Clay with Flat Oyster Shells.

The Golt Brickearth at the edge of the Fens between this Section and Cambridge contains in its upper part the same Fossils as in other more elevated situations.

Ickworth on the Summit of the heavy Lands about three Miles South of this Section has an extensive view of the surrounding Country.

The vast District of Blowing Sands which extends into Norfolk appears to have been drifted from the Stratum of Sand over the Chalk.

Woodland Clay Soil with retentive s[...] like high Suffolk and therewith conn[...]

GEOLOGICAL VIEW AND SE

B.

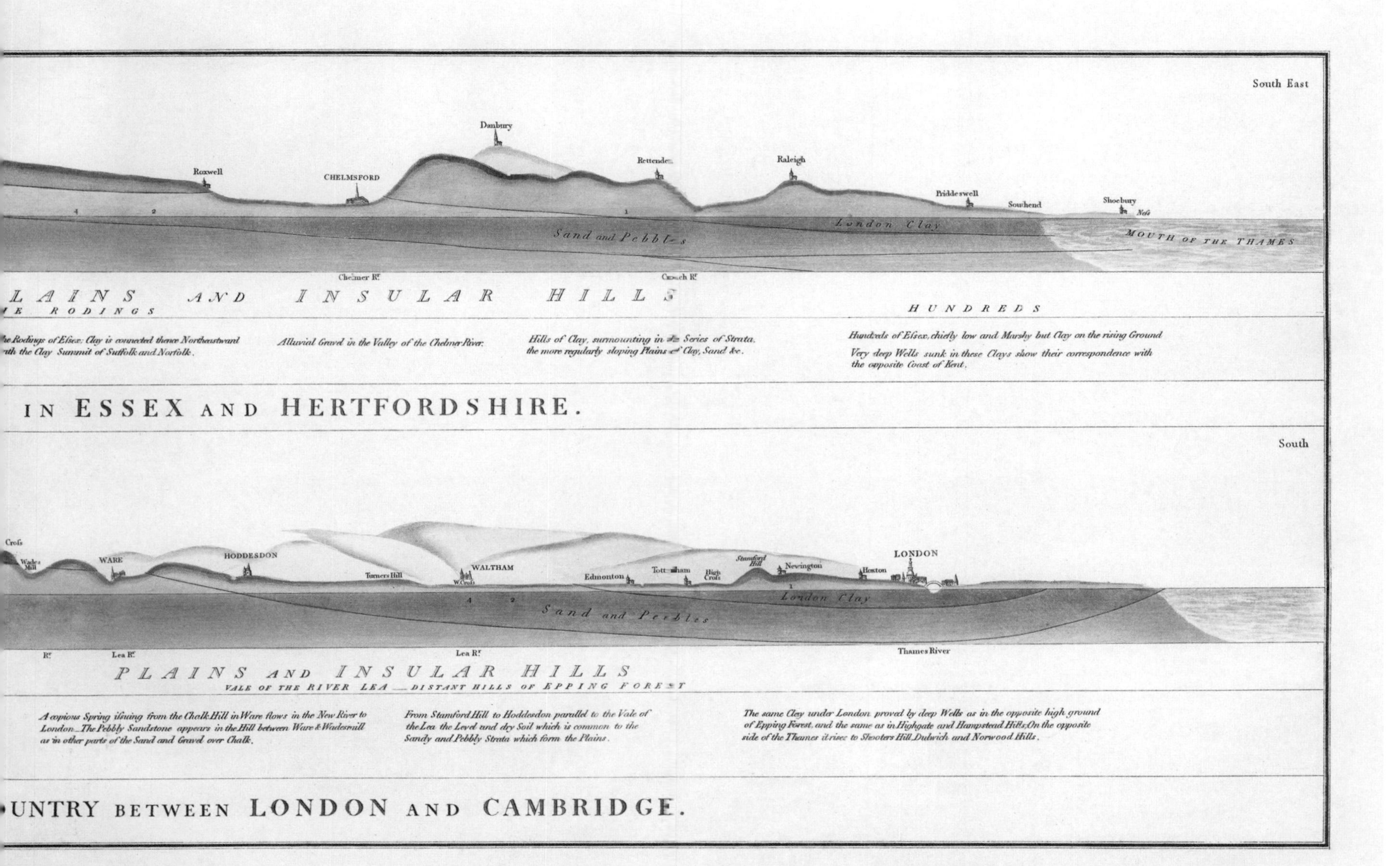
South East
Danbury
Roxwell
CHELMSFORD
Rettendon
Raleigh
Prittlewell
Southend
Shoebury
Ness
Sand and Pebbles
London Clay
MOUTH OF THE THAMES
Chelmer R.r
Crouch R.r
LAINS AND INSULAR HILLS
E RODINGS
HUNDREDS
he Rodings of Essex. Clay is connected thence Northeastward
ith the Clay Summit of Suffolk and Norfolk.
Alluvial Gravel in the Valley of the Chelmer River.
Hills of Clay, surmounting in a Series of Strata,
the more regularly sloping Plains of Clay, Sand &c.
Hundreds of Essex, chiefly low and Marshy but Clay on the rising Ground
Very deep Wells sunk in these Clays show their correspondence with
the opposite Coast of Kent.
IN ESSEX AND HERTFORDSHIRE.
South
Cross
Wades Mill
WARE
HODDESDON
Turners Hill
WALTHAM
W. Cross
Edmonton
Tottenham
High Cross
Stamford Hill
Newington
Hoxton
LONDON
London Clay
Sand and Pebbles
Lea R.r
Lea R.r
Thames River
PLAINS AND INSULAR HILLS
VALE OF THE RIVER LEA _ DISTANT HILLS OF EPPING FOREST
A copious Spring issuing from the Chalk Hill in Ware flows in the New River to
London _ The Pebbly Sandstone appears in the Hill between Ware & Wadesmill
as in other parts of the Sand and Gravel over Chalk.
From Stamford Hill to Hoddesdon parallel to the Vale of
the Lea the Level and dry Soil which is common to the
Sandy and Pebbly Strata which form the Plains.
The same Clay under London proved by deep Wells as in the opposite high ground
of Epping Forest, and the same as in Highgate and Hampstead Hills. On the opposite
side of the Thames it rises to Shooters Hill, Dulwich and Norwood Hills.
UNTRY BETWEEN LONDON AND CAMBRIDGE.

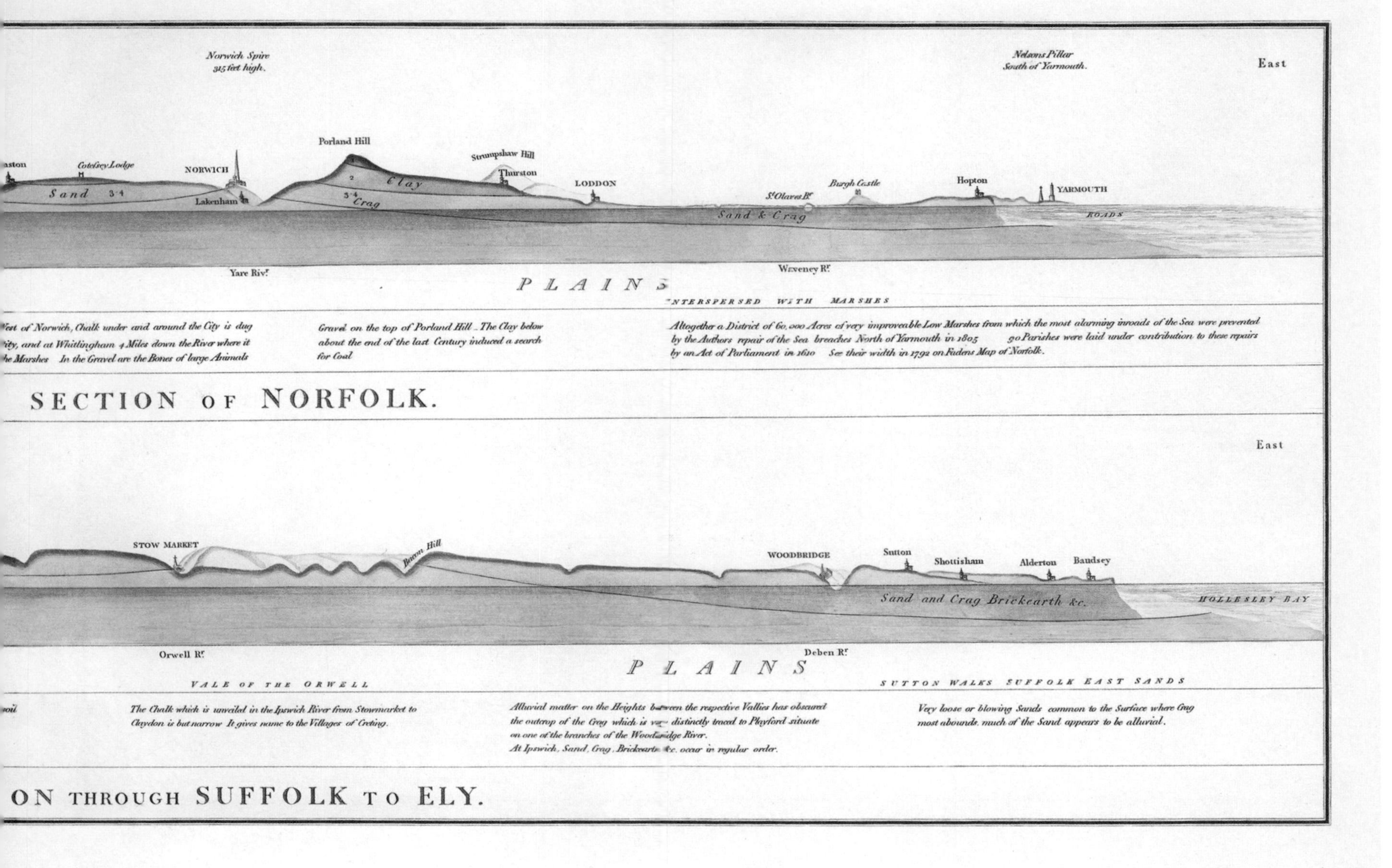
Norwich Spire
315 feet high.
Nelsons Pillar
South of Yarmouth.
East
Cote Grey Lodge
NORWICH
Lakenham
Sand
Porland Hill
Clay
Crag
Strumpshaw Hill
Thurston
LODDON
St. Olaves Br.
Burgh Castle
Hopton
YARMOUTH
Sand & Crag
ROADS
Yare Riv.r
Waveney R.r
PLAINS
INTERSPERSED WITH MARSHES
Gravel on the top of Porland Hill _ The Clay below
about the end of the last Century induced a search
for Coal
Altogether a District of 60,000 Acres of very improveable Low Marshes from which the most alarming inroads of the Sea were prevented
by the Authors repair of the Sea breaches North of Yarmouth in 1805 90 Parishes were laid under contribution to these repairs
by an Act of Parliament in 1610 See their width in 1792 on Fadens Map of Norfolk.
SECTION OF NORFOLK.
East
STOW MARKET
WOODBRIDGE
Sutton
Shottisham
Alderton
Bawdsey
Sand and Crag Brickearth &c.
HOLLESLEY BAY
Orwell R.r
Deben R.r
PLAINS
VALE OF THE ORWELL
SUTTON WALKS SUFFOLK EAST SANDS
The Chalk which is unveiled in the Ipswich River from Stowmarket to
Claydon is but narrow It gives name to the Villages of Creting.
Alluvial matter on the Heights between the respective Vallies has obscured
the outcrop of the Crag which is very distinctly traced to Playford situate
on one of the branches of the Woodbridge River.
At Ipswich, Sand, Crag, Brickearth &c. occur in regular order.
Very loose or blowing Sands common to the Surface where Crag
most abounds, much of the Sand appears to be alluvial.
ON THROUGH SUFFOLK TO ELY.

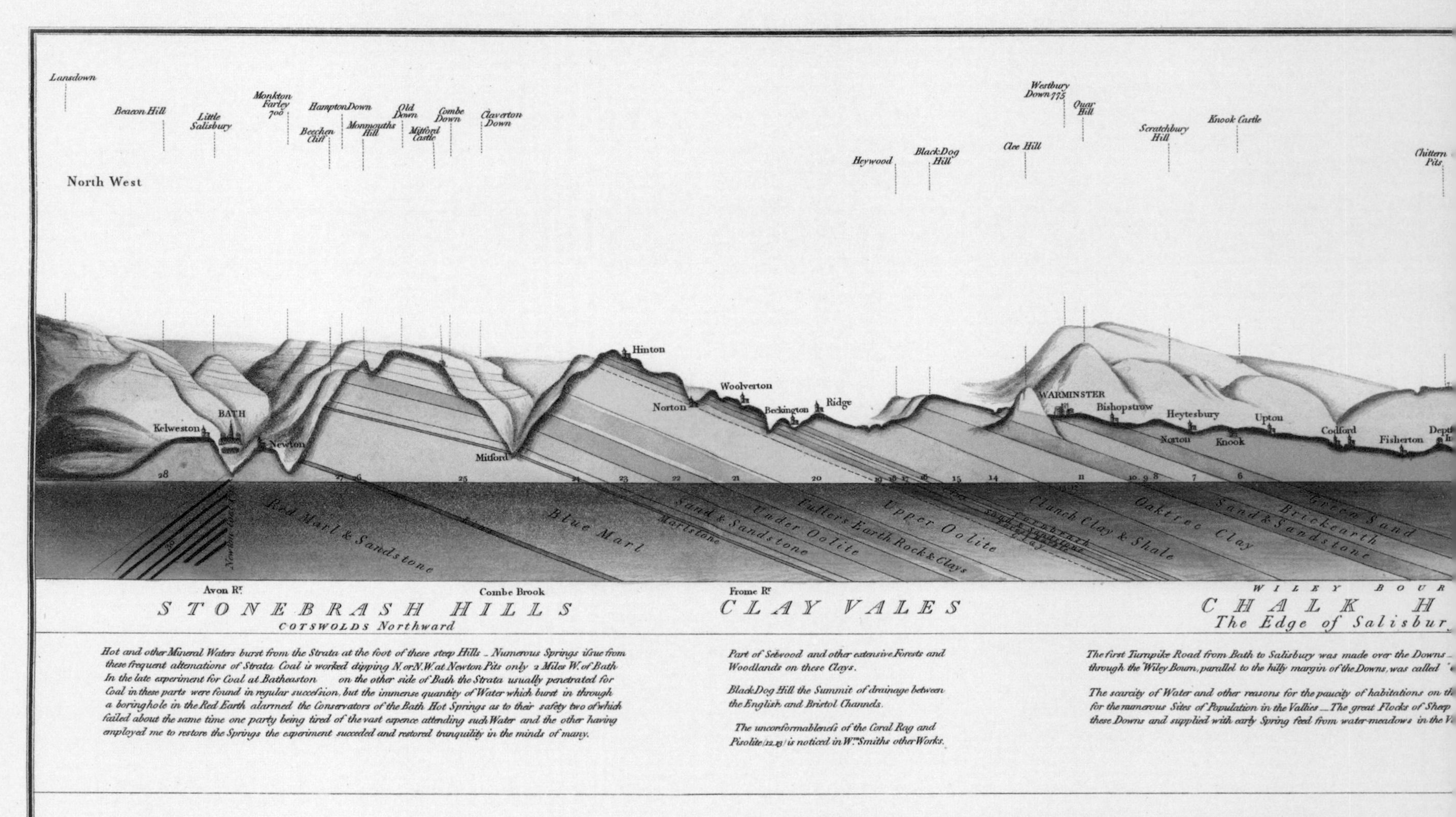

Lansdown
Beacon Hill
Little Salisbury
Monkton Farley 700
Hampton Down
Old Down
Combe Down
Claverton Down
Beechen Cliff
Monmouths Hill
Mitford Castle
Westbury Down 775
Heywood
Black Dog Hill
Clee Hill
Scratchbury Hill
Knook Castle
North West
Kelweston
BATH
Newton
Mitford
Hinton
Norton
Woolverton
Beckington
Ridge
WARMINSTER
Bishopstrow
Heytesbury
Upton
Norton
Knook
Codford
Fisherton
Red Marl & Sandstone
Blue Marl
Marlstone
Sand & Sandstone
Under Oolite
Fullers Earth Rock & Clays
Upper Oolite
Clunch Clay & Shale
Oaktree Clay
Sand & Sandstone
Brickearth
Green Sand
Avon R.
Combe Brook
Frome R.
STONEBRASH HILLS
COTSWOLDS Northward
CLAY VALES
WILEY BOUR
CHALK H
The Edge of Salisbur
Hot and other Mineral Waters burst from the Strata at the foot of these steep Hills _ Numerous Springs ifsue from these frequent alternations of Strata Coal is worked dipping N. or N.W. at Newton Pits only 2 Miles W. of Bath In the late experiment for Coal at Batheaston on the other side of Bath the Strata usually penetrated for Coal in these parts were found in regular succefsion, but the immense quantity of Water which burst in through a boringhole in the Red Earth alarmed the Conservators of the Bath Hot Springs as to their safety two of which failed about the same time one party being tired of the vast expence attending such Water and the other having employed me to restore the Springs the experiment succeded and restored tranquility in the minds of many.
Part of Selwood and other extensive Forests and Woodlands on these Clays.
Black Dog Hill the Summit of drainage between the English and Bristol Channels.
The unconformablenefs of the Coral Rag and Pisolite (12.13) is noticed in Wm. Smiths other Works.
The first Turnpike Road from Bath to Salisbury was made over the Downs _
through the Wiley Bourn, parallel to the hilly margin of the Downs, was called
The scarcity of Water and other reasons for the paucity of habitations on th
for the numerous Sites of Population in the Vallies _ The great Flocks of Sheep
these Downs and supplied with early Spring feed from water-meadows in the V
SECTION OF THE STRATA THROUG
on the ROAD fr
LONDON; PUBLISHED BY

C.

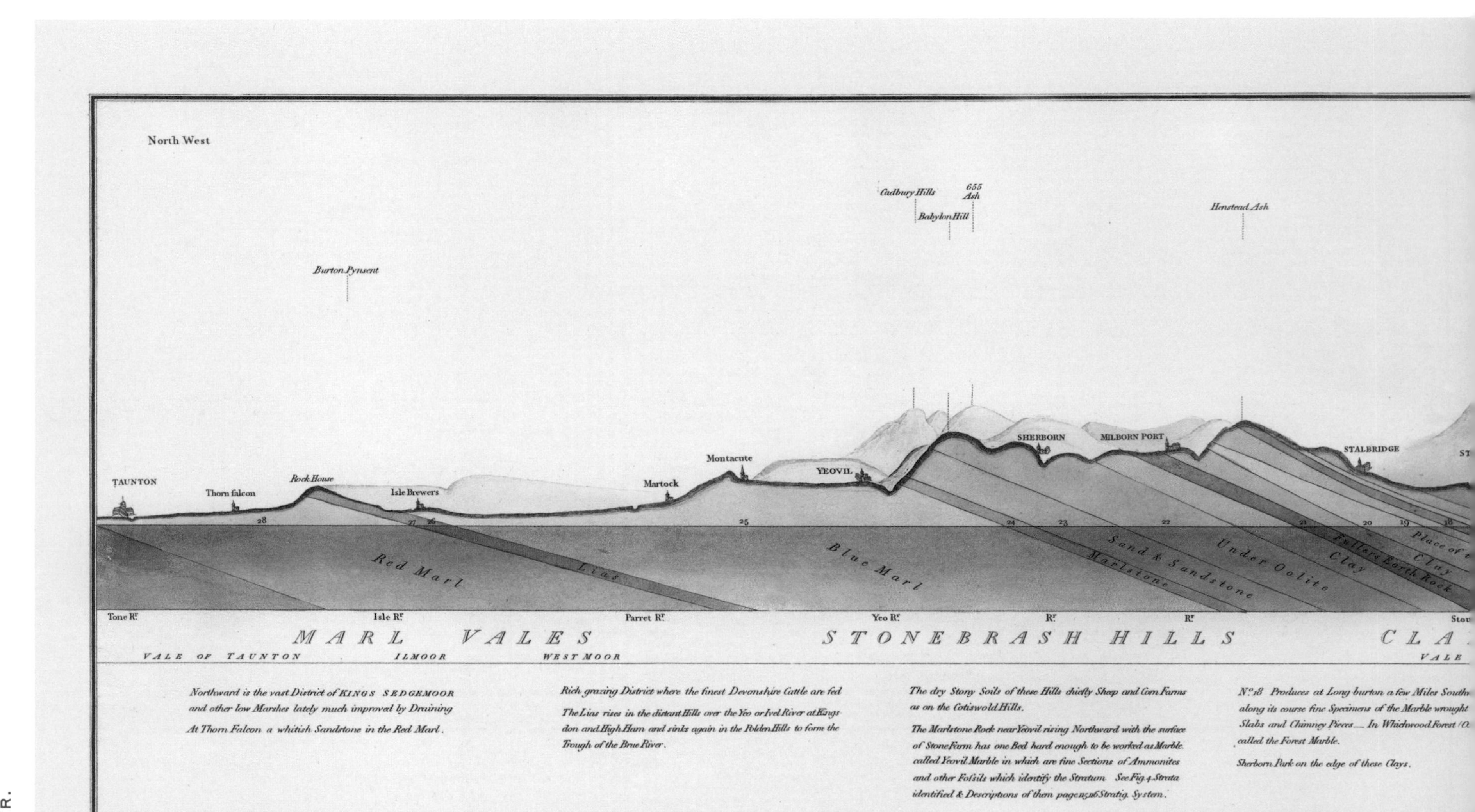

North West
Cadbury Hills
655 Ash
Babylon Hill
Henstead Ash
Burton Pynsent
TAUNTON
Thorn falcon
Rock House
Isle Brewers
Martock
Montacute
YEOVIL
SHERBORN
MILBORN PORT
STALBRIDGE
Red Marl
Lias
Blue Marl
Marlstone
Sand & Sandstone
Under Oolite
Clay
Fullers Earth Rock
Tone R.
Isle R.
Parret R.
Yeo R.
MARL VALES
VALE OF TAUNTON
ILMOOR
WEST MOOR
STONEBRASH HILLS
CLA
VALE
Northward is the vast District of KINGS SEDGEMOOR and other low Marshes lately much improved by Draining
At Thorn Falcon a whitish Sandstone in the Red Marl.
Rich grazing District where the finest Devenshire Cattle are fed
The Lias rises in the distant Hills over the Yeo or Ivel River at Kingsdon and High Ham and sinks again in the Polden Hills to form the Trough of the Brue River.
The dry Stony Soils of these Hills chiefly Sheep and Corn Farms as on the Cotiswold Hills.
The Marlstone Rock near Yeovil rising Northward with the surface of Stone Farm has one Bed hard enough to be worked as Marble. called Yeovil Marble in which are fine Sections of Ammonites and other Fofsils which identify the Stratum See Fig. 4 Strata identified & Descriptions of them page 15.16 Stratig. System.
No. 18 Produces at Long burton a few Miles South
along its course fine Specimens of the Marble wrought
Slabs and Chimney Pieces _ In Whichwood Forest (O
called the Forest Marble.
Sherborn Park on the edge of these Clays.
GEOLOGICAL VIEW AND SECTION THROUGH D
on the ROAD through YEOVI
LONDON; PUBLISHED BY J

D.

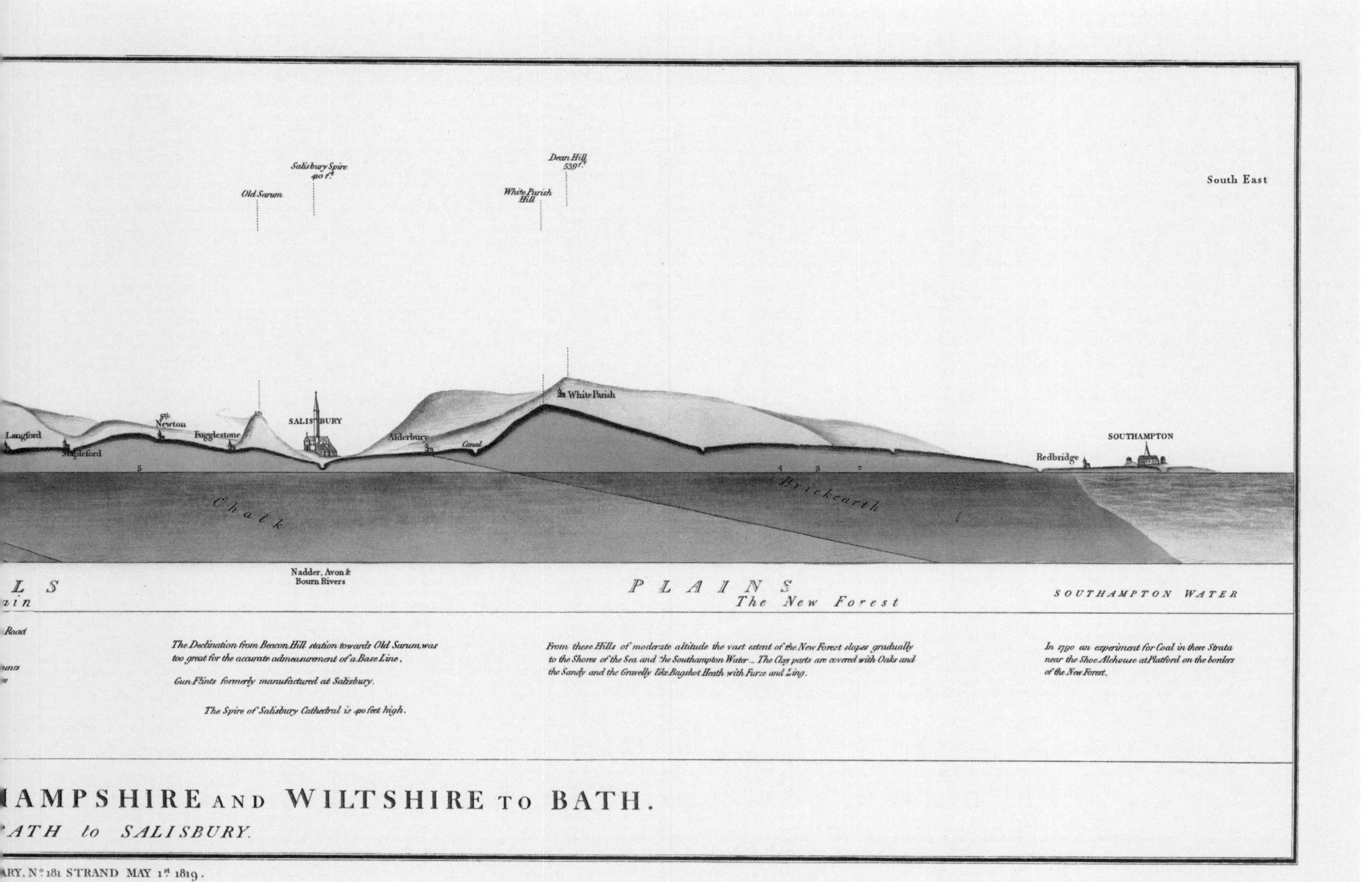
South East
Salisbury Spire 410 Ft.
Old Sarum
Dean Hill 539 Ft.
White Parish Hill
Langford
Stapleford
Newton
Fugglestone
SALISBURY
Alderbury
Canal
White Parish
Redbridge
SOUTHAMPTON
Chalk
Brickearth
Nadder, Avon & Bourn Rivers
L S
PLAINS
The New Forest
SOUTHAMPTON WATER
The Declination from Beacon Hill station towards Old Sarum, was too great for the accurate admeasurement of a Base Line.
Gun Flints formerly manufactured at Salisbury.
The Spire of Salisbury Cathedral is 410 feet high.
From these Hills of moderate altitude the vast extent of the New Forest slopes gradually to the Shores of the Sea and the Southampton Water _ The Clay parts are covered with Oaks and the Sandy and the Gravelly like Bagshot Heath with Furze and Ling.
In 1790 an experiment for Coal in these Strata near the Shoe Alehouse at Platford on the borders of the New Forest.
AMPSHIRE AND WILTSHIRE TO BATH.
ATH to SALISBURY.
RY, No. 181 STRAND MAY 1st 1819.

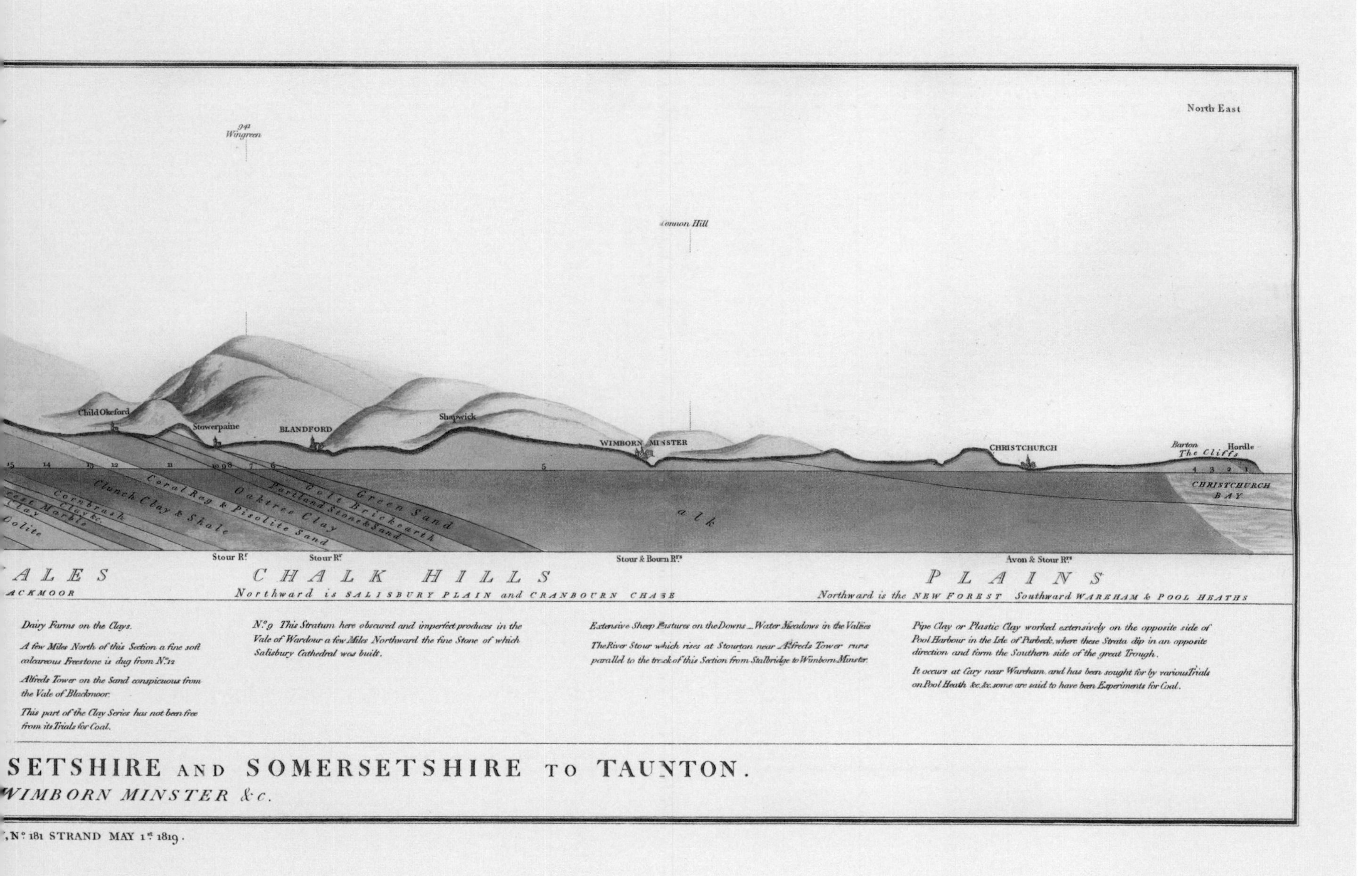
North East
941 Wingreen
Child Okeford
Stowerpaine
BLANDFORD
Shapwick
WIMBORN MINSTER
CHRISTCHURCH
Barton
Hordle
The Cliffs
CHRISTCHURCH BAY
Coral Rag & Pisolite
Clunch Clay & Shale
Oaktree Clay
Portland Stone & Sand
Pisolite Sand
Golt Brickearth
Green Sand
Cornbrash
Forest Marble
Clay
Oolite
Stour Rr.
Stour & Bourn Rrs.
Avon & Stour Rrs.
ALES
ACKMOOR
CHALK HILLS
Northward is SALISBURY PLAIN and CRANBOURN CHASE
PLAINS
Northward is the NEW FOREST Southward WAREHAM & POOL HEATHS
Dairy Farms on the Clays.
A few Miles North of this Section a fine soft calcareous Freestone is dug from No. 12
Alfreds Tower on the Sand conspicuous from the Vale of Blackmoor.
This part of the Clay Series has not been free from its Trials for Coal.
No. 9 This Stratum here obscured and imperfect produces in the Vale of Wardour a few Miles Northward the fine Stone of which Salisbury Cathedral was built.
Extensive Sheep Pastures on the Downs _ Water Meadows in the Vallies
The River Stour which rises at Stourton near Alfreds Tower runs parallel to the track of this Section from Stalbridge to Wimborn Minster.
Pipe Clay or Plastic Clay worked extensively on the opposite side of Pool Harbour in the Isle of Purbeck, where these Strata dip in an opposite direction and form the Southern side of the great Trough.
It occurs at Cary near Wareham, and has been sought for by various Trials on Pool Heath &c. &c. some are said to have been Experiments for Coal.
SETSHIRE AND SOMERSETSHIRE TO TAUNTON.
WIMBORN MINSTER &c.
, No. 181 STRAND MAY 1st 1819.

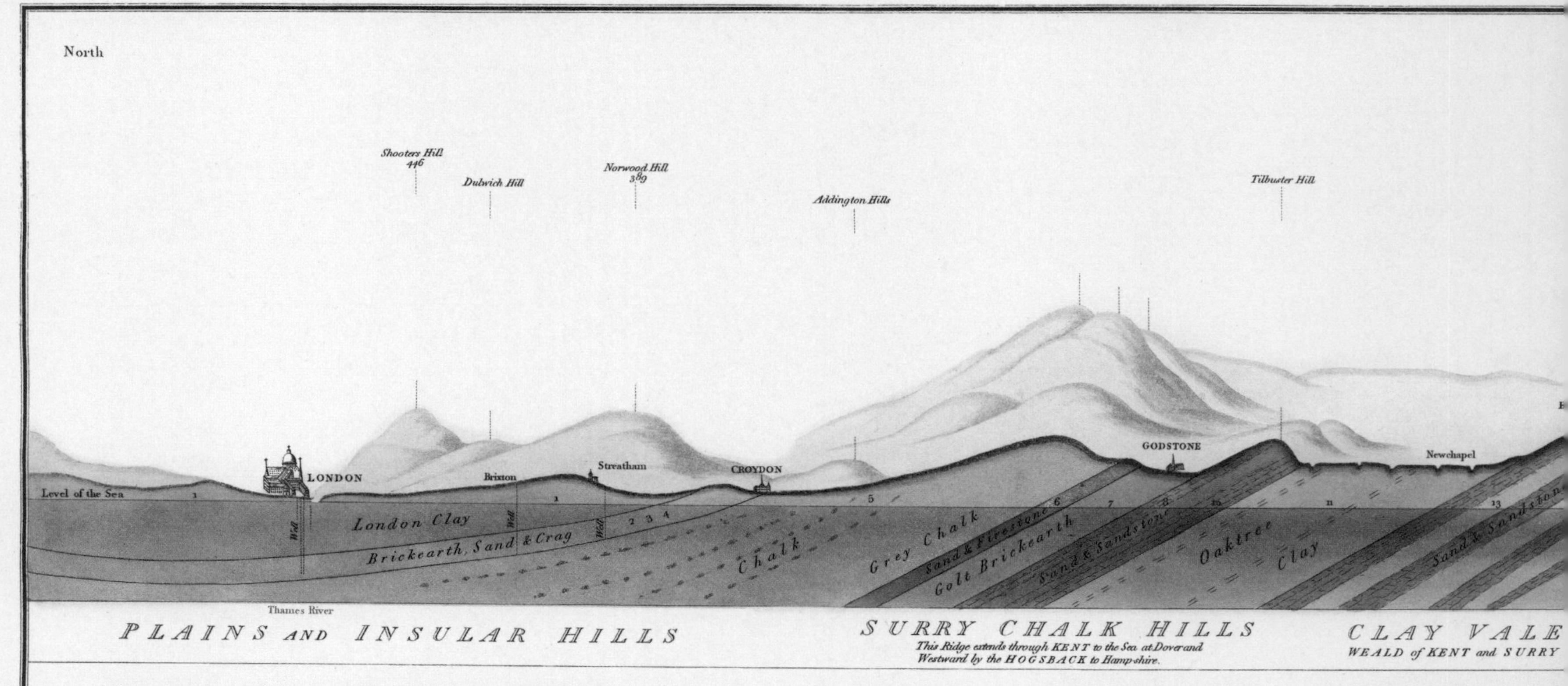
North
Shooters Hill
446
Dulwich Hill
Norwood Hill
389
Addington Hills
Tilbuster Hill
LONDON
Brixton
Streatham
CROYDON
GODSTONE
Newchapel
Level of the Sea
Well
London Clay
Brickearth, Sand & Crag
Chalk
Grey Chalk
Sand & Firestone
Golt Brickearth
Sand & Sandstone
Oaktree
Clay
Thames River
PLAINS AND INSULAR HILLS
SURRY CHALK HILLS
This Ridge extends through KENT to the Sea at Dover and
Westward by the HOGSBACK to Hampshire.
CLAY VALE
WEALD of KENT and SURRY
The little Water found in these Hills generally of a bad Quality _ In deep Wells sunk through the Clay Water rises from the Chalk
Shooters Hill higher than Norwood and other Clay Hills South of London shows at its base the Chalk at Greenwich, Sand & Loam
at Woolwich _ Higher in the Clay lie Organized Fossils similar to those in the Clay on Newhaven Castle Hill _ In deep Wells
at Brixton Causway and Streatham the same species are found. On the Plain of Blackheath which is Sand and Gravel & the
base of the more elevated Clay in the Hill, Wells are sunk into the Chalk, which shows the absurdity of the common
opinion that Blackheath contains Coal.
Fossil Oyster Shells at Addington
Hills and Sudbury
Fossil Echini occur in these
Hills as in other similar parts
of the Chalk.
The Golt Brickearth
is identified by a small
transparent belemnite.
On Copthorn Common near Newchapel
is the Summit Level of the proposed
Weald of Kent and Mesterham and
Portsmouth Canals 166 feet above the
Tunbridge Navigation.
VERTICAL SECTION OF STRATA IN SURRY DIPPING NORTHWARD
LONDON; PUBLISHED BY JO

E.

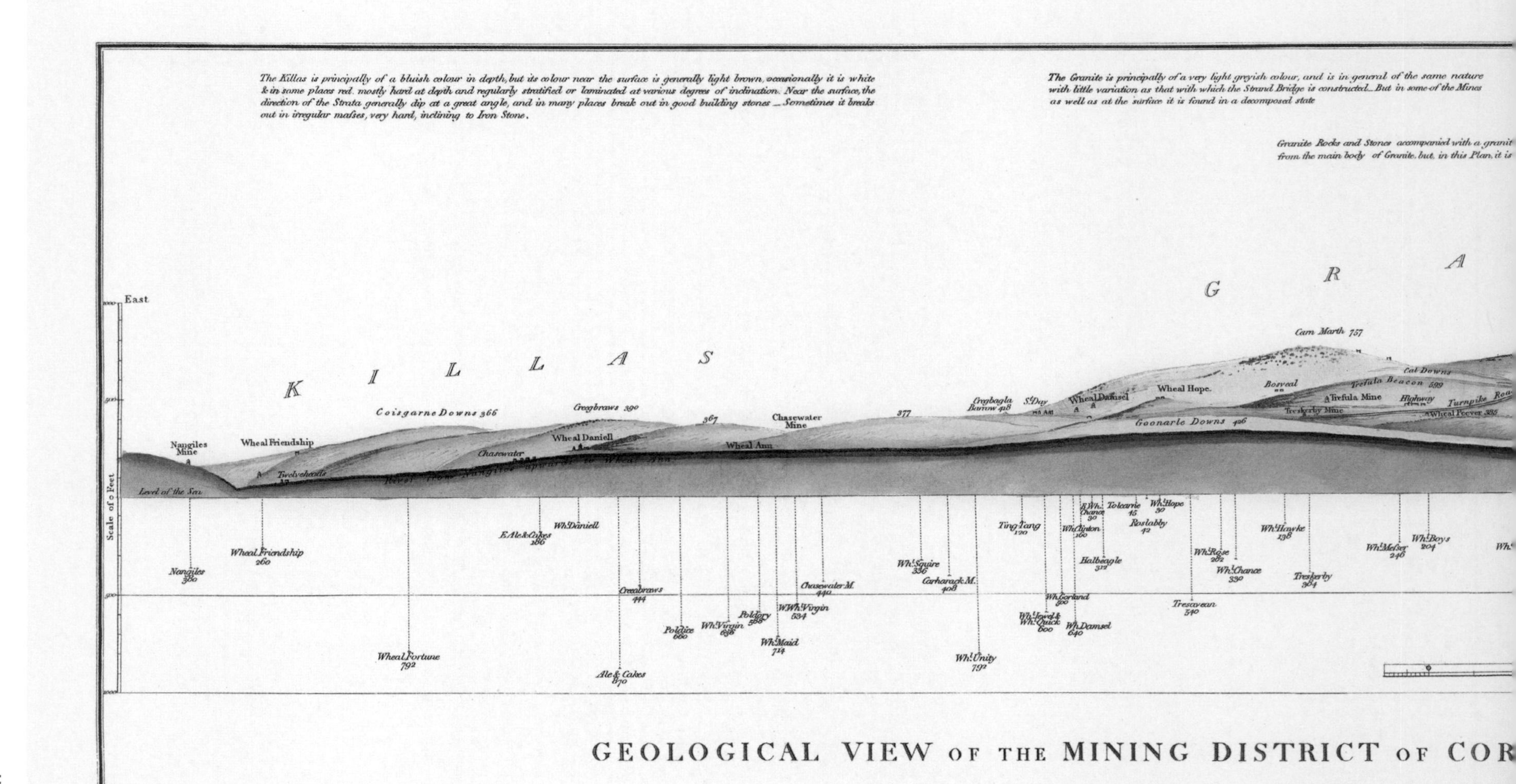
The Killas is principally of a bluish colour in depth, but its colour near the surface is generally light brown, occasionally it is white
& in some places red. mostly hard at depth and regularly stratified or laminated at various degrees of inclination. Near the surface, the
direction of the Strata generally dip at a great angle, and in many places break out in good building stones _ Sometimes it breaks
out in irregular masses, very hard, inclining to Iron Stone.
The Granite is principally of a very light greyish colour, and is in general of the same nature
with little variation as that with which the Strand Bridge is constructed _ But in some of the Mines
as well as at the surface it is found in a decomposed state
Granite Rocks and Stones accompanied with a granit
from the main body of Granite, but, in this Plan, it is
East
K I L L A S
G R A
Carn Marth 757
Coisgarne Downs 366
Goonarle Downs 416
Wheal Friendship
Wheal Daniell
Chasewater Mine
Wheal Hope
Level of the Sea
Scale of Feet
Wheal Fortune
792
Ale & Cakes
870
GEOLOGICAL VIEW OF THE MINING DISTRICT OF COR
shewing the Elevations of the Hills and the depth
LONDON; PUBLISH

F.

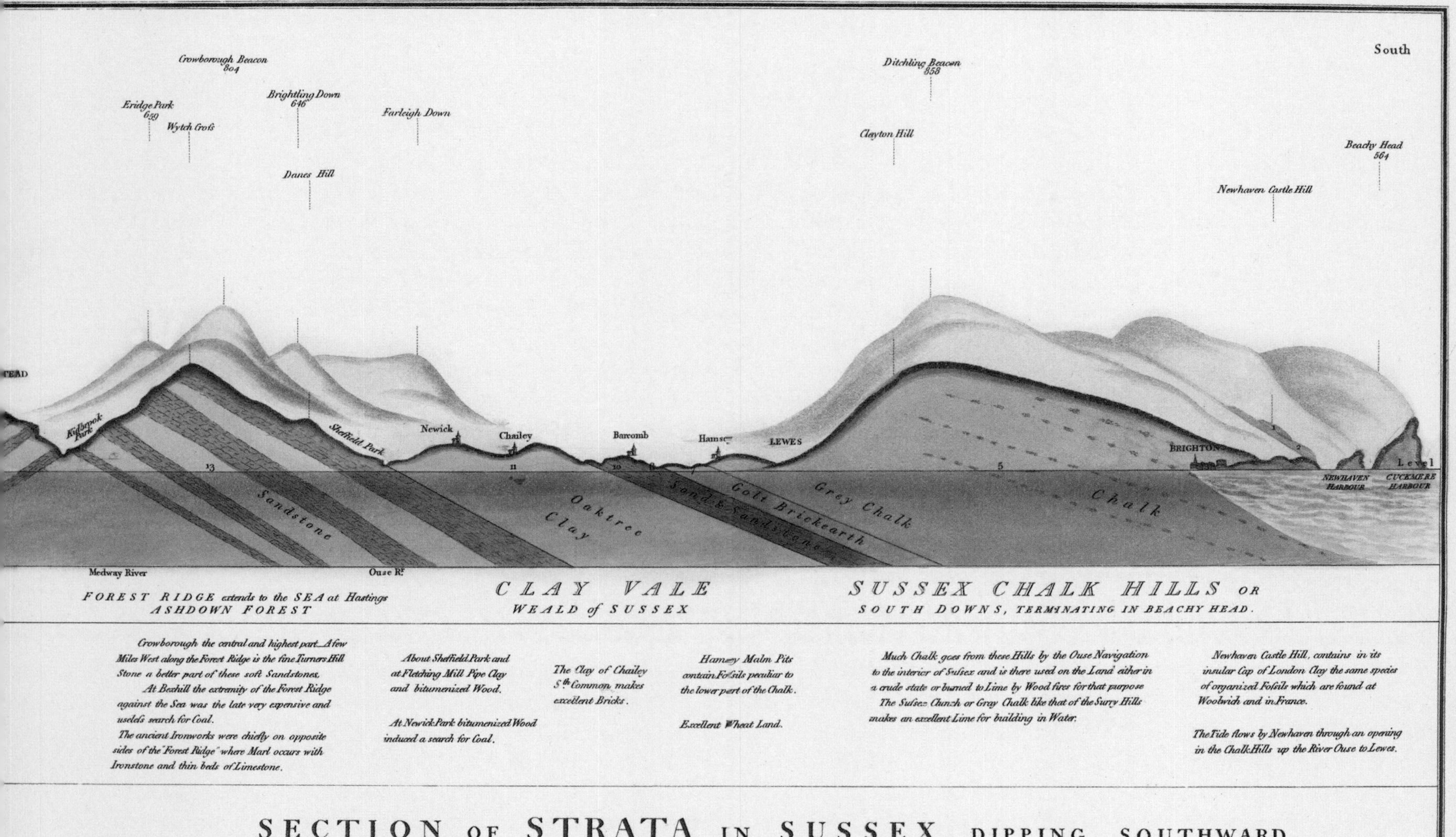
South
Crowborough Beacon 804
Eridge Park 659
Wytch Croſs
Brightling Down 646
Farleigh Down
Danes Hill
Ditchling Beacon 858
Clayton Hill
Beachy Head 564
Newhaven Castle Hill
Newick
Chailey
Barcomb
Hamsey
LEWES
BRIGHTON
Level
NEWHAVEN HARBOUR
CUCKMERE HARBOUR
Sandstone
Oaktree Clay
Golt Brickearth
Grey Chalk
Chalk
Medway River
Ouse R.
FOREST RIDGE extends to the SEA at Hastings
ASHDOWN FOREST
CLAY VALE
WEALD of SUSSEX
SUSSEX CHALK HILLS OR
SOUTH DOWNS, TERMINATING IN BEACHY HEAD.
Crowborough the central and highest part_A few Miles West along the Forest Ridge is the fine Turners Hill Stone a better part of these soft Sandstones.
At Bexhill the extremity of the Forest Ridge against the Sea was the late very expensive and useleſs search for Coal.
The ancient Ironworks were chiefly on opposite sides of the "Forest Ridge" where Marl occurs with Ironstone and thin beds of Limestone.
About Sheffield Park and at Fletching Mill Pipe Clay and bitumenized Wood.
At Newick Park bitumenized Wood induced a search for Coal.
The Clay of Chailey S.th Common makes excellent Bricks.
Hamsey Malm Pits contain Foſsils peculiar to the lower part of the Chalk.
Excellent Wheat Land.
Much Chalk goes from these Hills by the Ouse Navigation to the interior of Suſsex and is there used on the Land either in a crude state or burned to Lime by Wood fires for that purpose
The Suſsex Church or Gray Chalk like that of the Surry Hills makes an excellent Lime for building in Water.
Newhaven Castle Hill, contains in its insular Cap of London Clay the same species of organized Foſsils which are found at Woolwich and in France.
The Tide flows by Newhaven through an opening in the Chalk Hills up the River Ouse to Lewes.
SECTION OF STRATA IN SUSSEX DIPPING SOUTHWARD.
No. 181 STRAND MAY 1st 1819.

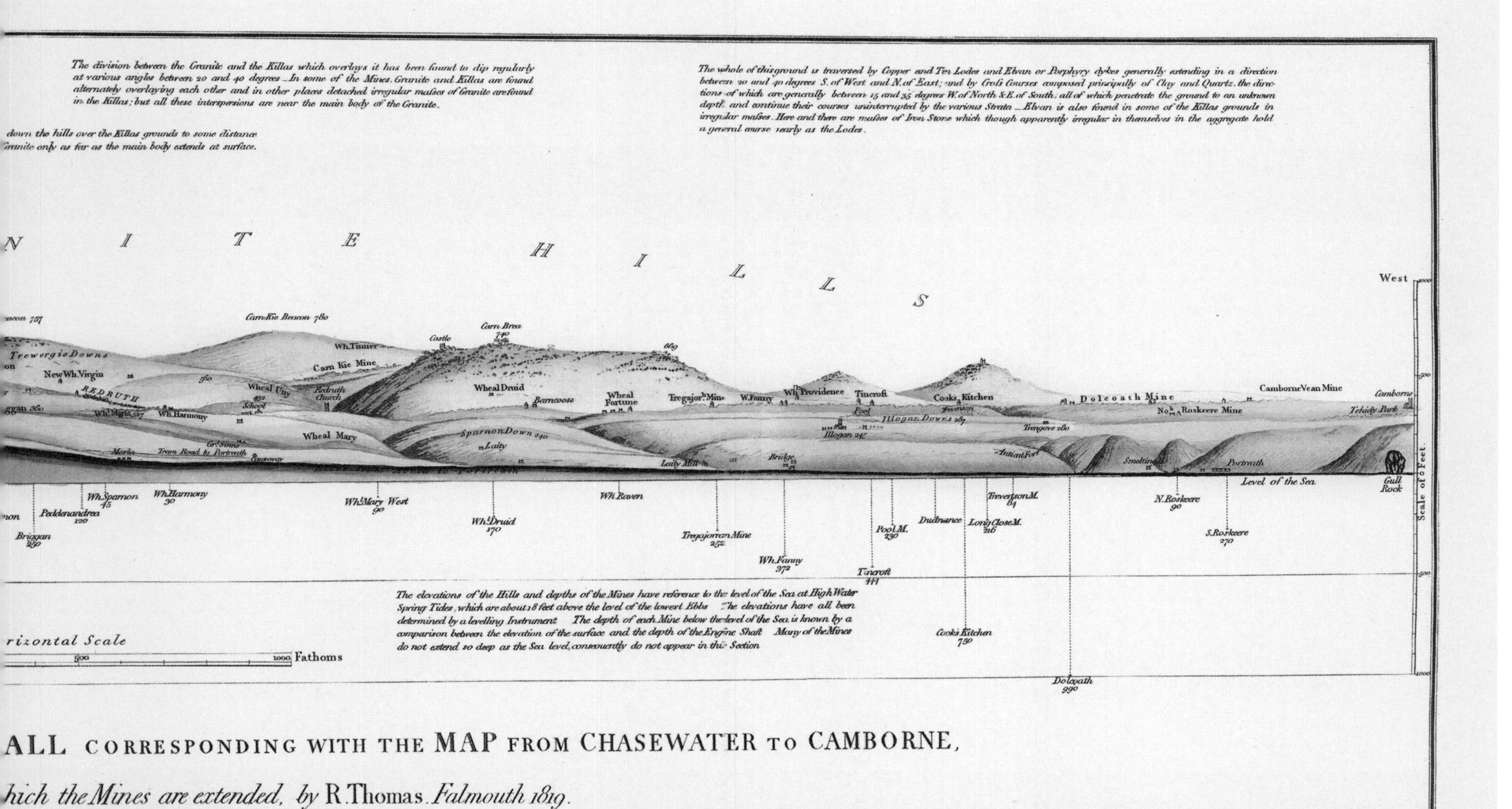
The division between the Granite and the Killas which overlays it has been found to dip regularly at various angles between 20 and 40 degrees _ In some of the Mines, Granite and Killas are found alternately overlaying each other and in other places detached irregular maſses of Granite are found in the Killas; but all these interspersions are near the main body of the Granite.
down the hills over the Killas grounds to some distance
Granite only as far as the main body extends at surface.
The whole of this ground is traversed by Copper and Tin Lodes and Elvan or Porphyry dykes generally extending in a direction between 20 and 40 degrees S. of West and N. of East; and by Croſs Courses composed principally of Clay and Quartz, the directions of which are generally between 15 and 35 degrees W. of North & E. of South; all of which penetrate the ground to an unknown depth and continue their courses uninterrupted by the various Strata _ Elvan is also found in some of the Killas grounds in irregular maſses. Here and there are maſses of Iron Stone which though apparently irregular in themselves in the aggregate hold a general course nearly as the Lodes.
N I T E H I L L S
West
Carn Kie Beacon 780
Carn Brea
Trewergie Downs
New Wh. Virgin
Carn Kie Mine
Wheal Druid
Wheal Mary
Sparnon Down
Wheal Fortune
W. Fanny
Wh. Providence
Tincroft
Cooks Kitchen
Dolcoath Mine
Camborne Vean Mine
Roskeere Mine
Level of the Sea
Gull Rock
Wh. Sparnon 45
Wh. Harmony 30
Peddenandrea 120
Briggan 260
Wh. Mary West 90
Wh. Druid 170
Wh. Raven
Tregajorran Mine 252
Wh. Fanny 372
Tincroft 444
Pool M. 230
Dudnance
Long Close M.
Treverson M.
Cooks Kitchen 750
Dolcoath 990
N. Roskeere 90
S. Roskeere 270
Scale of Feet
The elevations of the Hills and depths of the Mines have reference to the level of the Sea at High Water Spring Tides, which are about 18 feet above the level of the lowest Ebbs The elevations have all been determined by a levelling Instrument The depth of each Mine below the level of the Sea is known by a comparison between the elevation of the surface and the depth of the Engine Shaft Many of the Mines do not extend so deep as the Sea level, consequently do not appear in this Section
rizontal Scale
500
1000 Fathoms
ALL CORRESPONDING WITH THE MAP FROM CHASEWATER TO CAMBORNE,
hich the Mines are extended, by R. Thomas. Falmouth 1819.
CARY, No. 181 STRAND, JULY 1st 1819.

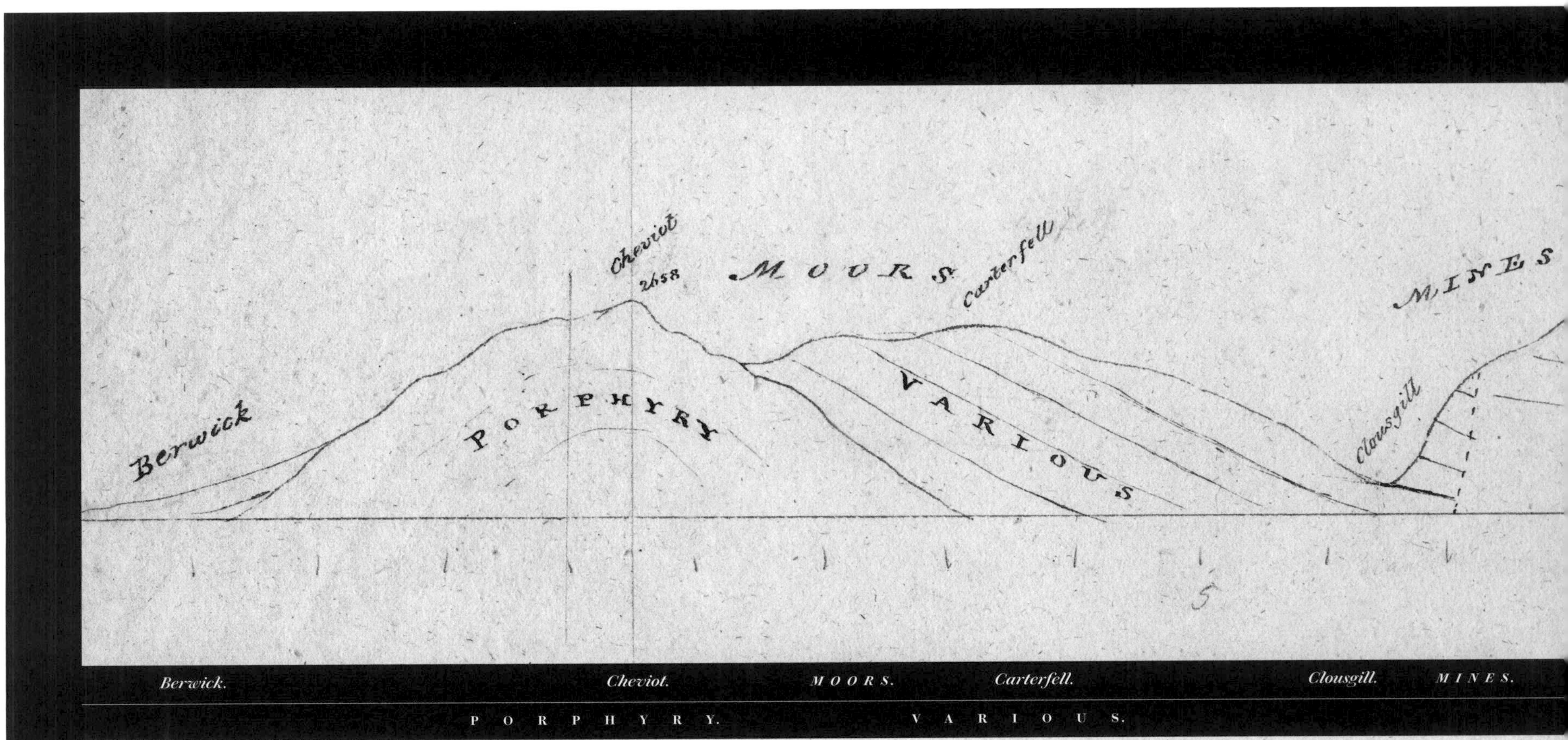

Berwick. Cheviot. MOORS. Carterfell. Clousgill. MINES.
PORPHYRY. VARIOUS.

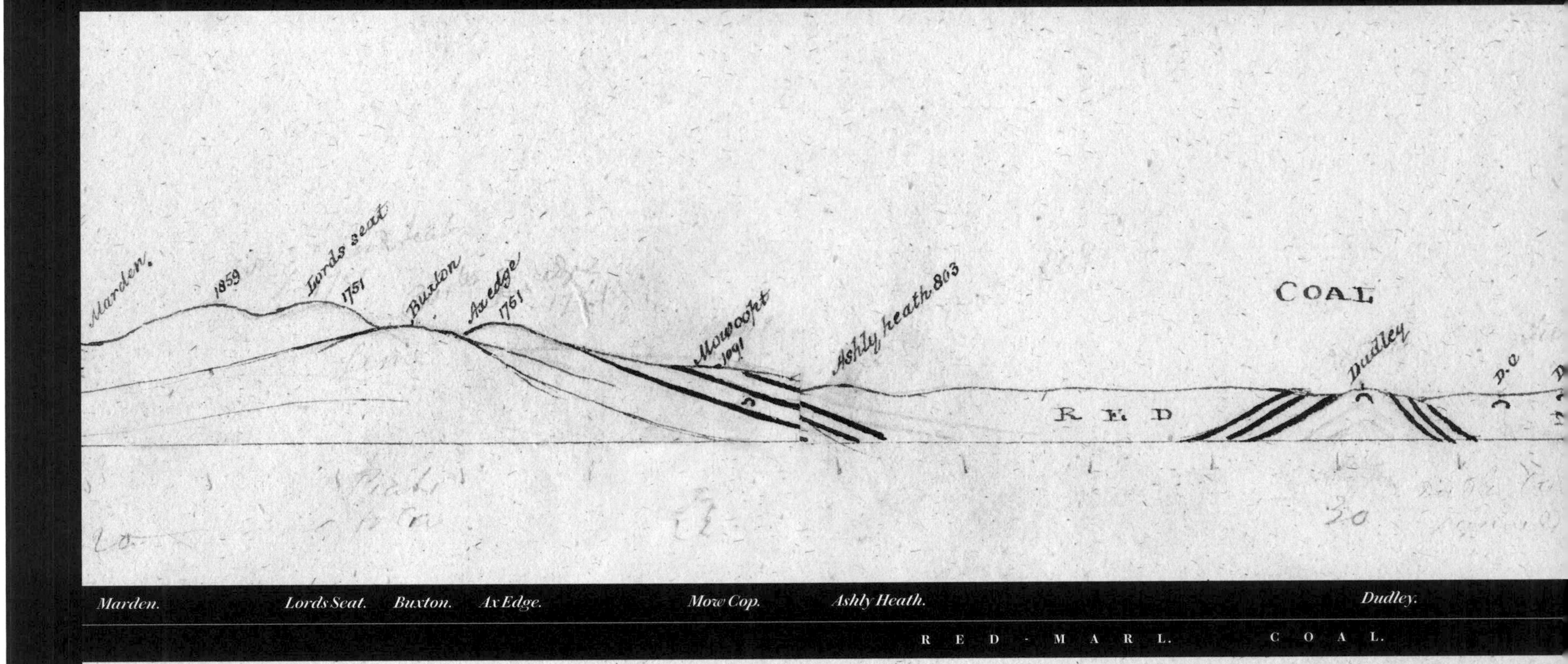

Marden. Lords Seat. Buxton. Ax Edge. Mow Cop. Ashly Heath. Dudley.
RED - MARL. COAL.

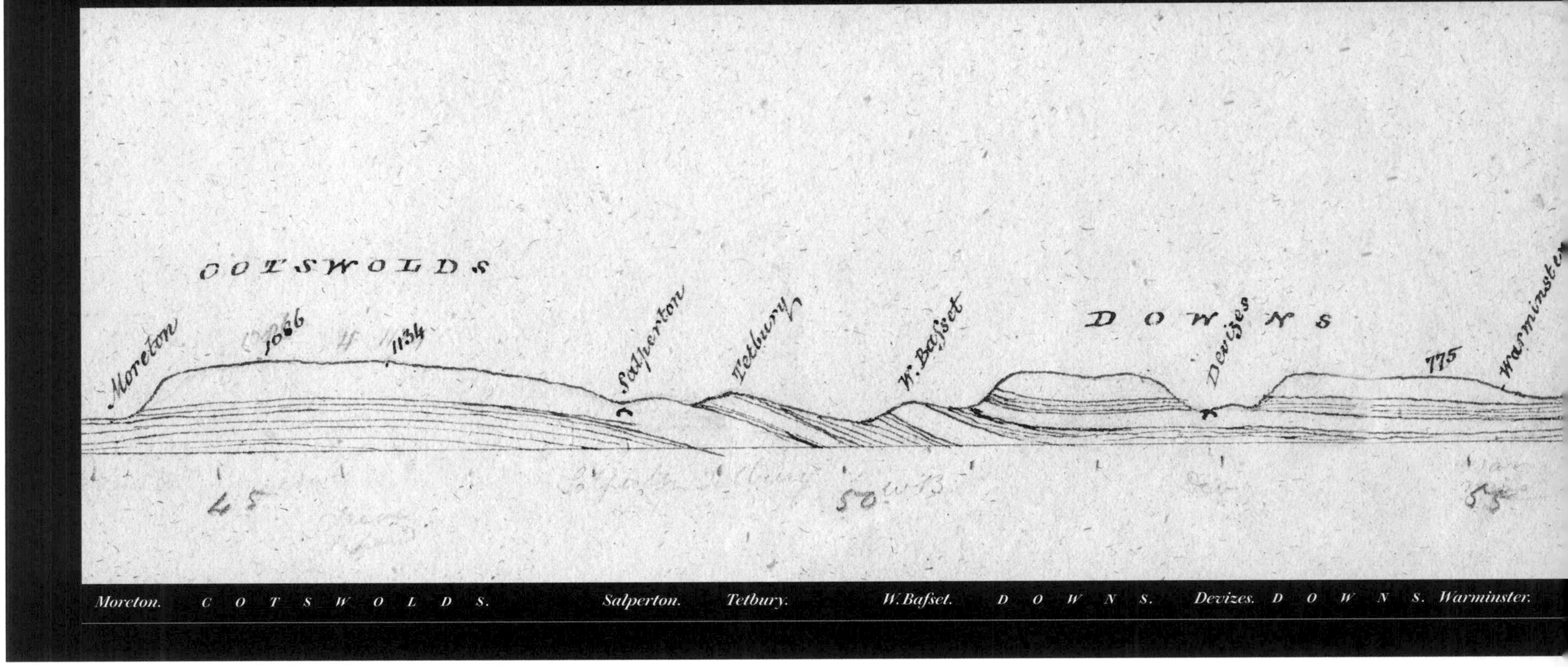

Moreton. COTSWOLDS. Salperton. Tetbury. W. Baſset. DOWNS. Devizes. DOWNS. Warminster.

PAGES 172–77: SECTIONS A–E.

Smith and Cary issued five large, hand-coloured, panoramic horizontal sections, dated 1 May 1819, which sold for five shillings each. They were also available in a bound quarto volume with several additional sections, including two mining sections of Cornwall by Richard Thomas (1779–1858), dated 1 July 1819.

PAGES 176–77: SECTION F.

Geological View of the Mining District of Cornwall corresponding with the Map from Chasewater to Camborne shewing the Elevations of the Hills and depths to which the Mines are extended, Richard Thomas, 1819. Produced to accompany his 1819 report on a survey of the mining district of Cornwall.

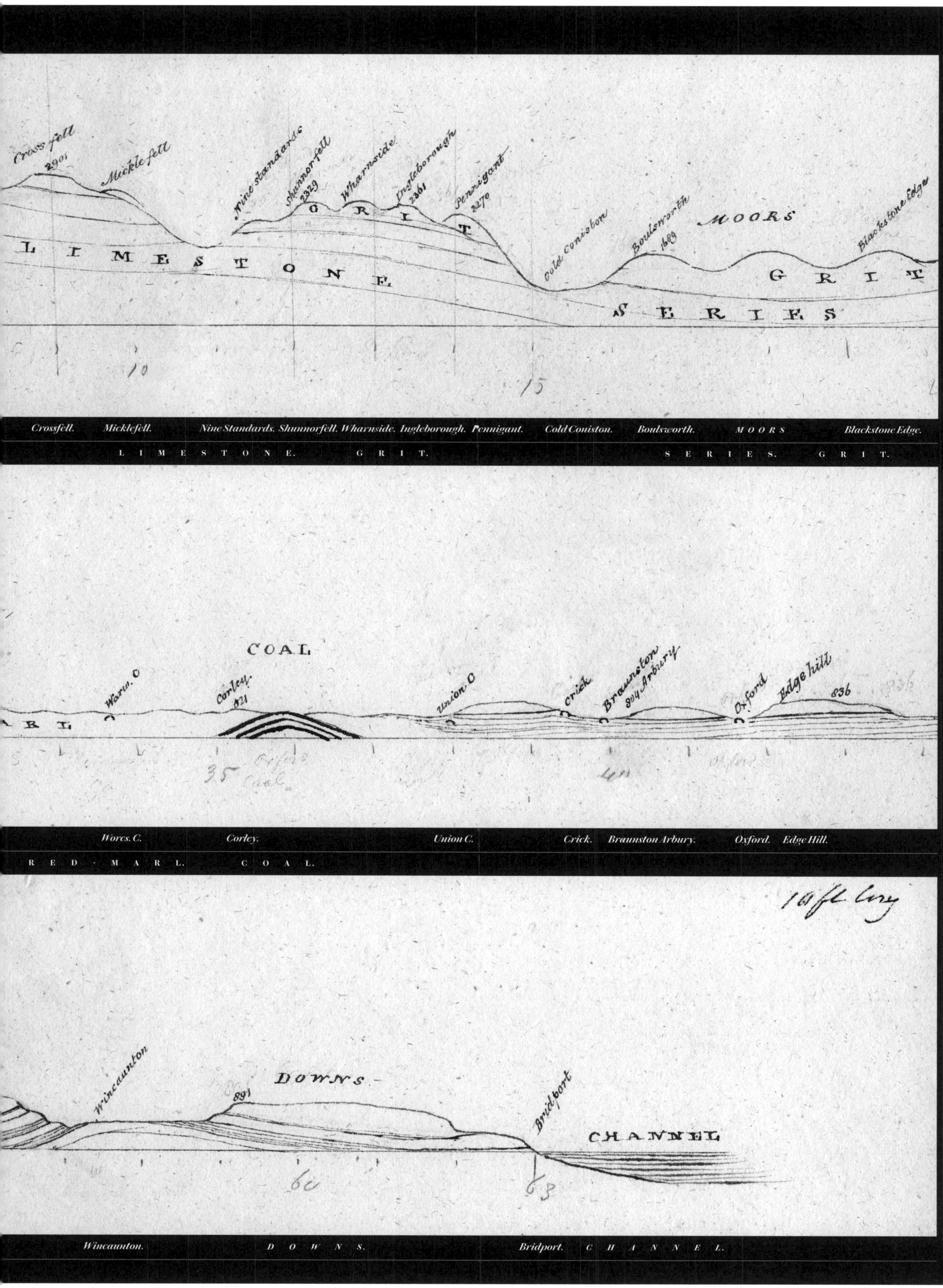

Crossfell. Micklefell. Nine Standards. Shunnorfell. Wharnside. Ingleborough. Pennigant. Cold Coniston. Boulsworth. MOORS Blackstone Edge.

L I M E S T O N E. G R I T. S E R I E S. G R I T.

Worcs. C. Corley. Union C. Crick. Braunston Arbury. Oxford. Edge Hill.

R E D - M A R L. C O A L.

Wincaunton. D O W N S. Bridport. C H A N N E L.

ABOVE: SUMMIT RIDGE OF ENGLAND, JOHN PHILLIPS, 1823.

Created by William Smith's nephew and assistant John Phillips, this long manuscript cross section shows the topography and the varied dips of the strata from Berwick and the 'Porphyry' (Granite) of the Cheviot in the north of England, through the Pennines and the Coal basins of the Midlands overlain by Red Marl, through the Great Oolite of the Cotswold Escarpment to the Chalk downs of Dorset in the west, and on to Bridport on the English Channel coast. The section is 92 cm (1 yd) in length.

Part I was published in 1819 and contained maps of Norfolk, Kent, Wiltshire and Sussex; Part II, published later in 1819, comprised maps of Gloucestershire, Berkshire, Surrey (spelled 'Surry') and Suffolk; Part III, published in 1820, had the maps for Oxfordshire, Buckinghamshire, Bedfordshire and Essex; the maps in Part IV, in 1821, covered the whole of Yorkshire; Part V contained maps for Nottinghamshire, Leicestershire, Huntingdonshire and Rutlandshire, was published in 1822, although the maps are dated 1821; and Part VI, containing Cumberland, Durham, Northumberland and Westmoreland, was published in 1824.

Each part was priced at £1 5s, but the maps could be bought separately for 5s 6d each. They were available as a flat sheet or mounted on cloth and folded into a case. Like the small England and Wales map of 1820, the county geological maps probably sold well, as production continued well into the 1830s. However, after Part VI was issued in 1824, no new county maps were published, even though three, Lincolnshire, Northamptonshire and Somersetshire, were close to completion, and at least eight others were in various stages of drafting. So Smith's *New Geological Atlas* can join *Strata Identified by Organized Fossils* and *A Stratigraphical System of Organized Fossils* in the list of his unfinished publications (Torrens, 2016).

LATER PUBLICATION OF THE COUNTY MAPS.

John Cary senior retired in 1820–21 and after his death in 1835, his map business continued as G. & J. Cary under his sons George and John until 1850. At some point in the late 1840s, the copper plates for Smith's county maps were acquired by the map-seller George Frederick Cruchley (1796–1880). A skilled map-maker in his own right, trained by the cartographer Aaron Arrowsmith, Cruchley discovered that there was easier money to be made by reprinting, with minor modification and on cheap paper, older maps from other publishers. With his acquisition of the Cary plates, from 1855 Cruchley began reprinting the county maps, overprinted – roughly and often inaccurately – with railway lines and stations. *Cruchley's Railway Map*, later *Cruchley's Railway and Telegraphic Map*, was printed on thin, poor quality paper, but sold for only sixpence. These railway maps often included Smith's geological lines, uncoloured key boxes and geological information, and the sheets advertised that the maps were available geologically coloured for 3s 6d.

An example was published in *The Wiltshire Archaeological and Natural History Magazine* in June 1858 to accompany an article on the geology of Wiltshire by George Poulett Scrope (1797–1876). The plate has been partly re-engraved and the date clumsily altered from 1855 to 1858. The printed map contains no mention that it is largely the work of William Smith, although Scrope does credit Smith in the text of his article. Likewise, the Smith-Cary map of Surrey ('Surry') was reissued in 1863 re-engraved with additional geological lines to accompany the *Flora of Surrey* by James Alexander Brewer (1818–1886). A separate map of the *Flora of Surrey* was also published by the Holmesdale Natural History Club. The Smith-Cary-Cruchley county maps, with their geological lines, continued production even into the early twentieth century after the plates passed to the Edinburgh map publishers Gall and Inglis in 1876, reaching almost a hundred years of publication (Davis, 1952).

STRATIFICATION IN HACKNESS HILLS.

Between 1828 and 1834, following his move with Phillips north to Yorkshire after his release from debtors' prison, Smith was employed by Sir John Vanden Bempde Johnstone (1799–1869) as land steward on his estate at Hackness near

FIG. 9.

Fig. 9. Smith's geological manuscript annotations on Cary's 1811 *New Map of Essex* in preparation for the engraving of the plate for their *Geological Map of Essex*, published in 1820.

FIG. 10.

Fig. 10. Smith's *Geological Map of Essex* first published on 1 February 1820 in Part 3 of his *Geological Atlas*.

Scarborough. Early on during his time there, Smith prepared a large-scale map of the estate, *Stratification in Hackness Hills*, which was exhibited at the first meeting of the British Association for the Advancement of Science in York in 1831 (Sheppard, 1917; Eyles, 1969). It was printed in London in 1832 by the lithographer William Day (1797–1845) at his premises at 17 Gate Street, Lincoln's Inn Fields. Holder of a Royal Warrant, Day was one of the leading lithographic printers of the time. On a scale of 12 chains to 1 inch (just over 6½ inches to the mile), this was one of Smith's largest-scale published maps, and is, perhaps, one of his rarest.

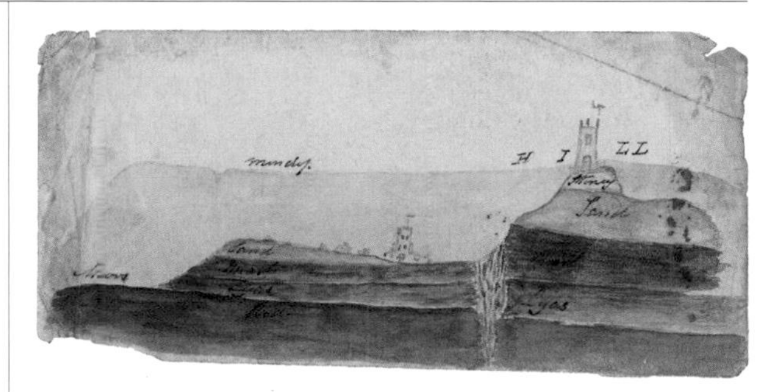

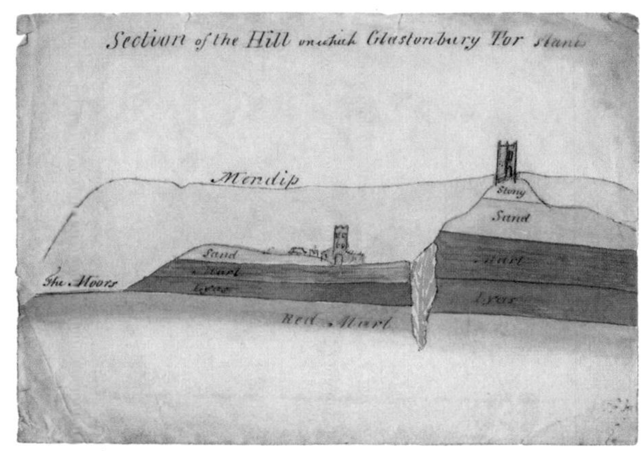

FIG. 11.

Fig. 11.
Section of the Hill on which Glastonbury Tor Stands. The strata shown on Smith's 1815 map were based on local observations made over a period of about twenty years. At Glastonbury Tor in Somerset, Smith has sketched a rough cross-section, perhaps done in the field, showing the strata making up the Tor and its surroundings with the Mendip Hills in the background. This would later be redrawn as a tidy copy.

ACHIEVEMENTS.

William Smith's geological mapping, and his large 1815 map of England and Wales in particular, is remarkable for a number of reasons. First, this was entirely the work of one man working single-handedly, not that of a consortium, society, organization, network or collaboration. As his contemporary, the French geologist and civil engineer Jean-François d'Aubuisson de Voisins (1769–1841) acknowledged in 1819: 'That which many distinguished mineralogists had done in a small part of Germany over half a century, one man (William Smith, mine engineer) had done for the whole of England; his work being as elegant in its results as it was astonishing in its extent.' ('*Ce que les minéralogistes les plus distingués ont fait dans une petite partie de l'Allemagne, en un demi-siècle, un seul homme [M. William Smith, ingénieur des mines] l'a entrepris et effectué pour toute l'Angleterre; et son travail, aussi beau par son résultat, qu'il est étonnant par son étendue.*') (d'Aubuisson de Voisins, 1819; Eyles, 1985.)

Secondly, the principles underlying the map – the recognition of the regularity of the succession of the strata and the value of fossils in distinguishing strata of similar appearance – were worked out by Smith himself. This is not to say that Smith was the only person to develop such ideas, but he was the first to apply them systematically and scientifically to the production of such a large-scale map over a whole country. Smith's principles allowed later workers, such as Roderick Impey Murchison (1792–1871) and Adam Sedgwick in the 1830s to begin to unravel the complexities of the older rocks of Wales which Smith had left undivided, leading to the establishment of the Cambrian and Silurian systems.

Thirdly, Smith originated the idea of using colour to indicate the sequence and orientation of the strata in a manner that can be appreciated at a glance, and came up with a considered selection of colours which broadly remains with us today. Fourthly, the mapping and fossil collecting had to be fitted around Smith's other surveying and drainage commitments, through which he had to earn a living. And fifthly, the mapping was conducted without any government support or financial assistance, but merely with the aid of a few enlightened individuals such as Sir Joseph Banks and the encouragement of friends such as Richardson and Townsend.

Smith's geological maps are today rare and valuable – perhaps fewer than 150 copies of his large map of 1815 still exist – but they are also things of beauty; they have an inherent attractiveness that calls for appreciation even before the background to their construction is considered. They are a testimony to Smith's dedication and perseverance over many decades, and he was clear as to their worth. This was not some mere academic exercise; Smith saw his maps as entirely utilitarian. 'The uses of such a discovery of regularity in the courses and order of the strata, and the facility of tracing them by means of their imbedded organic remains, are numerous.' He saw his map as being of value to the landowner, farmer, miner, brick-maker, architect, fuller, founder, glass-maker, potter, chemist, colourmen, 'vitriol, alum, and salt-makers', canal engineer and contractor, land drainer and irrigator, as well as to the geologist and natural philosopher (Smith, 1815a). The utility and economic value of geological maps was finally recognized by the government with the foundation, in 1835, four years before Smith's death, of today's British Geological Survey.

DERBY
Sandiacre
Basingfield
Hare Hill
Sutton on the Hill
Doveridge
Holbrook Ho.
Okbrook
Risley
BRAMCOTE HILLS
Chilwell
Clifton
Bridgford
Clipstone
Park
Dalbury
Alvaston
Elvaston
Ruddington
Sudbury
SINFIN MOOR
Hilton
EGGINGTON HEATH
Chellaston
Bradmore
Moorton
Dove R.
Tutbury
Twyford
Cavendish Br.
Trent R.
Stanton on the Woulds
NEEDWOOD FOREST
Castle Donington & Pa.
Stanton
Thorney Lane
Winchill
Melbourn
Kegworth
Widmere Pool
Burton
Sinai Park
Bredon
Diseworth
Hall Park
Tatenhill
Trent and Mersey Canal
Langley
Hatherne
Wimeswould
Wood Houses
Drinklow
Hartshorn
Blackfordby
Worthington
Lime Works
Prestwould
Hoton
Burton on the Woulds
Dalby on the Woulds
WOLDS
Ashby de la Zouch
Cole Orton
Sheepshead
Thrinkston
Rail Road
Loughborough
Walton on the Woulds
Sex or Segs Hill
Coton Park
Barrat Pool
CHARNWOOD
Quorndon
Ragdale
Whitwick
Beau Manor
Bardon Hill 853
Woodhouse
Measham
Ravenstone
FOREST
Mount Sorrel
Brooksby
HEATH
Grand Trunk Ca.
Normanton on the Heath
Mease R.
Stretton in the Field
Ulverscroft Abbey
Swithland
Elmhurst
Elford
Stanton under Bardon
Heather
Markfield
Bradgate Park
Clifton Campvil
Ashby de la Zouch Ca.
Newtown Linford
Groby Pool
Fisherwick Pa.
Statford
Gopsal
Bagworth
Thornton
Twyford
Tamworth
Coventry Canal
Orton on the Hill
Ratby
Glenfield
Weeford
Market Bosworth
LEICESTER
Keame
Bourne R.
Sence R.
Desford
Houghton on the Hill
Canwell Hall
Wilnecote
Pipehall
Tooley Pa.
Newal Park
Ratcliffe Culey
Upton
LEICESTER FOREST
Oadby
Atherstone
Fenny Drayton
Barwell
Glen Magna
Whetstone
Sutton
Higham
Potters Marston
Burton Overy
Hinckley
Stoney Stanton
Sharnford
Hurst Gr.
Cardsworth
Kilworth Beauchamp
Minworth
Tame R.
Maxtoke Castle
Barton Hastings
High Cross
Frowleworth
Willoughby Waterless
Foxton
Coleshill Hall
Coleshill
Arley
Wolvershill
Sharford
Wolvey
Gilmorton
Mowsley
Market Harborough
Corley 521
Withibrook
Over Claybrook
Bitteswell
Theddingworth
Newland
Lutterworth
Elmdon
Longford
Hops Ford
Husbands Bosworth
Bickenhill
Pickford Gate
Cottesbatch
Solihull
Hampton in Arden
Radford
Kilworth
Clipston
Mereden
Estern Gr.
Shawell
Brinklow
Cosford
Caves Inn
Stanford & Hall
Welford
Berkswell
COVENTRY
Swinford
Bretford Bridge
Knoll
Wolston
Lilbourn
Naseby
Redfern
Milburn Br.
Lawford
Hill Morton
Packwood
DUNSMORE HEATH
Rugby
Warren Ho.
Kenilworth
Stoneley Pa.
Avon R.
Lawford Lo.
Cottesbrook
Bourton
Dunchurch
Hollowell
Wapenbury
Liveridge Hill
Wedgnock Park
Wooton Leek
Blacklow Hill
Birdingbury
Holdenby
Henley in Arden
Leamington
WARWICK
Radford
Wolfhamcote
Braunston
Welton
Warwick & Napton Canal
Wolverton
CORNAVY
Southam
Flecknoe
Daventree
Brockhall
Tachbrook Bishops
Barford
Napton on the Hill
Dodford
NORTHAMPTON
Ladbroke
Harpole
Black Hill
Ashorne
Alveston
Weedon
Rodburne
Bugbrook
Red Hill
Stratford on Avon
Morton Morrell
Wellesburne Hastings
Charwelton
Farthingstone
Cold Higham
Byfield
Kineton
Penny Common
Aston in the Walls
Woodford
Maidford
Tiffield
Claydon
Edgcot
Radway
Ratley
Wood End
Bidmarsh
HEATH
Grand Junction Can.

'The greatest difficulties in understanding such an extensive branch of natural history arose from the want of some method of generalizing the information, which could only be supplied by a map that gives, in one view, the locality of thousands of specimens. By strong lines of colour, the principal ranges of strata are rendered conspicuous, and naturally formed into classes, which may be seen and understood at a distance from the map, without distressing the eye to search for small characters.'

SMITH'S *MEMOIR*, 1815.

DETAIL FROM SHEET VIII, SHOWING THE MIDLANDS.

To indicate the base of a stratum or layer and that it lies above the adjacent layer, Smith used a deeper shade of the same colour. Here the deeper blue shading visible on the west edge of the blue Lias Limestone stratum shows that this is the base of the Lias. Smith devised this colouring convention to help readers visualize the order of the strata.

V. FOSSIL COLLECTOR.

A remarkable story of fossil collecting and study; the importance of fossils to Smith's unique stratigraphical theory; the superb fossil drawings of James Sowerby; the unfortunate yet imperative sale of Smith's beloved fossil collection to the British Museum; a dispiriting stint in the debtors' prison during 1819.

Fossils provided William Smith (1769–1839) with evidence for his map and they were part of his life from an early age. Growing up in Churchill in Oxfordshire in the Jurassic Cotswolds, where fossils were easily found, he was familiar with the attractive flat sea urchins known locally as 'poundstones' because most were of a similar weight and were used by the dairywomen to weigh out the butter. Nearly spherical brachiopods, which are robust and easily fall out of the rock, were abundant. The local children played a form of marbles with them and called them 'pundibs' (Phillips, 1844, p. 3). William Smith's fossil collection, which he amassed over the years, contains a number of specimens from Churchill, but it is not known when they were collected as he returned there often to visit family members.

A specimen labelled 'Mearns Pit, High Littleton' is probably one of the earliest fossils acquired by Smith. It is a plant specimen, *Cyperites*, from the Coal Measures, and the only fossil in Smith's collection from this location. Smith was sent there by his employer Edward Webb (1751–1828) to survey the estate of Lady Elizabeth Jones (1741–1800) in 1791. While there he worked on the Mearns Colliery in the Somerset Coal Field and lived at Rugbourne Farm. This was a very formative period for Smith: he later said 'if Bath is the cradle of geology, then Rugborne is the birthplace' (Cox, 1942, p. 72).

COLLECTING THE EVIDENCE.

It was during Smith's next employment, on the proposed Somersetshire Coal Canal, that he began to appreciate the importance of fossils as an aid to recognition of the different strata through which the canals were excavated. He was first sent with two directors to travel 1,450 km (900 miles) around the country to inspect canals already constructed or in progress. It was probably on these travels that Smith collected his specimens from the Thames & Severn Canal near the Sapperton tunnel. Smith wrote the fossil location on his specimens in Indian ink, a habit that helped him greatly later when he came to draw up a catalogue of his collection. The construction of the Somersetshire Coal Canal provided him with useful evidence

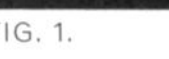

FIG. 1.

Fig. 1. 'Poundstone' from Naunton, not far from Churchill in Oxfordshire where Smith grew up. It is now known as the sea urchin *Clypeus ploti* NHMUK E538 from Smith's collection of 'Under Oolite' (inferior Oolite) fossils. It is 8 cm (3 in.) in length.

FIG. 2.

Fig. 2. Fossil plant stems from the Coal Measures *Cyperites* sp. NHMUK V741 Mearns Pit, High Littleton, where Smith worked as a surveyor in 1791–93. This is probably one of the first specimens in Smith's collection. It is 14 cm ($5\frac{1}{2}$ in.) in length.

to support his theories. Two branches of the canal were cut from the coal mines following two adjacent valleys which were nearly parallel for a stretch of almost 10 km (6 miles). This allowed Smith to verify his ideas on the regular order of the strata identified by the fossils they contained. Individual fossils helped him match the strata, but he also realized that 'suites of fossils' were found in particular strata and it was the assemblage that was the key.

When Smith came to list the locations of the specimens he chose for the plate illustrating his 'Oak Tree Clay' (Kimmeridge Clay) in his *Strata Identified by Organized Fossils*, all nine came from the North Wilts Canal (although he only actually figured three of them). Smith had visited this canal while it was under construction, probably sometime between 1814 and 1817. A large number of specimens in his collection come from other canals in areas where he worked or visited, and he was able to match the specimens from the North Wilts Canal with those from other locations. For example, he found *Ostrea delta* on the Kennet & Avon Canal at Seend, a well near Swindon, the Wilts & Berks Canal, Even Swindon, Bagley Wood Pit and Wootton Bassett.

In 1799 Smith's work with the Somersetshire Coal Canal Company came to an end, partly because of the catastrophic flooding that year which contributed to the failure of the caisson lock. But that same flooding created work for Smith, the drainage expert. He took premises in Bath at Trim Street, where he laid out his fossils in boxes around the floor and impressed his visitors with his knowledge of the local 'Order of Strata'. Many of them had fossil collections of their own, and he took great pleasure in arranging these in the correct order. As early as 1796 he wrote (reproduced in Cox, 1942, p. 12, fig. 1):

> *Fossils have been long studied as great curiossities [sic] collected with great pains treasured up with great Care and at a great Expense and shown and admired with as much pleasure as a Child's rattle or his Hobbyhorse is shown and admired by himself and his playfellows – because it is pretty And this has been done by Thousands who have never paid the least regard to that wonderful order & regularity with which Nature has disposed these singular productions and assigned to each Class its peculiar Stratum*

Smith's work took him all over the country and he writes in his diary about riding on horseback, making descents into stone quarries where 'he could not refrain from loading his pockets with identifying fossils' (Phillips, 1844, p. 35). Most villages had their own quarry close to the road

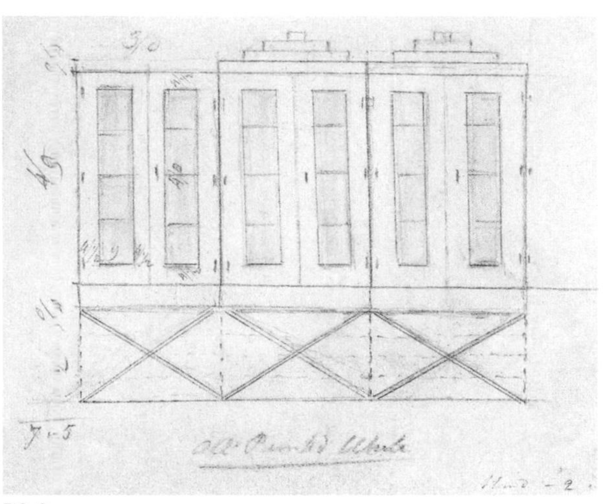

FIG. 3.

Fig. 3.
Plan for fitting up a Museum: a drawing of cabinets and accompanying text by Smith from *c.* 1804–5, detailing how the cabinets should extend around the room, with drawers underneath a table to house the fossils not on display. He also describes how to display the maps. The text underneath states: 'All painted white'.

for repairs, making collecting comparatively easy. By 1801 Smith had amassed enough material to consider publishing his ideas on the strata of England and Wales. His friend the Rev. Benjamin Richardson (1758–1832) persuaded him to publish a *Prospectus*, in which he also stated his intention to describe the 'animal remains' in each stratum.

PUBLICATION.

In 1804 Smith rented a property in London at 15 Buckingham Street, adjacent to the York Watergate leading to the Thames. It was ideally situated near the map-makers and publishing houses centred on the Strand, including John Cary (1754–1835), who a decade later became his cartographer. Smith arranged his fossil collection with the help of another friend, the Rev. Joseph Townsend (1739–1816) on sloping shelves according to the order of strata (Morton, 2001, p. 65) – a sketch and description, thought to be from this period, gives a flavour of the layout. He wanted to display his maps and large fossil collection to prospective clients in his private museum. Smith hoped to convince his visitors of the importance of fossils in differentiating between strata. Fellows of the Geological Society of London visited, but did not consider Smith for Fellowship because of his humble background. Much later, in 1831, they made amends by awarding him the first Wollaston Medal.

The year after William Smith's map of England and Wales was finally published in 1815, he began to publish the fossil evidence on which much of the map was based in *Strata Identified by Organized Fossils*. He wanted to convey to the reader the fundamental importance of the ordering of strata to understanding the lie of the land and availability of resources. The initial volume, which appeared on 1 June 1816, was intended to be the first of seven to be published monthly, but only four volumes were issued over

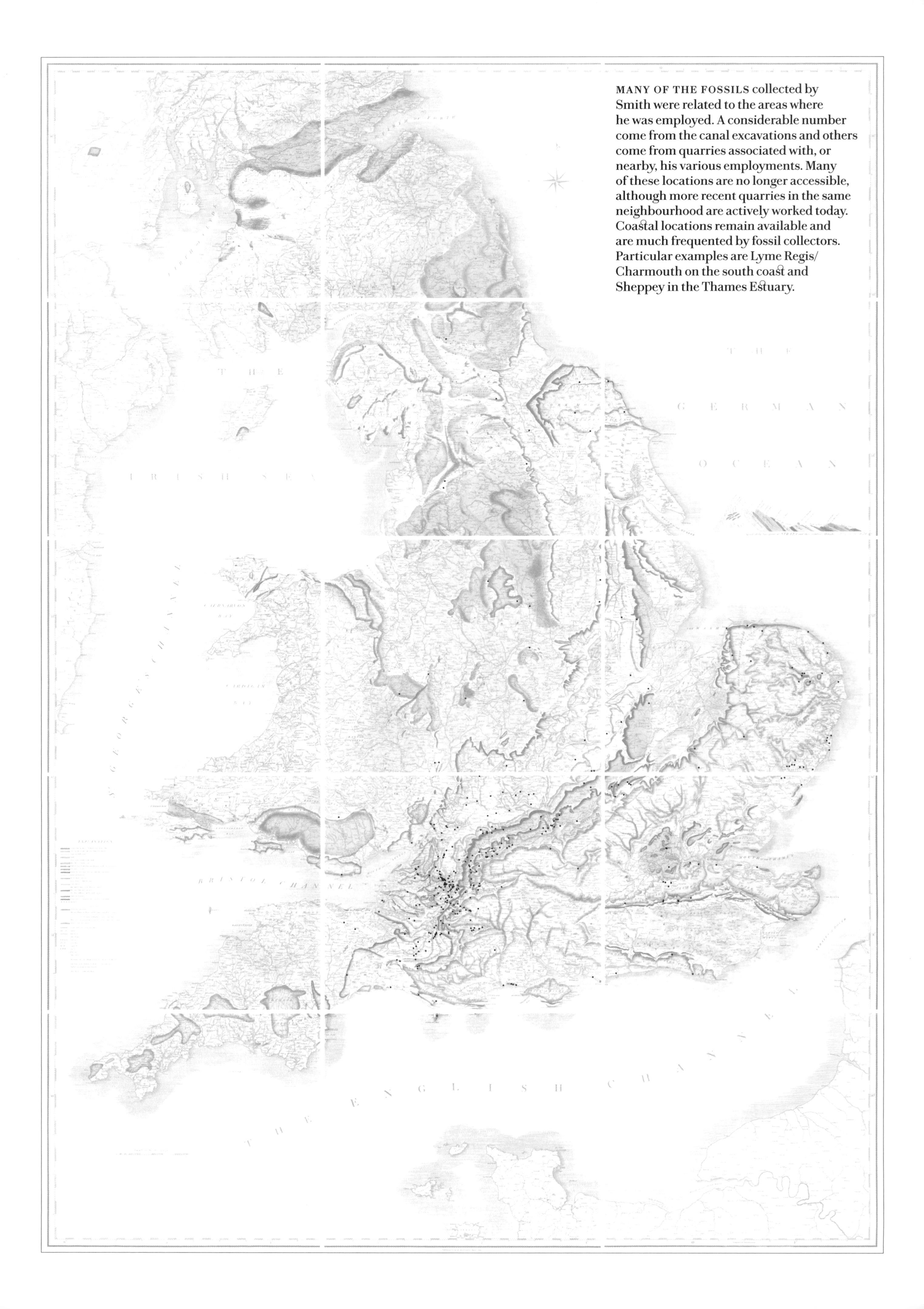

MANY OF THE FOSSILS collected by Smith were related to the areas where he was employed. A considerable number come from the canal excavations and others come from quarries associated with, or nearby, his various employments. Many of these locations are no longer accessible, although more recent quarries in the same neighbourhood are actively worked today. Coastal locations remain available and are much frequented by fossil collectors. Particular examples are Lyme Regis/Charmouth on the south coast and Sheppey in the Thames Estuary.

THE NORTH.

Clithero, *Lancashire.*
Danby Beacon (nr), *North Yorkshire.*
Kirby Moorside, *North Yorkshire.*
Malton, *North Yorkshire.*
Nenthead, *Yorkshire.*
Richmond, *North Yorkshire.*
Scarborough, *North Yorkshire.*
Spofforth, *North Yorkshire.*
Topcliffe, *North Yorkshire.*
Whaley, *Lancashire.*
Whitby, *North Yorkshire.*

WALES & CENTRAL ENGLAND.

Arbury, *Warwickshire.*
Bonsal, *Derbyshire.*
Critch, *Derbyshire.*
Newark, *Nottinghamshire.*
Northampton, *Northamptonshire.*
Ticknal, *Derbyshire.*
Towcester, *Northamptonshire.*
Buxton, *Derbyshire.*
Dudley, *West Midlands.*
Ilmington, *Warwickshire.*
Malvern, *Worcestershire.*
Mathon, *Herefordshire.*
Steeraway Wrekin, *Shropshire.*
Wrekin, *Shropshire.*

EAST ANGLIA & THE SOUTH EAST.

Aldborough, *Suffolk.*
Alderton, *Suffolk.*
Bagley Wood Pit, *Oxfordshire.*
Beddington Common, *Oxfordshire.*
Bentley, *Suffolk.*
Bognor, West Sussex.
Bracklesham Bay, *West Sussex.*
Bramerton, *Norfolk.*
Brightwell, *Suffolk.*
Burgh Castle, *Norfolk.*
Burnham Overy, *Norfolk.*
Bury St. Edmonds, *Suffolk.*
Carshalton, *London.*
Cherry Hinton, *Cambridgeshire.*
Chesterford, *Essex.*
Chipping Norton, *Oxfordshire.*
Churchill, *Oxfordshire.*
Clayton Hill, *West Sussex.*
Croydon, *London.*
Dry Sandford, *Oxfordshire.*
Ensham Bridge, *Oxfordshire.*
Enstone, *Oxfordshire.*
Foxhole, *Suffolk.*
Fulbrook, *Oxfordshire.*
Gagenwell (nr), *Oxfordshire.*
Garsington Hill, *Oxfordshire.*
Godstone (nr), *Surrey.*
Grimston (nr), *Norfolk.*
Guildford, *Surrey.*
Happisburgh, *Norfolk.*
Harwich, *Essex.*
Heddington Common, *Oxfordshire.*
Hickling, *Norfolk.*
Highgate Archway, *London.*
Highgate, *London.*
Hinton Waldrish, *Oxfordshire.*
Holkham Park, *Norfolk.*
Holywell, *Suffolk.*
Hordel Cliff, *Hampshire.*
Hunstanton Cliff, *Norfolk.*
Ipswich, *Suffolk.*
Kennington, *Oxfordshire.*
Keswick, *Norfolk.*
Leighton Beaudesert, *Bedfordshire.*
Leiston parish, *Suffolk.*
Leiston, Old Abbey, *Suffolk.*
Lexham, *Norfolk.*
Little Brickhill, *Buckinghamshire.*
Little Harwood, *Buckinghamshire.*
Marcham, *Oxfordshire.*
Minsmere Iron Sluice, *Suffolk.*
Moushold, *Norfolk.*
Near Godstone, *Surrey.*
Near May Place, *London.*
Near Stilton, *Cambridgeshire*
Newborn, *Suffolk.*
Newhaven Castle Hill, *East Sussex.*
Newton, *Nottinghamshire.*
Normanton Hill, *Lincolnshire.*
North of Norwich, *Norfolk.*
North of Reigate, *Surrey.*
North of Stamford, *Rutland.*
Northfleet, *Kent.*
Norwich and Yarmouth (bet), *Norfolk.*
Norwich, *Norfolk.*
Peterborough (nr), *Northamptonshire.*
Playford, *Suffolk.*
Prisley Farm, *Bedfordshire.*
Puddle Hill, near Dunstable, *Bedfordshire.*
Reading, *Berkshire.*
Reigate (north), *Surrey.*
Sandford Churchyard, *Oxfordshire.*
Sheppey, *Kent.*
Shippon, *Oxfordshire.*
Shotover Hill, *Oxfordshire.*
Sleaford, *Lincolnshire.*
Smitham Bottom, *London.*
Stanton, *Oxfordshire.*
Steppingley Park, *Bedfordshire.*
Stoke Hill, *Suffolk.*
Stony Stratford, *Buckinghamshire.*
Stunsfield, *Oxfordshire.*
Sunningwell, *Oxfordshire.*
Sutton, *London.*
Tattingstone Park, *Suffolk.*
Taverham, *Norfolk.*
Thame, *Oxfordshire.*
Thornbury, *Buckinghamshire.*
Thorpe Common, *Suffolk.*
Trimingsby, *Norfolk.*
Upton, *Oxfordshire.*
Well at Brixton Causeway, *London.*
Wellingborough, *Northamptonshire.*
Westerham, out of a deep well, *Kent.*
Westoning, *Bedfordshire.*
Whitlingham, *Norfolk.*
Wighton, *Norfolk.*
Woburn, *Bedfordshire.*
Woodford, *Northamptonshire*
Woolverton, *Buckinghamshire.*
Woolwich, *London.*

SOUTH WEST.

Abbotsbury (nr), *Dorset.*
Aberthaw, *Glamorgan.*
Alfred's Tower, *Somerset.*
Aston Somerville, *Worcestershire.*
Avoncliff, *Wiltshire.*
Banner's Ash, *Wiltshire.*
Barton, *Hampshire.*
Barton Cliff, *Hampshire.*
Bath, *Somerset.*
Bath (nr), *Somerset.*
Batheaston Pit, *Somerset.*
Bathhampton, *Somerset.*
Bayford (south), *Somerset.*
Bedminster Down, *Bristol.*
Bedminster Down, *Bristol.*
Below Combe Down Caisson, *Somerset.*
Berfield, *Wiltshire.*
Berkley, *Somerset.*
Between Nunny and Frome, *Somerset.*
Between Sherborn and Yeovil, *Dorset.*
Between Weymouth and Osmington, *Dorset.*
Black-Dog Hill, near Standerwick, *Wiltshire.*
Blackdown, *Devon.*
Blue Lodge, *South Gloucestershire.*
Bradford, *Wiltshire.*
Bradford Lock, *Wiltshire.*
Breadstone, *Gloucestershire.*
Brent, *Somerset.*
Brinkworth Common, *Wiltshire.*
Broadfield Farm, *Gloucestershire.*
Bruham Pit, *Somerset.*
Bubdown, *Dorset.*
Calne, *Wiltshire.*
Canal at Bradford, *Wiltshire.*
Canal at Seend, *Wiltshire.*
Castle Combe, *Wiltshire.*
Charlton Horethorn, *Somerset.*
Charmouth, *Dorset.*
Cheltenham, *Gloucestershire.*
Chicksgrove, *Wiltshire.*
Chillington, *Somerset.*
Chittern, *Wiltshire.*
Christian Malford, *Wiltshire.*
Chute Farm, *Wiltshire.*
Clifton, *Bristol.*
Closworth, *Somerset.*
Coal Canal, *Somerset.*
Coleshill, *Wiltshire.*
Combe Down, *Somerset.*
Combhay, *Somerset.*
Crescent Fields, *Somerset.*
Crewkerne, *Somerset.*
Crickley Hill, *Gloucestershire.*
Crockerton, *Wiltshire.*
Cross Hands, *South Gloucestershire.*
Damerham, *Hampshire.*
Dauntsey House, *Wiltshire.*
Derry Hill, *Wiltshire.*
Devizes, *Wiltshire.*
Devizes (nr), *Wiltshire.*
Devonshire Buildings, Bath, *Somerset.*
Didmarton, *Gloucestershire.*
Dilton, *Wiltshire.*
Dinton Park, *Wiltshire.*
Dowdswell Hill, *Gloucestershire.*
Down Ampney, *Gloucestershire.*
Draycot, *Wiltshire.*
Dudgrove Farm, *Gloucestershire.*
Dundry, *North Somerset.*
Dunkerton, *Somerset.*
Dun's Well, Silton Farm, *Dorset.*
Dursley, *Gloucestershire.*
Elmcross, *Wiltshire.*
Even Swindon, *Wiltshire.*
Evershot, *Dorset.*
Farley, *Somerset*
Farley Castle, *Somerset.*
Fonthill, *Wiltshire.*
Foss Cross, *Gloucestershire.*
Frocester Hill (foot), *Gloucestershire.*
Frocester Hill (top), *Gloucestershire.*
Glastonbury, *Somerset.*
Gloucester & Berkley Canal, *Gloucestershire.*
Great Ridge, *Wiltshire.*
Grip Wood, *Wiltshire.*
Hardington, Somerset.
Heytesbury, *Wiltshire.*
Highworth, *Wiltshire.*
Hilmarton, *Wiltshire.*
Hinton, Somerset.
Hogwood Corner, Somerset.
Holt, *Wiltshire.*
Horningsham, *Wiltshire.*
Hotwells, Bristol.
Hutton Hill, *North Somerset.*
John St, Bath, *Somerset.*
Kelloways, *Wiltshire.*
Kennet & Avon Canal, *Wiltshire.*
Kennet & Avon Canal at Seend, *Wiltshire.*
Kennet & Avon Canal, near Trowbridge, *Wiltshire.*
Keynsham, *Somerset.*
Knook, *Wiltshire.*
Lady Down, *Wiltshire.*
Ladydown Farm, *Wiltshire.*
Lansdown, *Somerset.*
Latton, *Gloucestershire.*
Latton (north), *Wiltshire.*
Laverton, *Somerset.*
Leigh upon Mendip, *Somerset.*
Liliput, *South Gloucestershire.*
Limpley Stoke, *Wiltshire.*
Longleat Park, *Wiltshire.*
Lullington, *Somerset.*
Maisey Hampton, *Gloucestershire.*
Marston Magna, *Somerset.*
Marston, near Frome, *Somerset.*
Mazen Hill, *Wiltshire.*
Melbury, *Dorset.*
Mells, *Somerset.*
Merthyr Tydvil, *Mid Glamorgan.*
Minching Hampton Common, *Gloucestershire.*
Mitford, *Somerset.*
Monkton Combe, *Somerset.*
Muddiford, *Dorset.*
Mudford, *Somerset.*
Nailsworth, *Gloucestershire.*
Naunton (nr), *Gloucestershire.*
Near Bath, *Somerset.*
Near Devizes, *Wiltshire.*
Near Lansdown, *Somerset.*
Near Shrivenham, *Oxon.*
Near Warminster, *Wiltshire.*
North Cheriton, *Somerset.*
North Wiltshire Canal, *Wiltshire.*
Norton, *Wiltshire.*
Norton Bavant, *Wiltshire.*
Oldford near Frome, *Somerset.*
Orchardleigh, *Somerset.*
Penard Hill, *Somerset.*
Pewsey, *Wiltshire.*
Pickwick, *Wiltshire.*
Pipehouse, *Somerset.*
Portland, *Dorset.*
Pottern, *Wiltshire.*
Poulton, *Gloucestershire.*
Poulton Quarry, Bradford, *Wiltshire.*
Purton Passage, *Gloucestershire.*
Redland (nr Bristol.), *Bristol.*
Redlynch, *Somerset.*
Redlynch (nr), *Somerset.*
Road, *Somerset.*
Road, Coal Experiment, *Somerset.*
Rowley Bottom, *Somerset.*
Rundaway Hill, *Wiltshire.*
S. W. of Tellisford, *Somerset.*
S.W. of Wincanton, *Somerset.*
Sattyford, *Somerset.*
Sheldon, *Wiltshire.*
Sherborn, *Dorset.*
Siddington, *Gloucestershire.*
Silton Farm, *Dorset.*
Slabhouse, Mendips, *Somerset.*
Smallcombe Bottom, *Somerset.*
Sodbury, South *Gloucestershire.*
Somerton, *Somerset.*
South of Bayford, *Somerset.*
Stanton near Highworth, *Wiltshire.*
Steeple Ashton, *Wiltshire.*
Stoford, *Somerset.*
Stoke, *Somerset.*
Stone Farm Yeovil, *Somerset.*
Stoney Littleton, *Somerset.*
Stourhead, *Wiltshire.*
Stow on the Wold, *Gloucestershire.*
Stowey, *Somerset.*
Stratton, *Wiltshire.*
Swindon, *Wiltshire.*
Teffont, *Wiltshire.*
Tellisford, *Somerset.*
Thames & Severn Canal, *Gloucestershire.*
Tinhead, *Wiltshire.*
Tisbury, *Wiltshire.*
Tracey Park, South *Gloucestershire.*
Trowle, *Wiltshire.*
Tucking Mill, *Somerset.*
Turnpike near Bratton, *Somerset.*
Tytherton Lucas, *Wiltshire.*
Ubby Hill, Mendips, *Somerset.*
Vinyard Down, *Somerset.*
Warminster, *Wiltshire.*
Watchet, *Somerset.*
Well at Seagry, *Wiltshire.*
Well near Swindon, Wiltshire & Berks Canal, *Wiltshire.*
Wells (nr), Somerset.
Westbrook, *Wiltshire.*
Weston, *Somerset.*
Westwood, *Wiltshire.*
Wick Farm, *Somerset.*
Wiltshire & Berks Canal, *Wiltshire.*
Wiltshire & Berks Canal near Shrivenham, *Oxon.*
Wiltshire & Berks Canal, near Chippenham, *Wiltshire.*
Wincanton, *Somerset.*
Wincanton (north side), *Somerset.*
Winsley, *Wiltshire.*
Wotton Basset, *Wiltshire.*
Wotton Basset (nr), *Wiltshire.*
Wraxhall, *Wiltshire.*
Writhlington, *Somerset.*
Yeovil, *Dorset.*

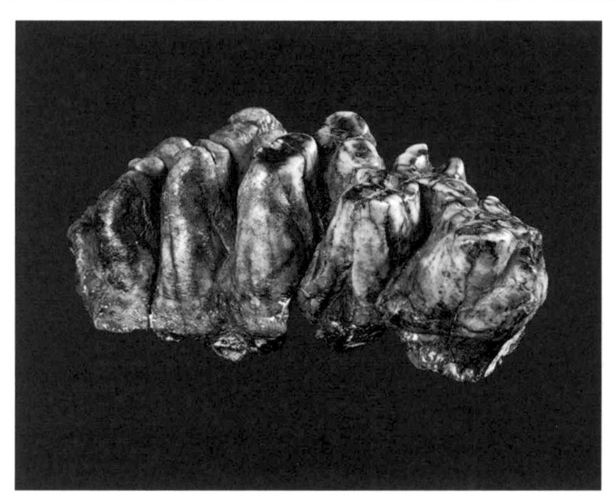

FIG. 4.

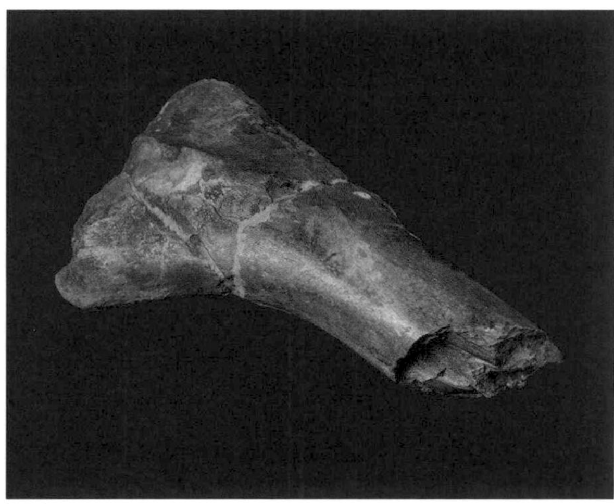

FIG. 5.

Figs. 4 & 5. Two vertebrate fossils acquired by Smith. Fig. 4 is a mastodon tooth *Anancus arvernensis* NHMUK M1983, from the gravels at Whitlingham in Norfolk and given to Smith. Smith described it as a 'tooth of some extinct monstrous animal'. It is 17 cm (6⅔ in.) in length. Fig. 5 is the lower end of the right shin bone, or tibia, of the dinosaur loosely referred to as *'Iguanodon'* NHMUK R526. It was given to Smith in 1809 from the Lower Cretaceous Wealden strata at Cuckfield, West Sussex, and described by Smith as 'bones of gigantic dimensions'. It is 38 cm (15 in.) in length.

a period of three years. Smith's collaborator on this project was the pre-eminent fossil illustrator of the day, James Sowerby (1757–1822). The two men agreed on the importance of illustrating British fossils, even though their focus was different: Sowerby concentrated on the fossils themselves, while Smith was concerned with the rock strata they were found in and their consequent geological order. Smith and Sowerby first met in 1808 when the latter had already embarked on the first volume of his *The Mineral Conchology of Great Britain* (1812), and he was able to help Smith identify his fossils, while Smith encouraged Sowerby to collect fossils for himself.

Neither Smith nor Sowerby made financial gain from their joint venture, but the eighteen plates are masterpieces. The fossil assemblages chosen were representative of particular strata and the plates were printed on coloured paper to represent the strata in which they are found using the same scheme as in Smith's map. In his 'Plan of the Work' dated 1800 Smith writes 'these as well as the strata, to make them more striking, should be coloured' (Morton, 2001, p. 42).

Remarkably one man roaming the countryside of England and Wales, by stagecoach, on horseback or on foot, sometimes covering as much as 16,000 km (10,000 miles) in a single year, accurately mapped nearly all the strata and in the correct order. Curiously, the exception is the first stratum that William Smith figured: the London Clay. Smith had no idea of the actual ages in years of his strata and did not speculate on their date. He was more concerned with establishing the regular order in which they occurred, but did say 'that the earth had been an infinite time in arriving at perfection' (Torrens, 2003, p. 190). We now know that the London Clay is dated to 55–50 million years ago whereas the Crag, figured by Smith as Plate 2, is only about 2 million years old. Smith's order is understandable as he found a clay resting on the Crag in East Anglia and mistook it for the London Clay.

Among the London Clay fossils Smith illustrated is a Jurassic ammonite. What Smith could not be aware of at the time was that it had been transported from Jurassic rocks by an ice sheet and deposited within glacial till above the Crag. Smith continually updated his ideas and in a hand-annotated copy of his fossil catalogue published in 1819, he amended his entry of *Ammonites communis* with 'Lias fossil diluvial' (Wigley, 2019, p. 89). He realized that the clay at Happisburgh in Norfolk that contained the ammonite was not London Clay, but he believed instead that it was deposited by the Deluge – the idea of biblical flood as an actual event was widely believed for centuries.

TEETH AND BONES.

Most of the fossils selected by Smith and illustrated by Sowerby are of marine invertebrates although the plates sometimes included shark and fish teeth. Other vertebrates were rare and included a large Mastodon tooth from Whitlingham in Norfolk used for the frontispiece, which was given to Smith while working in the area. A small toe bone in the Crag plate has been tentatively identified as that of a gazelle. If correct, this would be the oldest record of gazelles in the UK (*c.* 2 million years old). Although Smith does not include Wealden Strata within his fossil descriptions he did include them on his map. In these terrestrial rocks fossils are comparatively rare. However, while working in Sussex in 1809, Smith visited the quarry in Cuckfield where Mary Ann Mantell (1795–1869), the wife of Gideon Mantell (1790–1852), found the first dinosaur tooth some thirteen years later (Charig, 1983, p. 50). Smith acquired 'several bones of gigantic dimensions' (Phillips, 1844, p. 63) long before the term 'dinosaur' was invented.

In his two Chalk plates, Smith chose not to follow his convention of matching the colour of the background on which the fossils were illustrated with the colour used for the respective stratum on his 1815 map. For the map he chose green to depict the rolling slopes of the Downs, making it easier to read and more attractive. The background to the fossils in the plates was white, a more realistic colour for the Chalk. For the Upper Chalk, the majority of the fossils illustrated were collected in Norfolk, where he was employed on several drainage projects, and are labelled 'mostly Norwich' or 'N. of Norwich', whereas most of the Lower Chalk fossils figured come from the Warminster area of Wiltshire, which he first encountered in 1799 (Phillips, 1844, p. 39).

Smith's first 'Green Sand' plate is one of the most striking, depicting a large funnel-shaped sponge (termed *Alcyonite* by Smith). The colour of the background brings the fossils into sharp contrast and its greenish hue reflects the mineral glauconite. Fossils pictured in this plate were collected from 'Blackdown', at the edge of an outlier of Upper Greensand in the Blackdown Hills in Devon, whereas the majority of the fossils depicted on the second 'Green Sand' plate come from Warminster or nearby Chute on the western edge of Salisbury Plain, close to the Lower Chalk locations.

Smith made two plates with specimens for his 'Coral Rag & Pisolite' (Coralline Oolite), probably because the three corals depicted are some of the larger specimens in his collection and make a pleasing display on their own. On the second plate he included the large gastropod he called *Melania striata*, a distinctive, long-lived species that is not a good indicator for a particular stratum. The best match in Smith's collection for this specimen in fact comes not from the Coral Rag, but from much older rocks of the Inferior Oolite. An even larger specimen in his collection does come from the Coral Rag, but neither its size nor colour match Sowerby's drawing on the Coral Rag Plate.

On his 'Kelloways Stone' (Kellaways Rock) plate Smith figures the famous species of *Gryphaea* that he identified at the coal trial at Bruham (Brewham), near Bruton, which proved to him that the miners were far too high in the succession ever to encounter coal. Repeated layers of clay were often confusing and it was only with reference to the enclosed fossils that the correct strata could be ascertained (Torrens, 2003, p. 175).

It was not until 1819 that Smith published the fourth and last of his *Strata Identified by Organized Fossils* featuring the Cornbrash, Forest Marble, Clay over the Oolite, Upper Oolite and Fuller's Earth. Smith recognized early on that this suite of strata followed a band right across England from Dorset to Yorkshire, and they form a colourful central strip to his 1815 map. His collection contains fossils from the entire length. Cornbrash is Smith's name for the lower slopes of the 'Stonebrash Hills' where he observed rich fields of corn growing on the rubbly soil, and is still used today. Other names Smith adopted were ones used by local quarrymen.

THE SALE OF SMITH'S FOSSILS.

Unfortunately Smith had problems with raising the finance for his large map from subscribers, mainly because fourteen years had passed since the pledges to him had first been made following his Prospectus. By the time of publication in 1815 several subscribers had died and others were untraceable. This was in addition to Smith's financial woes caused by the failure of a quarry for Bath stone he had acquired at Kingham Field, near the home he had bought at Tucking Mill. To resolve his financial difficulties, Smith resolved to sell his fossil collection to the British Museum. Negotiations began in 1815, facilitated in the first instance by his friend and supporter, Sir Joseph Banks (1743–1820), who was a Trustee at the Museum. Smith was aware that recent acquisitions of mineral collections by the British Museum had attracted large sums, so he was

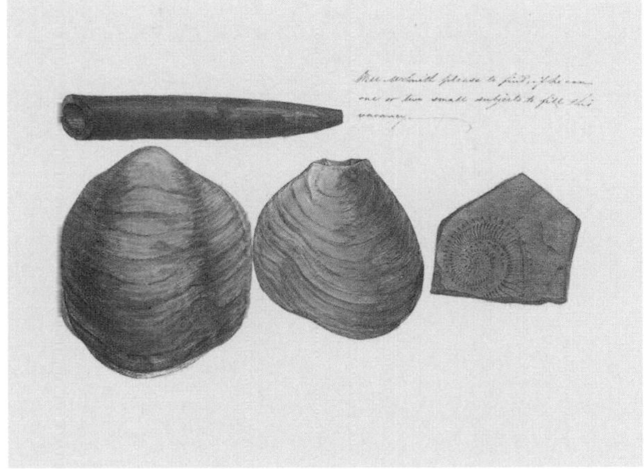

FIG. 6

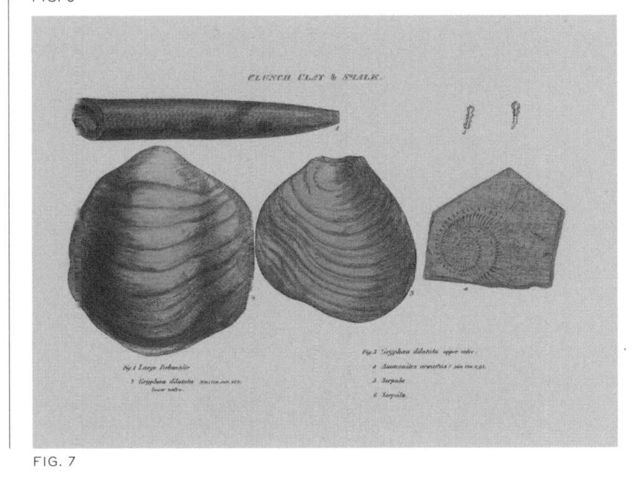

FIG. 7

Figs. 6 & 7.
An early version of Sowerby's drawings for the 'Clunch Clay' plate (Fig. 6) has a note appended 'Will Mr Smith please to find if he can find one or two small subjects to fill this vacancy'. Smith duly obliged by adding two small tube worms, common at certain horizons in the Oxford Clay. These two small serpulid worms, *Genicularia vertebralis*, appeared in the published versions (Fig. 7).

SMITH'S FOSSILS ORGANIZED BY STRATA.

PLAINS. NOS. 1–4.

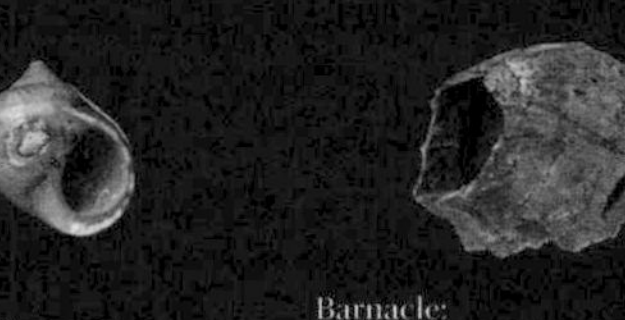
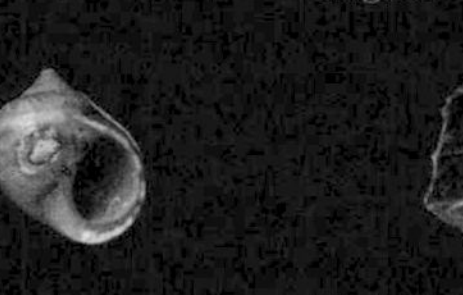

LONDON CLAY. 1

Gastropod: *Viviparus suessoniensis.* Well, Brixton causeway.

Block with bivalves: *Abra splendens.* Sheppey.

Bivalve: *Glycymeris brevirostris.* Bognor.

Gastropod: *Volutospina denudata.* Bognor.

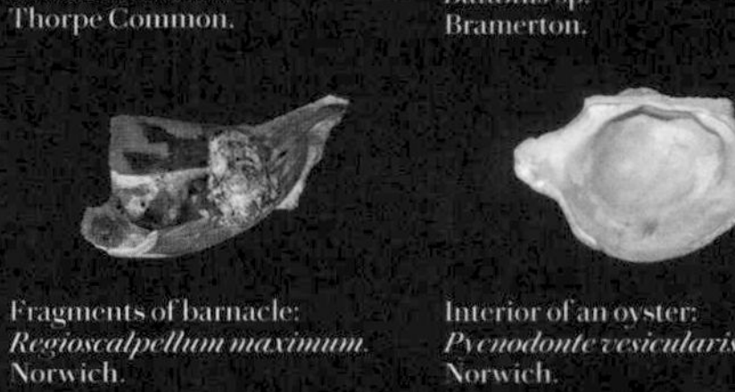

CRAG. 3

Gastropod: *Neptunea angulata.* Alderton.

Gastropod: *Nucella incrassata.* Bramerton.

Gastropod: *Littorina littorea.* Thorpe Common.

Barnacle: *Balanus* sp. Bramerton.

CHALK HILLS. NOS. 5–6.

UPPER CHALK. 5

Sponge: *Sporadoscinia alcyonoides.* Wighton.

Sponge: *Toulminia catenifer.* Chittern.

Fragments of barnacle: *Regioscalpellum maximum.* Norwich.

Interior of an oyster: *Pycnodonte vesicularis.* Norwich.

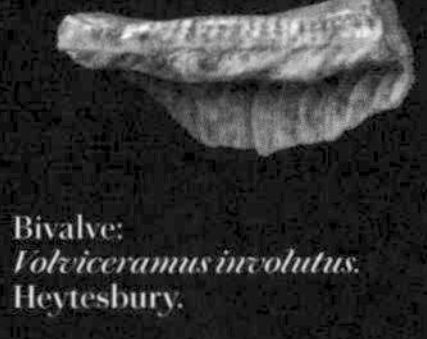

LOWER CHALK. 5

Bivalve: *Volviceramus involutus.* Heytesbury.

Bivalve: *Mytiloides labiatus.* Warminster.

Gastropod: *Bathrotomaria* sp. Mazen Hill.

Ammonite: *Schloenbachia subtuberculata.* Rundaway Hill.

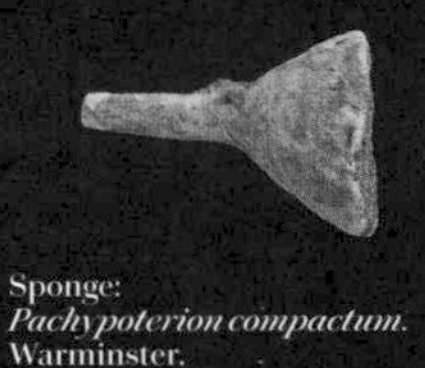

GREEN SAND. 6

Sponge: *Pachypoterion compactum.* Warminster.

Sponge: *Siphonia tulipa.* Pewsey.

Bivalve: *Epicyprina angulata.* Blackdown.

Gastropod: *Cretaceomurex calcar.* Blackdown.

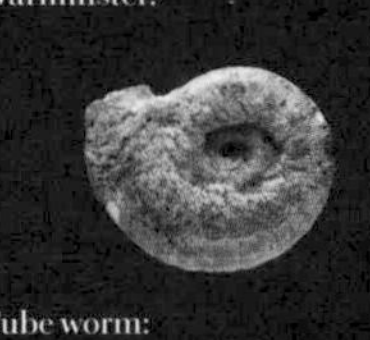

GREEN SAND. 6

Tube worm: *Rotularia concava.* Near Warminster.

Gastropod: *Nummogaultina fittoni.* Rundaway Hill.

Bivalve: *Merklinia scabra.* Chute Farm.

Brachiopod: *Dereta pectita.* Chute Farm.

CLAY VALES. NOS. 7–15.

BRICKEARTH. 7

Ammonite: *Hoplites dentatus.* Steppingley Park.

Ammonite: *Idiohamites* sp. Near Grimstone.

Echinoid: *Holaster* sp. Near Devizes.

Belemnite: *Neohibolites minimus.* Near Grimstone.

PORTLAND ROCK. 9

Gastropod: *Neritoma sinuosa.* Swindon.

Gastropod: *Aptyxiella portlandica.* Portland.

Bivalve: *Eomiodon* sp. Swindon.

Bivalve: *Myophorella incurva.* Swindon.

OAKTREE CLAY. 11

Gastropod: *Bathrotomaria reticulata.* North Wilts Canal.

Bivalve: *Nanogyra nana.* North Wilts Canal.

Bivalve: *Nanogyra nana.* North Wilts Canal.

Bivalve: *Deltoideum delta.* North Wilts Canal.

CORAL RAG & PISOLITE. 12

Coral: *Isastrea explanata.* Stanton near Highworth.

Coral: *Complexastrea depressa.* Steeple Ashton.

Coral: *Thecosmilia annularis.* Steeple Ashton.

Gastropod: *Oolitica muricata.* Derry Hill.

CLUNCH CLAY & SHALE. 14

Belemnite: *Cylindroteuthis puzosiana.* Dudgrove Farm.

Bivalve: *Gryphaea dilatata.* Derry Hill.

Bivalve: *Gryphaea dilatata.* Derry Hill.

Ammonite: *Kosmoceras spinosum.* Thames & Severn Canal.

KELLOWAYS STONE. 15

Ammonite: *Cadoceras sublaevis.* Kellaways.

Ammonite: *Sigaloceras calloviense.* Kellaways.

Ammonite: *Proplanulites koenigi.* Kellaways.

Bivalve: *Gryphaea dilobotes.* Ladydown.

STONEBRASH HILLS. NOS. 16–24.

CORNBRASH. 16

Gastropod: cf. *Ampullospira* sp. Road.

Ammonite: *Clydoniceras discus.* South-west of Wincanton.

Bivalve: *Modiolus imbricatus.* Wick Farm.

Bivalve: *Trigonia cruris.* North side of Wincanton.

Gastropod:
Brotia melanioides.
Woolwich.

Shark's tooth:
Otodus obliquus.
Sheppey.

Bivalve:
Striarca wrigleyi.
Highgate.

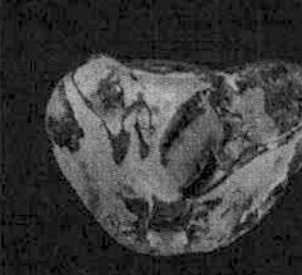

Crab:
Zanthopsis sp.
Sheppey (front).

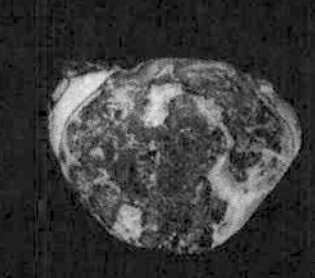

Crab:
Zanthopsis sp.
Sheppey (back).

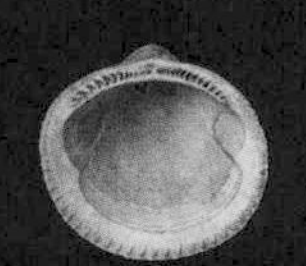

Bivalve:
Glycymeris variabilis.
Tattingstone Park.

Bivalve:
Cerastoderma hostei.
Tattingstone Park.

Bivalve:
Mya arenaria.
Bramerton.

Short vertebra
of a teleost indet.
Thorpe Common.

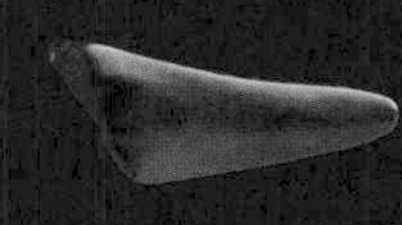

Large shark's tooth:
'Isurus' sp.
Stoke Hill.

Toe phalange of a mammal:
possibly gazelle.
Tattingstone Park.

Oyster attached to a belemnite:
(*Belemnitella mucronata*).
Norwich.

Bivalve:
Mimachlamys mantelliana.
Norwich.

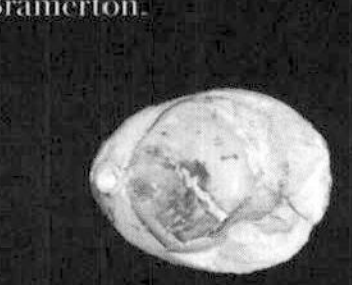

Brachiopod:
Concinnithyris subundata.
Norwich.

Echinoid:
Echinocorys scutata.
Norwich.

Palate of a fish:
Ptychodus mammillaris.
Near Warminster.

Fish tooth:
Enchodus sp.
North of Norwich.

Gastropod:
Bathrotomaria sp.
Warminster.

Brachiopod:
Gibbithyris semiglobosa.
Heytesbury.

Brachiopod:
Orbirhynchia cuvieri.
Heytesbury.

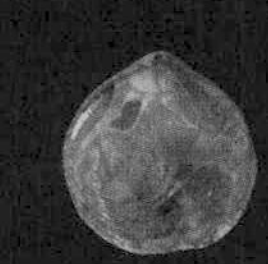

Brachiopod:
Gibbithyris semiglobosa.
Heytesbury.

Shark tooth: indet.

Gastropod:
Torquesia granulata.
Blackdown.

Bivalve:
Glycymeris sublaevis.
Blackdown.

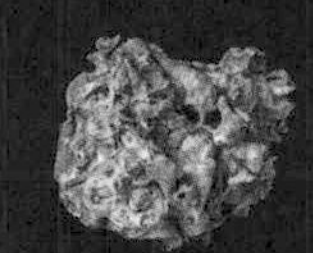

Bivalve:
'Mactra' angulata (and others).
Blackdown.

Echinoid:
Catopygus columbarius.
Chute Farm.

Echinoid:
Holaster laevis.
Chute Farm.

Brachiopod:
Cyclothyris latissima.
Chute Farm.

Bivalve:
Amphidonte obliquatum.
Alfred's Tower.

Bivalve:
Neithea gibbosa.
Near Warminster.

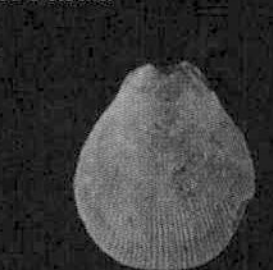

Bivalve:
'Chlamys' aff. *subacuta.*
Chute Farm.

Bivalve:
Amphidonte obliquatum.
Stourhead.

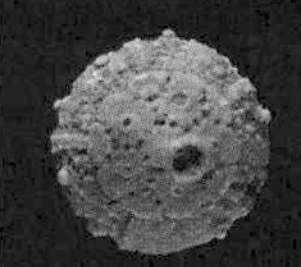

Echinoid:
Salenia petalifera.
Warminster.

Echinoid:
Discoides subuculus.
Warminster.

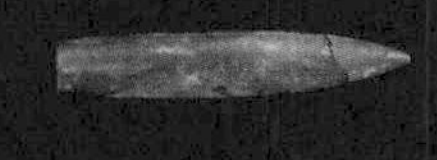

Belemnite:
Neohibilites minimus.
Prisley Farm.

Bivalve:
Protocardia dissimilis.
Swindon.

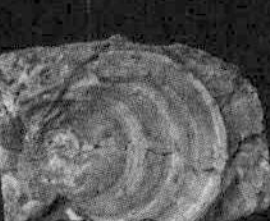

Bivalve:
Camptonectes lamellosus.
Swindon.

Section of larger piece
of conifer.
Woburn.

Ammonite:
Pictonia baylei.
Wootton Bassett.

Bivalve:
Neocrassina ovata.

Brachiopod:
Torquirhynchia inconstans.
Bagley Wood Pit.

Gastropod:
Ampullospira sp.
Longleat Park.

Gastropod:
Bourguetia saemanni.
Caisson, Wilts & Berks Canal.

Bivalve:
Actinostreon gregarium.
Wiltshire.

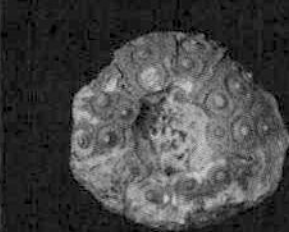

Echinoid:
Paracidaris smithii.
Hilmarton.

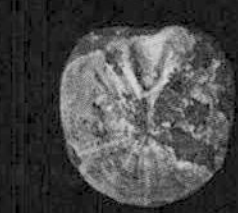

Echinoid:
Nucleolites clunicularis.
Meggot's Mill, Coleshill.

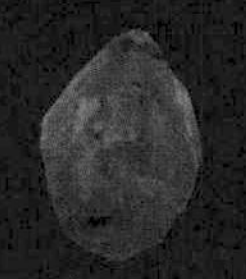

Brachiopod:
Ornithella ornithocephala.
Wilts & Berks Canal, near Chippenham.

Bivalve:
Protocardia buckmani.
Trowle.

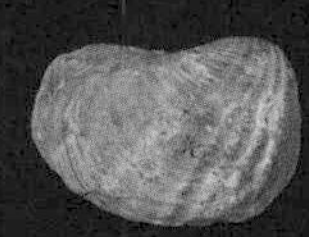

Bivalve:
Pholadomya deltoidea.
Road.

Bivalve:
Pleuromya uniformis.
Road.

Bivalve:
Meleagrinella echinata.
Draycot.

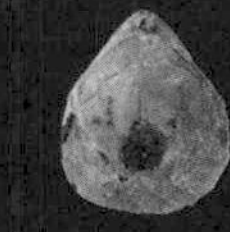

Brachiopod:
Digonella siddingtonensis.
Latton.

STONEBRASH HILLS. NOS. 16–24.

18 — **FOREST MARBLE.**

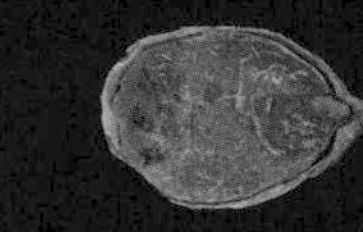

Gastropod: *Cylindrites archiaci.* Farley.

Gastropod: *Rostellaria* in block.

Bivalve: *Catinula* sp. Wincanton.

Bivalve: *Plagiostoma subcardiiformis.* Siddington.

19 — **CLAY OVER THE UPPER OOLITE.**

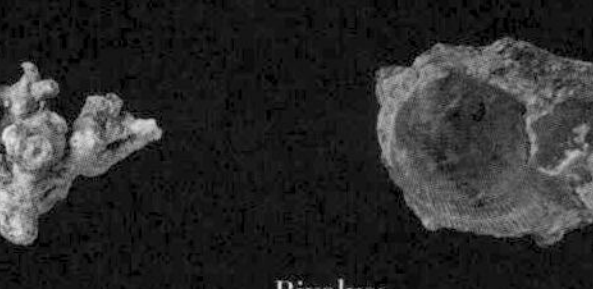

Crinoid: *Apiocrinites elegans.* Bradford.

Crinoid: *Apiocrinites elegans.*

Bryozoan: *Terebellaria ramosissima.* Hinton.

Bivalve: *Plagiostoma cardiiformis.* Combhay.

20 — **UPPER OOLITE.**

Coral: *Cladophyllia babeana.* Castle Coombe.

21 — **FULLER'S EARTH & ROCK.**

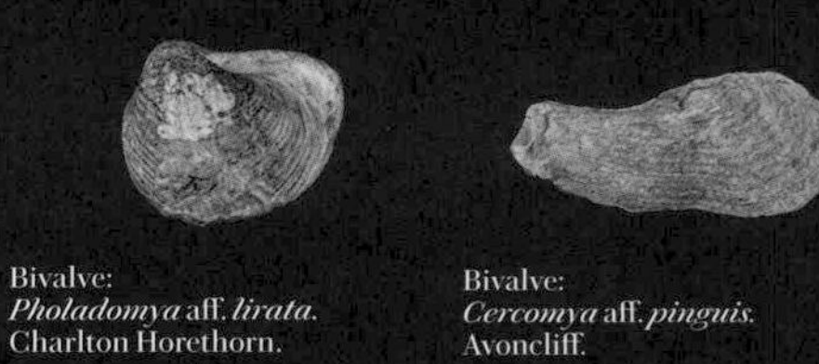

Bivalve: *Modiolus anatinus.* Avoncliff.

Bivalve: *Ceratomya* aff. *striata.* Hardington.

Bivalve: *Pholadomya* aff. *lirata.* Charlton Horethorn.

Bivalve: *Cercomya* aff. *pinguis.* Avoncliff.

22 — **UNDER OOLITE.**

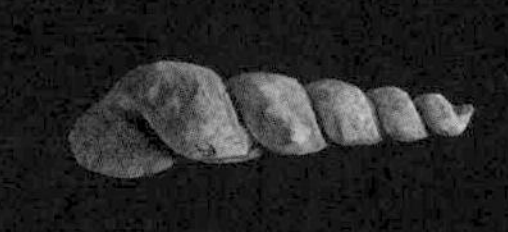

Gastropod: *Pseudomelania* sp. Tucking Mill.

Gastropod: *Pyrgotrochus conoideus.* Near Bath.

Gastropod: *Pyrgotrochus* sp. Somersetshire Coal Canal.

Gastropod: *Pleurotomaria granulata.* Sherborne.

22 — **UNDER OOLITE.**

Bivalve: *Trigonia costata* (internal mould). Cotswold Hills.

Bivalve: *Trigonia costata.* Tucking Mill.

Bivalve: *Astarte elegans* (internal mould). Somersetshire Coal Canal.

Bivalve: *Variamussium* cf. *pumilum.* Churchill.

23 — **SAND AND SANDSTONE.**

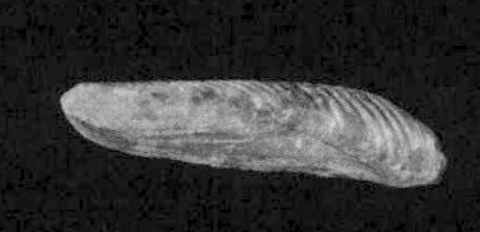

Belemnite: *Belemnopsis* sp. Tucking Mill.

Ammonite: *Pleydellia burtonensis.* Yeovil.

Ammonite: *Witchellia* sp. Penard Hill.

24 — **MARLSTONE.**

Ammonite: *Tragophylloceras undulatum.* Somersetshire Coal Canal.

Ammonite: *Zugodactylites sp.* Somersetshire Coal Canal.

Witchellia sp.

Ammonite: *Grammoceras striatulum.* Glastonbury.

MARL VALES. NOS. 25–29.

25 — **BLUE MARL.**

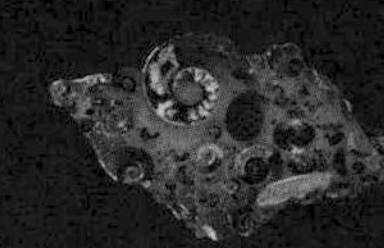

Block with ammonites: *Promicroceras planicosta.* Marston Magna.

Ammonite: *Asteroceras smithi.*

Bivalve: *Cardinia listeri.* Mudford.

26 — **LIAS.**

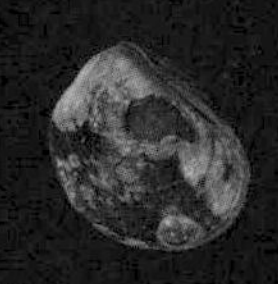

Bivalve: *Plagiostoma giganteum.* Near Bath.

Bivalve: *Gryphaea arcuata.* Gloucester & Berkley Canal.

Brachiopod: *Spiriferina walcotti.* Keynsham.

Shark fin spine: *Acrodus curtus.* Near Lyme.

26 — **LIAS.**

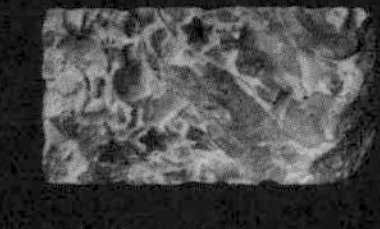

Crinoid, disarticulated: *Isocrinus psilonoti.* Gloucester & Berkley Canal.

Crinoid, articulated: *Pentacrinites fossilis.* Near Lyme Regis.

Gastropod: *Pleurotomaria* cf. *cognata.* Purton Passage.

Ammonite: *Euagassiceras sauzeanum.* Stony Littleton.

29 — **REDLAND LIMESTONE.**

Coral: *Amplexus coralloides.* Leigh.

Coral: *Actinocyathus floriformis.* Bristol.

Crinoid stems: indet. Mells.

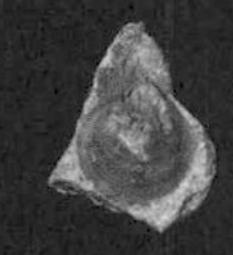

Bivalve:
Camptonectes auritus.
Farley.

Fish tooth:
Eomesodon sp. (splenial bone).
Stonesfield.

Shark palate:
Asteracanthus magnus.
Pickwick.

Shark palate:
Asteracanthus tenuis.
Pickwick.

Bivalve:
Oxytoma costata.
Bradford.

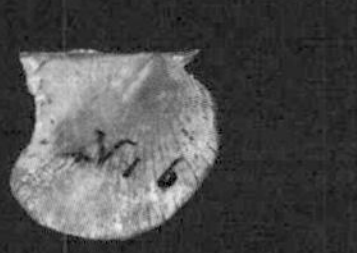

Bivalve:
Oxytoma costata.
Hinton.

Brachiopod:
Digonella digona.
Farley.

Brachiopod:
Dictyothyris coarctata.
Farley.

Bivalve:
Actinostreon marshii.
Cotswold Hills.

Brachiopod:
Rhynchonelloidella media.

Gastropod:
Nerinea sp.
Smallcombe Bottom.

Gastropod:
Nerinea sp.
Churchill.

Gastropod:
Ampullina sp.
Bath.

Nautilus:
Cenoceras excavatus.
Sherborne.

Ammonite:
Teloceras calix.
Sherborne.

Bivalve:
Fragment of large
Trichites ploti.
Bath.

Brachiopod:
Acanthothyris spinosa.
Churchill.

Echinoid:
Clypeus ploti.
Near Naunton.

Bivalve:
Pseudopecten equivalvis.
Kennet & Avon Canal.

Ammonite:
Zugodactylites braunianus.
Bath.

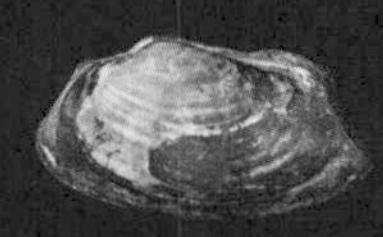

Bivalve:
Pleuromya aff. *uniformis.*

Fig. 8.
Smith's rock collection was sold to the British Museum with his fossils in 1818, although they were not catalogued and are often not well documented. (Top) 'Killas' from Exmoor in Devon; (centre) pink granite from Great Strickland, similar to the Lake District Shap Granite nearby; (bottom) chipped-off sarsen stone from one of the large standing stones at Stonehenge. Sarsens are still in abundance in their natural form in nearby valleys.

FIG. 8.

extremely disappointed when he was offered only £500 for his collection of over 2,000 specimens.

Correspondence exists in the British Museum archives about the catalogue Smith prepared to accompany the fossils, and also a second batch of specimens which was not accepted until 1818. The documents make depressing reading and it is clear that the chief negotiator, Charles König, the assistant to the Principal Librarian, Joseph Planta, did not appreciate the scientific significance or importance of displaying such a collection arranged in strata as stipulated by Smith. Eventually Smith was paid a further £100 for his catalogue, which was published in 1817, though, like his *Strata Identified by Organized Fossils*, it was not completed. It ended before the Lower Jurassic strata and the remaining Lower Jurassic and Carboniferous strata were handwritten as Part 2, but lack of finance prevented publication.

Support from Banks eventually persuaded the Trustees to accept the 260 additional fossils, for which Smith was paid a further £100, bringing the total number of specimens, according to John Phillips (1800–1874), to 2,657, including a number of rock specimens in addition to the fossils ('supposedly 693 species') (Phillips, 1844, p. 79).

Smith's catalogue, *A Stratigraphical System of Organized Fossils*, is a comprehensive list of all the fossils sold to the British Museum arranged first stratigraphically and then systematically within each stratum, beginning with the most primitive and ending with the vertebrates. Smith devised a unique cataloguing system to ascribe scientific attributes to the fossils. Each genus was given a capital letter, starting with A. Different species were then given consecutive numbers; lower case letters, a, b, c etc, distinguished different localities. Many of Smith's specimens, but not all, have this code inscribed in Indian ink on them and were probably added as the catalogue was prepared. It was only when the fossils were laid out in their correct order at 15 Buckingham Street that they were in position to have the identifying letters and numbers added.

Smith's nephew, John Phillips, who had accompanied him on his expeditions and was quick to learn, helped Smith draw up his catalogue and his contribution cannot be overstated. It is unknown whether the writing on the specimens is Smith's or that of his apprentice, but it is probable that the handwritten Part 2 of the catalogue is in Phillips's hand (Wigley, 2019, p. 8). Phillips went on to have a distinguished geological career himself, finally becoming Chair of Geology at Oxford University, President of the Geological Society and Keeper of the Ashmolean Museum and the Oxford University Natural History Museum.

It is clear from the published catalogue that Smith intended to continue with the unfinished volumes of *Strata Identified*. For his Under Oolite he listed the fossils he wanted to illustrate in two plates. From the manuscript catalogue of Part 2 it is possible to match the fossils he had in mind with unpublished sketches in the archives of the Oxford University Museum of Natural History for his proposed three Lias plates and the Redland Limestone plate.

The total of £700 that Smith received from the British Museum was insufficient to repair his finances and in June 1819 he was committed to the King's Bench debtors' prison in Southwark for almost ten weeks. The financial crisis was a tragedy for Smith personally, but saved his seminal fossil collection in its entirety for future generations. What happened to any fossils he collected during the remaining twenty years of his life remains a mystery.

DISPLAYING THE COLLECTION.

The surviving correspondence shows the British Museum initially had no space to house Smith's collection, delaying its transfer. This had the advantage that Smith was able to arrange and catalogue his collection while displayed in his premises in Buckingham Street. Documents in the archives relate to initial storage space, including an agreement from the Museum's Trustees to alter a glazed cabinet to mount the display.

There is no later correspondence and it seems Smith's collection was never even unpacked. In 1827, more than ten years later, a handwritten inventory of the exhibition rooms details Smith's design of sloping shelves within the glazed cabinet. The text implies that the collection was not yet installed and that the Trustees would not in any case approve either of the way Smith wanted to arrange his fossils, or of his nomenclature.

> *The upright glass cases at the North side of this Room, were fitted up with sloping shelves for the reception of the greater part of Mr. William Smith's collection of Secondary Fossils, intended to illustrate the succession of Strata in England and Wales, and purchased in 1817 by Parliament. This method of arranging a collection of Fossils is not likely to meet with the approbation of the Trustees. It does not appear to possess any advantage over the common way of arranging rocks and fossils; neither are Geologists agreed with regard to several supposed facts intended to be illustrated and explained by that arrangement; not to mention that the nomenclature used by W. Smith, has not been generally adopted by writers on this subject.*

FIG. 9.

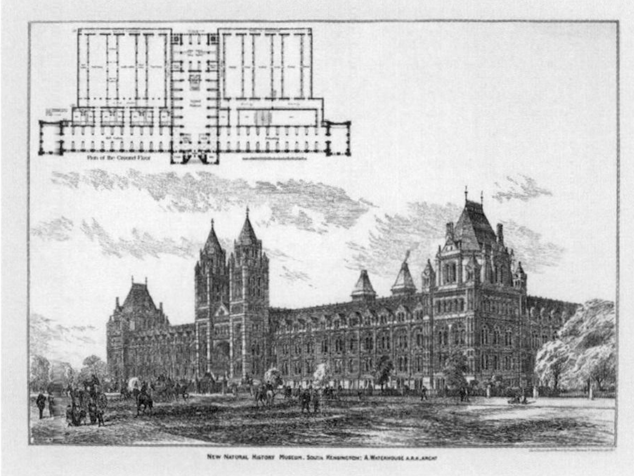

FIG. 10.

The absence of his collection on display is borne out by the observation twenty years later by Thomas Hoblyn of the Treasury, whom Smith had corresponded with about the sale, that 'the specimens remain in the Boxes at the British Museum unopened' (Wigley, 2019, p. 6). In 1838, Smith himself observed that 'the collection lay in obscurity' and in 1844, after Smith's death, John Phillips did not know what had happened to it (Torrens, 2016, p. 35).

In the 1880s the fortunes of Smith's collection changed. It was transferred to the newly built Natural History Museum in South Kensington and an early guide to the displays (1886) mentions the William Smith exhibit which included a copy of his map, his bust and a cabinet displaying his fossils. A more detailed version is found in the 1897 edition. That was how the situation remained until the 1930s, when the displayed fossils were relocated to an A-frame cabinet in a different gallery. In both locations the residue of the collection was stored in drawers underneath the display. In 1976 the Smith collection was removed from display to be housed in the newly opened east wing of the museum, dedicated to the storage of the palaeontological collections. The fossils and rocks are kept together in Smith's stratigraphical order and remain today available for study. Fossils continue to play an important part in the Museum's work, providing evidence in the ongoing search for resources and dating strata.

To mark 200 years since the publication of Smith's famous map in 2015, a temporary display, *Mapping a Nation*, was mounted alongside the permanent life-size facsimile of the 1815 map. Four of the original Sowerby hand-coloured plates were selected and the actual fossils illustrated were displayed in the same arrangement alongside. Following this exhibition, the authors located the fossils for the remaining plates, which were photographed for *William Smith's Fossils Reunited* (Wigley, 2019).

Fig. 9.
Smith did not figure his specimens for his 'Under Oolite' (Inferior Oolite), but he did list them and selected specimens. Smith's code is inscribed on most of the fossils in this plate, most easily seen in the gastropod *Nerinea* sp. NHMUK G1661 E 1 a (above), which also has the location (Smallcombe Bottom Bath). This specimen is 3.4 cm (1⅓ in.) in length. The nautiloid, *Cenoceras excavatus* NHMUK C666 (below), is labelled Sherborne H 3 a. This specimen is 6.5 cm (2½ in.) in length.

Fig. 10.
Plan of the ground floor and drawing of the façade of the new Natural History Museum. Guide books from 1886, 1897 and 1907 detail Smith's collection on display in Gallery XI of 'Historical & Type collections' (extreme right gallery, on the north side). It occupied Table-cases 10–12 in the middle of the right-hand side along with a bust of Smith and a copy of his map.

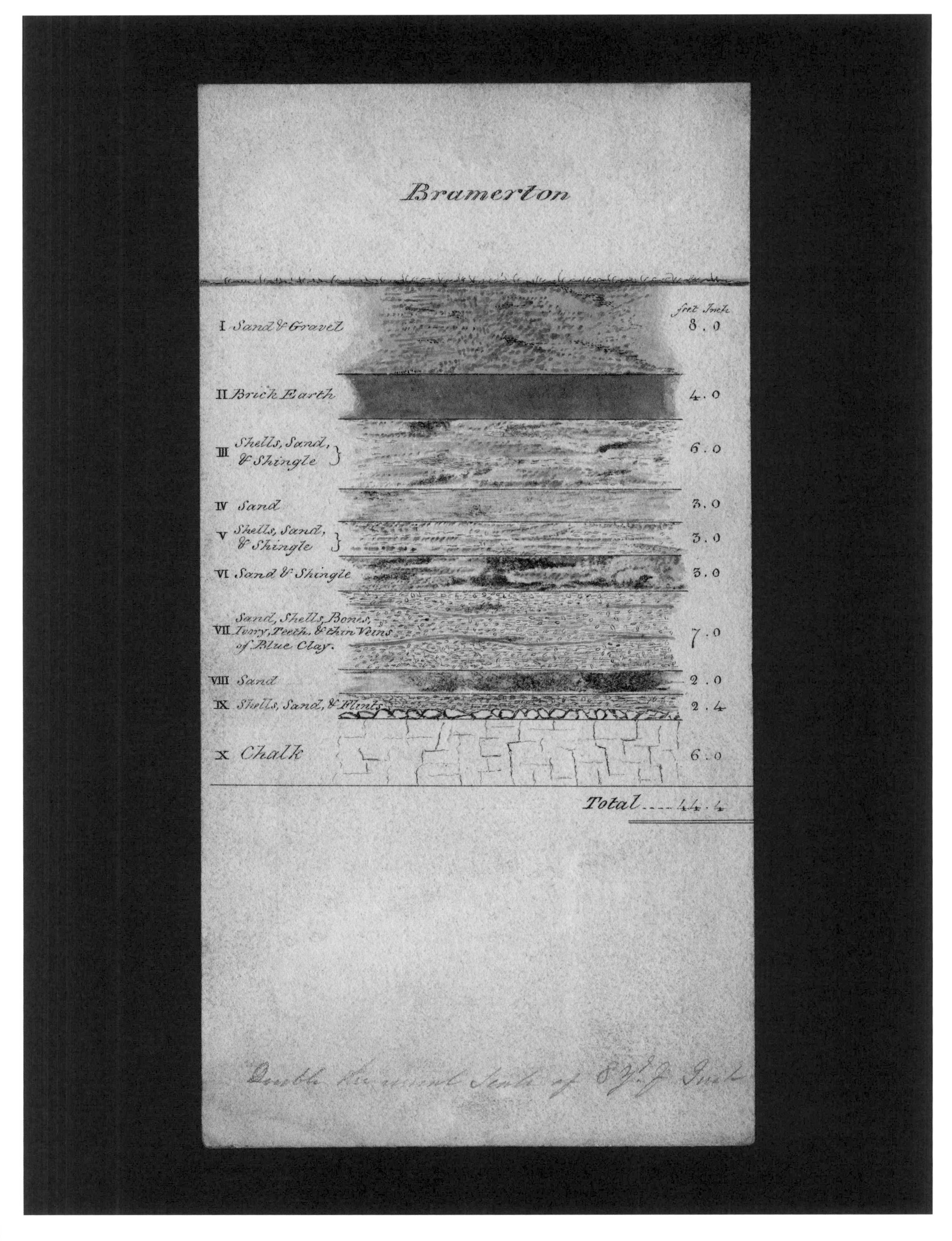

BRAMERTON LOG SECTION, NORFOLK.

Smith will have drawn this section while working on drainage projects in the area; he was resident in Norwich for about three years. During that time he collected many specimens from Bramerton including the barnacle, gastropod encrusted with barnacles and the large bivalve, *Mya lata*, that he illustrated in his Crag plate. He lists many others in *A Stratigraphical System of Organized Fossils*: Gastropods: *Murex striatus* and *Murex* species 3, *Turbo littoreus* and *Turbo* species 2 and 3, *Turritella trilineata*, *Scalaria similis*, *Trochus* species 2; Bivalves: *Cardium*, *Mactra*, *Tellina bimaculata*. They will have come from the shell layers between the Brickearth and Chalk. All of the fossils illustrated by Sowerby can be found in the Red Crag but some of them extend upwards into the Norwich Crag, and some down to the Coralline Crag beneath. Since 1985 Bramerton Pits, north of the village of Bramerton, on the southern banks of the River Yare, has been recognized as a Site of Special Scientific Interest as the type section for the Norwich Crag Formation and nearby Blakes Pit is the type site of the Bramertonian Stage within that.

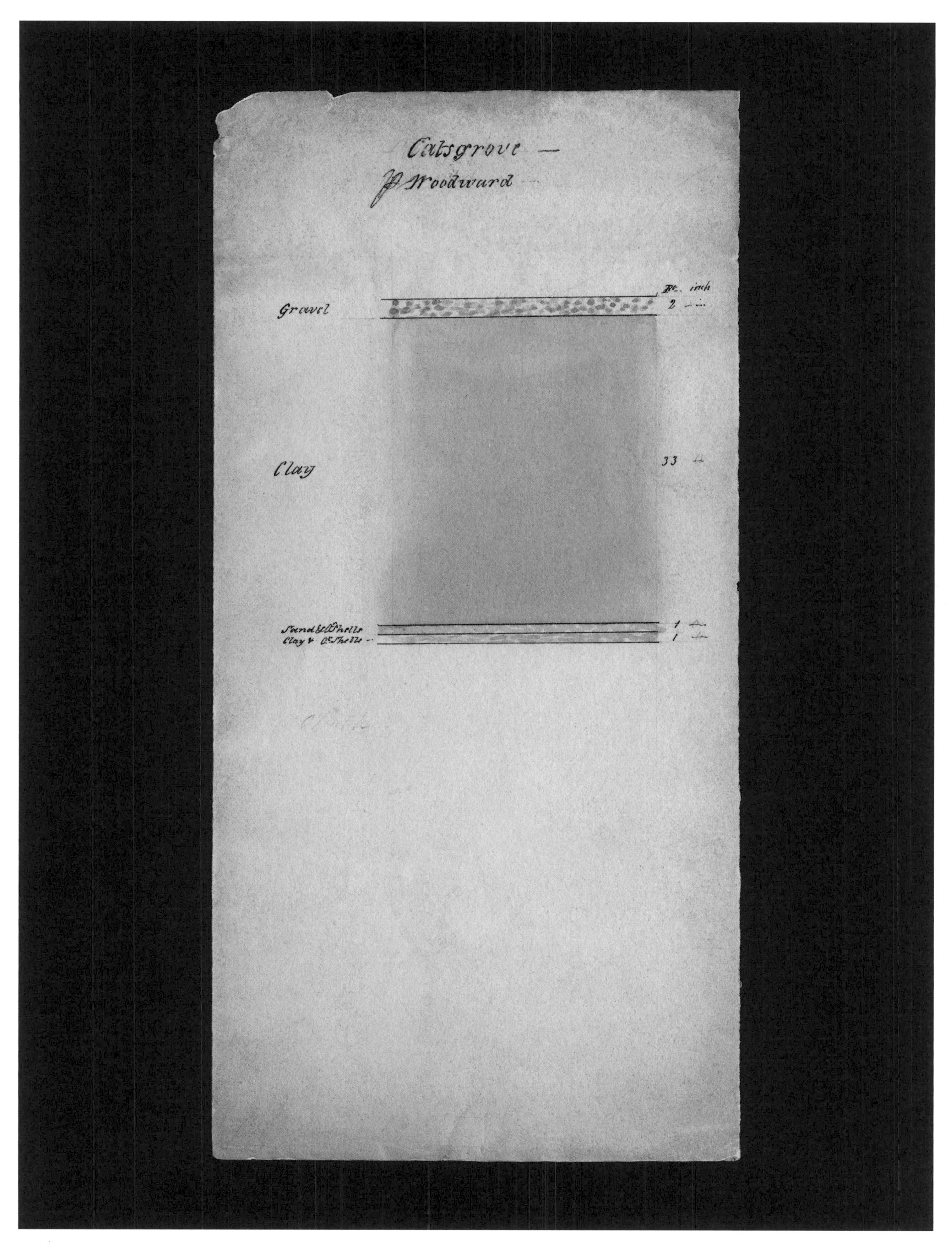

CATSGROVE LOG SECTION, NEAR READING, BERKSHIRE.

Catsgrove Hill was famous for its brick pits described in detail both by Ure (1814, p. 295–7) and Buckland (1817). Smith did not collect fossils from there, nor are the Reading Beds described in either *Strata Identified by Organized Fossils* or *A Stratigraphical System of Organized Fossils*. However, it is likely that Smith was familiar with the brick pits as he describes the layer between the London Clay and Chalk as 'Brickearth'. At the top of the section it says 'P. (or p.) Woodward'. This is undoubtedly the section that Dr. John Woodward described in his *Catalogue of Fossils* (vol. ii, p. 41) as the measurements match exactly.

The catalogue was deposited, along with the fossils, in Cambridge Museum in 1728 and referred to by Sedgwick 100 years later in 1822 (Sedgwick, 1822, p. 340). Both are still housed in the Sedgwick Museum and the oyster from the base has been identified as *Ostrea bellovacina* from the Upnor Formation underneath the Reading Formation. It is probable that Smith illustrated the section from Woodward's description but as no details of the Reading Beds were given ('thin clay of various colours, purple, blue, red, liver-colour'), he chose to paint them uniform grey. The 'P' probably stands for 'per' or something similar.

'I have, with immense labour and expense, collected specimens of each stratum, and of the peculiar extraneous fossils, organic remains, and vegetable impressions, and compared them with others from very distant parts of the island, with reference to the exact habitation of each, and have arranged them in the same order as they lay in the earth; which arrangement must readily convince every scientific or discerning person, that the earth is formed as well as governed, like the other works of its great Creator, according to regular and immutable laws, which are discoverable by human industry and observation, and which form a legitimate and most important object of science.'

SMITH'S *MEMOIR*, 1815.

DETAIL FROM SHEET XI, SHOWING FOSSIL-RICH REGIONS.
The strata of Dorset and Hampshire, from the Lias in the west to the London Clay in the east, are rich with fossils of fish, giant marine reptiles, ammonites and numerous other invertebrates.

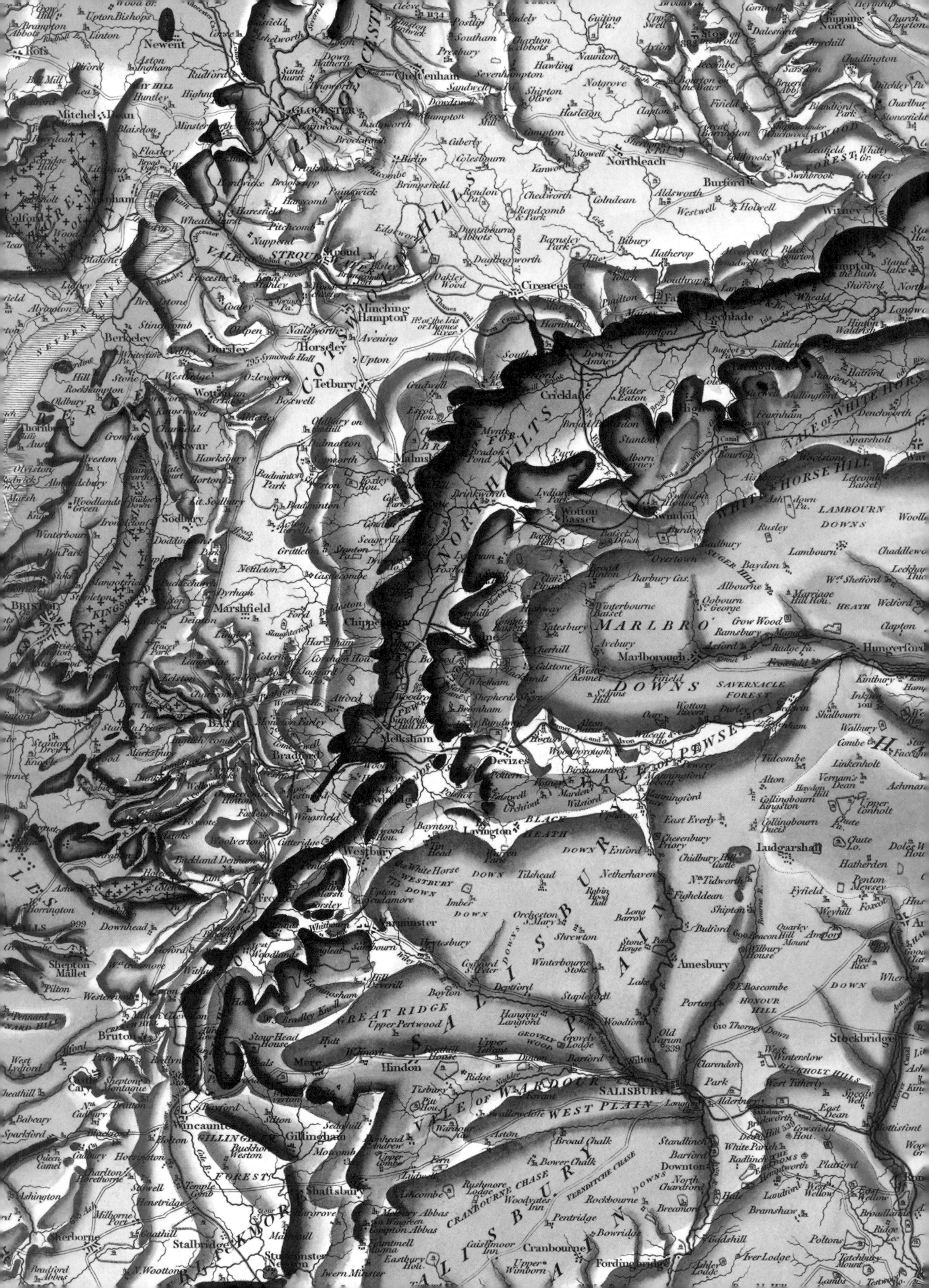

Newent
Upton Bishops
Linton
Cheltenham
Gloucester
Northleach
Burford
Witney
Chipping Norton
Stroud
Minching Hampton
Cirencester
Tetbury
Berkeley
Dursley
Horseley
Avening
Upton
Wotton under Edge
Cricklade
Lechlade
Sodbury
Marshfield
Chippenham
Malmsbury
Bath
Bradford
Melksham
Devizes
Wotton Basset
Swindon
Marlborough
Hungerford
Westbury
Warminster
Frome
Shepton Mallet
Bruton
Wincaunton
Mere
Hindon
Shaftsbury
Gillingham
Salisbury
Amesbury
Ludgarshall
Stockbridge
Cranbourne
Fordingbridge
Sherborne
Stalbridge
Milborne Port
COTSWOLD HILLS
NORTH WILTS
VALE OF WHITE HORSE
WHITE HORSE HILL
LAMBOURN DOWNS
MARLBRO' DOWNS
SAVERNACLE FOREST
VALE OF PEWSEY
SALISBURY PLAIN
GREAT RIDGE
VALE OF WARDOUR
WEST PLAIN
CRANBOURNE CHASE
VERNDITCHE CHASE
SEVERN RIVER
VALE OF STROUD
WHICHWOOD FOREST
BLACK HEATH
BUCKHOLT HILLS
HONOUR HILL
Stour Head House
Stone Henge
Old Sarum

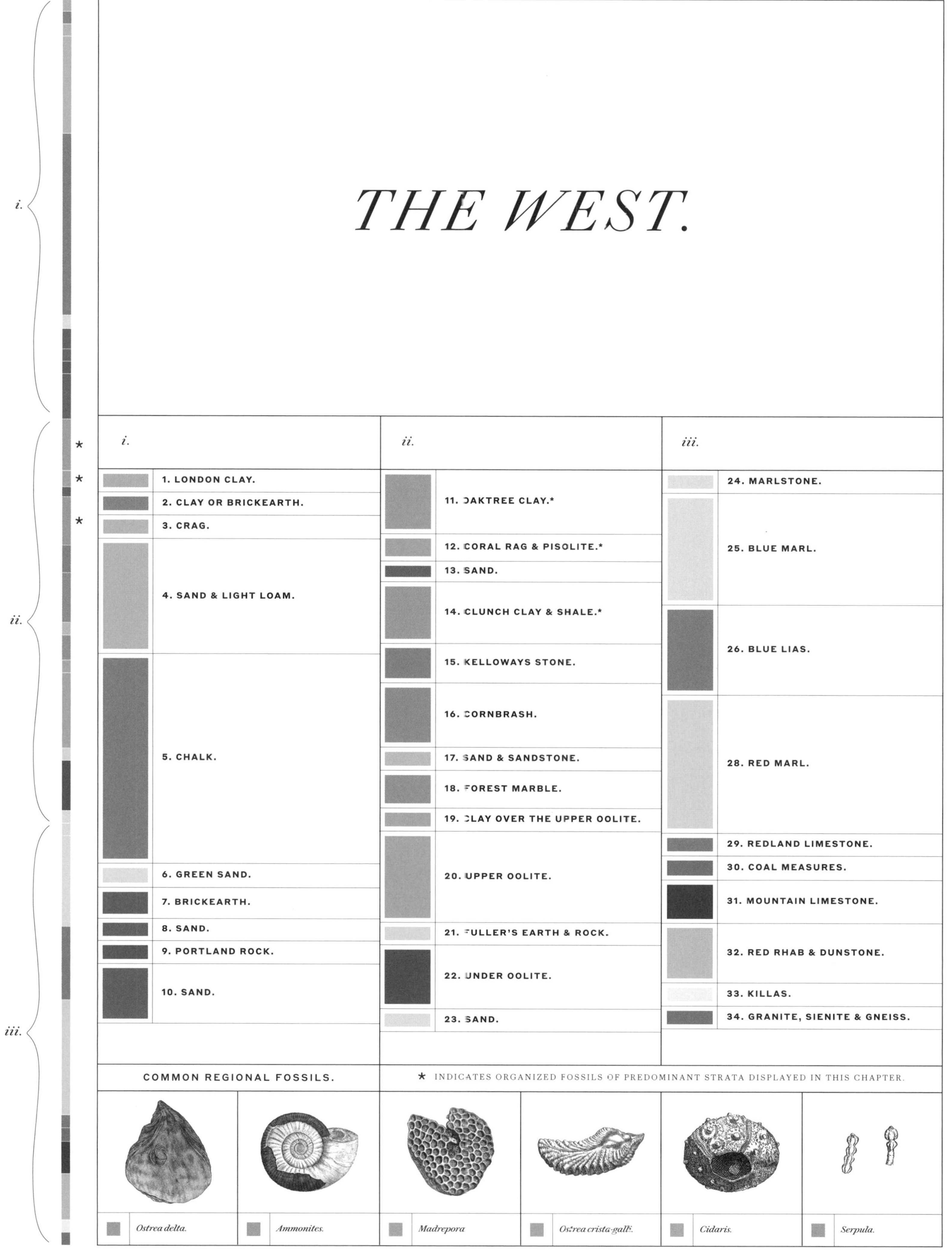

i.
ii.
iii.
*
*
*
THE WEST.
i.
ii.
iii.
1. LONDON CLAY.
2. CLAY OR BRICKEARTH.
3. CRAG.
4. SAND & LIGHT LOAM.
5. CHALK.
6. GREEN SAND.
7. BRICKEARTH.
8. SAND.
9. PORTLAND ROCK.
10. SAND.
11. OAKTREE CLAY.*
12. CORAL RAG & PISOLITE.*
13. SAND.
14. CLUNCH CLAY & SHALE.*
15. KELLOWAYS STONE.
16. CORNBRASH.
17. SAND & SANDSTONE.
18. FOREST MARBLE.
19. CLAY OVER THE UPPER OOLITE.
20. UPPER OOLITE.
21. FULLER'S EARTH & ROCK.
22. UNDER OOLITE.
23. SAND.
24. MARLSTONE.
25. BLUE MARL.
26. BLUE LIAS.
28. RED MARL.
29. REDLAND LIMESTONE.
30. COAL MEASURES.
31. MOUNTAIN LIMESTONE.
32. RED RHAB & DUNSTONE.
33. KILLAS.
34. GRANITE, SIENITE & GNEISS.
COMMON REGIONAL FOSSILS.
* INDICATES ORGANIZED FOSSILS OF PREDOMINANT STRATA DISPLAYED IN THIS CHAPTER.
Ostrea delta.
Ammonites.
Madrepora
Ostrea crista-galli.
Cidaris.
Serpula.

Bar chart of strata indicates the comparative area covered by each stratum in this region.

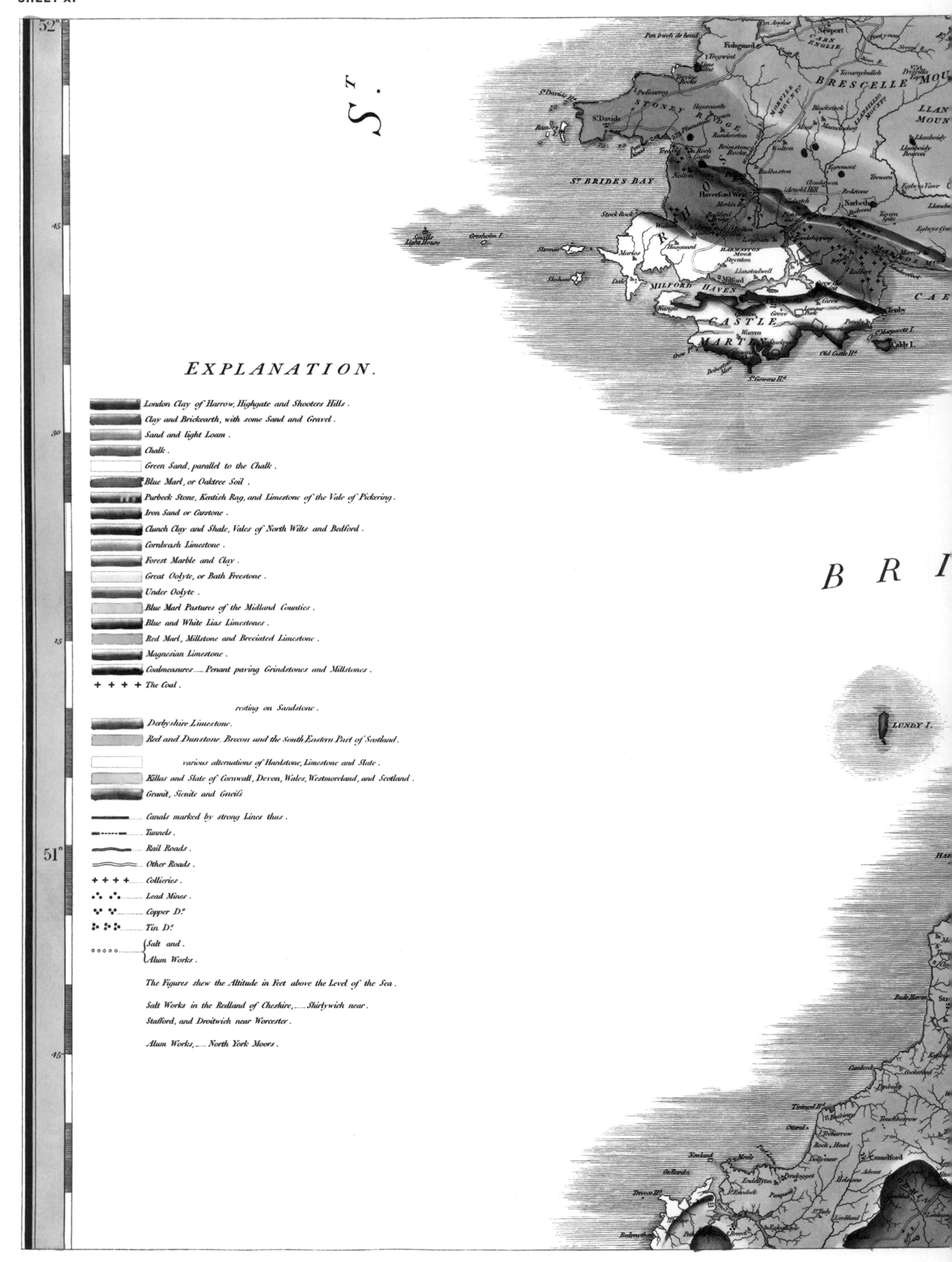
EXPLANATION.
London Clay of Harrow, Highgate and Shooters Hills.
Clay and Brickearth, with some Sand and Gravel.
Sand and light Loam.
Chalk.
Green Sand, parallel to the Chalk.
Blue Marl, or Oaktree Soil.
Purbeck Stone, Kentish Rag, and Limestone of the Vale of Pickering.
Iron Sand or Carstone.
Clunch Clay and Shale, Vales of North Wilts and Bedford.
Cornbrash Limestone.
Forest Marble and Clay.
Great Oolyte, or Bath Freestone.
Under Oolyte.
Blue Marl Pastures of the Midland Counties.
Blue and White Lias Limestones.
Red Marl, Millstone and Breciated Limestone.
Magnesian Limestone.
Coalmeasures___Penant paving Grindstones and Millstones.
+ + + + The Coal.
resting on Sandstone.
Derbyshire Limestone.
Red and Dunstone, Brecon and the South Eastern Part of Scotland.
various alternations of Hardstone, Limestone and Slate.
Killas and Slate of Cornwall, Devon, Wales, Westmoreland, and Scotland.
Granit, Sienite and Gneiss
Canals marked by strong Lines thus.
Tunnels.
Rail Roads.
Other Roads.
+ + + + Collieries.
Lead Mines.
Copper D°.
Tin D°.
Salt and.
Alum Works.
The Figures shew the Altitude in Feet above the Level of the Sea.
Salt Works in the Redland of Cheshire,___Shirlywich near.
Stafford, and Droitwich near Worcester.
Alum Works,___North York Moors.
52°
45
30
15
51°
45
S.T
B R I
S.T BRIDES BAY
MILFORD HAVEN
CASTLE MARTIN
STONEY RIDGE
BRESCELLE
Fishguard
Newport
S.t Davids
Haverford West
Narberth
Tenby
Caldy I.
LUNDY I.
Camelford

SHEET XI.

(see pages 204–205)

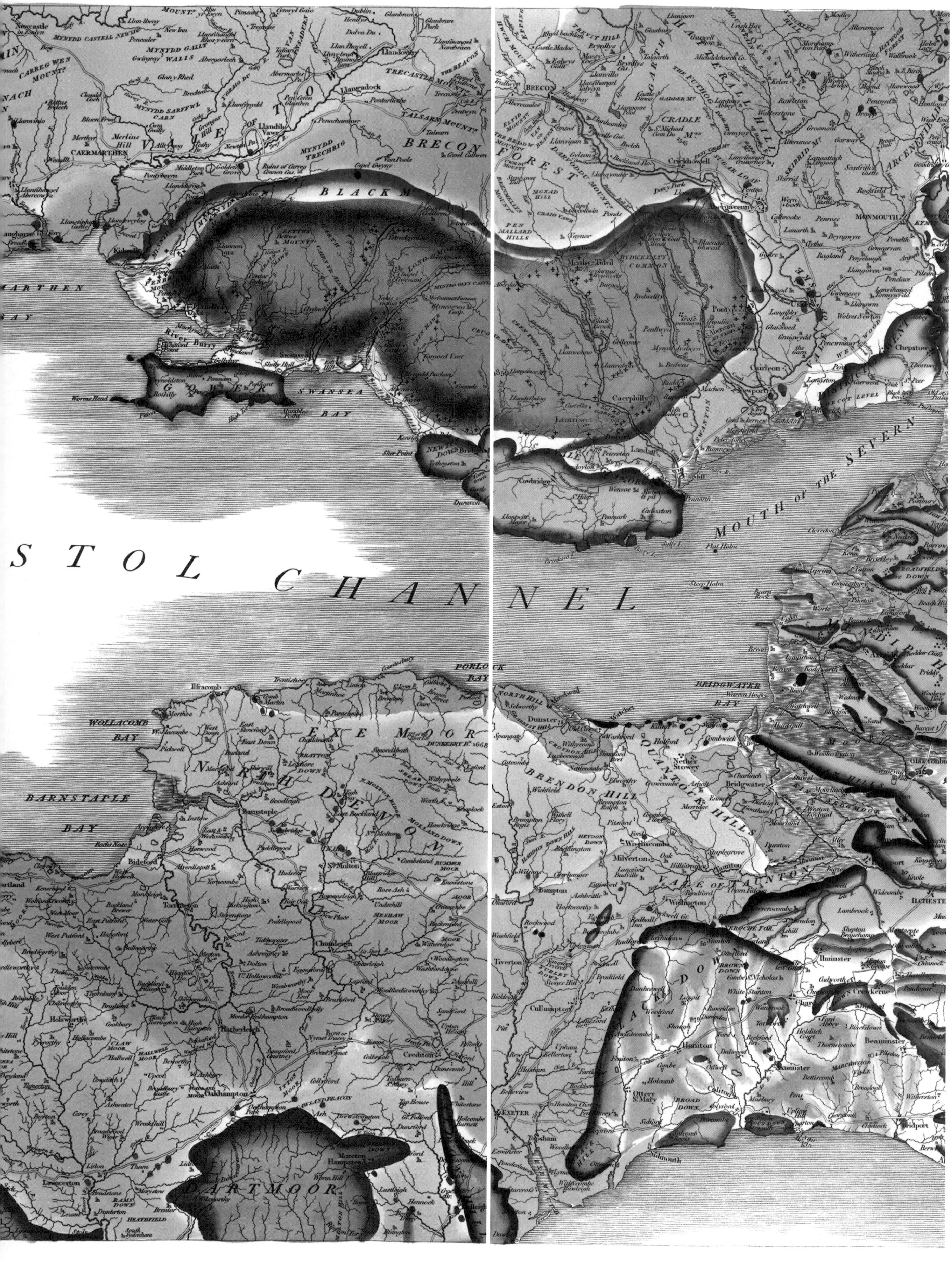

MOUTH OF THE SEVERN
NEL
BRIDGWATER
BAY
BRECON
FOREST
CRADLE
Merthyr Tidvil
Pontypool
Caerphilly
Lantrisent
Crickhowell
MONMOUTH
Chepstow
Newport
Cardiff
Cowbridge
Flat Holm
Steep Holm
Clevedon
BRISTOL
BATH
Marshfield
Tetbury
Cirencester
Gloucester
Cheltenham
Tewksbury
Ledbury
Newent
Stroud
Minching Hampton
Berkeley
Sodbury
Chippenham
Melksham
Devizes
Marlborough
Westbury
Warminster
Frome
Wells
Shepton Mallet
Glastonbury
Bridgwater
Minehead
Dunster
Watchet
Nether Stowey
QUANTOCK HILLS
BRENDON HILL
Milverton
Wellington
Taunton
VALE OF TAUNTON
ILCHESTER
Yeovil
Sherborne
Ilminster
Crewkerne
Chard
Honiton
Cullumpton
Tiverton
Bampton
EXETER
Topsham
Sidmouth
Ottery St. Mary
Axminster
Lyme
Bridport
Beaminster
DORCHESTER
Cerne Abbas
Blandford Forum
Wimborn Minster
Poole
Wareham
Weymouth
Melcomb Regis
STUDLAND BAY
Sandwich Bay
WAREHAM AND POOLE HEATHS
DORSETSHIRE DOWNS
Shaftsbury
Gillingham
Wincaunton
Bruton
Stalbridge
SALISBURY
GREAT RIDGE
VALE OF WARDOUR
WEST PLAIN
Hindon
Cranbourne
Fordingbridge
MENDIP HILLS
COTSWOLD HILLS
NORTH WILTS
Malmsbury
Wotton Basset
Cricklade
Northleach
Tilshead
Netheravon
Broad Chalk

(see pages 138–139)

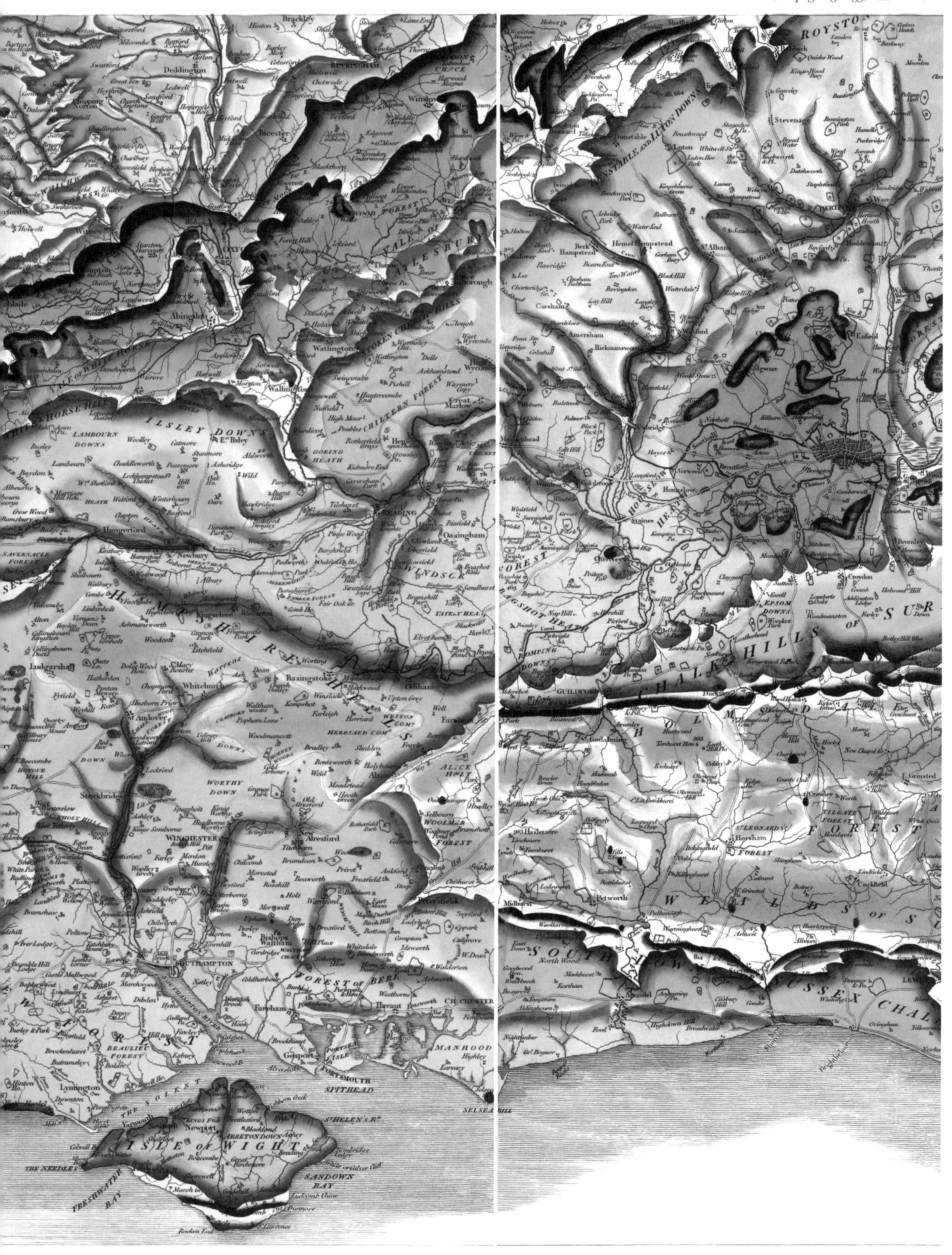

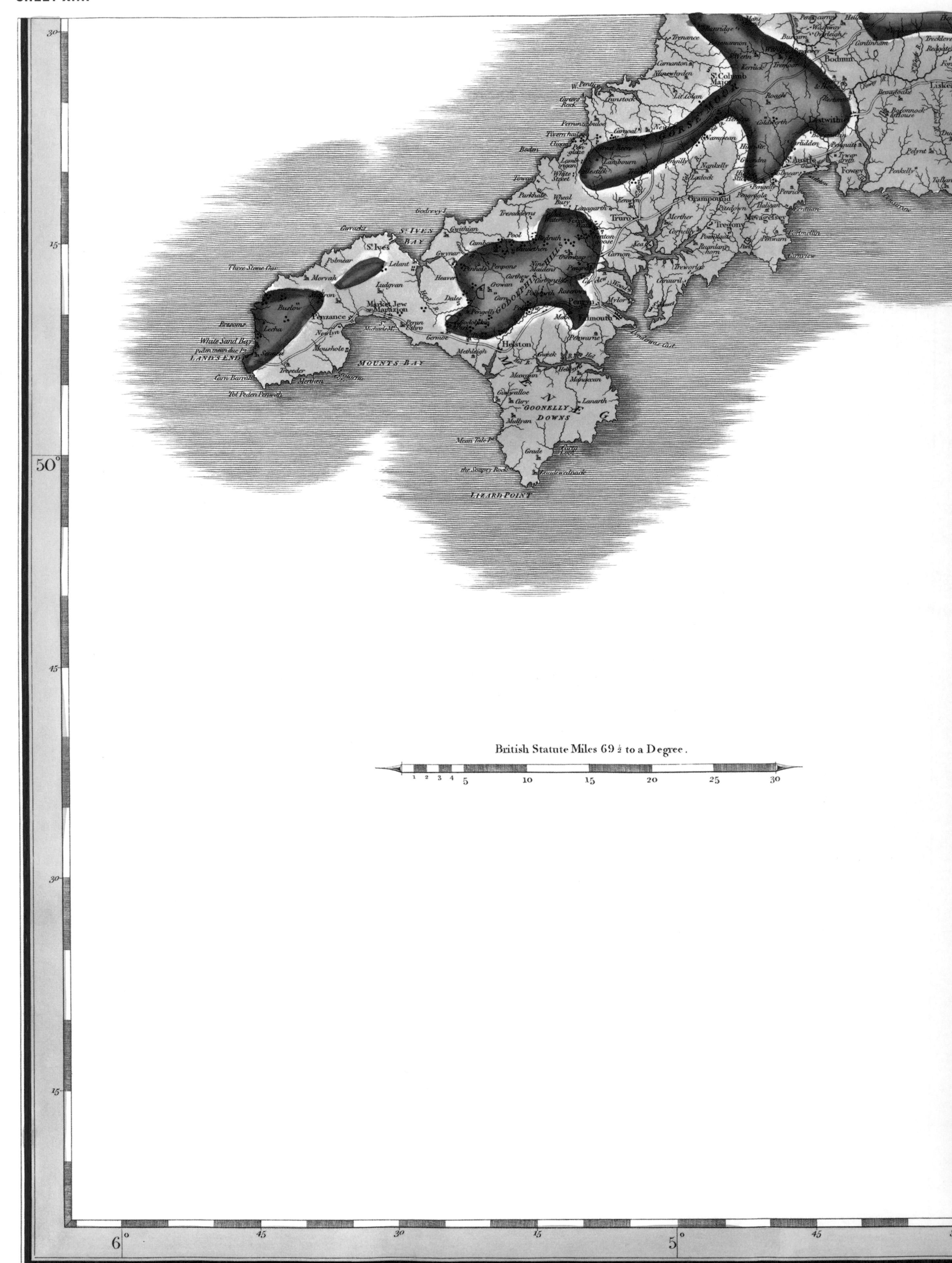
St. Columb Major
Bodmin
Lostwithiel
Fowey
St. Austle
Truro
Grampound
Tregony
Mevagissey
Penryn
Falmouth
Helston
Penzance
Newlyn
Mousehole
Market Jew Marazion
Ludgvan
Lelant
St. Ives
St. IVES BAY
Morvah
Polmear
Three Stone Oar
Burlow
Lecha
White Sand Bay
LAND'S END
Treveder
Merthen
MOUNTS BAY
Godrevy I.
Gwithian
Camborne
Redruth
Pool
Perranzabuloe
GOODOLPHIN HILLS
Gwennap
Mylor
Mawgan
Gunwalloe
Cury
Mullyan
GOONELLY DOWNS
Lanarth
Manaccan
Grade
Mean Tale Pt.
the Soapey Rock
LIZARD POINT
Lizard Point
British Statute Miles 69 ½ to a Degree.
1
2
3
4
5
10
15
20
25
30
50°
45
30
15
6°
5°

(see pages 208–209)

(see pages 206–207)

T H E E N

15 4° 45 30 15 3° 45

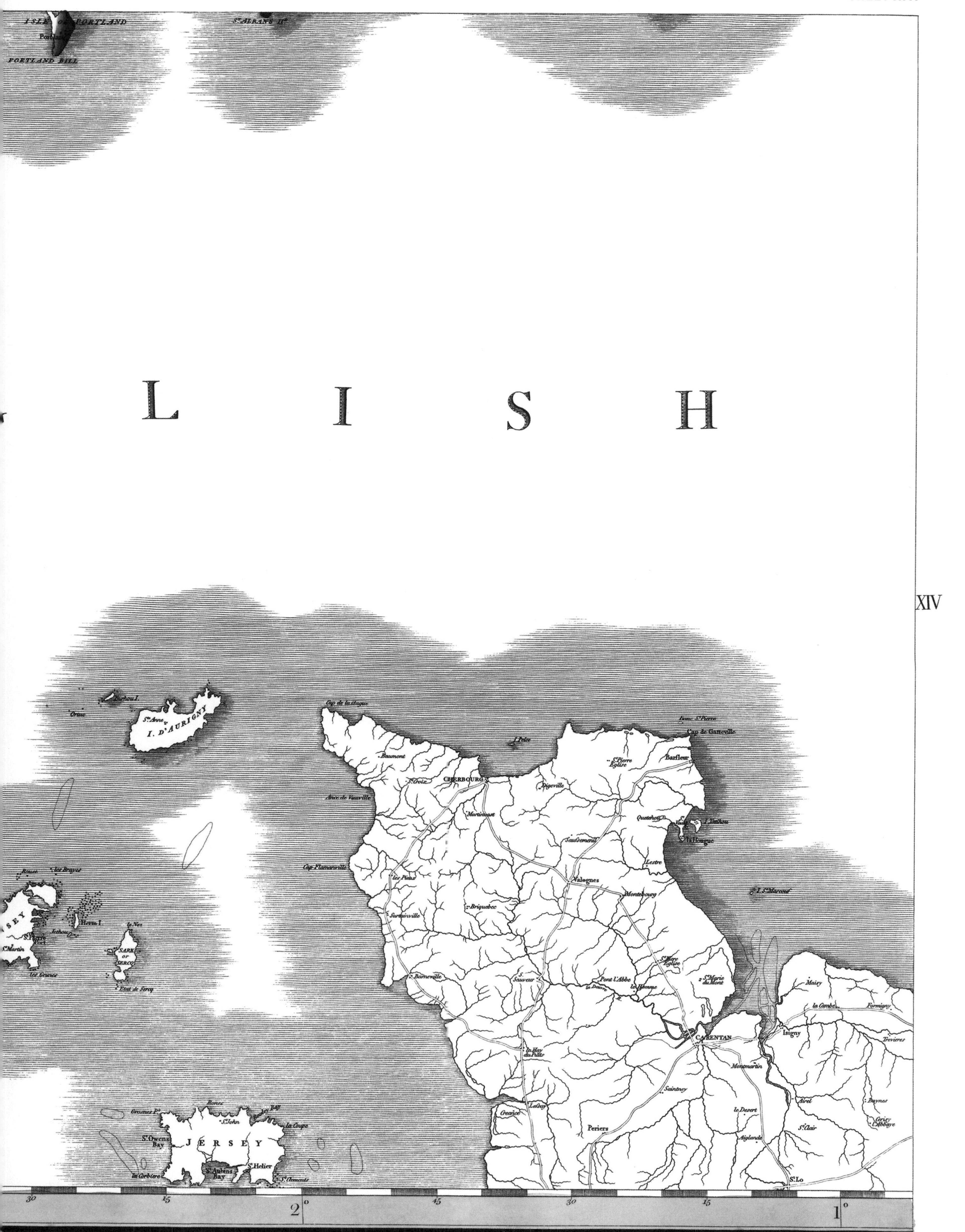

Published by J. Cary, 181 Strand, London: August 1st. 1815.

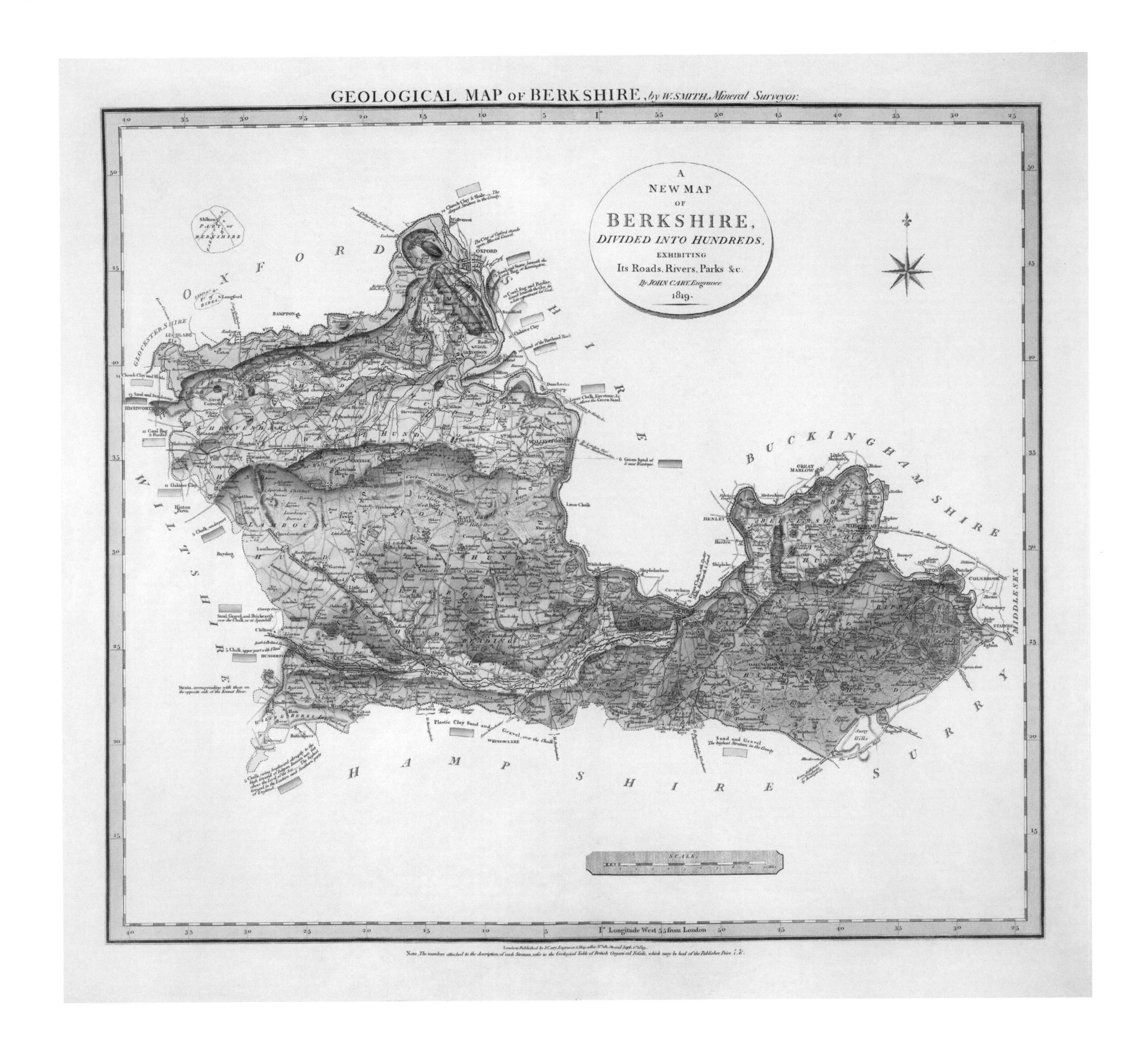

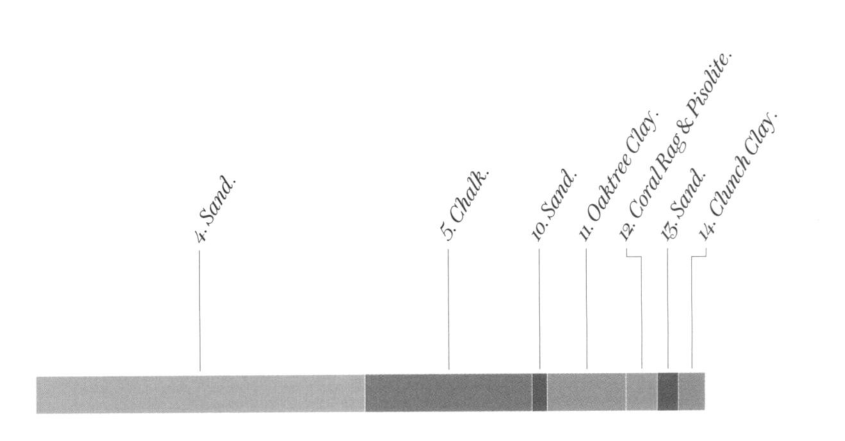

GEOLOGICAL MAP OF BERKSHIRE, 1819.

First published in Part II of *A New Geological Atlas of England and Wales* with the counties Gloucestershire, Surry (Surrey) and Suffolk. The top stratum is composed of Clay or Sand (18) over Brickearth (2) and the bottom stratum Clunch Clay and Shale (14).

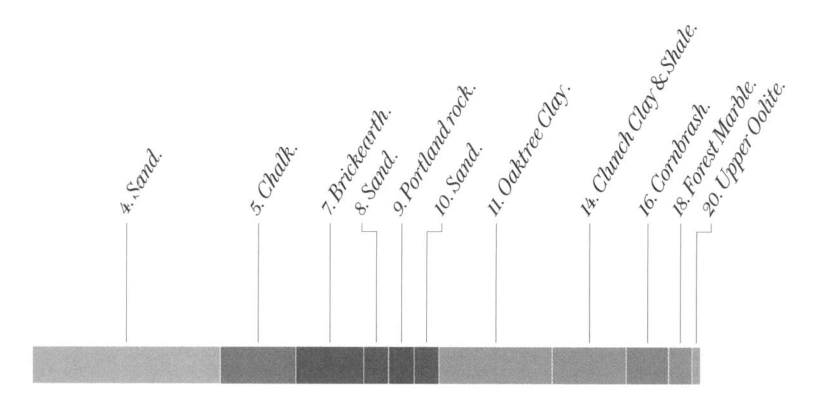

GEOLOGICAL MAP OF BUCKINGHAMSHIRE, 1820.

First published in Part III of *A New Geological Atlas of England and Wales* with the counties Oxfordshire, Bedfordshire and Essex. The upper stratum is composed of Brickearth (7), Sand (2, 4) and small pebbles over Chalk (5). Smith shows the deepest strata in the county to be the Upper Oolite.

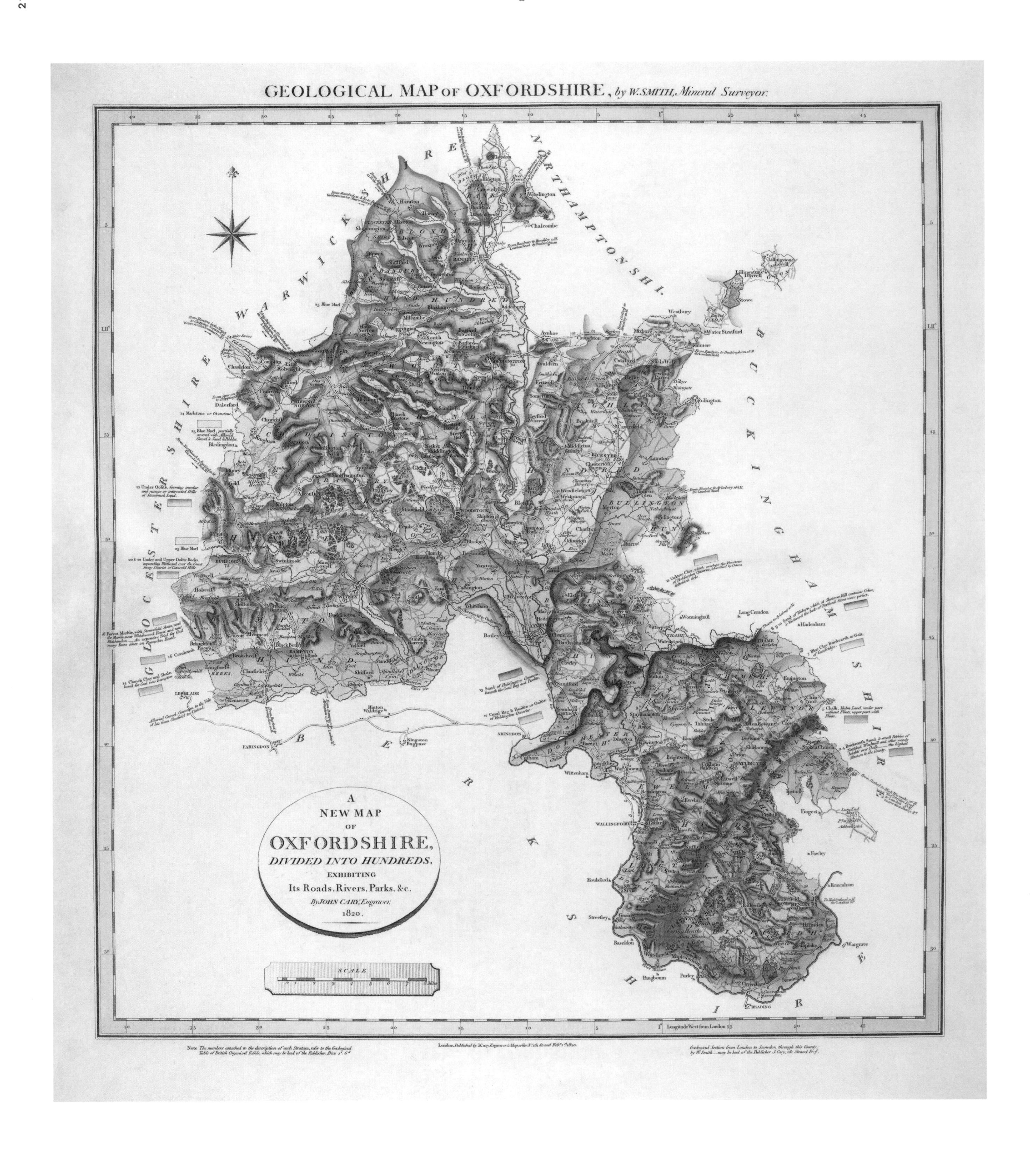

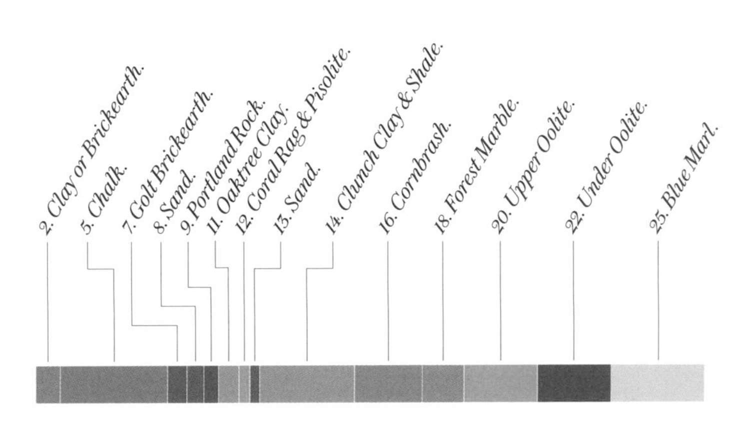

GEOLOGICAL MAP OF OXFORDSHIRE, 1820.

First published in Part III of *A New Geological Atlas of England and Wales* with the counties Buckinghamshire, Bedfordshire and Essex. The upper stratum is composed of Brickearth (2), Sand (4) and small pebbles over Chalk (5) and the lowest stratum comprises Blue Marl(25) partially covered with alluvial gravel, sand and pebbles.

GEOLOGICAL MAP OF GLOUCESTERSHIRE. *by W. SMITH, Mineral Surveyor.*

WORCESTERSHIRE

HEREFORDSHIRE

WARWICKSHIRE

OXFORDSHIRE

MONMOUTHSHIRE

SOMERSETSHIRE

WILTSHIRE

A NEW MAP OF GLOUCESTERSHIRE, *DIVIDED INTO HUNDREDS,* EXHIBITING Its Roads, Rivers, Parks &c. *By JOHN CARY Engineer.* 1819.

Longitude West from Greenwich

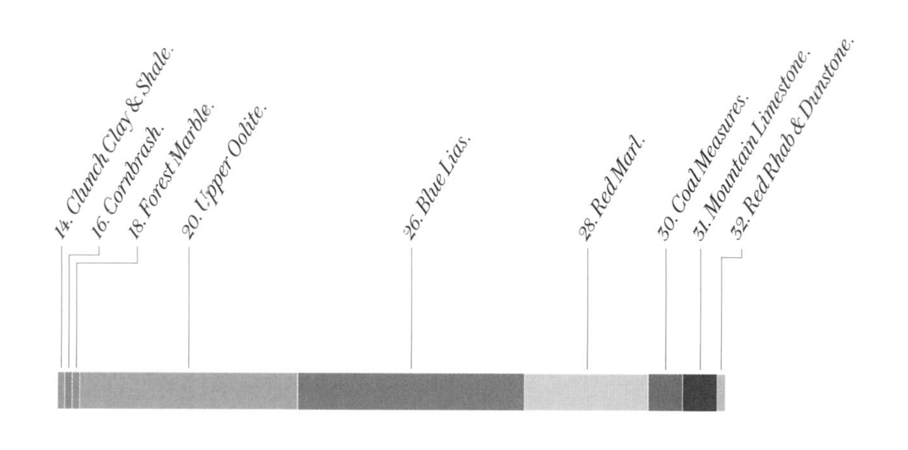

GEOLOGICAL MAP OF GLOUCESTERSHIRE, 1819.

First published in Part II of *A New Geological Atlas of England and Wales* with the counties Berkshire, Surry (Surrey) and Suffolk. The top stratum is composed of Clunch Clay and Shale (14) and the bottom stratum Red Rhab and Dunstone (32) with brecciated Sandstone.

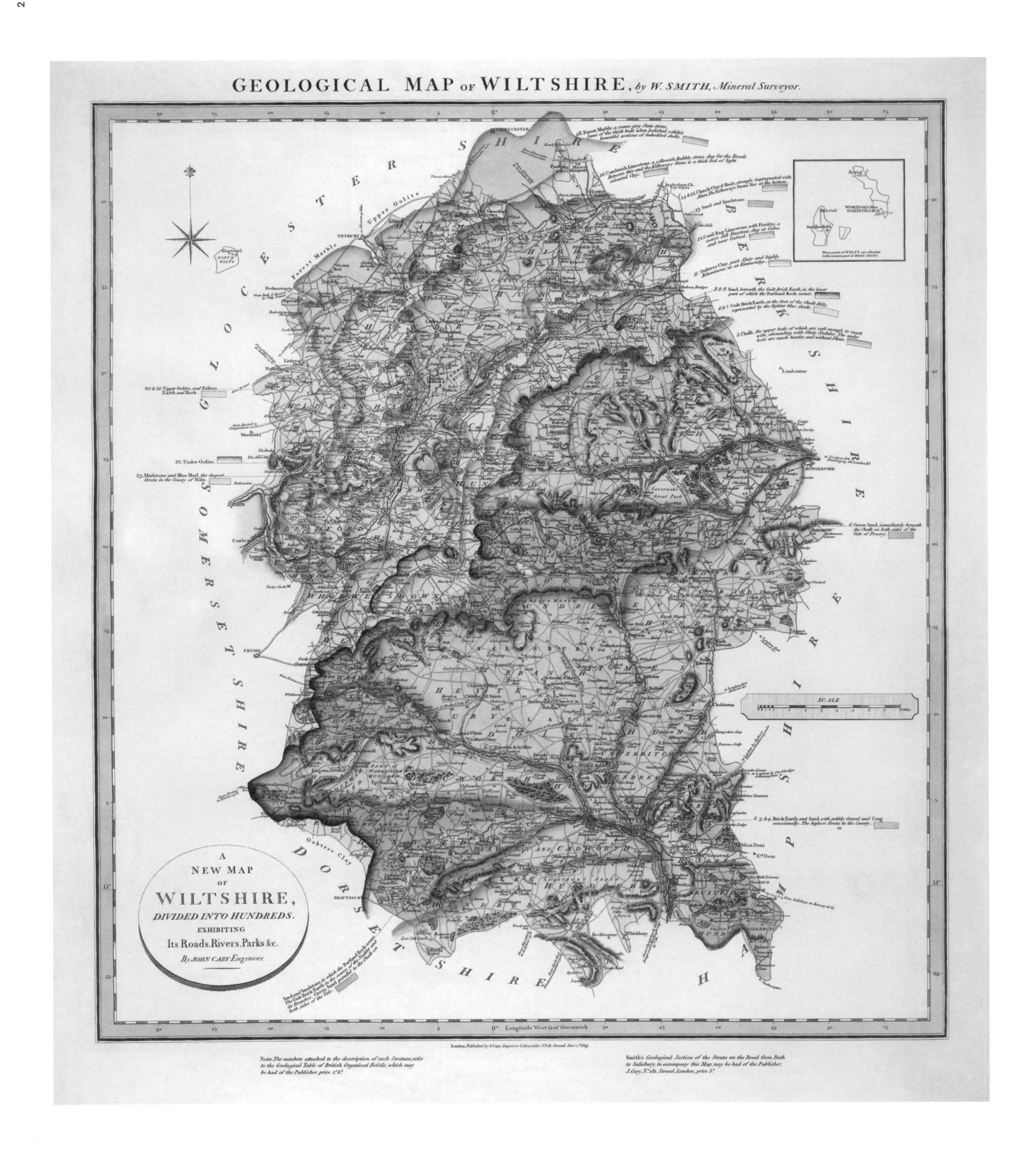

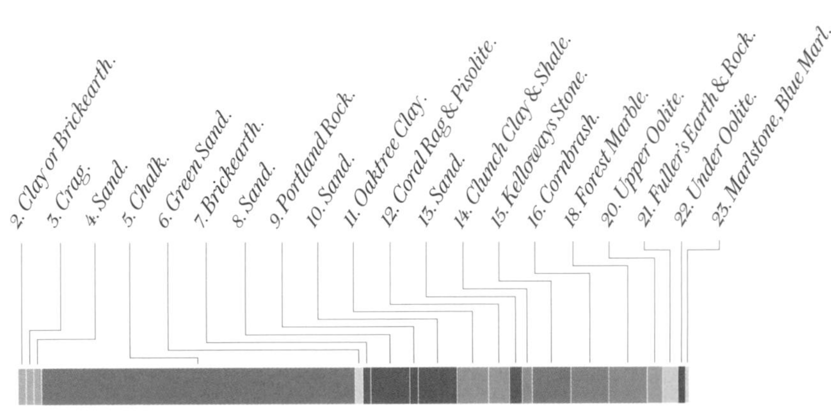

GEOLOGICAL MAP OF WILTSHIRE, 1819.

First published in Part I of *A New Geological Atlas of England and Wales* with the counties Norfolk, Kent and Sussex. The highest strata are composed of Brickearth (2) and Sand (4) with pebbly gravel and Crag occasionally (3) over Chalk (5), the upper beds of which abound with flinty nodules. The lowest stratum in the county is made up of Marlstone and Blue Marl (25).

GEOLOGICAL MAP OF SOMERSETSHIRE, *by W. SMITH, Mineral Surveyor.*

A NEW MAP OF SOMERSETSHIRE, *DIVIDED INTO HUNDREDS.* EXHIBITING Its Roads, Rivers, Parks &c. *By JOHN CARY, Engraver.* 1829

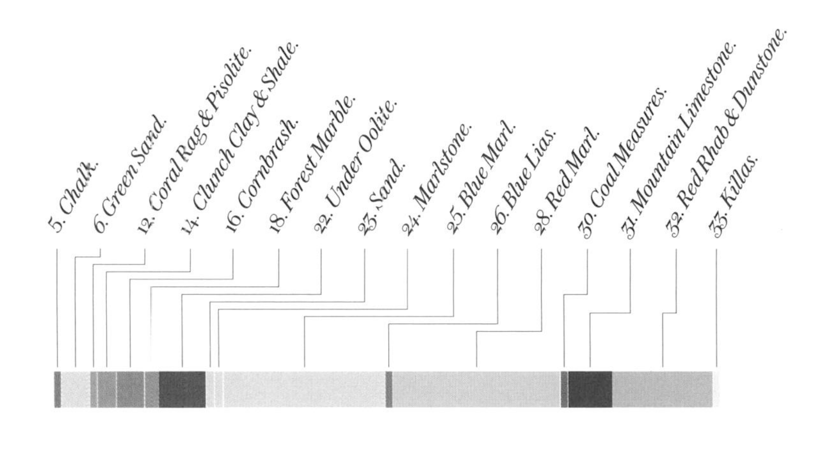

GEOLOGICAL MAP OF SOMERSETSHIRE.

A digitally reconstructed Somersetshire county map based on Smith's line-work and John Cary's *New Map of Somersetshire*, 1829. The upper stratum is composed of Chalk (5) and the lowest stratum Killas (33).

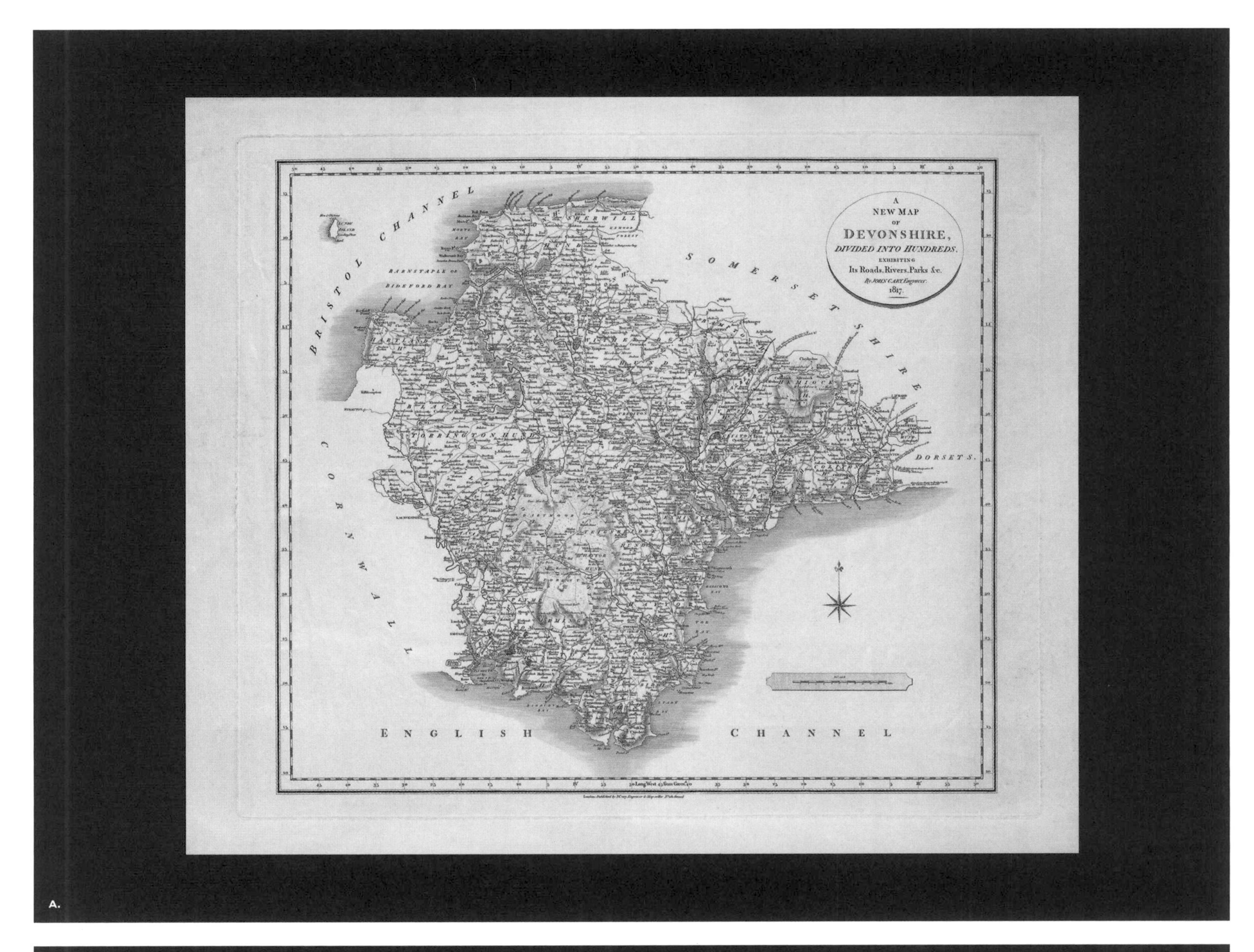

A.

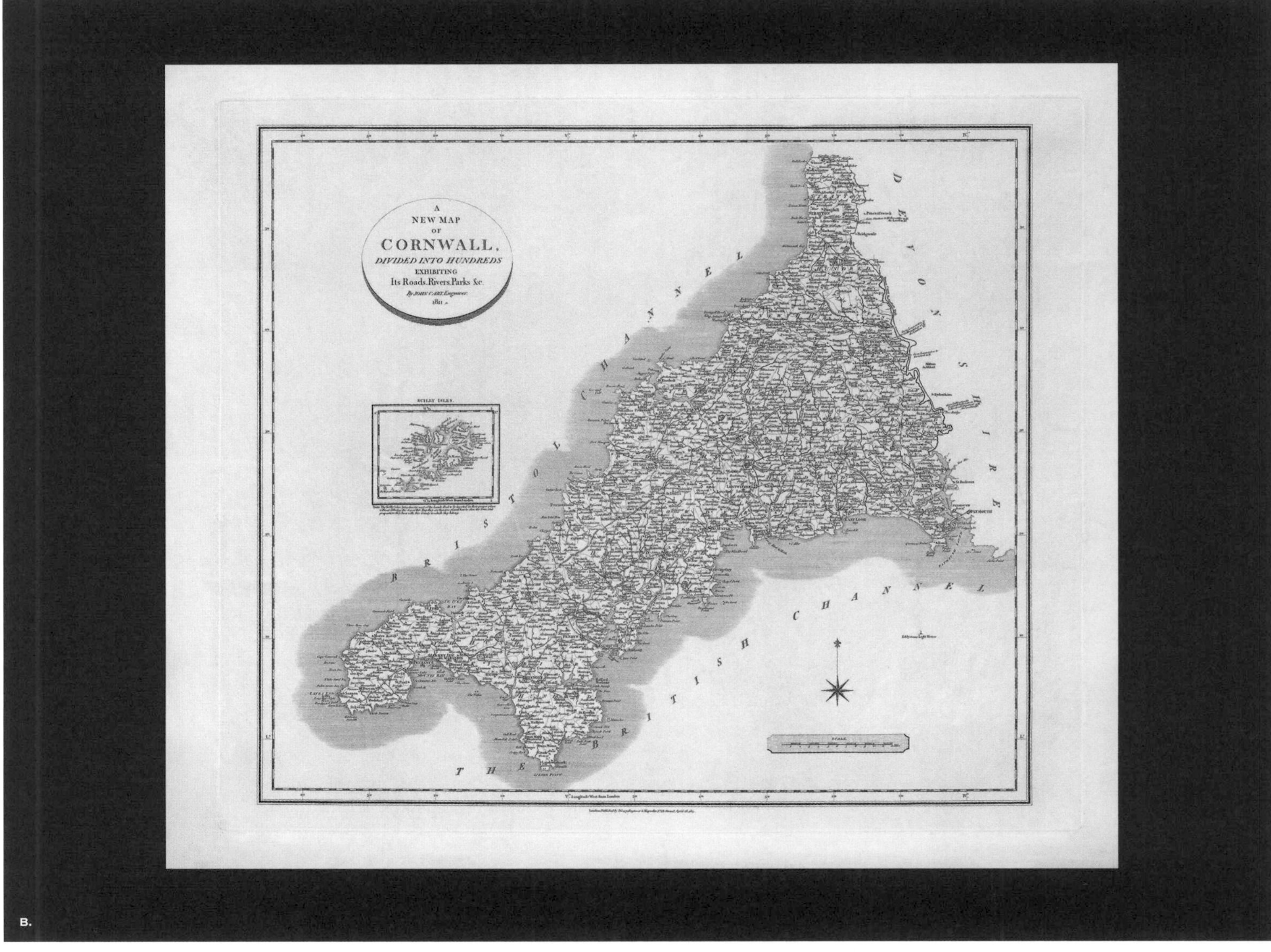

B.

A. UNPUBLISHED MAP OF DEVONSHIRE.

There is no geology visible on this map, however there is some light pencil shading around the Dartmoor Granite. Smith never published a completed map of Devonshire.

B. UNPUBLISHED MAP OF CORNWALL.

There is no geology visible on this map. Smith never published a completed map of Cornwall.

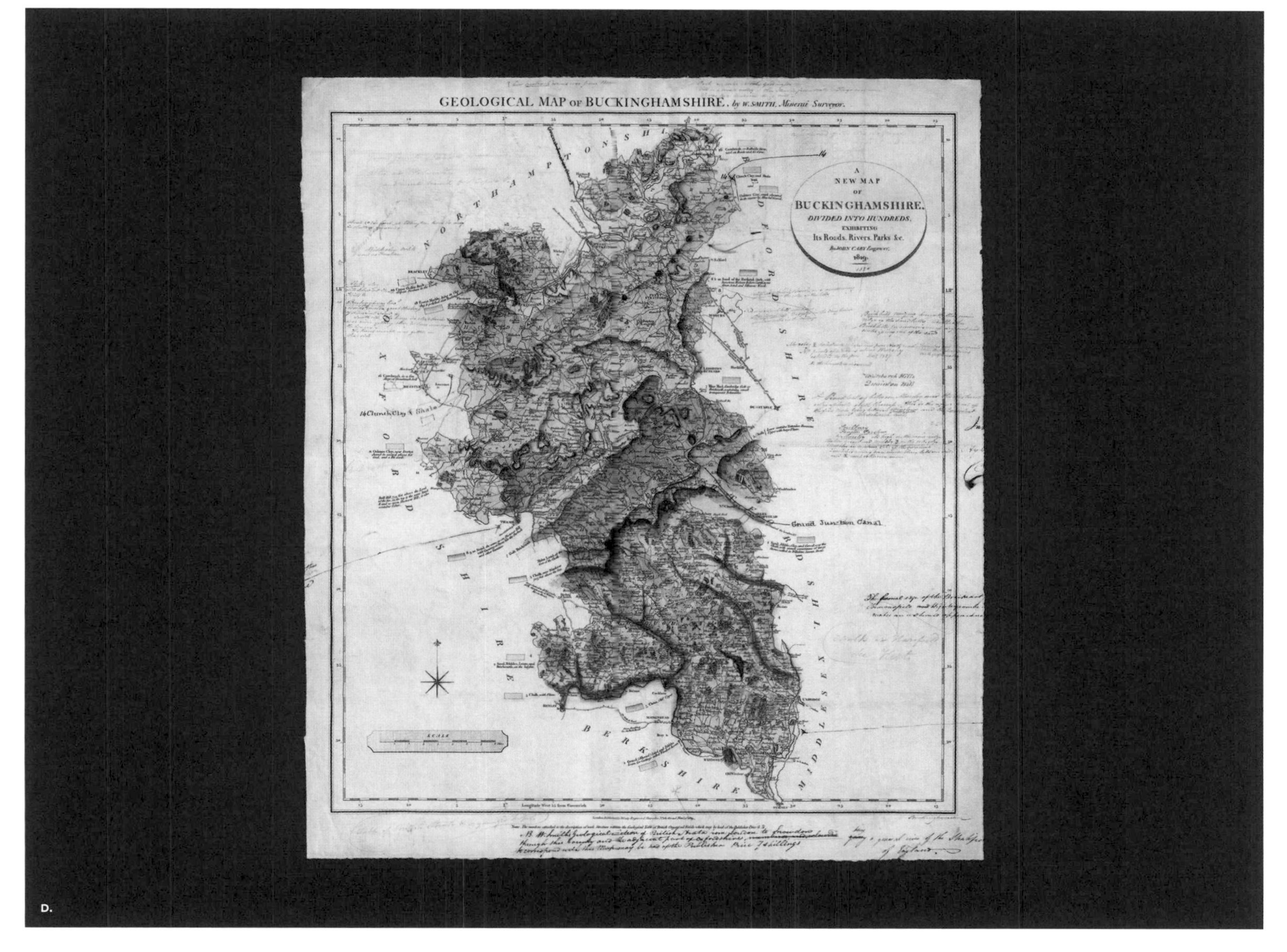

C. UNPUBLISHED MAP OF HAMPSHIRE.

An early draft map, showing no annotation other than 'Ham' in the lower right, and two bands of colouring.

D. UNPUBLISHED MAP OF BUCKINGHAMSHIRE.

A draft map with heavy annotation. It is likely to be a final proof copy.

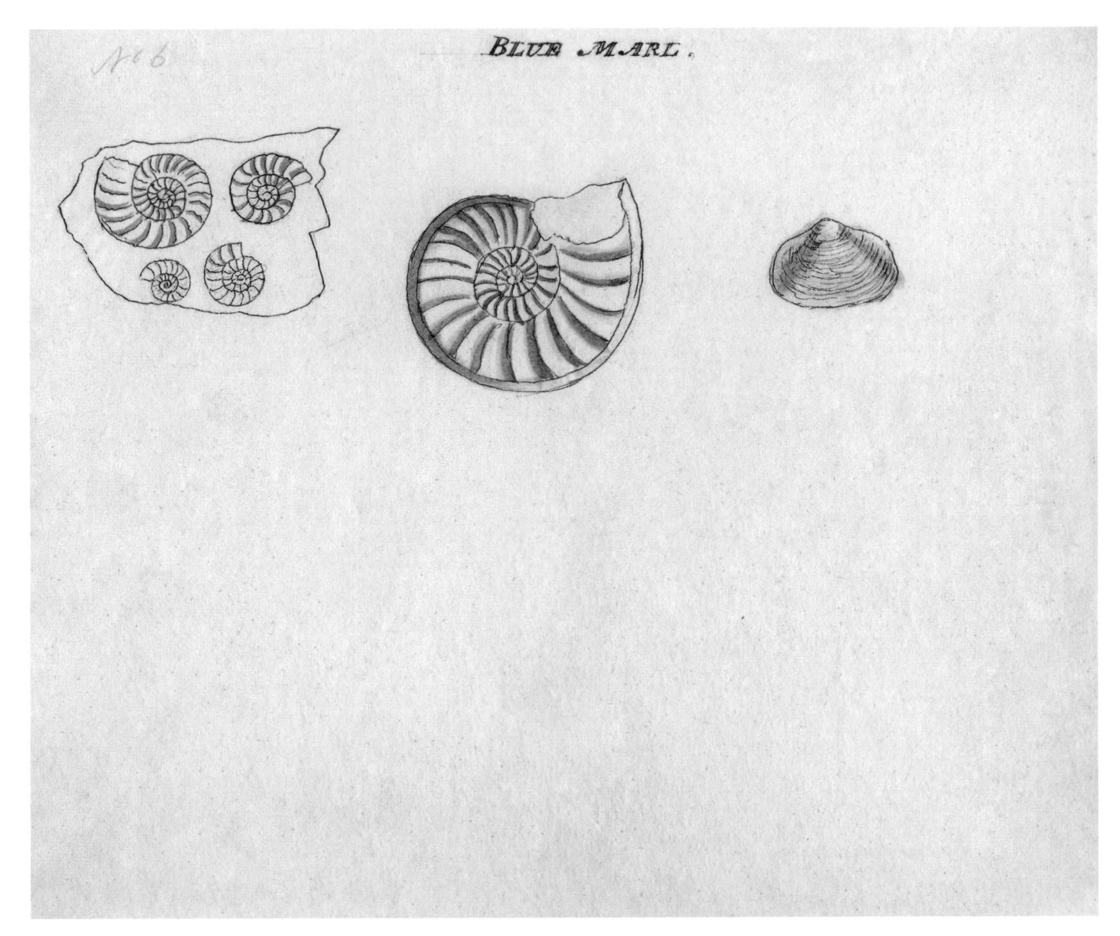

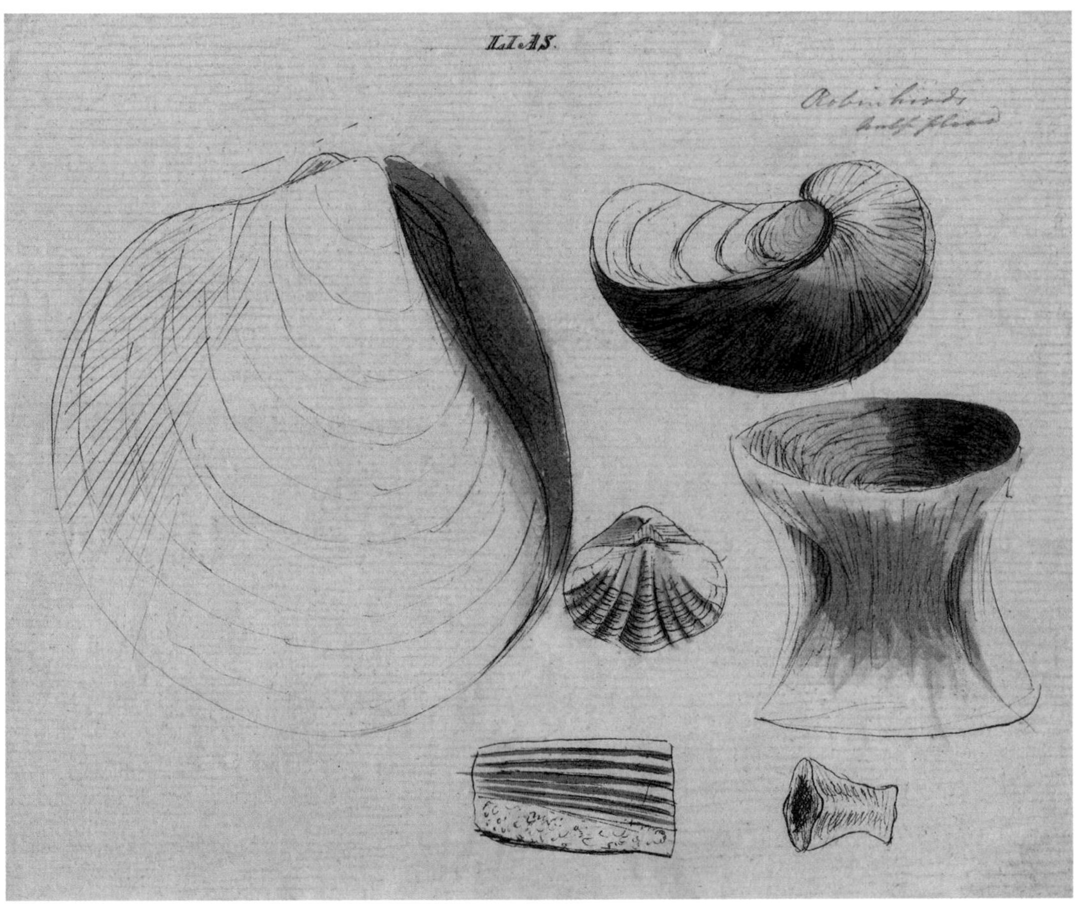

FOSSILS FOUND IN BLUE MARL [ABOVE].

Smith's *Strata Identified by Organized Fossils* was never completed because of his financial difficulties but four sketches, three of Lias and the Redland Limestone were found among his papers at the Oxford Museum of Natural History. His Blue Marl plate features a block of the famous Marston Magna Stone. One of the ammonites on this block was named *Asteroceras smithi* after Smith by Sowerby.

FOSSILS FOUND IN LIAS [BELOW].

The layered rocks around Lyme Regis, Dorset, were known as Lias from the way it is layered. Many of the specimens that Smith collected from the Lias relate to his work on the canals, including the Coal Canal. Nowadays the famous area for collecting is Lyme Regis/Charmouth and some of the specimens come from that area.

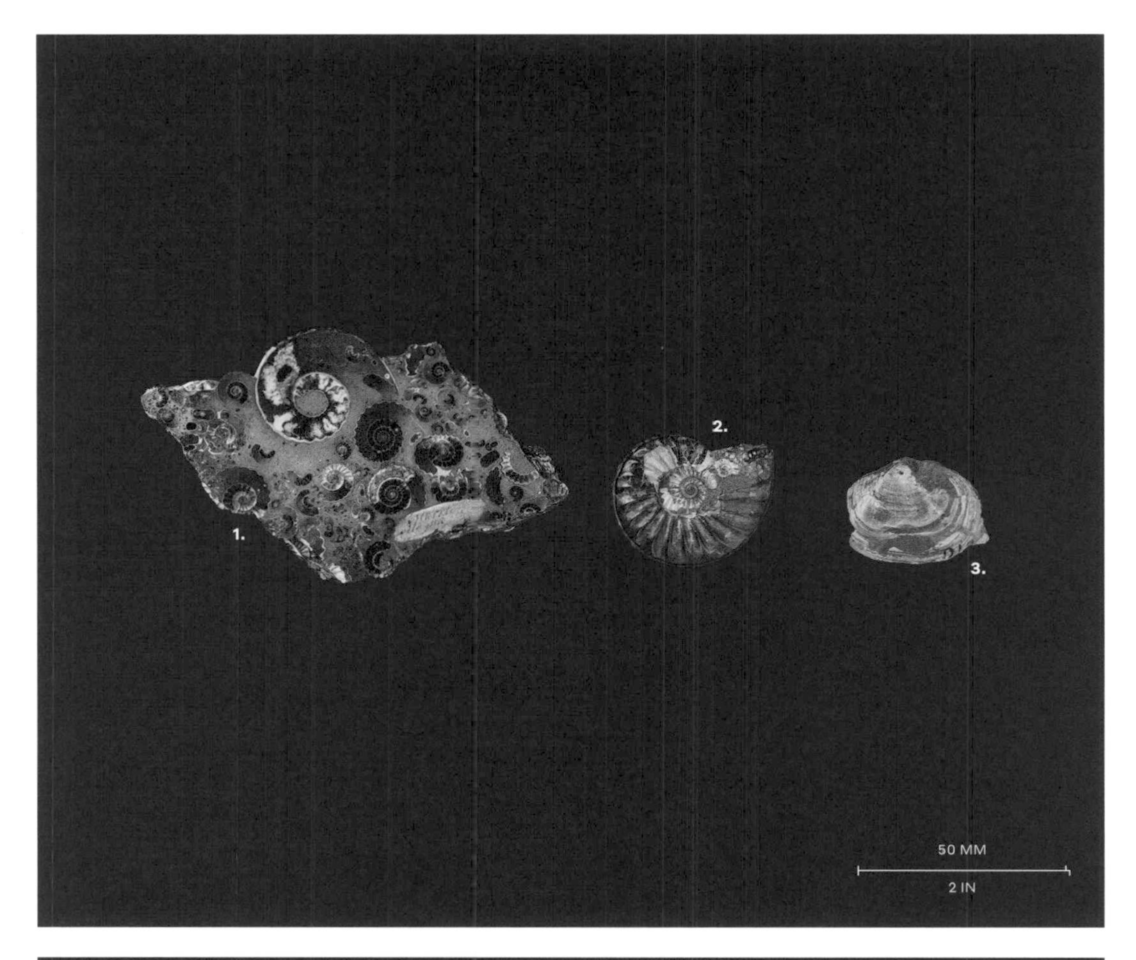

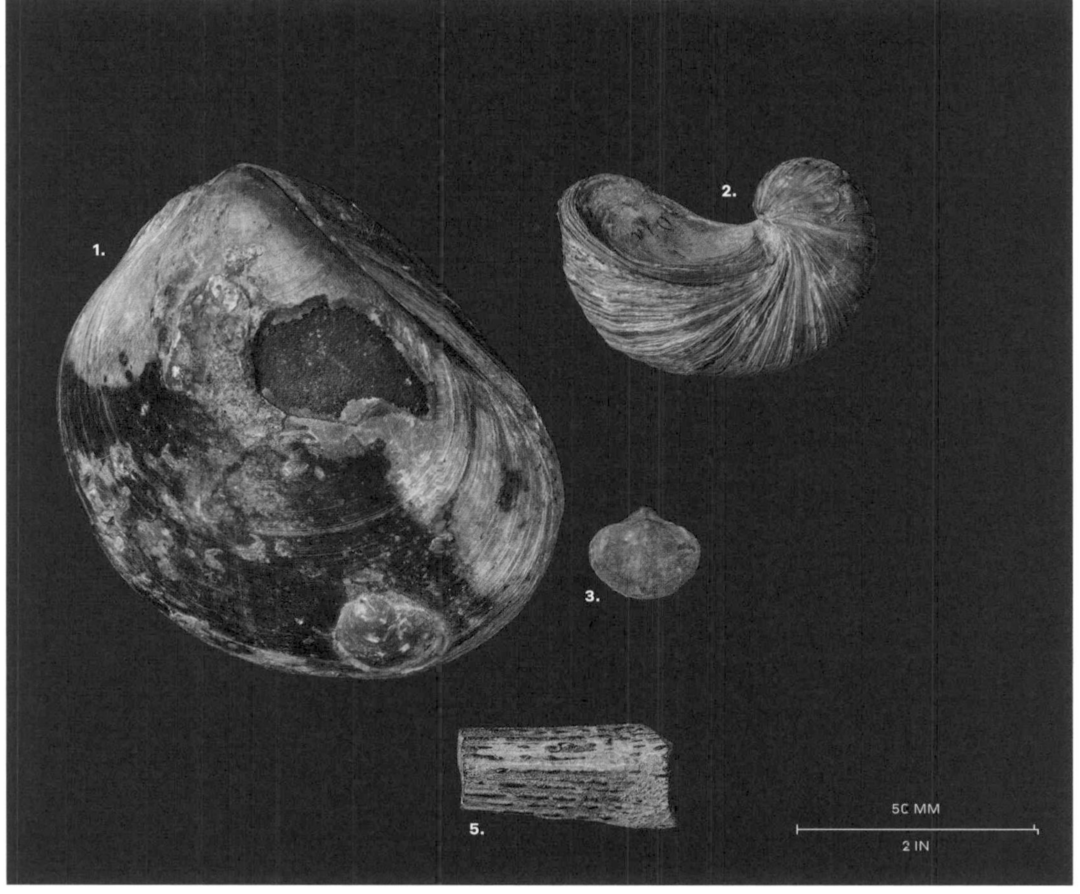

FOSSILS FOUND IN BLUE MARL [ABOVE].

1. *Promicroceras planicosta* C736 Marston Magna; **2.** *Asteroceras smithi* C737 (no locality); **3.** *Cardinia listeri* L1765 Mudford.

FOSSILS FOUND IN LIAS [BELOW].

1. *Plagiostoma giganteum* L1780 near Bath; **2.** *Gryphaea arcuata* L1737 Gloucester & Berkley Canal; **3.** *Spiriferina walcotti* B1504 Keynsham; **5.** Shark fin spine *Acrodus curtus* P4841 near Lyme.

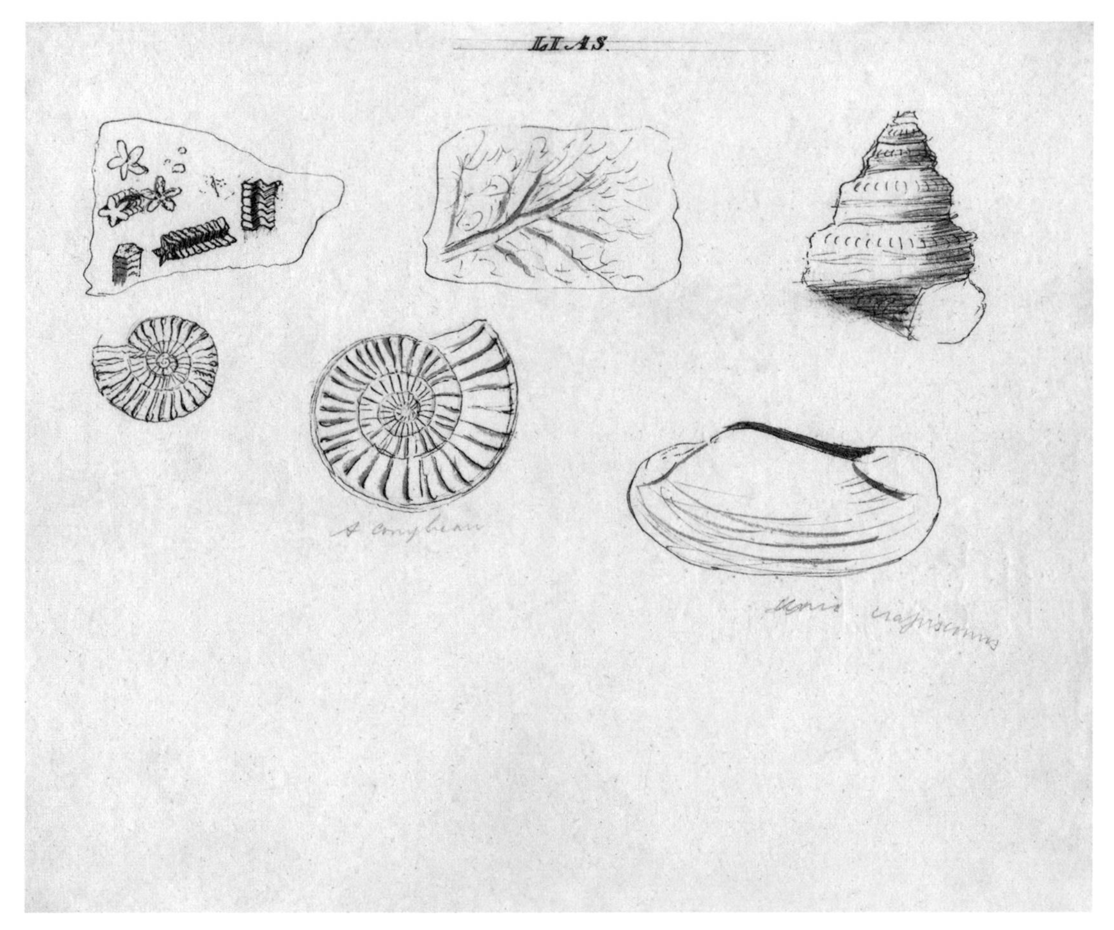

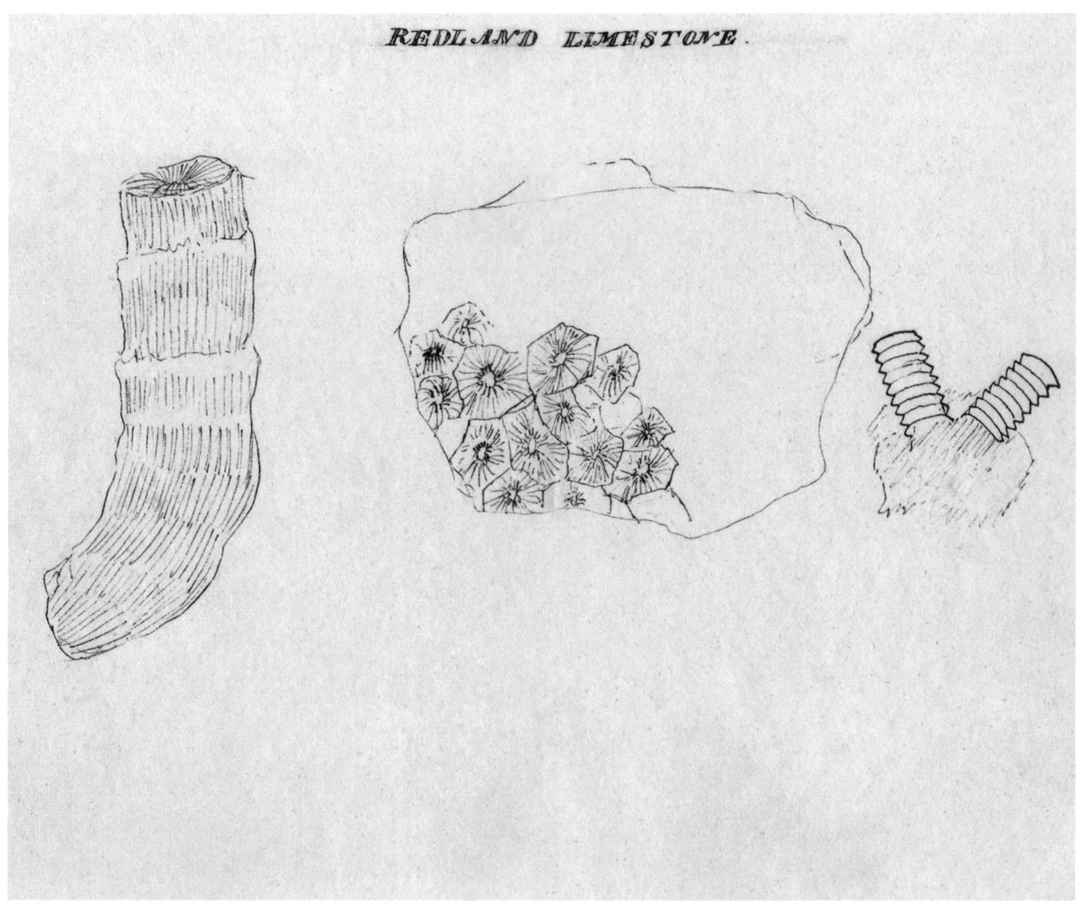

FOSSILS FOUND IN LIAS [ABOVE].

In Smith's strata he divided the Lias into White Lias (below) and Blue Lias (above) – terms still used today. In his two Lias plates he did not make this distinction. The upper middle fossil (**2.**) is a rare example of an articulated crinoid that is still found from a certain bed at Lyme Regis.

FOSSILS FOUND IN REDLAND LIMESTONE [BELOW].

Smith correlated the Redland Limestone on his 1815 map with his 'Magnesium Limestone' in Yorkshire although the fossils shown are all of Carboniferous age, rather than Permian. By the time he came to publish his 1819 map of Gloucestershire, he had revised the age of these limestones around Bristol, indicating that they belonged to his Mountain/Derbyshire Limestone, i.e. Carboniferous.

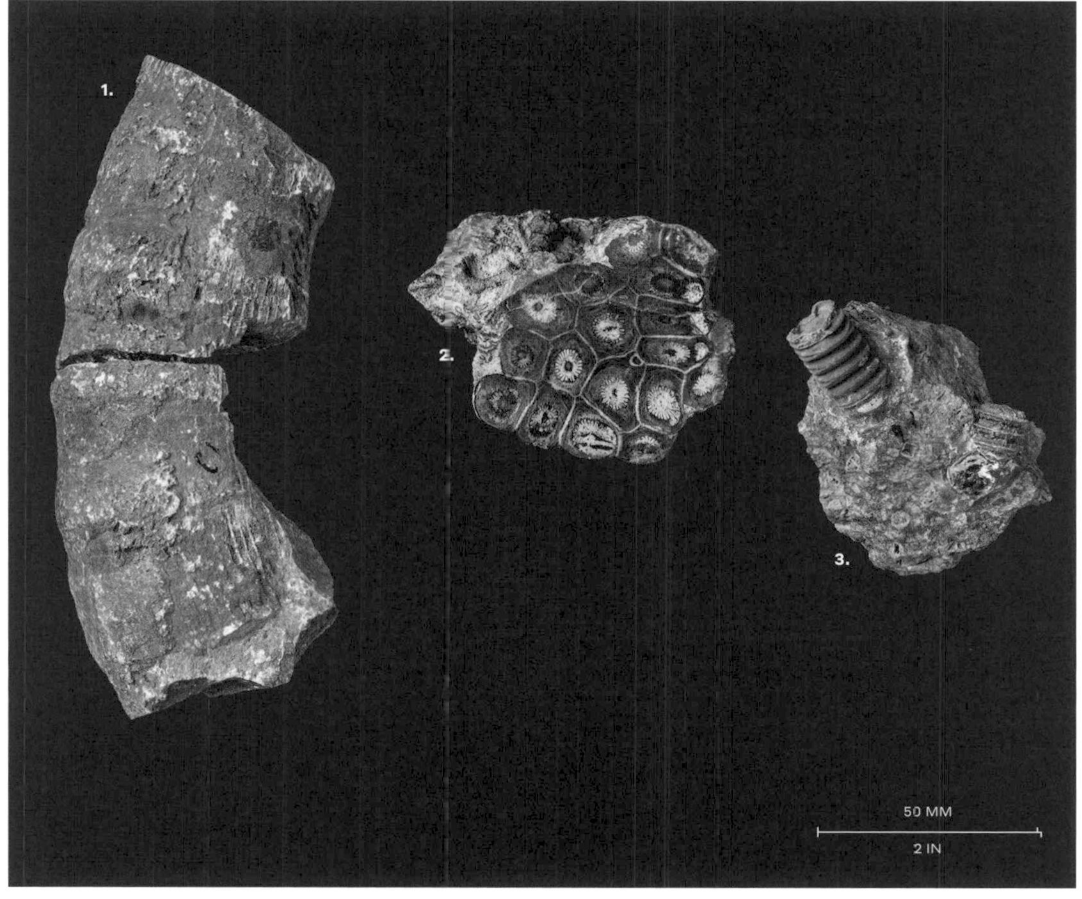

FOSSILS FOUND IN LIAS [ABOVE].

1. *Isocrinus psilonoti* E561 Gloucester & Berkley Canal; **2.** *Pentacrinites fossilis* E542 near Lyme Regis; **3.** *Pleurotomaria* cf. *cognata* G1670 Purton Passage; **4.** *Euagassiceras sauzeanum* C83396 Stony Littleton; **5.** *Zugodactylites braunianus* C723 Bath; **6.** *Pleuromya* aff. *uniformis* L1769 (no location).

FOSSILS FOUND IN CARBONIFEROUS LIMESTONE [BELOW].

1. *Amplexus coralloides* R1069 Leigh (substitute); **2.** *Actinocyathus floriformis* R1067 Avon Section, Bristol (substitute); **3.** Crinoid stems indet. E548 Mells.

VI. WELL SINKER.

Contemplating the likely source and nature of underground water supplies; the ancient art of dowsing; proposing a scientific method for locating water underground; the invaluable assistance of John Farey; the trials and tribulations of sinking deep wells; the peculiar difficulties of providing Scarborough with water.

In late eighteenth-century Britain, when William Smith (1769–1839) was formulating his ideas on stratigraphy, there were conflicting theories as to the origin of the underground waters that supplied springs and wells. Although most who looked into the matter considered that shallow wells and ephemeral springs, which dried up in summer months, were sourced from rainfall, many believed that the amounts of rain, and the available underground storage, were not sufficient to maintain the year-round flow of perennial springs and deeper groundwaters, and that an additional source of recharge must exist. A popular explanation was that seawater seeped through the Earth, losing its salts by filtration en route. Thus William Pryce (1725–1790), a Cornish physician, identified both temporary and perennial water flows in local mines; the temporary flows were related to seasonal rainfall while the perennial ones were thought to be derived from the sea, by ducts and cavities running through the Earth like veins and arteries in the human body.

Another theory, credited to the ancient Greek philosopher Aristotle (*c.* 384–322 BC), held that the additional water was formed by condensation in subterranean caverns of vaporous air generated by the Earth's heat. Erasmus Darwin (1731–1802), the grandfather of Charles, had progressive views on many topics, but was unable to accept that, for the most part, groundwater was derived from rainfall. He considered that springs originated from the night-time condensation of atmospheric water on the summits of hills, which were colder than the surrounding plains.

FIG. 1.

Fig. 1. A waterpump in Cornhill, City of London. There was a well recorded here in 1282. However, it was rediscovered and deepened in 1799, and this pump erected through the contributions of the Bank of England, the East India Company and bankers and traders from the ward of Cornhill. Among the contributors were Fire Insurance Companies, who valued access to water for fire-fighting. A pump like this takes water from the gravels which overlie the London Clay rather from a deep source in the Sands and Chalk beneath.

SINKING WELLS.

The concept of separate origins for deep and shallow groundwaters received apparent support from deep wells dug in the Thames Basin. In 1781–1782 the military engineer Thomas Hyde Page (1746–1821) supervised the sinking of a well at the Sheerness Garrison on the Isle of Sheppey, where the River Medway enters the Thames Estuary. Water-bearing sands were struck at 98 m (some 320 ft) depth, with water rising to within half a metre of the surface. Some ten years later, water struck at a depth of 79.3 m (260 ft) in a well at Notting Hill in west London eventually overflowed at the surface. To the water engineers involved, the presence of water under pressure at such depths was easier to understand on the basis of recharge from below by filtered seawater or condensing vaporized air than by rainfall from above. Their scepticism was understandable as there was no obvious mechanism by which rain could pass through more than 80 m (262 ft) of clay to saturate an underlying layer of sand.

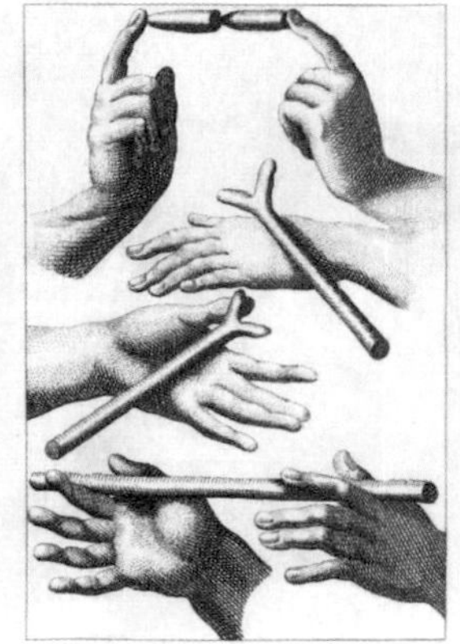

FIG. 2.

Fig. 2.
A water diviner from the eighteenth century (left) and a type of divining rod used to search for water (right). Smith was sceptical of the water diviners' capability; he considered the method unscientific, the practitioners charlatans and their clients gullible.

Guidance on finding water at shallow depth was provided as early as the first century BC by the Roman architect and engineer Vitruvius (*c.* 75–*c.* 15 BC). His methods involved close examination of the landscape and identification of water-loving plants, followed by a series of experiments in a small shallow pit. One test involved placing a fleece of wool in the pit; if the next day water could be squeezed from it, water was thought to be abundant there.

The expense involved in sinking wells to prospect for deep groundwaters was considerable as they were traditionally dug by hand. Drilling techniques introduced from Flanders reduced costs, and drilled wells started to proliferate by the beginning of the nineteenth century. The Sheerness well had proved that fresh water was present at depth, and many boreholes were drilled in the expectation that fresh water was available anywhere if one drilled deep enough, a justifiable assumption if the source was migrating seawater or condensation in subterranean caverns. Borehole locations were selected on a random basis or on the advice of water diviners or dowsers, whose ability to locate water was widely accepted. Failures were common, leading to frustration and sometimes litigation.

CONFLICT WITH DOWSERS.

Vitruvius made no mention of dowsing and it seems that the use of divining rods, in the way they were employed in William Smith's day, dates from the sixteenth century. Smith had problems with competing dowsers and one encounter on the Mendip Hills between Smith and a miner with a dowsing rod was described by his nephew and biographer, John Phillips (1800–1874). Consulted by a landowner, Smith found that he had been anticipated in giving an opinion by the dowser. As they traversed the ground, Smith dropped a small stone not native to the Mendips wherever the movement of the miner's twig indicated water would be found. Smith then 'asked the miner if he could rediscover the points indicated.' Unaware of Smith's subterfuge, he readily agreed, passing the spots where the stones lay and stopping at other localities indicated by his twig. Smith pointed out that, 'as the water had in so short a time changed its situation at all the points, it would be imprudent to spend money in following it' (Phillips, 1844, p. 132).

It was to be William Smith, aided by his pupil John Farey (1766–1826), who showed that an understanding of stratification could explain the presence of reservoirs of fresh water at depth, providing a scientific basis for predicting the presence of water-bearing formations without the need for the dowser's magical rod.

APPLICATION OF SMITH'S IDEAS.

John Farey first met Smith in 1801, becoming his pupil and friend. By this time Smith had already drawn up a list of strata for his Bath friends, which had been duplicated and circulated widely. Fearing that his innovative views would be claimed by others, who would gain the credit to which Smith was entitled, these same friends had encouraged him to publish, and his *Prospectus* was issued on 1 June 1801. This described the practical applications of his ideas, including the claim that they had enabled him to gain a complete knowledge of all springs.

That Smith was using his knowledge to advise on water-supply problems at this time is confirmed by Farey, to whom Smith demonstrated the order of strata while working in Buckinghamshire in the spring of 1802. The two men met with the rector of Newton Longville, a village in the Vale of Aylesbury. With no local springs, the rector was sinking a speculative well in the parsonage garden, but at a depth of 30 m (100 ft) was still in clay and was about to abandon the work. Smith referred to his map of the strata and pointed out that the clay was underlain by limestone, which cropped out in the River Ouse some 13 km (8 miles) to the north-west. He assured the clergyman that if he persevered he would certainly reach this limestone and obtain a plentiful supply of good fresh water. Well sinking was continued, and limestone was reached as predicted at 72 m (235 ft) below the surface. This upper limestone was thin and the water yield poor, but an auger hole bored through the underlying clay to a second limestone produced a plentiful jet of water. When visited by officers of the Geological Survey in 1904 the well was still in use, but has since been sealed.

Over the next few years Farey recognized that increasing numbers of people were profiting from Smith's ideas without acknowledgment and urged him to publish something, however

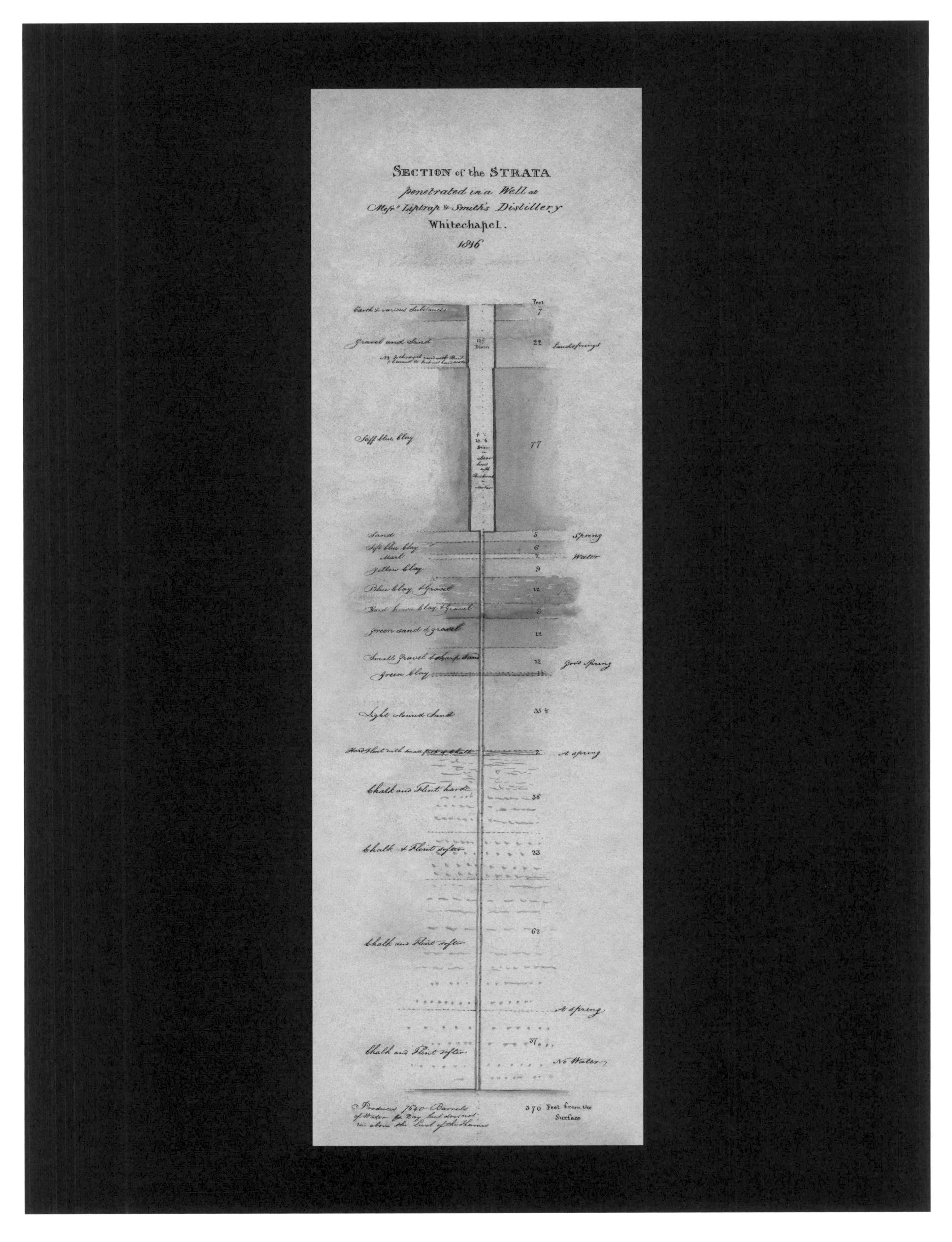

SECTION OF THE STRATA PENETRATED IN A WELL AT MESSRS. LIPTRAP & SMITH'S GIN DISTILLERY, WHITECHAPEL, LONDON, 1816.

William Smith collected information relevant to his consulting work, including this diagram prepared by the drilling contractor Mr R. Walker of a deep well excavated and bored for a distillery in Whitechapel, east London. It reached the Chalk at 64.6 metres (212 ft) below ground level and penetrated a further 48 metres (158 ft) within the Chalk encountering water inflows at several depths. The water entered the Chalk in the hills of the Chilterns and percolated over centuries into the London Basin. As the Chilterns are at a higher elevation than the Chalk beneath London, water was under considerable pressure and rose to the surface, or just below, in wells without the use of pumps. Distilleries and other industrial users drew large volumes of water from the Chalk, whilst the Metropolitan Water Companies relied mainly on river water. On this log, Smith has added the note that the daily yield was 7,640 barrels (approximately 1250 cu. metres). Apart from water data, Smith would have related the well driller's material descriptions to his stratigraphic model.

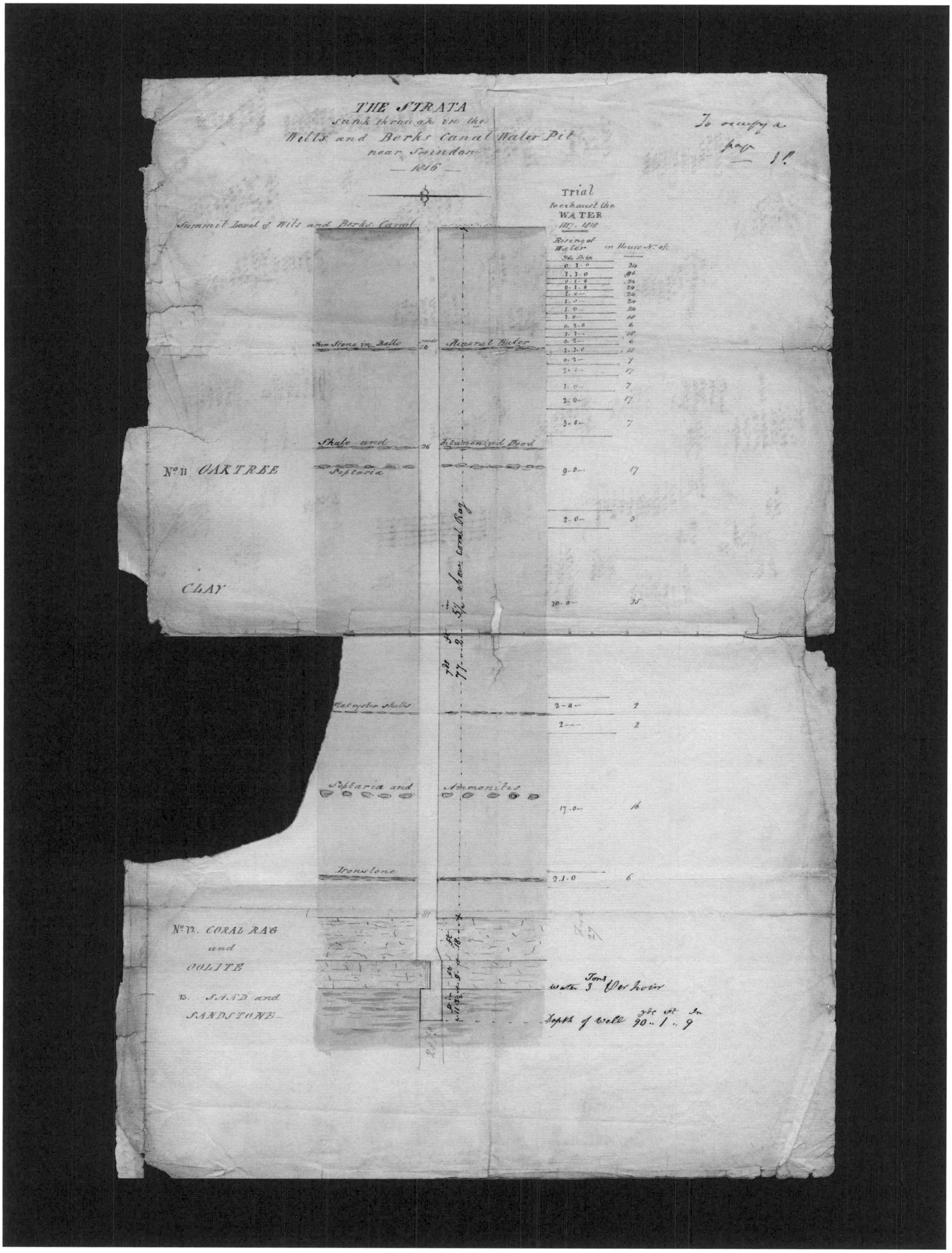

SECTION OF STRATA IN A WELL ON THE SUMMIT LEVEL OF THE WILTSHIRE & BERKSHIRE CANAL, NEAR SWINDON, 1816.

Canals require water input at their summit levels where water flows away in both directions. This can be supplied from storage reservoirs or by groundwater from wells. Engineers constructing the Wiltshire & Berkshire Canal between Swindon and Wootton Bassett consulted Smith when a well they were digging failed to find water. From his stratigraphic knowledge Smith recommended deepening the well and constructing lateral tunnels at the base. Although yields were initially good they fell away quickly and the well was abandoned (see p. 227).

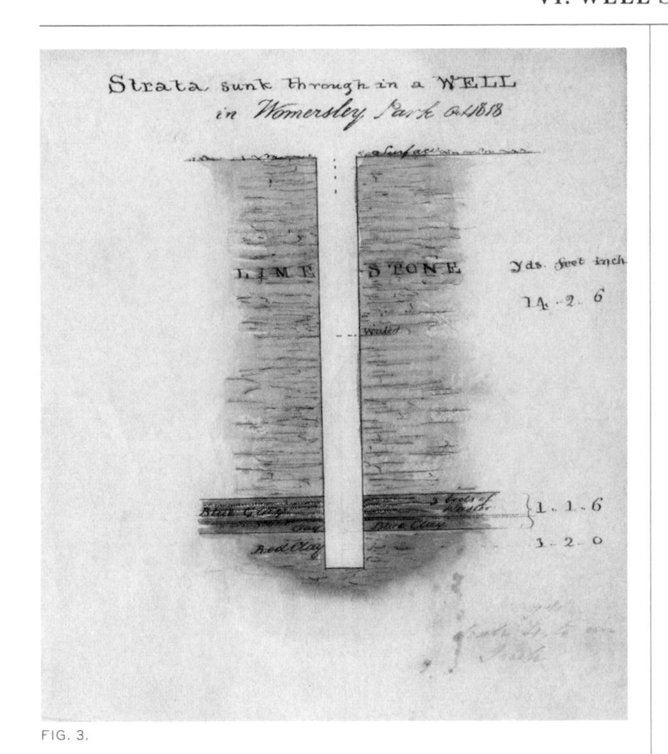

FIG. 3.

Fig. 3.
Smith's illustration of the strata sunk through for a well in Womersley Park, Yorkshire. Smith required information on local water resources and the standing water table for his feasibility report for the proposed Aire & Don Canal, which passed close to Womersley Park (see pp. 80–81). Water occurs in Limestone overlying Blue and Red Clays with the groundwater surface (water table) at approximately 7 metres (44 ft) below ground level. Coincidentally, at Womersley, Smith discovered a Limestone of lithographic standard, which his nephew, John Phillips, prepared and used to produce the map for this feasibility study.

short, to establish his priority. His pleas were ignored and eventually it was Farey himself who assumed the role of Smith's advocate and publicist. In December 1806 *The Monthly Magazine* published a letter from 'Aquarius' seeking advice on methods of obtaining water in times of scarcity. Farey seems to have been the only person to respond, in April 1807, and it gave him the perfect opportunity to show how Smith's theories could be applied to the origin and exploitation of springs and deeper groundwaters.

He began by insisting that all springs could be accounted for by the 'descent and filtration of the water supplied on the surface by rains, dews, etc' (Farey, 1807, p. 211). Using the London area as an example, he explained that water percolated through gravel, sand and soil close to the surface, until its downward passage was halted by clay. Here the water saturated the overlying porous material almost to the surface or moved sideways near the top of the clay until discharging in the form of springs at a lower elevation. Close to the River Thames, the gravel and sand were saturated and were also supplied by river water. This, he explained, was the origin of the water extracted from shallow wells in and around London.

With respect to the deep wells, Farey was clear that Smith's work on geological strata, which was to be published shortly, would explain the presence of the bodies of fresh water that were their source. He described how Smith had demonstrated that 'every stratum, whether of clay, sand, chalk, stone, etc. which we meet with in sinking a well ... forms part of an extended inclining plane' (Farey, 1807, p. 212), which at some distance from the well cropped out at the surface where rain, falling on porous strata, would percolate and soak down. The deep wells in London obtained their water from a thick and fully saturated stratum of loose sand below clay. Farey pointed out that the outcrop of this sand could be traced through a number of villages about 25 km (15 miles) to the north, and the water in the deep wells came from rainfall moving downslope from this outcrop. The relative height of the villages above the city, accounted for the pressure of water and the overflowing conditions in lower-lying areas such as Notting Hill and Richmond upon Thames. He concluded that the newly acquired knowledge of stratification led to the possibility not only of finding plentiful water anywhere, but also enabled its quantity to be assessed and the costs of exploitation to be estimated.

Farey publicized Smith's work in further articles, but little was written that related specifically to water supply and the exploitation of deep groundwater. In a further letter published in *The Monthly Magazine* in 1823, Farey complained that, despite his 1807 letter and William Smith's by now published maps and sections, 'which long ere this ought to have made the principles of deep well-sinking or boreing (sic.) for water sufficiently familiar to the British public, to have guarded it against quackery ... such is not the case' (Farey, 1823, 309). Instead, individuals with no geological knowledge 'have rashly undertaken ... to obtain what nature withholds'. This was partly because Smith and his pupils published no detailed guidance, perhaps preferring to restrict its dissemination since providing advice on water supply became a significant part of their own professional practice and income. Another factor was that many of the local practical well sinkers and drillers approached by a landowner wanting a water supply were probably illiterate, and preferred to rely on the mystical power of the dowser to locate a source rather than the, to them at least, equally mystical 'natural order of strata' propounded by Smith and his pupils.

WATER FOR THE WILTS & BERKS CANAL.

In 1816 Smith's advice was sought over the lack of water to feed the summit level of the Wilts & Berks Canal in the vale between Swindon and Wootton Bassett. He records in his diary that he walked from Swindon, with Mr Whitworth the canal engineer, to examine the site on Tuesday 9 April 1816. The Canal Company, without seeking geological advice, had first sunk a pit by the side of the canal to about 37 m (120 ft) depth, and had then bored the same depth again. The stratum exposed in the pit consisted of a tenacious clay containing layers of stony nodules. Smith recognized that no water could be obtained until the bore extended through the clay. A number of deep wells were known in the area, producing copious supplies of water which overflowed at the surface. This showed that the source was on higher ground and Smith's experience of the local strata suggested to him that the clay would be underlain by a series of limestones and sandstones which cropped out at Wootton Bassett to the south. Although at outcrop these strata were thin and interlayered with clay, he had evidence that they might be thicker at the canal well site. He was therefore confident that 'Water will most probably soon be found, which may be expected to rise to the surface, but ... the discharge will be slow unless it be assisted by machinery' (Phillips, 1844, p. 83).

The workmen proceeded to deepen the pit and Smith's predictions proved correct. The clay was bottomed and the underlying sandstones and limestones with their thin clays were penetrated with only moderate yields of water. Boring deeper, the quantity of water increased continually until a total depth of 80 m (263 ft) was reached, the lowest 4.5 m (15 ft) being bored through rock. At this depth water rushed into and filled the pit to a level very close to the surface. Smith concluded that the geological investigations had proved the continuity of the water-bearing rocks and had demonstrated that there was a sufficient outcrop to gather a good supply of water from rainfall. All that needed to be done to increase the yield was to drive a series of tunnels (headings) into the rock at the base of the pit to intersect further water-bearing joints and fractures.

Although the initial yield was encouraging, by 1820 the supply of water was 'very limited' and the well was abandoned. The failure was blamed

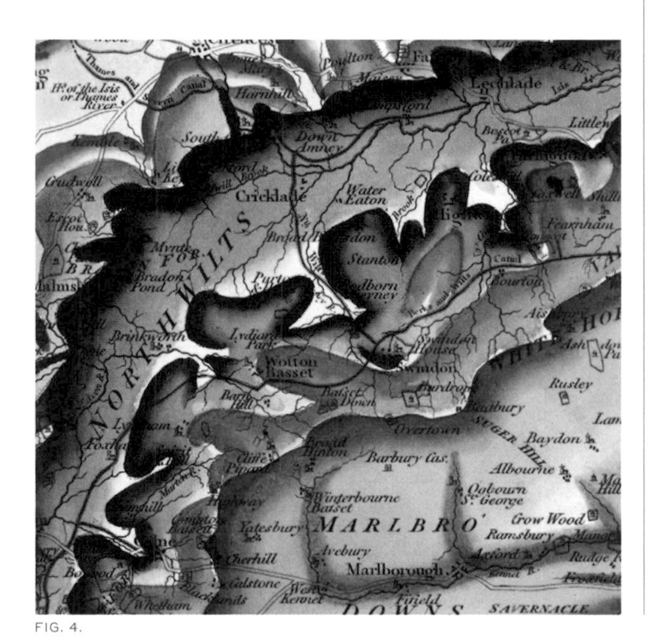

FIG. 4.

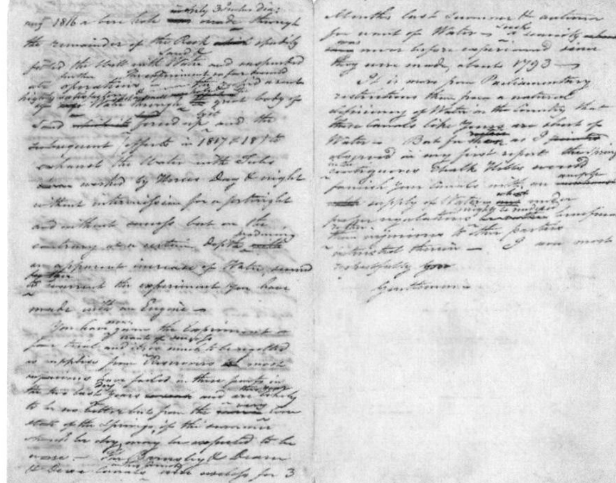

FIG. 5.

on either not following Smith's advice on the headings, the limited thickness of the sandstones and limestones, or on the presence of faults which reduced the effective volume of rock feeding water to the well. However, there is another reason why yields were lower than Smith anticipated. His estimation of the amount of water was based on observations of the water-bearing rock at outcrop, where water simply drained from pores and fractures as water levels fell. At the canal site, the water in the rock was confined under pressure beneath 71 m (233 ft) of clay, and water would have been released as a result of the decrease in pressure within the water and rock framework. The water available through the latter process is orders of magnitude less than that yielded by gravity drainage. The behaviour of such confined rocks was not understood until the earlytwentieth century and would not have been recognized or understood by Smith.

WATER SUPPLY IN SCARBOROUGH.

By the early nineteenth century Scarborough had become a fashionable spa town, attracting several thousand people during the season for sea bathing and 'taking the waters', which were thought to have health benefits. As numbers

Fig. 4.
An extract from Smith's 1815 map, showing the location of the Wiltshire and Berkshire Canal near Swindon and Wootton Bassett. Smith was employed to advise on the canal's water supply.

Fig. 5.
The draft of a letter of commiseration from Smith in Yorkshire to the commissioners for the Wiltshire and Berkshire Canal in 1820, on the diminishing yields from the 'Farindon [sic] Well Experiment' despite initial excellent results and the introduction of an 'Engine' to pump water. He remarks that the preceding two years have been dry and the 'want of water' is similar for the Barnsley Canal and the Dearne and Dove Canal, which related to his proposals for the Aire & Don Canal. He also intriguingly remarks that 'It is more from Parliamentary restrictions than from a natural deficiency of Water in this Country that these Canals like yours are short of Water'.

of people increased in the summer, additional sources of water were needed and in May 1826 the town's commissioners noted a small volume of water flowing from a borehole that had been drilled some years previously for draining land, about 2 km (1¼ miles) west of the town. Instead of acting as a drain, the borehole had penetrated confined beds where the water pressure was sufficient to produce overflowing conditions. On cutting an open channel up to the borehole the flow increased; which encouraged the commissioners to investigate further, at the same time seeking the advice of Smith, who was by then a resident in the town.

Smith described the rock from which the water flowed as a fine-grained crumbly sandstone and the excavated channel as being in a very tenacious clay. The face of the sandstone was almost vertical, like a wall, and a modern interpretation suggests a sandstone crag, which had been covered by glacial deposits plastered up against the valley side when ice covered this part of Yorkshire some 21,000 years ago. Deepening the open channel up to the borehole to a depth of around 4.7 m (15 ft) increased the flow to around 60 hogsheads (14.7 cu. m or 520 cu. ft) per hour, but this had decreased alarmingly over the summer of 1826 to less than 6 hogsheads (1.5 cu. m or 53 cu. ft) per hour. Water was collected in a basin, and cast iron pipes fed it, via an existing channel, into Scarborough. Smith suggested that by cutting off the flow during the coming winter, water could be dammed up within the sandstone and then released the following summer. He calculated that some 15,000 hogsheads (3,682 cu. m or 130,000 cu. ft) might be stored in this way. A system of puddled clay and pipes was installed to dam up or draw off water as required and water allowed to collect over the winter, with the success of the scheme determined by the yield the following summer. Unfortunately, little more is known about Smith's geological reservoir. John Phillips claimed that the scheme was a success and that 'the little spring became a great blessing' and a benefit to the town (Phillips, 1844, p. 113). However, there is no documentary evidence to support this and Smith's estimate of the sustainable yield was probably wildly over optimistic. A topographic map of 1850 shows an 'old conduit' at the site and it seems likely that, even if the spring supply was used at all, it was not for long.

FIG. 6.

Fig. 6. Two *Poetical Sketches of Scarborough* from 1813. Even before the arrival of the railway in 1845, Scarborough was an attractive location for the leisured classes, who came to enjoy the waters of the Spa (above) or the beach complete with bathing machines (below). The Scarborough Races were such a draw that the lecture series that Smith and his nephew gave was scheduled to coincide with them, in order to swell the audience.

MINERAL SPRINGS AT BATH AND HARROGATE.

The baths which give the city of Bath its name are fed directly by three hot springs. By 1808 there had been increasing problems with the flow of the hottest of these, the Hetling Spring, and the Hot Bath which it supplied was taking increasingly longer to fill. William Smith's advice was then sought as he was working nearby on the Batheaston coal trial, which some local people believed was connected to the failure of the spring at Bath. As Phillips noted: 'Not without much opposition, he was allowed to open the hot-bath spring to its bottom, and thus detect the lateral escape of the water. The spring had in no sense failed, but its waters flowed away in new channels.' (Phillips, 1844, p. 64.) Smith quickly recognized that the hot springs had a deep source and that the spring pipe was acting as a natural borehole bringing water to the surface. This flow was being obstructed by river gravels, in which a ruminant bone and rounded flint were found by workmen. Clearing away these sediments, which were probably mixed with clay that had disintegrated in contact with the thermal water, restored the flow to the Hot Bath. Smith succeeded where previous attempts had failed and a report to the Common Council of the City of Bath in 1811 commended his work, ascribing its success to his 'thorough knowledge of the strata through which the water passes on its approach to the surface of the earth'.

Some thirty years later Smith was involved in what became known as the Harrogate Well Case. Sulphur-rich waters were first discovered in Harrogate in the seventeenth century and were protected in 1770 by an Act of Parliament. Over the years, hotels were built close to the original public sulphur well and their entrepreneur owners exploited the waters by digging their

FIG. 7.

Fig. 7.
Smith's *Geological Map of Scarborough*, 1831. Smith presented this manuscript map to the Geological Society on the occasion of his receipt of the Wollaston Medal. The base map is a map of Scarborough published by Robert Knox in 1821, using a scale of one inch to the mile. Smith drew the boundaries very faintly in pencil and applied the water colour himself. The names of the strata and geological descriptions are in his extremely neat hand in the left margin, opposite their outcrop. His signature is within the frame in the lower right corner.

own wells. Problems came to a head in 1835 when the owner of the Crown Hotel, Joseph Thackwray, searching for fresh supplies, sank a well only 15 m (50 ft) from the public one. Rival hotel proprietors claimed that the public well was seriously affected; Thackwray was accused of violating the 1770 Act and a legal action was brought against him. The case was heard at York Assizes on 14 March 1837 and distinguished groups of scientific experts were recruited by both sides. William Smith was one of five experts assembled to appear on behalf of the prosecution.

He argued in a statement that the various springs all arose from 'one common store' and that mixing and interaction of water and rock as water found various routes to the surface accounted for the different compositions of individual springs. In contrast, the defence insisted that the springs did not rise from a common source, but that each was connected to a separate underground reservoir. The case was settled by a compromise which allowed Thackwray to retain ownership of the well but required him to allow the public free access. Much to the disappointment of the audience who had assembled, Smith and his fellow experts were never called upon to give evidence or be cross-examined.

Smith's advisory work at Bath and Harrogate demonstrates his good understanding of concepts such as porosity, permeability and hydraulic gradient. Coupled with his knowledge of stratigraphy, this provided the scientific basis for the discovery and exploitation of underground waters. On the chalk downs and limestone heaths, water could be found almost anywhere, but Smith was able to predict the depth of water-bearing rocks beneath the clay vales and fenlands, as well as the quality of the water that might be obtained. His one failure was adequately to estimate the sustainable yield of wells and boreholes, a skill which would not be acquired until the twentieth century when geologists and engineers began to understand how water-bearing rocks release stored water.

Dunstable
Breachwood
Luton
Whitwell St.
Broad Water
Wood Hall
Puckeridge
Standon
Bishops Stortford
Birchanger
Easton Lodge
Dunmow
Luton Hoe Park
the Hoo
Knebworth & Pl.
Sacomb & Pa.
Hatfield Forest
Datchworth
High Rothing
Kingsbourne Green
Lamer
Welwyn
Stapleford
Thundridge
Spel Brook
Widford
Broad Oak
Wheathempstead
Ware
THE
High Easter
Redburn
Brickhill
Digswell & Ho.
HERTFORD
Hertford Heath
Pincey Brook
Water End
Sandridge
RODINGS
Hemel Hempstead
St. Albans
Hoddesdon
Netteswell
Canal
Gorham Bury
Hatfield
Bayford
Bedwell Pa.
Broxbourn
Rye Hill
High Laver
Black Hill
N. Mims
Thornwood Com.
Nth. Weald
Forest
Chipping Ongar
Writtle Park
Waterdale
Waltham Abbey
Langley Bury
Ridge Hill
Potters Bar
Epping
Stapleford Tawney
Ingatestone
Ridge
ENFIELD
New R.
Hill Hall
Roding R.
Watford
Enfield
FOREST
Weald Hall
Chingford
Brentwood
Stanmore
Edgware
Chigwell
HENHAULT
Woodford
Thorndon Hall
Moor Pa.
Weald Stone
Tottenham
Romford
Harefield
Stamford Hill
Lower Forest
Langdon
Harrow on the Hill
Brent
Nth. Okendon
Rystfield
Northolt
Kilburn
Hampstead
Barking
Uxbridge
Grand
Junction
Canal
Regents Can.
LONDON
Sth. Okendon
Bell Ho.
Southall
Marshes
Ford Pl.
Hayes
Hanwell Pa.
Acton
Purfleet
Grays Thurrock
Brentford
Chiswick
Brompton
the Marshes
Norwood
Woolwich
Erith
Longford
Thames R.
Tilbury Fort
HOUNSLOW
Hounslow
Richmond
Wandsworth
Camberwell
Welling
Dartford
Green Hith
HEATH
Richmond Park
Wimbledon Park
Lewisham
Eltham
Crayford
Park Wood
Swanscomb
Staines
Old Riv.
Mount Maskall
Kempton Pa.
Bushey Park
Kingston
Norwood
Bromley
St. Ann's Hill
Mitcham
Masons Hill
St. Mary Cray
Oatlands Pa.
Morden
Beddington Pa.
Woodside
Cracking Hill
Ash
Luddesdon
Claygate
Croydon
Prats Bottom
Eynesford
Sutton
Coomb
Holwood Hill
Claremount Park
Ewell
Lamberts Oaks
Addington Lodge
KENT
Woodland
EPSOM DOWNS
Farley
SURRY
AND
Down
Knockholt
Marams Court Hill
Woodmanston
Okeshot
Woodcot Park
VALE
OF
Ockham & Park
Botley Hill 880
Wildernes
HILLS
Eastwick Pa.
Leatherhead
Westerham
Seven Oaks
Fair Lawn
Norbury Pa.
Kingswood Farm
Mereworth & Cas.
890
Bank Ho.
CHALK
Clandon Pa.
Dorking
Lynn Hurst
Tunbridge
Wood Hatch
Ridge Green
Pound Hill
Furnace
Leigh
Albury & Pa.
Harlswood
Crowhurst
Medway R.
Quarry Hill
Bramley
Homewood Com.
Leigh
Horne
Gabricks Ho.
Penhurst & Pl.
Speldhurst
Hurtwood
993
Horse Hill
Horley
WEALDS
OF
Tanhurst Hou
Leeth Hill Pla
New Chapel Gr.
Tunbridge Wells
Ashhurst
Charlwood
Ewhurst
Ockley
WATERDOWN
FOREST
Felbridge Pa.
E. Grinsted
Okewood Chap.
Ridge Pa.
County Oak
Turner Hill
Eridge Park
Okewood Hill
Crawley
Worth
Forest Row
Frant 659
Lit. Bookhurst
ASHDOWN

'Water of apparently different qualities percolates within the bowels of the earth through nearly all parts of this basin, and receives modifications in its passage through the different strata therein, so that the quality of the water in its course may vary considerably in different parts of this basin, from the character of the strata through which it passes to its principal issue – the public sulphur wells – where the waters appear to be concentrated by the peculiar cross declination of the rock contiguous thereto.'

FROM SMITH'S STATEMENT ON THE SUPPOSED DAMAGE TO HARROGATE'S OLD SULPHUR WELL, 1837.

DETAIL FROM SHEET XII, SHOWING LONDON AND SURROUNDS.
Although London is built on Clay, water from deep wells beneath the city came from a stratum of Sand, which is shown on Smith's map some 24 km (15 miles) to the north, around Hatfield, Ridge and Bushey, immediately south of Watford. Deeper wells in the chalk yielded substantial volumes of water that infiltrated in the Chilterns to the north and the North Downs to the south.

VII. MENTOR.

A retreat to Yorkshire; the conundrum of Mary Ann Smith; the sincere devotion of dutiful nephew John Phillips; the formation of the Yorkshire Philosophical Society; meandering public lectures on geology; a cunning cylindrical plan for the Rotunda Museum; our hero at last decorated with medals, prizes and honour.

Following the collapse of his finances in insolvency, William Smith (1769–1839) emerged from the King's Bench debtors' prison in late 1819. He had lost his property at Tucking Mill, his books and maps, and his leased house in London. His vast and professionally useful collection of fossils had been sold to the British Museum. In the company of his nephew, John Phillips (1800–1874), he left London for the north to escape his creditors and lick his wounds.

WORK IN THE NORTH.

Smith continued to consult geologically, prospecting for coal and locating water sources. Much of the employment he found was in Yorkshire. He had no fixed address, lodging where he found work and continuing his field investigations in preparing manuscript maps for *A New Geological Atlas*, which he had begun in 1819 with John Cary (1754–1835). From February to April 1821, Smith and Phillips, undertook an extensive 'walking excursion', during which they investigated the coalfields of the West Riding of Yorkshire west of Doncaster. Smith's diary was empty for this period, but he recorded their geological observations and interpretations in his field notes. In the summer of 1821, Smith's four-sheet geological map of Yorkshire was published by Cary.

In 1823, after completing several mineral evaluations in Lancashire, in what was then Cumberland and Durham for a Col. Braddyll, Smith returned to a favourite lodging in Kirkby Lonsdale on the west edge of Yorkshire adjacent to the Lake District. He found it a congenial location of considerable geological interest. It was here he first met Adam Sedgwick (1785–1873), the new professor of geology at Cambridge, on his own summer mapping project in the Lake District. The stonemasons of the town had observed Smith's 'singular habits of handling the stones and trying their hardness against his teeth' and, guessing by his geological hammer that Sedgwick 'was of the same trade', directed the professor to Smith (Phillips, 1844, p. 103).

Between projects, according to Phillips, Smith found Kirkby Lonsdale a 'sweet retirement ... feeding all the best qualities of his mind by calm meditations ... Frequent walks made all the neighbourhood familiar to us for a circle of fifteen miles, and gave the opportunity of completing the maps of Westmoreland and Lancashire' for the *New Geological Atlas* (Phillips, 1844, p. 104).

However, no more county maps were issued after 1824, when this cartographic project was ended by Cary's sons, George and John, who had taken over the business in 1820–21 on their father's retirement. The *Atlas* was only half completed, though some plates, including Somerset, were already partly engraved with geological boundaries. Among Smith's papers in the Oxford University Museum of Natural History

FIG. 1.

Fig. 1. *Map of South Britain* on which Smith had shaded the seventeen counties on twenty sheets, (Yorkshire taking four sheets) of his series of county maps, published between 1819 and 1822. His note on it was written at this stage when the last four northern counties were nearly ready. These were published in 1824; three other counties were in preparation.

archive is a small map of 'South Britain' with the twenty published sheets and 'four almost fit for publication' shaded in, and with an undated note (almost a plea perhaps) that 'many others in a forward state'. For Smith, this cancellation of yet another publishing project and the thwarting of his ambition for a larger-scale map of 'South Britain' must have been traumatic.

Phillips does not refer to this calamity directly, but does describe a period around that time of concentrated experiments with a small smelting furnace, which 'was kept in frequent use in a garden, surrounded by walls which were covered by fossils, rocks and minerals. Thus pleasantly passed days, months, almost years, in seclusion.' (Phillips, 1844, p. 106)

LECTURING.

At this nadir, Smith received an invitation from the recently formed Yorkshire Philosophical Society to give a series of public lectures on geology in York. Phillips noted that 'Mr. Smith, who though he had never *lectured*, had spent half his life *talking* on geology, immediately accepted the proposal. New maps were coloured, new sections drawn' (Phillips, 1844, p. 107), and fossils borrowed to illustrate eight lectures, each delivered in morning and evening meetings over four weeks. They were so well received that the Society paid Smith £60, which was £10 more than the original offer, and John Phillips received £20 for 'arranging the Geological Department of the Museum'.

Word of the lectures spread, and invitations were received from Scarborough, Hull and Sheffield. For the York lectures, Phillips had prepared figures and assisted his uncle, but at Scarborough the series was extended to nine lectures and Phillips delivered the middle three on 'Organic Remains'. The series was scheduled for late August and early September 1824 when Scarborough's summer population was at its peak, after the York Races. More invitations to lecture followed, from Wakefield, Leeds and again from York; by mutual agreement Phillips assumed more of the lecturing and eventually took over the entire series.

Smith's lecture scripts indicate careful planning, compiled from notes prepared over several years and pasted into small octavo booklets to refer to while speaking. It appears that he intended to publish a general text based upon his lecture notes. However, of Smith's style of lecturing Phillips observed (Phillips, 1844, pp. 109–10):

> *A certain abstractedness of mind, generated by long and solitary meditation, a habit of following out his own thoughts into new trains of research, even while engaged in explaining the simplest facts, continually broke the symmetry of Mr. Smith's lectures. Slight matters, things curious in themselves but not clearly or commonly associated with the general purpose of the lecture, swelled into excrescences, and stopped the growth of parts which were more important in themselves, or necessary to connect the observations into an intelligible and satisfactory system. But there was a charm thrown over these discourses by the novelty and appropriateness of the diagrams and modellings which exemplified the arrangement of rocks, the total absence of all technical trifling from the explanations, and the simplicity and earnestness of the man.*

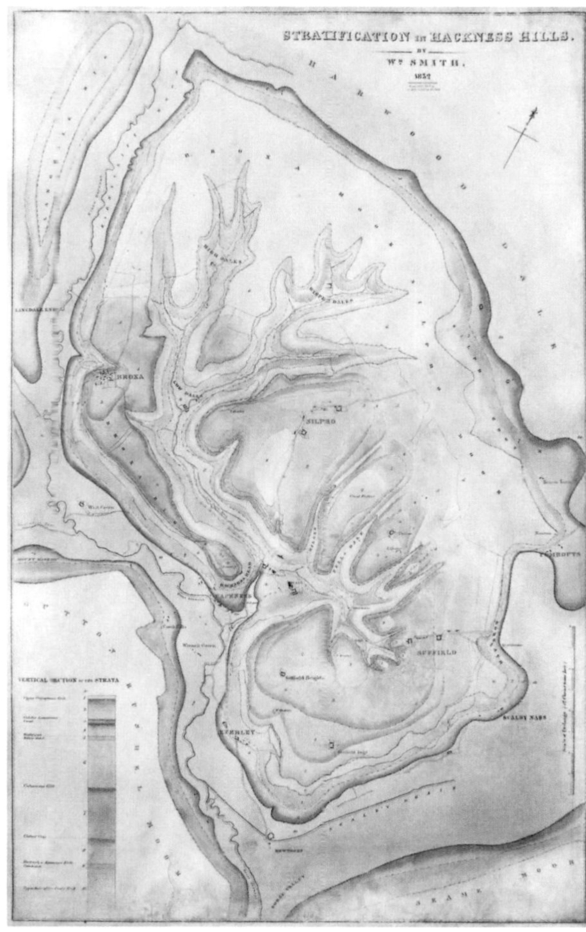

FIG. 2.

Fig. 2.
Stratification in Hackness Hills, 1832.
Smith's last major work, a limited edition of this map, was issued as a hand-coloured print in 1832. The base map was drawn by W. Day of Lincoln Inn Fields, London, and is of a rougher quality than the John Cary maps Smith used for earlier works.

MAPPING YORKSHIRE.

It was while he was based in Scarborough giving lectures that Smith met Sir John Vanden Bempde Johnstone (1799–1869), brother-in-law of the Rev. William Vernon Harcourt (1789–1871), President of the Yorkshire Philosophical Society. Johnstone engaged Smith to map his estate at Hackness, 10 km (6 miles) east-north-east of Scarborough.

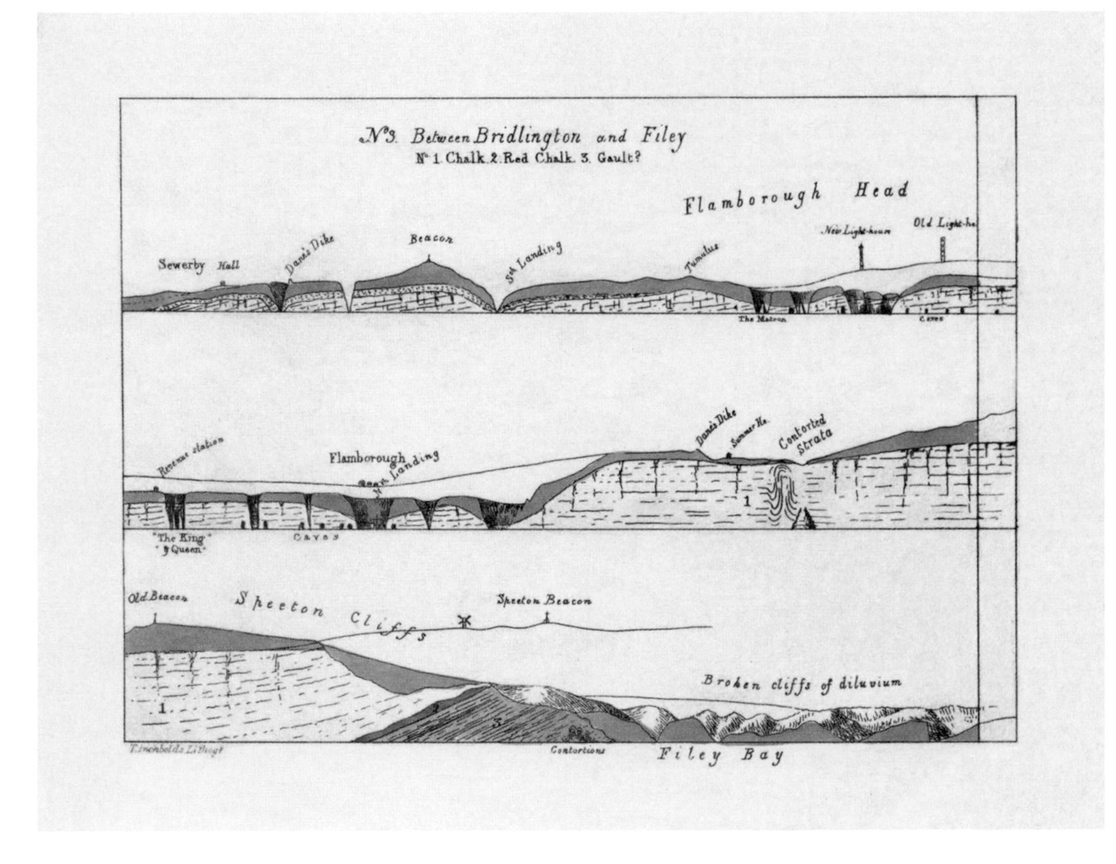

NO. 3. BETWEEN BRIDLINGTON AND FILEY, 1829.

This is the third of seven coastal sections included in John Phillips' *Illustrations of the Geology of Yorkshire, Part 1. The Yorkshire Coast* (1829), which he dedicated to his uncle William Smith, who had observed and discussed the cliff sections with him. The preceding sections are entirely coloured purple, without explanation as to what this colour denotes. At the bottom of this section, the annotation 'Broken cliffs of diluvium' provides the answer. Diluvium referred to the superficial deposits above the solid geology or bedrock. Not until after Smith's death did Louis Agassiz (1807–73) introduce the now accepted concept of continental glaciation and deposits by glaciers and their meltwaters. The chalk is not coloured in the figure as it is naturally white, but drawn with massive horizontal to gently dipping beds. The section shows the numerous caves and collapse features typical of the soluble chalk.

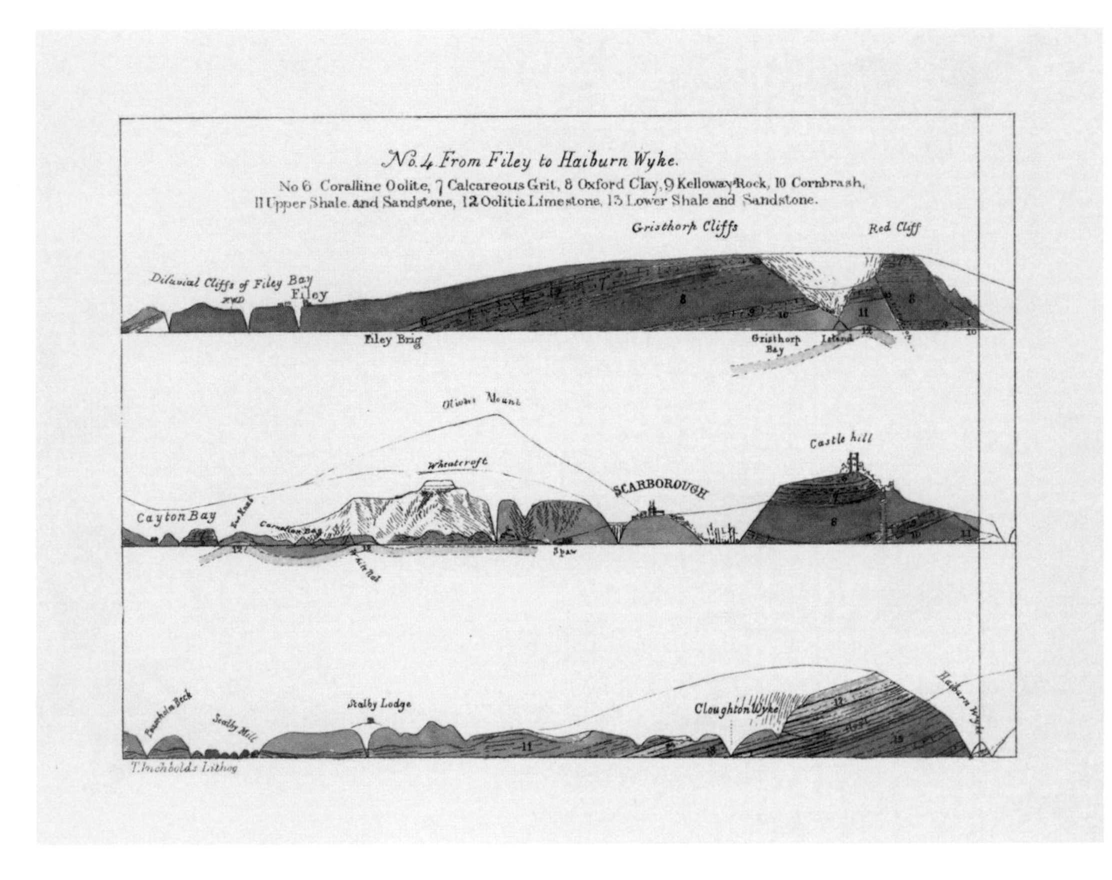

NO. 4. FROM FILEY TO HAIBURN WYKE, 1829.

The fourth coastal section from John Phillips' *Illustrations of the Geology of Yorkshire, Part 1. The Yorkshire Coast* (1829). In the middle section, from Carnelian Bay to Scarborough, the high cliff section is described as thin Grits, Sandstones and Shales in 'irregular often grassy cliffs', with 'frequent and remarkable curvatures of the beds'. As this confused section did not fit the strata to the south and north, Phillips left it blank and indicated the irregularity diagrammatically. In the text, he notes that from the 'Spaw' to Scarborough the proportion of diluvium increases and purple is again used to indicate this. In the diluvium Phillips recognized pebbles from the Lake District and Scotland, though these remained a mystery until the work of Louis Agassiz. The iron footbridge which connects Scarborough Spa to St Nicholas Cliff is visible to the left of Scarborough. The typical strata resume in Castle Hill.

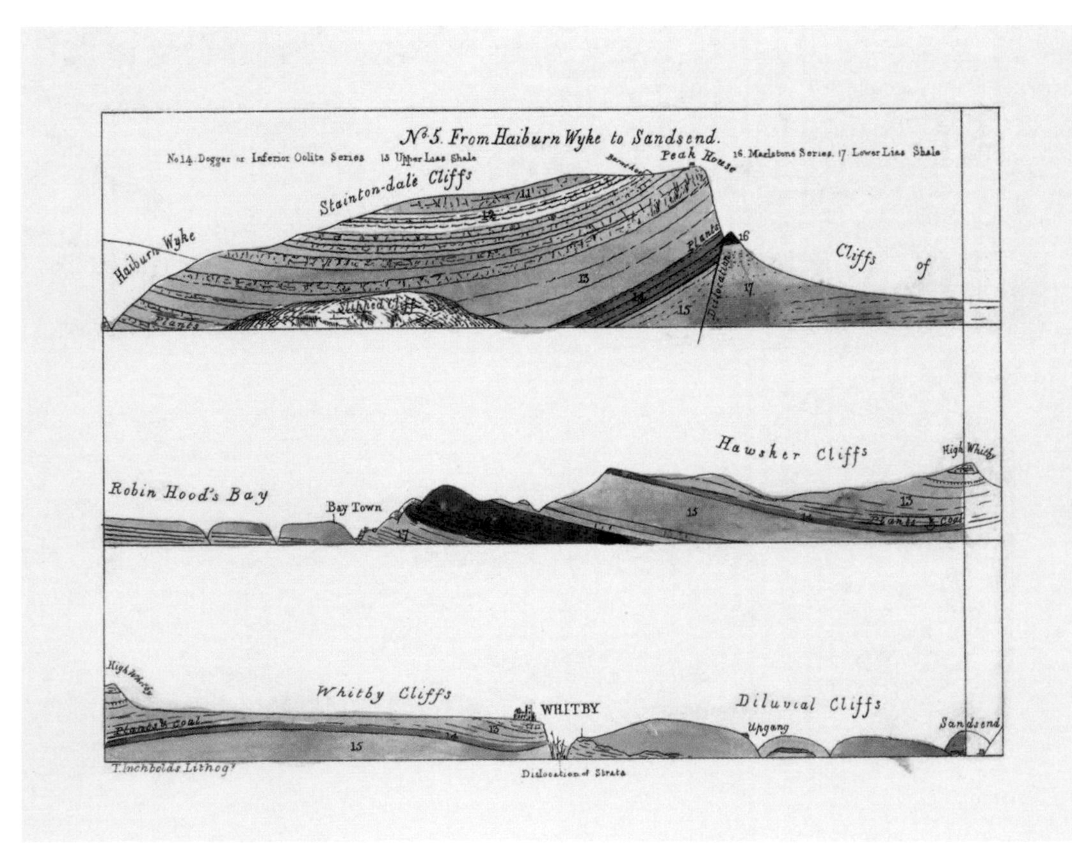

NO. 5. FROM HAIBURN WYKE TO SANDSEND, 1829.

In this fifth coastal section from John Phillips' *Illustrations of the Geology of Yorkshire, Part 1. The Yorkshire Coast* (1829), the strata are shown continuing to rise northward so that the Oolitic Limestone (yellow, numbered 12), first observed at sea level in section No. 4, rises and outcrops at the top of Staintondale Cliffs. Older, deeper strata constitute the cliff moving northward. In the top section Phillips drew and labelled a dislocation at Peak House; we know these now as faults. Another occurs at Whitby where the ships crowd the River Esk.

NO. 6. BETWEEN SANDSEND AND SALTBURN, 1829.

In the sixth section of John Phillips' *Illustrations of the Geology of Yorkshire, Part 1. The Yorkshire Coast* (1829) the cliffs are relatively more uniform in the Lias formations of Shales, Marlstones and Ironstones. Unremarked on dislocations or faults provide the few minor havens along this shore at Runswick, Staithes and Skinningrove. The highest cliffs at Boulby reveal all but the base of the Lias series. The Upper Lias rocks contained Alum Shale which was a nationally important resource for decades when Alum was essential to the textile and tanning industries as a mordant to fix or bind colour dyes to fabrics and leathers.

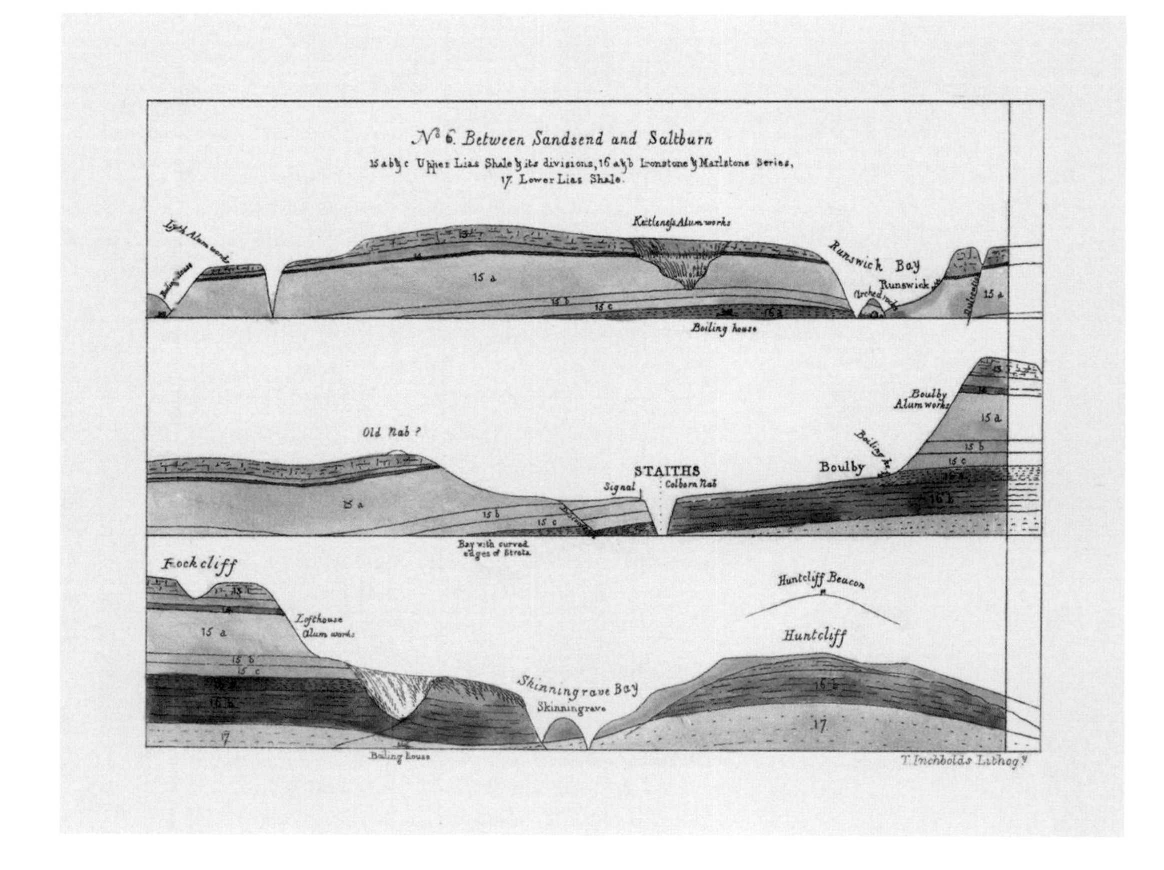

NO. 7. COAST NEAR REDCAR WITH ENLARGED SECTIONS, 1829.

The seventh section of John Phillips' *Illustrations of the Geology of Yorkshire, Part 1. The Yorkshire Coast* (1829) shows that north of Saltburn, the solid geology drops below sea level and is covered by diluvium. In the 'Enlarged Sections' below Phillips illustrated details of contortions, subtle dislocations and local variations in strata with page references to descriptions and discussion in his text.

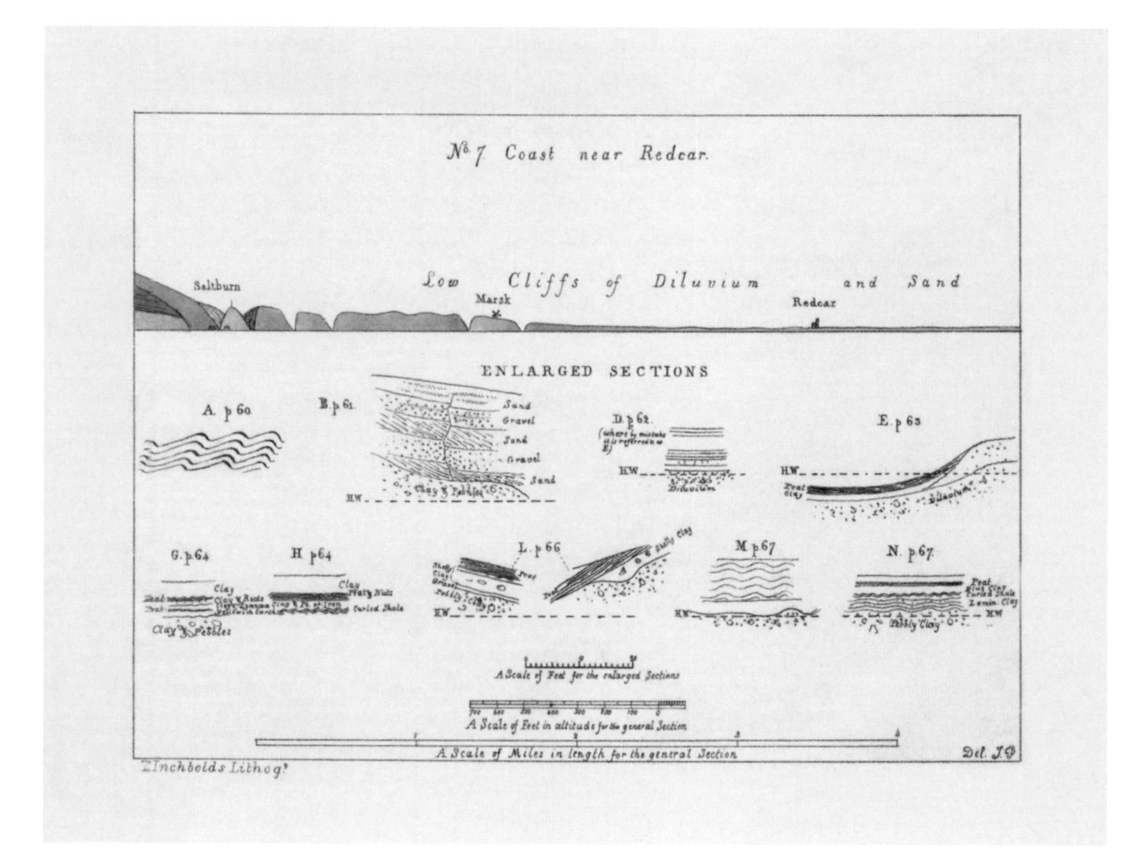

NO. 9. ENLARGED SECTIONS, 1829.

Following the seven coastal sections in *Illustrations of the Geology of Yorkshire, Part 1. The Yorkshire Coast* (1829), John Phillips provided two pages of scaled sections and general context. At Staithes, the strata are dislocated by 27 metres (90 ft), so that the strata of Colborn Nab on the right side lie beneath the exposed strata of Signal Cliff on the left side. Colborn Nab is now Cow Nab. The section through the Yorkshire Wolds is parallel to the coast and illustrates an anticlinal or dome structure that was eroded, then subsided to be buried by tropical marine deposits, which were uplifted to form the Chalk uplands of the Yorkshire Wolds. The boundary between the early deposits and the Chalk deposits marks a gap in the stratigraphic record of time and is known as an unconformity. The bottom section shows the sequence of rocks from the south-east to the north-west of Yorkshire.

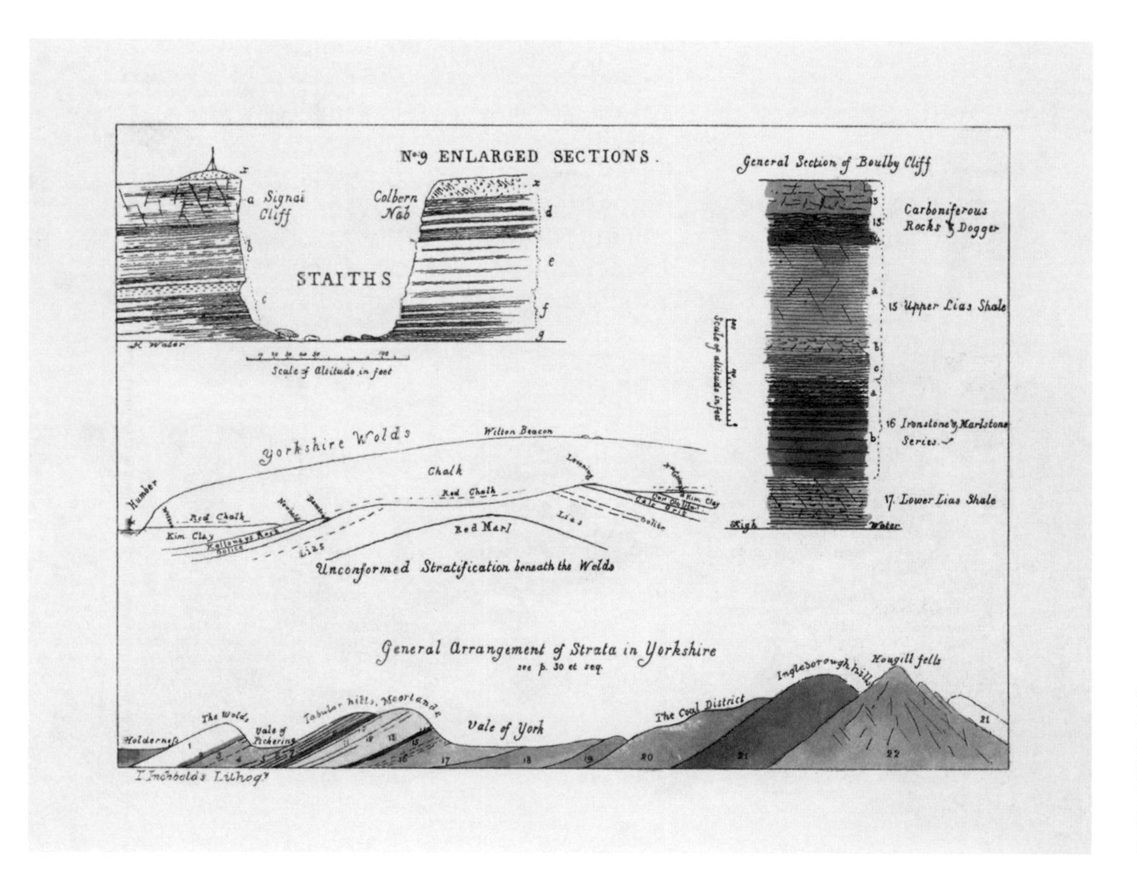

Several commissions to survey and prospect provided income and information to update his map of Yorkshire, and a unique revised version was provided to the Yorkshire Philosophical Society, now displayed in its restored library within the Yorkshire Museum. With Phillips, Smith recorded geological sections of the Yorkshire coastline, which were compiled and published by Phillips in the first part of his *Illustrations of the Geology of Yorkshire* in 1829, dedicated to his uncle (see pp. 234–5). So impressive was this pioneering work that the eminent London publisher John Murray undertook to publish a reissue in 1835. Murray invested in engravings of Phillips' lithographed map, sections and illustrations.

Johnstone had appointed Smith as land steward of the Hackness estate in the hope, as Phillips wrote 'that the retirement which seemed so well suited to Mr. Smith's age and taste, would have been memorable for the production of the results of a life of scientific toil'. However, 'Mr Smith meditated and wrote, but did not arrange his papers, and excepting a beautiful geological map of the Hackness estate ... nothing of importance came from his hands to the public.' Like many geologists after him, Smith found that report writing was less attractive than field work. But Phillips did record that Smith 'passed six of the calmest and happiest years of his declining life' at Hackness (Phillips, 1844, p. 113).

APPRECIATION.

While in Scarborough, Smith was consulted by the council to deal with the water supply for the increasing number of summer visitors and developed a novel solution involving recharging the aquifer in the winter months to store water for the summer. He also prepared a geological map centred on Scarborough at the scale of 1 inch to the mile. This was not published but remains as a manuscript in the Geological Society archives.

In the enlightened and welcoming environment of the town, Smith was invited by members of the Scarborough Philosophical Society to advise on a museum in 1827. He recommended the cylindrical form of the Rotunda, with the fossils displayed in stratigraphic order, the earliest on the lowest levels and the youngest on the top floor. Visitors would view the sequence while ascending a spiral staircase. Smith specified a unique system of sloping shelves that he had previously designed for his London house on Buckingham Street, intended as a museum to impress potential clients, often the owners of large estates who were visiting London for the winter social season.

For the Scarborough museum, John Phillips painted a large geological section of the coastal cliffs on the ring beam supporting the cupola crowning the Rotunda. Sir John Johnstone, who was then president of the Society, supplied the building stone at cost from his quarry at Hackness, which Smith supervised.

Smith's new home in Yorkshire, amid the circle of local savants, clients and his patron, brought stability to his life. One of the most heartening developments was his long overdue recognition in 1831 by a new generation of geologists at the Geological Society. Smith was awarded the first Wollaston Prize, 20 guineas, the Society's highest honour. At the award ceremony, the President, Adam Sedgwick, addressed Smith as the 'Father of English Geology' (Sedgwick, 1831, p. 278):

> *I for one can speak with gratitude of the practical lessons I have received from Mr. Smith: it was by tracking his footsteps, with his maps in my hand, through Wiltshire and the neighbouring counties, where he had trodden nearly thirty years before, that I first learned the subdivisions of our oolitic series, and apprehended the meaning of those arbitrary and somewhat uncouth terms, which we derive from him as our master, which have long become engrafted into the conventional language of English geologists, and through their influence have been, in part, also adopted by the naturalists of the Continent.*

On this occasion Smith donated to the Society the first geological map he had produced, of Bath, his original *Table of Strata*, both of 1799, and his

FIG. 3.

Fig. 3. The Scarborough Museum was completed in 1829 as a simple rotunda designed by the architect Richard Sharp of York to reflect Smith's idea, with wings added in 1861. The museum is shown here with Scarborough Crescent (above) and Scarborough Spa Bridge (below).

1801 map sketching the outline the geology of east and south England, the precursor of his first national map. Following the event, Smith wrote at length to his niece Anne Phillips (1803–1862), herself a knowledgeable amateur geologist, who then lived in York with her brother, that

> *while I was out of the room, Dr. Fitton moved that the President's address on the occasion be printed which was seconded by Mr. Greenough ... 90 merry Philosophical faces glowed over a most sumptuous dinner at the Crown and Anchor. The new President, Mr. Murchison then took the Chair – on his right sat Mr. Herschel, Sir John Johnstone, Professor Sedgwick, myself ... on his left, Davies Gilbert, the late president of the Royal Society ... Mr. Greenough, etc, etc ... After drinking success to their fellow associates in Science such as the Royal, Astronomical, Horticultural, Geographical ... coupled with the 'numerous Geological Societies which now spot the Great Oolitic Series was given with three times three' which was truly drunk with enthusiasm. From the pleasant manner in which it was given from the chair, I rose elated to thank them for their kind attention to me and others along the oolitic range which had also to boast of the birthplace of Newton and with a happy but unpremeditated allusion to benefits which Geology might have conferred on the County long before my time if Newton, even in his own fields, had but for once looked upon the ground. It produced a general laugh ... and John Cary has thrown new light on Geology by presenting me with a new pair of silver mounted spectacles.*

From the reference to George Bellas Greenough (1778–1855) seconding the motion, it appears that both he and Smith intended to draw a line under their past differences over accusations of plagiarism of Smith's 1815 map.

In 1832, at Oxford, the newly cast first Wollaston gold medal was presented to Smith at the second meeting of the British Association for the Advancement of Science (BAAS), which was founded in York in 1831. Smith was also awarded an annual government pension of £100, which Phillips believed was due to lobbying by members of the BAAS. In 1835 at the Dublin meeting of the Association, Smith was granted an honorary Doctor of Laws by Trinity College, there being no science degrees at this time. In his final years, Smith attended the annual week-long conferences of the BAAS, of which his nephew was Secretary. As Phillips observed (Phillips, 1844, p. 118):

FIG. 4.

Fig. 4.
Smith was the first person to be awarded the Wollaston Medal, the highest award of the Geological Society, in 1831. It was established by William Hyde Wollaston (1766–1828) to promote research concerning the mineral structure of the Earth. The gold medal was not ready at the time of the award, and was presented at the second British Association meeting in Oxford in 1832. From 1845, the medal was made of the much rarer metal palladium, part of the platinum group of metals and discovered by Wollaston in 1803.

> *To Mr. Smith the periodical return of these meetings was like the revival of spring to the vegetable world ... and for many years his presence ... was hailed with delight by those who, occupying the highest places in public opinion, were proud to call him the 'Father of Geology,' though to some of the views which they advocated he was disposed to show anything but parental affection. But he rarely took part in discussions at these meetings, and indeed seldom spoke, except to mention some striking fact, because his mind had been little trained to intellectual gymnastics, and an infirmity of age ... deafness began to rob him of half the wit and eloquence which he most admired and most longed to hear.*

MARY ANN SMITH.

Smith's time in Scarborough also appears to have brought a period of relative calm in a sometimes troubled domestic life. Smith's diaries for 1790–1801 and 1808 are missing, probably destroyed by Phillips. It appears Smith married Mary Ann (1790/92–1844) in 1808, the year his sister Elizabeth died, leaving her children in the care of Smith and his brother John. Initially they stayed with John in Somerset, and only in November 1815 did Phillips move to London to be with William.

Very little is known about Mary Ann, and a near total silence surrounds her. Phillips mentions a stay in Scarborough in 1820 'in hopes to soothe the mental aberration of his wife, which became very manifest in this year' (Phillips, 1844, p. 94). In a letter to Sedgwick in 1831, he noted 'Smith's achievements had been despite long and heavy afflictions. Poverty, disappointment and neglect forced seclusion from the world of science – these have been heightened by a still more severe and invincible torment: a mad, bad, wife.'

In 1877, long after the deaths of all concerned, Professor William Crawford Williamson (1816–1895), the son of Smith's landlord in Scarborough in 1826–28, recalled (Williamson, 1877, p. 65):

> *When Smith came to Scarborough he was accompanied by his wife, an eccentric little round-faced woman, who was about as unsuited for being the partner of a meditative philosopher as she could well be ... She seemed to have neither relative or friend, and during the many years in which she survived her husband the only persons who exhibited the slightest care for her were the professor [Phillips] and his sister. Of a small and somewhat stunted figure, oddly attired, with her cheeks rouged up to the highest point of which they were capable, and with short, dark-coloured, girlish curls gathered closely round her brow, she was daily to be seen walking a few yards in the rear of her husband, who plodded steadily on his way, apparently too much immersed in his geological meditations to give a thought to her who followed behind. ... She was occasionally subject to violent outbursts of temper, which, for the time, disturbed the even tenor of her husband's way. He never wrangled with her on such occasions, but would quietly walk out of the room, locking the door as he did so, and leave the house for a solitary stroll. On more than one occasion ... I have known her to dash some object through the window of her temporary prison as he passed outside of it, with a view of bringing the truant back. Yet notwithstanding these occasional outbursts, and their intellectual un-congeniality, their domestic life was, on the whole, a happy one.*

In 1835, Smith wrote to Anne Phillips: 'Mrs S. is very well, cheerful and active and has more than once reproached me for not writing to you.' Following Smith's death in 1839 and the cessation of his pension, Mary Ann had to leave their rented home. She was not recorded on the 1841 census and was admitted to York Lunatic Asylum on 15 February 1842, where she died on 27 June 1844.

This is the sum of the truncated marital record – a great difference in age and temperaments, an initial motivation to care for his dead sister's children, and long periods away leaving his wife alone in a large empty museum of a house, plagued by creditors.

STONE FOR THE HOUSES OF PARLIAMENT.

In 1838 Smith's final professional consultancy was an invitation to join Henry De la Beche (1796–1855) and the architect, Charles Barry (1795–1860) on a parliamentary commission to select a suitable building stone for the new Houses of Parliament, following the fire in 1834 that had destroyed most of the medieval parliamentary buildings. In fact, Phillips had been requested by De la Beche for this, but was too busy with other Geological Survey work and recommended his uncle, whose experience was far greater. Following the Newcastle meeting of the British Association in 1838, the commissioners travelled to more than a hundred quarries over three months in Scotland, England and Wales.

Stone was evaluated for its appearance, durability, the size of the largest blocks, the volume available and cost. Each stone was subject to laboratory testing and assessed by stone masons for its workability. Examples of existing buildings using each stone were included for readers of the report, which was published on 16 March 1839. The recommendation was for a Magnesian limestone from the Anston quarry in South Yorkshire. In the event, the Anston stone selected rapidly decayed in London's pollution, although similar stone from the same quarry performed well in the Geological Survey's own Museum of Economic Geology off Whitehall.

INFLUENCE AND LEGACY.

In August 1839, while on his way from London to the annual British Association meeting in Birmingham, William Smith caught a chill and died, aged seventy, in Northamptonshire. His legacy is substantial on both a personal level

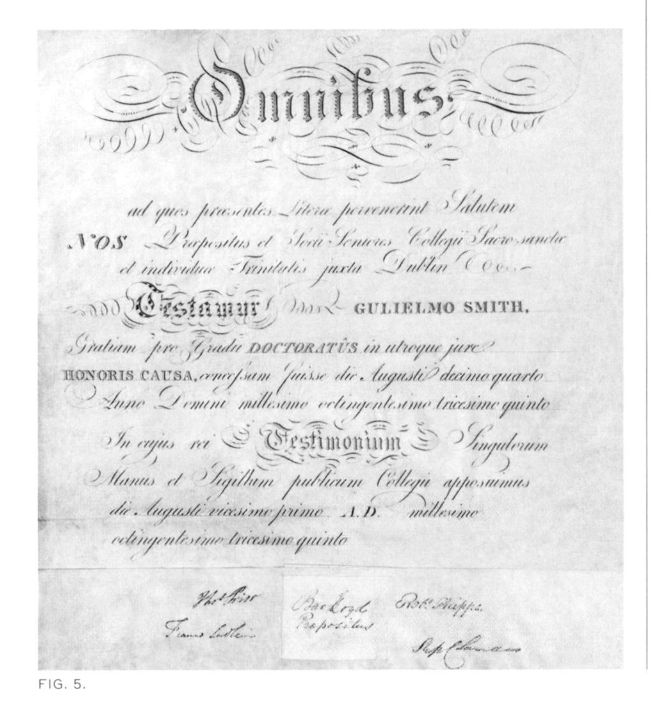

Omnibus

ad quos praesentes Literae pervenerint Salutem

NOS Praepositus et Socii Seniores Collegii Sacro-sanctae et individuae Trinitatis juxta Dublin

Testamur GULIELMO SMITH,

Gratiam pro Gradu DOCTORATÛS in utroque jure HONORIS CAUSA, concessam fuisse die Augusti decimo quarto Anno Domini millesimo octingentesimo tricesimo quinto

In cujus rei Testimonium Singulorum Manus et Sigillum publicum Collegii apposuimus die Augusti vicesimo primo A.D. millesimo octingentesimo tricesimo quinto

FIG. 5.

Fig. 5. The Honorary Doctorate of Laws degree awarded to William Smith by Trinity College, Dublin, during the 1835 meeting of the British Association.

and professionally to the discipline of geology. He was the single strongest influence on his nephew, John Phillips, the first geological apprentice. As a teenager Phillips helped describe and illustrate Smith's fossil collection for the catalogue that accompanied its sale to the British Museum. He drew the lithographed maps for Smith's consulting reports and he observed and mapped alongside Smith. When the opportunity to lecture arose in York he rapidly succeeded his uncle, and was appointed curator of the Yorkshire Museum. His early publications on the geology of York were assisted by Smith, but were far more detailed and technical. A series of overlapping careers as Curator at the Yorkshire Museum, Secretary of the newly formed British Association for the Advancement of Science in York, palaeontologist for the embryonic Geological Survey, lectureships at University College London and Trinity College, Dublin, followed. Phillips ultimately became professor of geology at Oxford and founder and curator of the Oxford University Museum of Natural History. Through Phillips's career and writing, Smith's influence expanded. His 1844 biography of Smith is our major source of information on William's life.

In a letter dated 30 March 1829 included with the copy of the *Illustrations of the Geology of Yorkshire, Part 1, The Coast* which he gave to Smith, Phillips wrote:

> *My Dear Sir,*
> *Many and changeful years have passed since your kindness was first bestowed on me. You took to your arms a friendless orphan, gave him the means of instruction, filled his mind with your own experience, and encouraged him to rival yourself. Others will praise your ardent zeal and great discoveries in science, but I am bound by a holier bond – the tie of increasing gratitude.*
> *Whatever I have done, whatever I may do, is in truth your own work, and can yield me but poor satisfaction unless it satisfies you. Receive then my first publication as a proof that I have not slighted your instructions, nor neglected your favourite science; accept it as a pledge of my unalterable affection, and let me now at least boast my happy connection with one who will be remembered when I am forgotten.*

Professionally, Smith's geological legacy is immense. Without precedents to follow, he invented a scientific field – biostratigraphy – through his observation that sedimentary rocks were deposited in a sequence of strata that is recognizable over extensive areas and definable by their fossil content. Characteristic

FIG. 6.

Fig. 6.
A group portrait of geological worthies at the Newcastle meeting of the British Association in 1838. John Phillips in the foreground is examining a fossil jawbone; William Smith is in profile on the extreme right. Others from left to right are: Roderick Murchison (second director general of the Geological Survey), Richard Owen (founder of the Natural History Museum in London), Henry De la Beche (founder of the Geological Survey), unknown with hearing trumpet, John Taylor (mining magnate), Charles Lyell (author of *Principles of Geology*), John Morris (professor at University College of London), William Buckland (professor at Oxford University), and unknown.

fossils enabled him, and all subsequent geologists, to identify where rocks were located in the stratigraphic sequence and to predict the presence and depth of a stratum of economic interest or its absence due to removal by erosion. This fundamental technique of fossil recognition and sequence interpretation was not recognized before Smith and today underpins exploration in sedimentary rocks for water, minerals, and fossil fuels which sustain life and our material world.

Smith's maps were hugely influential. The 1815 map reveals the big picture and rock relations over extensive areas. It is visible to an entire lecture hall, while his 1820 poster-sized single-sheet map accomplishes the same at a domestic scale. His choice of colours is still largely followed in standardized geological maps around the world, while the scales he adopted are practical and close to typical modern scales. The individual county maps show regional geology at various medium scales selected to fit the different county areas into a standard large atlas format, but they approximate the modern survey maps at quarter-inch or 1:250,000 scale. His Scarborough map shows local detail at the 1-inch scale that was commonly used by national geological surveys in English-speaking countries until metrication in the last century. And finally, Smith's map of the Hackness Estate nearly replicates the detailed scale of 6 inches to the mile, now 1:10,000, employed by all Geological Surveys in the British Isles. In his choice of legend colours and useful scales, Smith thus led and anticipated subsequent geological cartographers.

Smith was fortunate to live long enough to receive the appreciation that he so justly deserved – medals, prizes, an honorary degree, official appointment and a pension. Although this is not recorded, in his last appointment he would have been able to see the first official maps of the new Geological Survey of England and Wales, and if so, would surely have recognized the extent of his influence and legacy.

Scalby
Hackness
Scarborough
East Ayton
Wheat Croft
Cayton
Filey
Filey Beck
FILEY BAY
Brompton
Sherburn
Potter Brompton
Hunmanby
Fordon
Buckton Hall
Would Newton
Foxholes
Wm Heslerton
Grindale
Flamborough
FLAMBOROUGH HEAD
Kirby Grindilyth
Bridlington
Rudston
Beacon
BRIDLINGTON BAY
Groom
Sledmere
Kilham
Thornholm
Towthorp
Gransmoor
Garton
Great Driffield
Wetwang
Wansford
Huggate
Kirkburn
Beeford
Dringhoe
Emmotland
Bewholme
Nth Dalton
Hempholm
Kilnwick
Boshill
Hayholm
Hornsea
Holme on the Wolds
Walsand
Mere
Sigglesthorn
Goxhill
Sth Dalton
Leconfield
Goodmanham
Mappleton
Cherry Burton
Rise
Hatfield Magna
Routh
Lambwith Stream
Beverley
Benningholme
Sancton
Bentley
Burton Constable
Risby
Grimston
Beacon
Coniston
Danthorp
Hilston
Bilton
Bagwell
Kirk Ella
South Cave
Kingston upon Hull
Headon
Rimswell
Owthorne
Hessle
Ferriby
Winestead Park
Ryhill
Hollym
HUMBER RIVER
Whitton
Marsh
Salt Marshes
Patrington
Barton
Coxhill
Wellwick
Winterton
Sunk I.
Killingholme
Salt Marshes
Kilnsea
High Riseby
Wooton
MOUTH OF THE HUMBER
Elsham
Brocklesby Park
Thornham
SPURN HEAD
G. Grimsby
Barneby le Wold
Great Limber
Blow Wells
Clee Ness
Clee
Irby
Bradley
YORKSHIRE WOLDS
HOLDERNESS

'It will appear as unnecessary, as it would be difficult, to enumerate all the advantages, when it is considered what numerous coincidences and indisputable facts have occurred (in the course of so many years constant observation and experiment on strata, in different parts of the country) to found this extensive investigation, which must lead to accurate ideas of all the surface of the earth, if not to a complete knowledge of its internal structure, and the progress and periods of formation.'

SMITH'S *MEMOIR*, 1815.

DETAIL FROM SHEET VI, SHOWING THE YORKSHIRE WOLDS
Hackness and Scarborough, where Smith spent the latter part of his life, are visible at the top of this map segment at the edge of the Limestone strata of the Vale of Pickering. Sir John Johnstone supplied the stone from his quarry at Hackness for the Rotunda Museum in Scarborough, which was built in 1829 to a design suggested by William Smith.

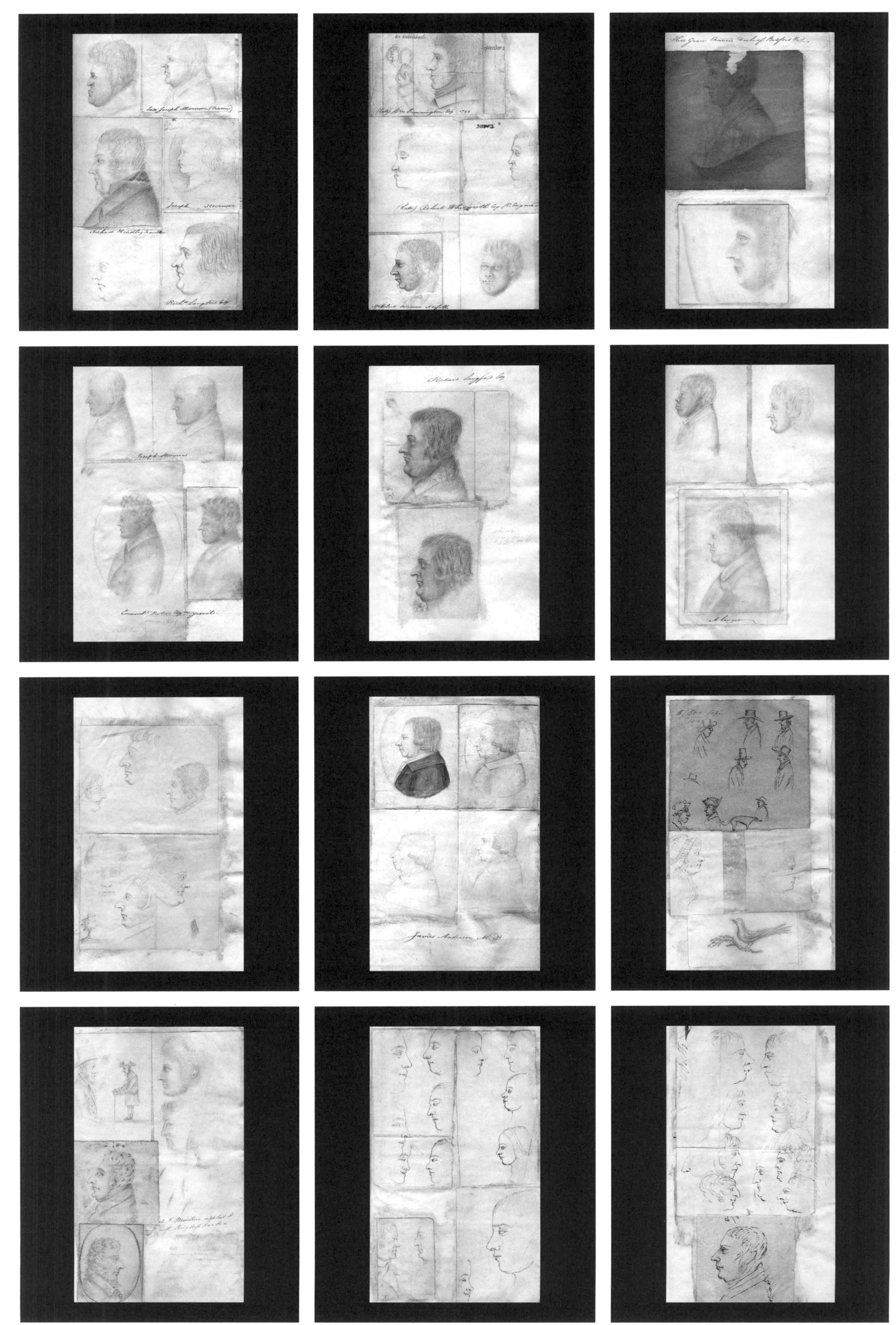

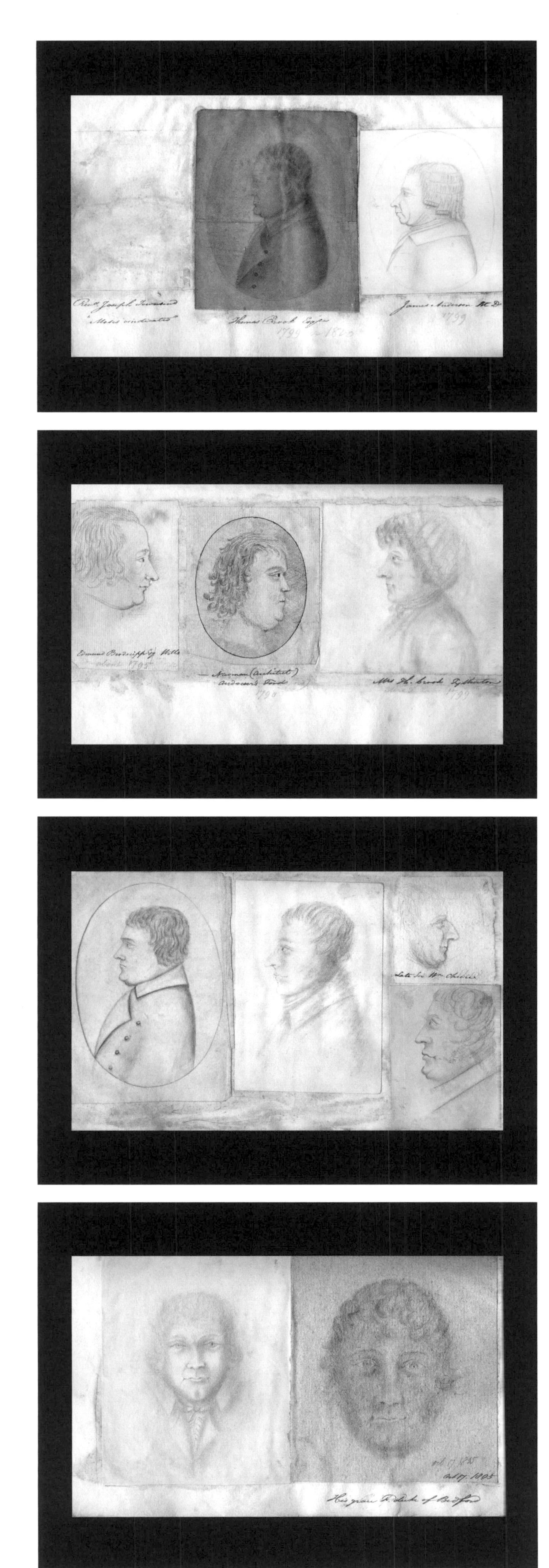

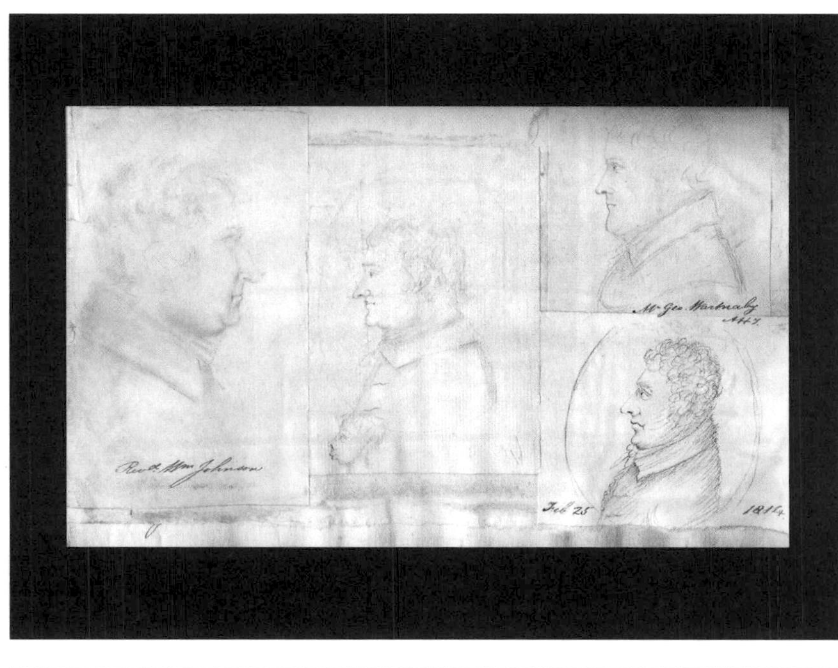

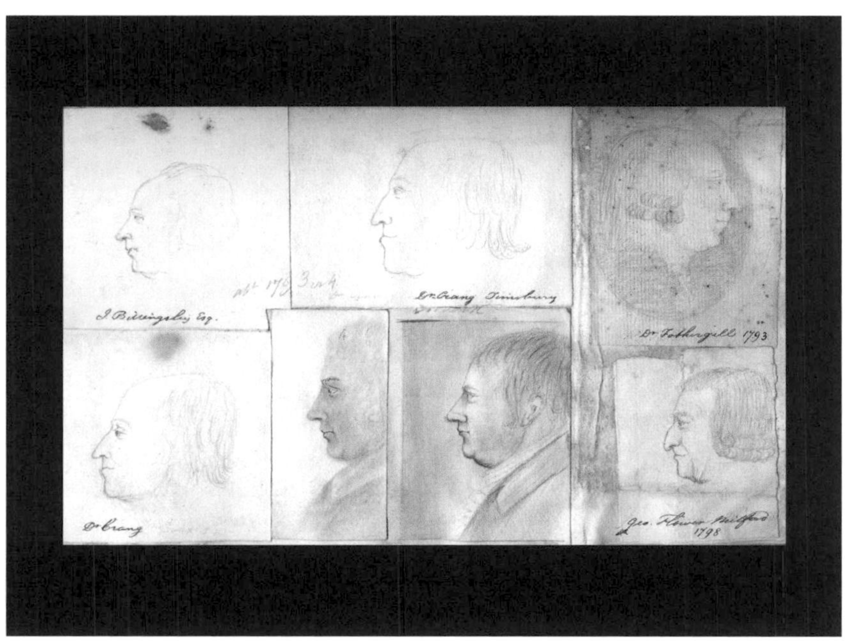

PORTRAITS DRAWN BY WILLIAM SMITH, 1798–1805.

Smith was a keen draughtsman, as this sketchbook of portraits of his various acquaintances shows. Many of Smith's friends from the scientific world are represented, including the Scottish agriculturalist James Anderson (third row, centre portrait opposite), canal engineer Robert Whitworth (top row, centre portrait opposite) and archaeologist William Cunnington (top row, centre portrait opposite). Smith's aristocratic employer Francis Russell, 5th Duke of Bedford also features (top row, right-hand portrait opposite and bottom portrait this page).

DRAMATIS PERSONÆ.

Sir Joseph Banks, 1st Baronet.
1743–1820.

Banks made his name as a botanist on an expedition to Newfoundland and Labrador in 1766, before joining Captain Cook on his famed expedition to the South Pacific two years later. He was elected President of the Royal Society in 1778, and also advised King George III on the Royal Botanic Gardens, Kew. He promoted the careers of many other scientists, including financially supporting William Smith.

Reverend William Vernon Harcourt.
1789–1871.

After serving in the navy and then studying at Oxford, Harcourt moved to Yorkshire to begin his career as a clergyman. Here he also pursued his scientific interests, constructing a laboratory at Bishopthorpe Palace and becoming the first president of the Yorkshire Philosophical Society. He encouraged investigation into Yorkshire's geology, and invited William Smith to give a course of lectures in the county.

William Buckland.
1784–1856.

Buckland was both a pioneering geologist and minister, known for his efforts to reconcile geological discoveries with the Bible. He was a lively and charismatic lecturer, and became well known and respected in the scientific community, becoming the first reader in Geology at Oxford University, and later the president of the Geological Society as well as the Dean of Westminster.

Charles König.
1744–1851.

Born in Germany, König came to England to organize Queen Charlotte's collections. Subsequently, he assisted Joseph Banks' librarian, then became keeper of natural history at the British Museum where he was responsible for evaluating Smith's fossil collection. As a mineralogist, he failed to appreciate the fossils' significance and greatly undervalued them at £700.

John Cary.
1754–1835.

After serving his apprenticeship, Cary established his own engraving business in 1783, and quickly gained a reputation for cartography. His *New and Correct English Atlas* (1787) became a standard reference work, and in 1794 the Postmaster General commissioned him to survey England's roads. He collaborated closely with William Smith, who primarily used Cary's maps as the base for his geological works.

Sir John Vanden-Bempde-Johnstone, 2nd Baronet.
1799–1869.

Vanden-Bempde-Johnstone was MP for Yorkshire in 1830 and then Scarborough in 1833. He employed William Smith as land steward of Hackness Hall when Smith was out of work and homeless in 1828, and donated the stone for the building of the Scarborough Rotunda Museum. He was a member of the Yorkshire Philosophical Society and married a sister of the Reverend William Vernon Harcourt.

Thomas Coke, 1st Earl of Leicester.
1754–1842.

Coke served as the MP for Norfolk from 1776 but lost his seat in 1784 due to his support for the American colonists during the American Revolutionary War. In his time away from parliament Coke applied himself to improving his Holkham estate with many agricultural reforms, during which he commissioned the help of William Smith. He returned to his parliamentary career in 1790.

George Bellas Greenough.
1788–1855.

Greenough was the Geological Society of London's first president, and led the production of their geological map of England and Wales, published in 1820. The map drew heavily on William Smith's work, although this was uncredited. Unlike Smith, Greenough was sceptical of the usefulness of fossils in correlating strata. However, the supposed rivalry between the two men is likely overstated.

John Farey.
1766–1826.

Farey met William Smith in 1801, when the former was employed on the Duke of Bedford's Woburn Estate as land agent and the latterly for irrigation and drainage works. The following year, Farey accompanied Smith on an exploration of Woburn's chalk hills and learnt about mineral surveying from him. He later surveyed estates across the country and wrote extensively on many scientific topics.

Sir Roderick Murchison, 1st baronet.
1792–1871.

After leaving the army, Murchison became a geologist and highly active in the Geological Society of London His most influential work was on the England–Wales border, in which he proposed the Silurian System in 1839. Smith guided him along the cliffs of Yorkshire in 1826, demonstrating the significance of fossils for locating strata in context. Murchison related his subsequent field work in the Brora coalfield to Smith's method.

Anne Phillips.

1803–62.

The niece of William Smith, Phillips worked closely with her brother John on his geological investigations. Her most important discovery was a rock conglomerate in the Malvern Hills, now known as Miss Phillips' Conglomerate. This provided crucial evidence for her brother's theory that Roderick Murchison's Silurian System was incorrect.

Adam Sedgwick.

1785–1873.

Sedgwick was a pioneering geologist and cleric, who along with Roderick Murchison applied himself to discovering the order of strata below the Old Red Sandstone. As president of the Geological Society of London he conferred the Wollaston medal on William Smith in 1831. As Professor of Geology at Cambridge he strongly supported advances in geology against conservative churchmen, but opposed Darwin's theory of evolution.

John Phillips.

1800–74.

After the death of his parents, Phillips' custody was assumed by his uncle William Smith. He became Smith's assistant, accompanying him on his geological investigations around the country and contributing particularly to an understanding of the geology of Yorkshire. In 1841 he published the first global geologic time scale, helping to standardize the terms Palaeozoic and Mesozoic.

Mary-Ann Smith

c. 1792–1844.

Mary Ann is believed to have married William Smith in 1809, when he was aged 40 and she was aged around 17. Very little is known about her. However, William Crawford Williamson, the son of the first curator of the Scarborough Rotunda Museum, described her as 'oddly attired' and prone to bursts of temper, but happily married. She had no children, and died in York Lunatic Asylum.

Joseph Planta.

1744–1827.

Born in Switzerland of noble descent, Planta succeeded his father as assistant librarian at the British Museum, before becoming the Principal Librarian, i.e. director, in 1799. In this role he increased the capacity of the reading rooms and made the galleries more accessible to the public. It was also during his tenure that the British Museum acquired the fossil collections of William Smith.

James Sowerby.

1757–1822.

Sowerby was an engraver and naturalist who made an important contribution to the field of natural history through his major illustrated works, including *English Botany* (1791–1814) and *Mineral Conchology of Great Britain* (1812). He collaborated with William Smith on *Strata Identified by Organized Fossils* (1816–19), for which he created nineteen illustrated plates. He had a large fossil collection of his own.

John Rennie (the elder).

1761–1821.

Rennie was one of the great civil engineers of his day. He designed a large number of canals during their 'golden age', and employed William Smith to execute surveys for the proposed Somersetshire Coal Canal. He also worked on drainage in the Norfolk Fens and Lincolnshire, designed many bridges, including Waterloo Bridge in London, and built docks in cities including Hull Liverpool and London.

Joseph Townsend.

1739–1816.

Townsend was a doctor, geologist and vicar known for his opposition to state provision for the poor, instead advocating for compulsory membership of friendly societies. It was while visiting at Townsend's house in Bath with the Reverend Benjamin Richardson that William Smith first explained his theory of stratigraphy, which Townsend recorded and later transcribed as Smith's *Order of Strata*.

Reverend Benjamin Richardson.

1758–1832.

After his ordination in 1782, Richardson was first a curate in Bradford-on-Avon, and then rector of Farleigh Hungerford in Somerset. He had a keen interest in geology and natural history, and was a friend and supporter of William Smith. He was present when Smith dictated his *Order of Strata* to the Reverend Joseph Townsend in 1799, and helped educate Smith's nephew John Phillips.

Reverend Richard Warner.

1763–1857.

As well as a clergyman, Warner was an enthusiastic antiquarian and the author of a wide range of books, including multiple histories of Bath, a Gothic novel, several accounts of his walking holidays and a reissued ancient cookery book. He was a friend of William Smith, and his histories of Bath clearly drew on Smith's geological work there. Smith also helped Warner with the subsidence below his house.

Francis Russell, 5th Duke of Bedford.

1765–1802.

As well as being a Whig politician, Russell took a great interest in agriculture and employed John Farey and William Smith to work on his Woburn estate in Bedfordshire. He was the first president of the livestock society, the Royal Smithfield Club, and a member of the original Board of Agriculture. It was at one of Russell's agricultural shows that Smith first promoted his proposed geological map in 1802.

Edward Webb.

1751–1828.

Webb was one of the foremost land surveyors of his day, and it was as his assistant in Stow-on-the-Wold that William Smith began his career. In this position Smith was able to travel the surrounding counties and begin to build up his knowledge of geology. Webb gave Smith plenty of encouragement and eventually sent him to survey an estate in Somerset, where Smith would develop his theory on the order of strata.

TABLE DETAILING WILLIAM SMITH'S FOSSILS FEATURED AS PHOTOGRAPHIC PLATES IN THIS BOOK.

This table gathers together information relating to the photographs of Smith's fossils that appear in the book, grouped according to the stratum in which they are commonly found. The first column indicates the photographic plate and page on which you can find each fossil grouping together with the stratum in which they were found. Where a fossil specimen is listed in Smith's publication *Strata Identified by Organized Fossils* (SIOF, 1816–19), with illustrations by Sowerby, the identification and locations where he found that species is given in the second column. In this book, Sowerby's illustration of each grouping is displayed opposite the photographic plate. Where a fossil specimen is listed in *A Stratigraphical System of Organized Fossils* (SS, 1817), published to accompany the collection of fossils Smith sold to the British Museum (BM), the identification and locations are given in the third column. A code is hand-written on many of the Smith specimens. These are added in brackets after the location. Occasionally the location is written. Smith's codes are shown in SS where the numeral is the species number and the lower-case letter the location; only in Part 2 of SS does Smith use the capital letter to denote the group. The capital lettering system begins again with each stratum. Notes in square brackets [ ] in that column refer to annotations in Smith's hand discovered by Hugh Torrens in an edition in Oklahoma. It is interesting to compare the two publications. With only two to three years between them Smith was able to add extra locations and often update the identifications. Some locations in SIOF do not appear in SS, so perhaps the specimens had been lost in the interim.

Smith's collection was sold to the British Museum in 1818 and moved to the Natural History Museum (NHM) when it opened in 1881, where it joined other natural history specimens. The fourth column provides the specimen label details on the figured specimens in the NHM collection. All NHM registration numbers should have the prefix NHMUK, which has not been added to the numbers given here. Where the photographed specimen is not the original but has been substituted for a similar specimen (usually the same species and location), the registration number is given in brackets. We have only ever substituted with Smith's own specimens. Where no specimen has been found in Smith's collection, there is no registration number. Finally, the fifth column provides additional notes by the authors of this table.

Both SIOF and SS were re-published in full, along with photographs of the actual specimens in *William Smith's Fossils Reunited* (Wigley, 2019).

KEY TO ABBREVIATIONS.	
Min. Conch.	*Mineral Conchology* (Sowerby's publication of fossils)
Linn.	Linnaeus (author of the fossil name)
Lam.	Lamarck (author of the fossil name)
indet.	indeterminable

PLATE. (PAGE)	STRATA IDENTIFIED BY ORGANIZED FOSSILS.	A STRATIGRAPHICAL SYSTEM OF ORGANIZED FOSSILS.	SPECIMEN LABEL DETAILS ON SMITH'S SPECIMENS IN THE NHM COLLECTION.	NOTES.
FRONTISPIECE. (188, 256)	*Mastodon arvernensis* / Norfolk		Mastodon tooth: *Anancus arvernensis* M1983 / Norfolk [Whitlingham]	The Mastodon tooth is from Norfolk. Its locality was first recorded by Smith's pupil, R. C. Taylor in April 1827 *Philosophical Magazine* (series 2, vol. 1 pages 283–84) as 'from Whitlingham near Norwich'. This was then confirmed by Smith himself in March 1836. Right upper third molar.
1. (62/3)	LONDON CLAY.	LONDON CLAY.	EOCENE, INCLUDING LONDON CLAY.	Within this plate, there are specimens that we would not now include under 'London Clay', although all but one are of Eocene age. The specimens from Woolwich come from the Woolwich Formation and are slightly older than the London Clay. The specimens from Barton, Bracklesham and Hordle Cliff are somewhat younger.
				Smith places the London Clay plate before his Crag plate. He mistook the glacial till at Happisburgh and elsewhere for London Clay and therefore thought it was younger than the Crag: the Crag is approximately 2–4 million years old, while the London Clay is around 55–50 million years old. In an annotated copy of SS found by Hugh Torrens in Oklahoma, Smith annotates the Happisburgh entries as Diluvial. Peter Riches discusses the confusion more fully (Riches, 2016–17).
1.	*Vivipara fluviorum* / Well, Brixton causeway / Hordwell Cliff	Vivipara: *Vivipara fluviorum* Min. Conch. / Brixton causeway, out of a deep well	Gastropod: *Viviparus suessoniensis* G1675 / Well, Brixton causeway	The plate drawing only features a portion of the block photographed and the photograph has been similarly trimmed.
2.	*Tellina* etc. / Sheppey / Happisburgh	*Tellina* (only 1 species) / a. Sheppey / b. Happisburgh Cliff	Block with bivalves: *Abra splendens* L1426 / Sheppey (Sheppey)	The block from Sheppey is the one photographed (and trimmed). It is very similar to the figured specimen. There is no London Clay at Happisburgh (it is glacial till). The specimen from there has not been found and Smith may have realized that it was 'Diluvial' before delivering his fossils to the BM.
3.	Arca Linn. *Pectunculus* Lam. / Bognor	*Pectunculus* Sp. 2: *Pectunculus* / Bognor	Bivalve: *Glycymeris brevirostris* L1430 / Bognor (Q 3)	Q 3 is written on the specimen yet there are only 2 species of *Pectunculus* listed in SS. We believe this to be a mistake on Smith's part and it should be Q 2.
4.	*Chama* / Hordwell Cliff	*Chama*: *Chama squamosa* Brander / Hordel Cliff	Bivalve: *Chama squamosa* / (not found)	
5.	*Voluta spinosa* / Barton	*Voluta*: *Voluta spinosa* Min. Conch. / Bognor	Gastropod: *Volutospina luctator* / (not found)	Figs 5 & 6 are amalgamated in SS with the locality of Bognor. This agrees with Fig. 6 in SIOF, but there Fig. 5 is listed from Barton. Certainly Smith's figured specimen is a typical Barton Fossil. We have only located and photographed Fig. 6, which is labelled from Bognor and is a London Clay Fossil.
6.	*Voluta* / Bognor	*Voluta*: *Voluta spinosa* Min. Conch. / Bognor	Gastropod: *Volutospina denudata* G1563 / Bognor (A 1)	
7.	*Cerithium* / Woolwich / Bracklesham Bay	*Cerithium* Sp. 1: *Cerithium melanioides* Min. Conch. / a. Woolwich / b. Bracklesham Bay	Gastropod: *Brotia melanioides* G1567 / Woolwich (D 1 a)	*Brotia melanioides* is well known from the Woolwich Formation in the Woolwich area, but is not found in the younger beds of Bracklesham Bay.
8.	Large Shark's Tooth / Sheppey / Highgate	Teeth: Large / Isle of Sheppey	Shark's tooth: *Otodus obliquus* P4829 / Sheppey	
9.	Small Shark's Tooth / Highgate	Teeth: Thin / Happisburgh Cliff	Shark's tooth: indet. / (not found)	SIOF lists Highgate as the locality (as many others). SS lists the locality as Happisburgh, which is less likely as the clay encountered there is glacial till coming from the north. No London Clay is known at Happisburgh (see above) and indeed Smith revised the entry to Diluvial.
10.	*Pectunculus decussatus* / Highgate	*Pectunculus* Sp. 1: *Pectunculus decussatus* Min. Conch. / Highgate archway	Bivalve: *Striarca wrigleyi* L1435 / Highgate (Q 1) /	
11.	*Ammonites communis* / Happisburgh	*Ammonites communis* Min. Conch. / Happisburgh Cliff	Ammonite: *Dactylioceras commune* / (not found)	Smith later corrected to Lias Fossil Diluvial [transported in the glacial till]. There is no London Clay at Happisburgh.
12.	*Calyptraea* Lam. / Barton Cliff	(none listed)	Gastropod: *Sigapatella aperta* / (not found)	There is no entry for Fig. 12 in SS and neither *Calyptraea* nor its relatives are listed. We have not found this specimen and it is possible that it never reached the BM.
13.	Crab / Sheppey / Highgate	Crab / Sheppey mentioned but not specified as Fig.	Crab: *Zanthopsis* sp. (1749 – substitute) / Sheppey	Only Sheppey is mentioned by name in SS. The photographed specimen is not the one figured in SIOF but it does come from Sheppey.
	Other London Clay sites listed in SS: Selsey Bill, Stubbington, Ryde, Isle of Wight, Muddiford, Alum Bay, Emmsworth, Pagham, Newhaven Castle, Harwich, Bexley Heath, Brentford, Richmond.			
	Sowerby London Clay additional sites (*Mineral Conchology*) listed in SS: Bracklesham Bay, Brentford, Minster Cliff, Regent's Canal, Croydon Canal, Streatham, in Surry [sic].			
2. (62/3)	CRAIG (CRAG IN TEXT).	CRAG.	PLIOCENE & PLEISTOCENE CRAG.	Smith places his 'Craig' plate second, even though the specimens are about 50 million years younger than the London Clay fossils (see above and Riches, 2016–17).
				The spelling 'Craig' on the plate is the spelling Sowerby uses in his *Mineral Conchology*. Smith uses Crag in the texts.
1.	*Murex contrarius* / Alderton, Suffolk / Thorpe Common, Harwich / Holywell near Ipswich / Tattingstone Park	*Murex* Sp. 4: *Murex contrarius* Linn. Min. Conch. / a. Alderton / b. Suffolk / c. Tattingstone Park / Foxhole / Playford / Sutton / Newborn / Brightwell	Gastropod: *Neptunea angulata* G1546 / Alderton (B 4 a)	
2.	*Murex striatus* / Bramerton / Holywell / Alderton / Aldeborough	*Murex* Sp. 2 / e. Bramerton / a. Between Norwich and Yarmouth / b. In the parish of Leiston / c. Thorpe Common / d. Tattingstone Park / Foxhole (covered in large *Balani*) / Playford / Sutton	Gastropod: *Nucella incrassata* with encrusting barnacles G1536 / Bramerton (B 2 e)	Apart from Bramerton, the locations listed in SIOF and SS are different.
3.	*Turbo littoreus* / Thorpe Common / Between Norwich & Yarmouth / Leiston old Abbey / Bramerton / Trimingsby	*Turbo* Sp. 1: *Turbo littoreus* Min. Conch. / c. Thorpe Common / a. Between Norwich and Yarmouth / b. Leiston old Abbey / d. Bramerton	Gastropod: *Littorina littorea* G1558 / Thorpe Common (F 1 c)	
4.	*Turbo* Linn. *Turritella* Lam. / Thorpe Common	*Turritella* Sp. 1 / Thorpe Common	Gastropod: *Potamides tricinctus* / (not found)	
5.	*Patella Fissura* Linn. *Emarginula* Lam. / Bramerton / Harwich / Holywell	*Emarginula reticulata* Min. Conch. / Bramerton	Gastropod: *Emarginula fissura* / (not found)	Only Bramerton is listed in SS. It is possible that the figured specimen was from a locality not listed and never came to BM.
6.	*Balanus tessellatus* / Bramerton	*Balanus tessellatus* Min. Conch. / b Bramerton on a murex / a. Aldborough / c. Burgh Castle / d. Tattingstone Park / e. Keswick / Foxhole on a murex	Barnacle: *Balanus* sp. 1747 / Bramerton	
7.	*Arca* Linn. *Pectunculus* Lam. / Tattingstone Park / Thorpe Common	*Pectunculus glycimeris*. *Arca glycimeris* Linn. / b. Tattingstone Park / a. Thorpe Common / Foxhole / Aldborough / Sutton / Newborn / Brightwell / Bentley	Bivalve: *Glycymeris variabilis* L1409 / Tattingstone Park (Q 2 b)	
8.	*Cardium* Linn. / Tattingstone Park / Bramerton / Happisburgh (or Hasbro') / Trimingsby	*Cardium* / a. Tattingstone Park / c. Bramerton / b. Happisburgh Cliff / Foxhole / Sutton / Newborn / Brightwell / Bentley	Bivalve: *Cerastoderma hostei* L1410 / Tattingstone Park (R 1 a)	
9.	*Mya lata* / Bramerton / Trimingsby	*Mya lata* Min. Conch. / Bramerton, hinges / Trimingsby, hinges / Aldborough, a cast of the inside	Bivalve: *Mya arenaria* (hinge only, L1413 – substitute) / Bramerton	Apart from the cast, only hinges are listed in SS. It is possible no whole specimens reached the BM.
10.	Vertebrae / Thorpe Common	(none listed)	Short vertebra of a Fish: indet. P4832 / Thorpe Common	Figs 10–21 are not listed in SS.
11.	ditto	(none listed)	Fish: Hourglass vertebra: *Platax woodwardi* P4839 / Thorpe Common	The descriptive names are from the edition of SIOF in NHM and are included with the photographed specimens to help distinguish.

PLATE...	STRATA IDENTIFIED...	A STRATIGRAPHICAL SYSTEM...	SPECIMEN LABEL DETAILS...	NOTES.
12.	ditto	(none listed)	Fish vertebra (worn): indet. P75892 (formerly P4839) / Thorpe Common	
13.	ditto	(none listed)	Fish vertebra showing the 6 ridges forming: half a vertebra indet. P4833 / Thorpe Common	
14.	ditto	(none listed)	A sort of star (same specimen as Fig. 13) / Thorpe Common	
15.	Palate / Tattingstone Park	(none listed)	Tooth palate of an Eagle Ray: possibly *Aetobatus* sp. (very worn) P4834 / Tattingstone Park	
16.	Tooth / Stoke Hill	(none listed)	Large shark's tooth: *'Isurus'* sp. (very worn) P4835 / Stoke Hill	
17.	Teeth / Reading, Ipswich	(none listed)	Shark's tooth (very worn): ? Lamnid P4836 / Reading, Ipswich	In the NHM edition of SIOF Smith notes: 3 others that are found in the London Clay in a more perfect state but they are characteristic here from being worn very smooth. (These will be derived from London Clay).
18.	ditto	(none listed)	Shark's tooth: ?Lamnid / (not found)	ditto
19.	ditto	(none listed)	Shark's tooth (very worn): 'Odontaspid' type P4837 / Reading, Ipswich	ditto
20.	Quadruped's bone / Tattingstone Park	(none listed)	Toe phalange of a mammal: possibly gazelle M1590 / Tattingstone Park	If the identification of gazelle is correct it would be the earliest occurrence in the UK.
21.	Stalactite / Burgh Castle	(none listed)	Stalactite (not fossil animal) (not found)	
	Sowerby Crag additional sites (*Mineral Conchology*): Newhaven Castle, Walton Nase			
3. (64/5)	**UPPER CHALK.**	**UPPER CHALK.**	**UPPER CRETACEOUS, UPPER CHALK.**	
1.	*Alcyonium* Flint, others in Chalk / Wighton / Wilts	*Alcyonium* Sp. 2 / Wighton / near Warminster / Guildford	Sponge: *Sporadoscinia alcyonoides* S9863 / Wighton (Norfolk on specimen)	
2.	ditto / Wighton /	*Alcyonium* Sp. 1 / Chittern	Sponge: *Toulminia catenifer* S9866 / Chittern (Chittern) /	
3.	*Serpula* / Norwich	*Serpula* / Norwich	serpulid worm: ?*Filogranula* sp. / (not found)	
4.	Valves of *Lepas* Linn. / Norwich	*Balanus* / Norwich	fragments of barnacle: *Regioscalpellum maximum* I 750 / Norwich /	
5.	Hollow Valve of *Ostrea* / Norwich	*Ostrea* / Norwich)	Interior of an oyster: *Pycnodonte vesicularis* L1446 / Norwich	
6.	Flatter valve of ditto / Norwich	*Ostrea* / Norwich)	Interior of an oyster: *Pycnodonte vesicularis* / (not found)	
7.	Ditto attached to a Belemnite / Norwich	(Fig. 7 not listed)	Belemnite: oyster attached to a belemnite / (*Belemnitella mucronata*) L1446 / Norwich (G 1) /	Fig. 7 is not mentioned in SS, probably because the oyster is the same sp. and location as Figs 5 & 6.
8.	*Pecten* / Norwich	*Pecten* Sp. 1 / Norwich	Bivalve: *Mimachlamys mantelliana* L1441 / Norwich	
9.	*Terebratula subundata* (long variety) / Norwich	*Terebratula* Smooth Sp. 2 VAR / *Terebratula subundata* Min. Conch. / Norwich	Brachiopod: *Concinnithyris subundata* B1392 / Norwich	
10.	*Echinus* / North of Norwich / Taverham / Croydon / Wilts	*Galea* Sp. 1 *E. ovatus* Leske / a. Norwich / b. Taverham / c. Croydon / Guildford / Bury	Echinoid: *Echinocorys scutata* (552 – substitute) / Norwich (D 1 a)	The figured specimen E551, (near Norwich, Norfolk) is on exhibition in the Enlightenment Gallery (King's Library), BM (2019). Substitute is from the same location but is not the same species. N.B. *Echinocorys scutata* is very varied in shape. The sub-species include var. *ovata*.
11.	Palate of a fish / Warminster	Fishes palates Sp. 1 / Near Warminster	palate of a fish: *Ptychodus mammillaris* P4813 / Near Warminster	
12.	Part of *Echinus* / North of Norwich	(none listed)	Echinoid: basal plate of a cideroid with boss for spine / (not found)	The specimen has not been located, nor is it listed in SS. It is possible that it never came to the BM.
13.	*Echinus* spine / North of Norwich	Spines of *Echini* Sp. 1 / Norwich	Echinoid: spine of a cideroid / (not found)	
14.	Shark's tooth with 2 ridges / North of Norwich	Teeth Sp. 1 / b. Norwich / a. Near Warminster	Fish tooth: *Enchodus* sp. P4816 / North of Norwich	
15.	Shark's tooth serrated / North of Norwich	Teeth Sp. 3 / a. Norwich / b. Near Warminster	Shark's tooth: 'Corax' tooth / (not found)	
16.	Vertebrae / North of Norwich	Vertebrae of fish / North of Reigate	Vertebra: indet. (possibly shark) / (not found)	
	Ovate Echini and Zoophites, without enumerating localities, may be found any where on the surface of Upper Chalk.			
4. (64/5)	**LOWER CHALK.**	**LOWER CHALK.**	**UPPER CRETACEOUS, LOWER CHALK.**	
1.	*Inoceramus Cuvieri* / Heytesbury / Knook Castle & Barrow / Hunstanton Cliff	*Inoceramus* Sp. 1 / *Inoceramus Cuvieri* Sowerby / Heytesbury / Knook Castle & Barrow / Hunstanton Cliff / Bury St Edmonds	Bivalve: *Volviceramus involutus* L1444 / Heytesbury	
2.	*Inoceramus* / Wilts (Warminster)	*Inoceramus* Sp. 2 / Near Warminster / Guildford	Bivalve: *Mytiloides labiatus* MB1147 / Warminster, Wiltshire	
3.	Cast of a *Trochus* / Mazen Hill	*Trochus* Sp. 3 / Mazen Hill	Gastropod: *Bathrotomaria* sp. G1571 / Mazen Hill	Mazen Hill was a quarry site near the Park of Sir William Pierce Ashe-A'Court, Bart 1795 (*c.* 1748–1817) at Heytesbury House, Wiltshire. His second wife was a fossil collector. Heytesbury House is about 32 km (20 miles) from Warminster (pers. comm. Hugh Torrens).
4.	*Ammonites* / Norton Bevant / Mazen Hill	*Ammonites tuberculatus* / Norton	Ammonite: *Schloenbachia subtuberculata* (C619 – substitute) / (Rundaway Hill, near Devizes, Wiltshire, H 2 a)	H 2 a is the Greensand code for this specimen. However, *Schloenbachia* is a Lower Chalk specimen. Often it is very difficult to distinguish the glauconitic Upper Greensand and the Glauconitic Marl at the base of the Lower Chalk. Substitute is not identical but is the same species.
5.	*Cirrus depressus* / Warminster	*Cirrus* / Near Warminster	Gastropod: *Bathrotomaria* sp. G1573 / Warminster	
6.	*Terebratula* / Heytesbury	*Terebratula* smooth Sp. 2 / *Terebratula subundata* Min. Conch. / a. Heytesbury / b. Near Warminster / c. Mazen Hill / Guildford	Brachiopod: *Gibbithyris semiglobosa* B1387 / Heytesbury I 2 a	SS not listed as figured, but description fits I 2 a.
7.	*Terebratula* / Heytesbury / Warminster / Mazen Hill	*Terebratula* plicated Sp. 4 / b. Heytesbury / a. Near Warminster / c. Mazen Hill / d. Norwich	Brachiopod: *Orbirhynchia cuvieri* (B1396 – substitute) / Heytesbury	This is not the figured specimen embedded in chalk, but is believed to be the same species and from one of the locations listed in both SS and SI.
8.	*Terebratula subundata* / Heytesbury / Warminster / Mazen Hill	*Terebratula* smooth Sp. 2 / *Terebratula subundata* Min. Conch. / a. Heytesbury / b. Near Warminster / c. Mazen Hill / Guildford	Brachiopod: *Gibbithyris semiglobosa* BF107 (re-registered from B1387 / Heytesbury I 2 a	SS not listed as figured, but description fits I 2 a.
9.	Shark's teeth / Warminster	Teeth sp. 2 (not listed as figured) / b. Warminster / a. Wilts	Shark tooth: indet. (P4819 – substitute) / (no location)	SS not listed as figured for Lower Chalk teeth. The specimen chosen is not the one figured by Smith and no location is given with the specimen.
5. (66/7)	**GREEN SAND (1).**	**GREEN SAND.**	**LOWER CRETACEOUS, UPPER GREENSAND (1).**	
1.	Alcyonite (funnel form) / Warminster / Pewsey / Devizes / Dinton Park	*Alcyonium* Sp. 1 / Warminster / Pewsey / Devizes /	Sponge: *Pachypoterion compactum* P3030 / Warminster	
2.	Alcyonite (doliform) / Pewsey / Warminster	*Alcyonium* Sp. 2 / Pewsey	Sponge: *Siphonia tulipa* P3018 / Pewsey	
3.	*Venus angulata* / Blackdown	*Venus* Sp. 2 *Venus angulata* Min. Conch. / Blackdown	Bivalve: *Epicyprina angulata* L1455 / Blackdown (P 2)	
4.	*Murex* Linn. / Blackdown	(none listed)	Gastropod: *Cretaceomurex calcar* G1584 / Blackdown	*Murex* is not listed at all in SS. The location of Blackdown on the label matches SIOF.
5.	*Turritella* / Blackdown	*Turritella* / Blackdown	Gastropod: *Torquesia granulata* G1581 / Blackdown (C 1)	
6.	*Pectunculus* Lam. / Blackdown	*Pectunculus* / Blackdown	Bivalve: *Glycymeris sublaevis* L1448 / Blackdown (K 1)	
7.	*Cardium* / Blackdown	*Cardium* / Blackdown	Bivalve: *'Mactra' angulata* – on block G1583 / Blackdown	None of the specimens on block are specified as figured in SS, except Fig. 10.
8.	*Rostellaria* Lam. / in a mass from Blackdown	(none specifically listed)	Gastropod: *Drepanocheilus calcaratus* – on block G1583	On same block.

PLATE...	STRATA IDENTIFIED...	A STRATIGRAPHICAL SYSTEM...	SPECIMEN LABEL DETAILS...	NOTES.
9.	*Trigonia alaeformis* / Blackdown	*Trigonia* / (none specifically listed)	Bivalve: *Pterotrigonia* cf. *aliformis* – on block G1583	On same block.
10.	*Cucullaea* Lam.	*Cucullaea*	Bivalve: ?*Idonearca* sp. – on block G1583	On same block.
6. (66/7)	GREEN SAND (2).	GREEN SAND (2D).	LOWER CRETACEOUS, UPPER GREENSAND (2).	
1.	*Vermicularia* (chambered) / Horningsham, Wilts.	*Vermicularia* Sp. 1 / *Vermicularia concava* Min. Conch. / Near Warminster	Tube worm: *Rotularia concava* A121 / Near Warminster (E 1) /	Fig. 1 not listed in SS, but specimen matches *Vermicularia concava* Sp. 1.
2.	*Solarium* Lam. / Rundaway	*Solarium* / Rundaway Hill	Gastropod: *Nummogaultina fittoni* G1580 / Rundaway Hill (B 1) /	The photographed specimen is a close match to the figured one and comes from the only location mentioned in SS.
3.	*Pecten* (echinated) / Chute Farm / Rundaway	*Pecten* Sp. 4 / Chute Farm	Bivalve: *Merklinia scabra* L1470 / Chute Farm	
4.	*Terebratula pectinata* / Chute Farm / Warminster	*Terebratula* plicated Sp. 4 / *Terebratula pectinata* Min. Conch. / a. Chute Farm / b. Warminster	Brachiopod: *Dereta pectita* B1407 / Chute Farm	
5.	*Terebratula Lyra* / Chute Farm	*Terebratula* plicated Sp. 4 / *Terebratula Lyra* Min. Conch. / Chute Farm	Brachiopod: *Terebrirostra lyra* / (not found)	
6.	*Terebratula* / Chute Farm / Warminster	*Terebratula* plicated Sp. 5 / a. Chute Farm / b. Warminster	Brachiopod: *Cyclothyris latissima* B1406 / Chute Farm	
7.	*Chama haliotoidea* / Alfred's Tower / Dilton / Black Dog Hill / Teffon / Evershot Stourton / Blackdown	*Chama haliotoidea* Min. Conch. / f. Alfred's Tower / a. Dilton / b. Black Dog Hill, near Standerwick / c. Teffont / d. Evershot / e. Stourton / g. Blackdown	Bivalve: *Amphidonte obliquatum* L1462 / Alfred's Tower (R 1 f)	
8.	*Pecten quadricostata* / Warminster / Chute Farm / Blackdown	*Pecten* Sp. 1 *Pecten quadricostata* / Near Warminster	Bivalve: *Neithea gibbosa* L1468 / Near Warminster	
9.	Pecten / Chute Farm	*Pecten* Sp. 5 / Chute Farm	Bivalve: *'Chlamys'* aff. *subacuta* L1473 / Chute Farm	
10.	*Ostrea* (*Gryphea* Lam.) / Stourton / Dinton Park / Tinhead	*Ostrea* Sp. 2 / a. Stourhead / b. Dinton Park / c. Tinhead	Bivalve: *Amphidonte obliquatum* L1464 / Stourhead (S 2 a)	Stourton (SIOF) and Stourhead (SS) are the same location.
11.	*Echinus* with a singular anal appendage / Warminster / Chute Farm	*Cidaris* Sp. 4 *Cidaris diadema* / b. Near Warminster / a. Chute Farm	Echinoid: *Salenia petalifera* E476 / Warminster (C 1)	
12.	*Echinites* Leske / Warminster / Chute Farm	*Conulus* / Near Warminster / Chute Farm	Echinoid: *Discoides subuculus* E487 / Warminster	
13.	*E. lapis caneri* / Chute Farm &c.	*Spatangus* no furrows Sp. 3 / *Echnites lapis caneri* Leske / Chute Farm	Echinoid: *Catopygus columbarius* E485 / Chute Farm (C 1 No. 3)	
14.	*Spatangus* Leske / Chute Farm / Warminster / Rundaway	*Spatangus* no furrows Sp. 3 / a. Chute Farm / near Warminster	Echinoid: *Holaster laevis* E483 / Chute Farm (B 3)	
15.	*Cyclolites* Lam. / Chute / Puddle hill, near Dunstable	*Madrepora* / Chute Farm	Coral: *Microbacia* sp. / (not found)	
16.	*Madreporite* / Chute Farm	*Millepora* / Chute Farm	Indet. / (not found)	
17.	*Alcyonite* / Chute Farm	*Alcyonium* / Chute Farm	Sponge *Barroisia* sp. / (not found)	
7. (96/7)	BRICKEARTH.	BRICKEARTH [GOLT].	LOWER CRETACEOUS, GAULT CLAY.	
1.	*Ammonites* / Steppingley Park / Near Godstone / Prisley Farm Bedfordshire	*Ammonites* with furrow Sp. 2 / a. Steppingley Park / b. Near Godstone / c. Prisley Farm	Ammonite [fragments only]: *Hoplites dentatus* (C627 – substitute) / Steppingley Park	Another specimen is in fragments that may have been the figured one. Pyrite decay is a common problem of Gault fossils.
2.	*Hamites* / North-west part of Norfolk	*Hamites* / Near Grimston	Ammonite: *Idiohamites* sp. C628 / Near Grimstone	Grimston is in the north-west part of Norfolk, so SS may just be more specific than SIOF. The correct spelling is without the final e.
3.	*Echinus* Linn. *Spatangus* Leske / Near Devizes	*Spatangus* / Near Devizes	Echinoid: *Pliotoxaster* sp. E489 / Near Devizes	
4.	*Belemnites* / Norfolk (north-west part) / North of Riegate / near Godstone / Steppingley Park / Prisley Farm /	*Belemnites* / c. Near Grimston / a. North of Reigate / b. near Godstone / d. Steppingley Park / e. Prisley Farm / Leighton Beaudesert / Westoning	Belemnite: *Neohibilites minimus* (C626 – substitute) / Near Grimstone	Smith's figured specimen is slightly larger than the photographed one. The location fits SS. Grimston is in the NW part of Norfolk.
5.	*Belemnites* / (no text for Fig. 5)	*Belemnites* / e. Prisley Farm / a. North of Reigate / b. Near Godstone / c. Near Grimston / d. Steppingley Park / Leighton Beaudesert / Westoning	Belemnite: *Neohibilites minimus* C625 / Prisley Farm, Bedfordshire	There is no entry for Fig. 5 in SIOF but both Figs 4 and 5 are referred to the same species in SS. The locations match.
8. (96/7)	PORTLAND STONE.	PORTLAND ROCK.	UPPER JURASSIC, PORTLAND STONE.	
1.	Cast of *Natica* Lam. / Swindon	*Natica* / Swindon	Gastropod: *Neritoma sinuosa* G1587 / Swindon (Swindon)	
2.	*Turritella* inside cast / Portland / Swindon	*Turritella* / a. Portland / b. Swindon	Gastropod: *Aptyxiella portlandica* G1585 / Portland	
3.	Cast of *Venus* Linn. / Swindon	*Astarte cuneata* Min. Conch. / Swindon	Bivalve: *Eomiodon* sp. L1486 / Swindon (Swindon E 2)	
4.	*Trigonia* / Swindon / Chicksgrove / Fonthill / Telfont	*Trigonia* Sp. 2 / Swindon	Bivalve: *Myophorella incurva* (L1484 – substitute) / Swindon (2)	This is not the correct specimen although it is a trigonid form Swindon. The locations in SIOF fit better with Sp. 1 in SS but the description of the specimen fits better with Sp. 2 and '2' is written on the specimen.
5.	*Venus* inside cast / Swindon / Chicksgrove	*Venus* Sp. 2 / Swindon / Chicksgrove	Bivalve: *Protocardia dissimilis* L1487 / Swindon (Swindon)	
6.	*Pecten* / Swindon / Chicksgrove / Portland / Thame	*Pecten* / Swindon / Chicksgrove	Bivalve: *Camptonectes lamellosus* L1492 / Swindon	
7.	Fossil Wood / Woburn / Fonthill	Wood / Woburn / Fonthill / Swindon	Section of larger piece of conifer (V475 ?substitute) / Woburn	The photographed image is just a small portion of a larger piece.
9. (98/9)	OAK-TREE CLAY.	OAK-TREE CLAY.	UPPER JURASSIC, KIMMERIDGE CLAY.	For the most part Smith indicates in SS which species are figured but does not number the figures.
1.	*Melania Heddingtonensis* / North Wilts Canal	*Melania Heddingtonensis* Min. Conch. / North Wilts Canal	Gastropod: *Pseudomelania heddingtonensis* / (not found)	*Melania Heddingtonensis* is named in SS (p. 41) but no mention that it is figured.
2.	*Turbo* / North Wilts Canal	*Turbo* / North Wilts Canal	Gastropod: indet. / (not found)	Smith confirmed at a later date that this is Fig. 2.
3.	*Trochus* / North Wilts Canal	*Trochus* / North Wilts Canal	Gastropod: *Bathrotomaria reticulata* 24817 / North Wilts Canal	Smith confirmed at a later date that this is Fig. 3.
4.	*Ampullaria* / North Wilts Canal	(none listed)	Gastropod: *Ampullina* sp. / (not found)	
5.	*Chama* / North Wilts Canal / Bagley Wood Pit / Well near Swindon, Wilts & Berks Canal	*Chama* Sp. 1 / North Wilts Canal, Well near Swindon, Wilts & Berks Canal, Bagley Wood Pit / Chama Sp. 2 *Chama striata* / North Wilts Canal / Bagley Wood pit	Bivalve: *Nannogyra nana* L53452-3 / North Wilts Canal (2 b)	We believe the photographed specimens with the code 2 b to be the figured one, yet the locations listed in SS for Sp. 1 includes the Well near Swindon as in SIOF. N.B. No Fig. is detailed against *Chama* in SS (p. 45).
6.	*Ostrea delta* / North Wilts Canal / Canal at Seend / Well near Swindon, Wilts & Berks Canal / Bagley Wood Pit / Near Shivenham / Even Swindon / Wootton Bassett	*Ostrea* Sp. 1 / Ostrea deltoidea Min. Conch. / North Wilts Canal / Kennett & Avon Canal at Seend / Well near Swindon, Wilts & Berks Canal / Bagley Wood Pit / Wilts & Berks Canal near Shivenham / Even Swindon / Wootton Bassett	Bivalve: *Deltoideum delta* L1495 / North Wilts Canal (D 1 b)	Smith confirmed at a later date that this is Fig. 6.
7.	*Ammonites* / North Wilts Canal / Well near Swindon, Wilts & Berks Canal	*Ammonites* Sp. 3 / North Wilts Canal / Well near Swindon, Wilts & Berks Canal	Ammonite: *Pictonia baylei* 37847 / Wootton Bassett	This is the correct specimen but Wootton Bassett is not mentioned in either SIOF or SS for this species. Smith confirmed at a later date that this is Fig. 7.
8.	*Venus* / North Wilts Canal	*Astarte ovata* / North Wilts Canal	Bivalve: *Neocrassina ovata* L256 / (no location)	The photographed specimen was from the Bright Collection but appears to be a match to the James Sowerby drawing in SIOF. We believe that Bright may have acquired it somehow.
9.	*Terebratula* / Bagley Wood pit / Well near Swindon, Wilts & Berks Canal / North Wilts Canal.	*Terebratula* plicated / Bagley Wood pit / Well near Swindon, Wilts & Berks Canal / North Wilts Canal.	Brachiopod: *Torquirhynchia inconstans* (B1409 – substitute) / Bagley Wood Pit	The photographed specimen is not the figured one but is the same species from the same location. Smith confirmed at a later date that this is Fig. 9. He adds '[on]ly one [rem]aining [wi]th septaria', so maybe the specimen figured in SIOF was lost at an early stage.
10. (98/9)	CORAL RAG & PISOLITE (1).	CORAL RAG & PISOLITE / [CORALINE OOLITE].	UPPER JURASSIC, CORALLINE OOLITE (1).	Smith indicates in SS which species are figured but does not number the figures.
1.	*Madrepora* / Stanton near Highworth / South of Bayford / Shippon / Bagley Wood Pit / Banner's Ash / Well near Swindon, Wilts & Berks Canal / Steeple Ashton	*Madrepora* Sp. 3 / Stanton near Highworth / South of Bayford / Shippon / Bagley Wood Pit / Banner's Ash / Well near Swindon, Wilts & Berks Canal / Steeple Ashton	Coral: *Isastrea explanata* R1076 / Stanton near Highworth	Smith confirmed at a later date that this is Fig. 1.

PLATE...	STRATA IDENTIFIED...	A STRATIGRAPHICAL SYSTEM...	SPECIMEN LABEL DETAILS...	NOTES.
2.	*Madrepora* / Steeple Ashton /	*Madrepora* Sp. 2 / Steeple Ashton /	Coral: *Complexastrea depressa* R1079 / Steeple Ashton	Smith lists this species in SS (p. 47) but does not mention that it is figured.
3.	*Madrepora* / Steeple Ashton / Longleat Park / Stratton / Ensham Bridge / Wootton Bassett / Banner's Ash / Well near Swindon, Wilts & Berks Canal / Shippon / Bagley Wood Pit / Stanton	*Madrepora* Sp. 1 / a. Steeple Ashton / b. Longleat Park / c. Stratton / d. Ensham Bridge / Wootton Bassett / Banner's Ash / Well near Swindon, Wilts & Berks Canal / Shippon / Bagley Wood Pit / Stanton near Highworth	Coral: *Thecosmilia annularis* 56336 / Steeple Ashton	Smith confirmed at a later date that this is Fig. 3.
11. (100/1)	**CORAL RAG & PISOLITE (2).**	**CORAL RAG & PISOLITE / [CORALINE OOLITE].**	**UPPER JURASSIC, CORALLINE OOLITE (2).**	Smith indicates in SS which species are figured but does not number the figures.
1.	*Turbo* / Derry Hill / Longleat Park / Steeple Ashton / Banner's Ash / Wootton Bassett / Bagley Wood Pit / Stratton	*Turbo*. Oaktree Clay / b. Derry Hill / a. Longleat Park / c. Steeple Ashton / Banner's Ash / Wootton Bassett / Bagley Wood Pit	Gastropod: *Ooliticia muricata* G1594 / Derry Hill (B 2 b)	
2.	*Ampullaria* / Longleat Park / Mersham / Kennington / Silton Farm / South of Bayford / Hinton Waldrish	*Ampullaria* / a. Longleat Park / b. Marcham / Kennington / Silton Farm / South of Bayford / Hinton Waldrish	Gastropod: *Ampullospira* sp. G1599 / Longleat Park (Longleat Park D 1 a)	
3.	*Melania striata* / Wilts & Berks Canal / Calne / Steeple Ashton / Silton Farm / Banner's Ash / Well near Swindon / South of Bayford	*Melania* Sp. 1 / *Melania striata* Min.Conch. / a Calne / b. Steeple Ashton / Silton Farm / Banner's Ash / Well near Swindon, Wilts & Berks Canal / South of Bayford	Gastropod: *Bourguetia saemanni* (G1646 probable substitute) / (Caisson, Wilts & Berks Canal) (B 3 Caisson)	The drawing of Fig. 3 Gastropod: *Bourguetia saemanni* G1646 is enigmatic. Smith's plate specimen is remarkably like the one found in the collection with the 'Under Oolite' specimens. 'B3 Caisson' is written on the specimen which fits with the numbering system and place for the Under Oolite in SS and the Caisson was constructed within the Inferior Oolite. As the range of this species extends from the Bajocian to the Upper Oxfordian we feel justified in photographing it for the Coral Rag & Pisolite plate. There is no similar specimen in the collection of that age.
4.	*Ostrea crista-galli* / Derry Hill / Shotover Hill / Westbrook / Longleat Park / South of Bayford / Wootton Bassett	*Ostrea* Sp. 1 *Ostrea crista galli* / a. Wilts / Derry Hill / b. Shotover Hill / c. Westbrook / d. Longleat Park / South of Bayford / Wootton Bassett	Bivalve: *Actinostreon gregarium* (L1533 – substitute) / Wilts (K 1 a)	There are several specimens under L1533 from Wilts. The code K 1 a matches SS but none of the specimens match Smith's plate image perfectly. SIOF makes no mention of Wilts as a location. Smith confirmed at a later date that this is Fig. 4.
5.	*Cidaris* / Hilmarton / Well near Swindon, Wilts & Berks Canal	*Cidaris* Sp. 3 / Hilmarton / Well near Swindon, Wilts & Berks Canal	Echinoid: *Paracidaris smithii* (E492 – substitute) / Hilmarton (?A 3)	Substitute is larger than Smith's but it is the same species from the same location. Smith confirmed at a later date that this is Fig. 5. The code on the specimen appears to read 'A 3' but it is difficult to decipher.
6.	*Clypeus* / Meggot's Mill, Coleshill / Longleat Park / Hinton Waldrish	*Clypeus* / a. Meggot's Mill, Coleshill / b. Longleat Park / c. Hinton Waldrish	Echinoid: *Nucleolites clunicularis* E495 / Meggot's Mill, Coleshill	Smith confirmed at a later date that this is Fig. 6. Smith notes (SIOF p. 20): 'This latter specimen, as shown in my "Stratigraphical Table of Echini" is one of the characteristic distinctions of the Pisolite part of the rock.'
12. (100/1)	**CLUNCH CLAY & SHALE.**	**CLUNCH CLAY & SHALE.**	**MIDDLE–UPPER JURASSIC, OXFORD CLAY.**	Smith indicates in SS which species are figured but does not number the figures.
1.	Large *Belemnites* / Dudgrove Farm /	*Belemnites* Sp. 1 / a. Dudgrove Farm / b. North Wilts	Belemnite: *Cylindroteuthis puzosiana* C640a / Dudgrove Farm (Dudgrove Farm near Lechlade C 1 a)	
2.	*Gryphaea dilatata* lower valve / Derry Hill / Meggot's Mill, Coleshill / Tytherton Lucas / Dudgrove Farm	*Gryphaea dilatata* Min. Conch. / lower valve / b. Derry Hill / a. Meggot's Mill Coleshill / c. Between Weymouth and Osmington / d. Tytherton Lucas / Dudgrove Farm	Bivalve: *Gryphaea dilatata* (L1518 – substitute) / Derry Hill F 1 b	None of the specimens from Derry Hill match exactly. Smith confirmed at a later date that this species is illustrated as Figs 2 & 3.
3.	*Gryphaea dilatata* upper valve / ditto	*Gryphaea dilatata* Min. Conch. / upper valve / ditto	Bivalve: *Gryphaea dilatata* (L1518 – substitute) / ditto	
4.	*Ammonites* / Thames & Severn Canal / Tytherton Lucas	*Ammonites* Sp. 3 / a. Thames & Severn Canal / b. Tytherton Lucas	Ammonite: *Kosmoceras spinosum* C636 / Thames & Severn Canal	Specimen label has 'Siddington, near Cirencester, Gloucestershire' added after 'Thames & Seven Canal'. Fig. 4 on Smith's plate is captioned *Ammonites arnatus*?, and the text description is *Ammonites*.
5.	*Serpula* / Wilts & Berks Canal, near Chippenham	*Serpula*, Sp. 2. Oaktree Clay / Steeple Ashton	Serpulid worm: *Genicularia vertebralis* / (not found)	No Fig. is detailed against Serpula in SS (p. 55). It is detailed as Serpula, Sp. 2 Oaktree Clay yet there is no Sp. 1. Locations are different in SIOF and SS.
6.	*Serpula* / ditto	ditto	Serpulid worm: *Genicularia vertebralis* / (not found)	Location is different in SS (Steeple Ashton).
13. (150/1, 9)	**KELLOWAYS STONE.**	**KELLOWAYS STONE.**	**MIDDLE JURASSIC, KELLAWAYS ROCK.**	Smith indicates in SS which species are figured but did not number the figures. Fig. numbers were added at a later date to all but Fig. 4.
1.	*Rostellaria* / Kelloways / Wilts & Berks Canal, near Chippenham	*Rostellaria* / Kelloways / Wilts & Berks Canal, near Chippenham	Gastropod: *Dicroloma* sp. / (not found)	
2.	*Ammonites sublaevis* / Kelloways / Ladydown Farm / Christian Malford	*Ammonites* Sp. 1 / *Ammonites sublaevis* Min. Conch. / a. Kelloways / b. Ladydown Farm / Christian Malford /	Ammonite: *Cadoceras sublaevis* C748 / Kellaways	
3.	*Ammonites Calloviensis* / Kelloways / Wilts & Berks Canal, near Chippenham	*Ammonites* Sp. 4 / *Ammonites Calloviensis* Min Conch. / a. Kelloways / b. Wilts & Berks Canal	Ammonite: *Sigaloceras calloviense* (C642b – substitute) / Kellaways	Smith's figured specimen is on tour in Singapore at the time of publication. This specimen is the same species from the same location (C642a).
4.	*Ammonites* / Kelloways / Dauntsey House / Wilts & Berks Canal, near Chippenham / Kennet & Avon Canal, near Trowbridge	*Ammonites* Sp. 2 / a. Kelloways / b. Dauntsey House / c. Wilts & Berks Canal / d. Kennet & Avon Canal	Ammonite: *Proplanulites koenigi* C643 / Kellaways	
5.	*Gryphaea incurva* / Ladydown / Kelloways / Wilts & Berks Canal, near Chippenham / Bruham Pit	*Gryphaea dilatata* Min.Conch. / c. Ladydown / a. Kelloways / b. Wilts & Berks Canal / d. Bruham Pit. Experiment for Coal	Bivalve: *Gryphaea dilobotes* L1778 / Ladydown (I 1 c)	This species was found by Smith at the Bruham Coal Trial and helped him prove that the miners were far too high in the succession ever to encounter coal.
6.	*Terebratula ornithocephala* / Wilts & Berks Canal near Chippenham / Thames & Severn Canal / Dauntsey House / Kelloways	*Terebratula ornithocephala* Min. Conch. / d. Wilts & Berks Canal / a. Thames & Severn Canal / b. Dauntsey House / c. Kelloways	Brachiopod: *Ornithella ornithocephala* B1417 / Wilts and Berks Canal, near Chippenham	
14. (150/1)	**CORNBRASH.**	**CORNBRASH.**	**MIDDLE JURASSIC, CORNBRASH.**	
1.	*Natica*? / Road / Sleaford / Wick Farm	*Natica*? / a. Road / b. Sleaford / Wick Farm	Gastropod: cf *Ampullospira* sp. G1605 / Road (Cornbrash E 1 a)	
2.	*Ammonites discus* / South-west of Wincanton / Closworth / Road / Chillington	*Ammonites* / *Ammonites discus* Min. Conch. / South-west of Wincanton / a. Closworth / b. Road	Ammonite: *Clydoniceras discus* C649 / South-west of Wincanton	
3.	*Modiola* / Wick Farm / Closworth / Holt	*Modiola* / b. Wick Farm / a. Closworth / c. Holt	Bivalve: *Modiolus imbricatus* L1536 / Wick Farm (F 1 b)	
4.	*Trigonia costata* / North side of Wincanton / Wick Farm	*Trigonia* Sp. 2 / *Trigonia costata* Min. Conch. Park. / North side of Wincanton / Wick Farm	Bivalve: *Trigonia crucis* L1531 / North side of Wincanton (Sp 2)	
5.	*Venus* Linn. / Trowle / Sheldon / South-west of Wincanton / Norton	*Venus*? Sp. 1 / a. Trowle / b. Sheldon / South-west of Wincanton	Bivalve: *Protocardia buckmani* L1556 / Trowle (Trowle L 1 a)	
6.	*Cardium* / Road / Elmcross / Wick Farm / Sleaford / Woodford / near Peterborough / near Stilton	*Cardium* / a. Road / b. Elmcross / c. Wick Farm / d. Sleaford / e. Woodford / f. near Peterborough /	Bivalve: *Pholadomya deltoidea* L1552 / Road (K 1)	
7.	*Unio*? / Road / North Cheriton / Draycote / Maisey Hampton / Sleaford / South-west of Tellisford / Sattiford / South-west of Wincanton	*Unio* Sp. 3 / b. Road / a. North Cheriton / c. Draycot / d. Maisey Hampton / e. Sleaford / f. South-west of Tellisford / g. Sattyford	Bivalve: *Pleuromya uniformis* L1543 / Road (M 3 b)	
8.	*Avicula echinata* / Draycot / Closworth / North Cheriton / Lullington / Trowle / Sheldon / Norton / Stony Stratford / South-west of Tellisford	*Avicula* Sp.1 *Avicula echinata* / f. Draycot / a. Closworth / b. North Cheriton / c. Lullington / d. Trowle / e. Sheldon / g. Norton / h. Stony Stratford / i. South-west of Tellisford / North side of Wincanton / Southwest of Wincanton	Bivalve: *Meleagrinella echinata* L1579 / Draycot	
9.	*Terebratula digona* (var. *globosa rotunda*) / Latton / Closworth / Redlynch / Trowle / Wick Farm / Sheldon / Woodford	*Terebratula* not plicated / *Terebratula digona* Min. Conch. / f. Latton / a. Closworth / b. Redlynch / c. Trowle / d. Wick Farm / e. Sheldon / g. Woodford	Brachiopod: *Digonella siddingtonensis* B1423 / Latton	
15. (152/3)	**FOREST MARBLE.**	**FOREST MARBLE.**	**MIDDLE JURASSIC, FOREST MARBLE.**	
1.	*Patella rugosa* / Minching Hampton Common / Hinton	*Patella rugosa* Min. Conch. / a. Minching Hampton Common / b. Hinton	Gastropod: *Symmetrocapulus tessoni* / (not found)	
2.	*Ancilla* / Farley Castle	*Ancilla* Sp. 1 / Farley	Gastropod: *Cylindrites archiaci* G1608 / Farley	
3.	*Rostellaria*? / Poulton / Farley Castle	*Rostellaria* / Poulton	Gastropod: *Rostellaria* in block L1596 / (no location, possibly Poulton)	
4.	*Ostrea* / Wincanton / Road Coal experiment	*Ostrea* Sp. 3 / a. Wincanton / b. Road, Coal Experiment	Bivalve: *Catinula* sp. L1589 / Wincanton	
5.	*Pecten* / Siddington / Foss Cross	*Pecten* Sp. 1 / a. Siddington / b. Foss Cross	Bivalve: *Plagiostoma subcardiiformis* L1591 / Siddington (I 1 a)	
6.	*Pecten* / Farley Castle	*Pecten* Sp. 4 / Farley	Bivalve: *Camptonectes auritus* L1594 / Farley (I 4)	

PLATE...	STRATA IDENTIFIED...	A STRATIGRAPHICAL SYSTEM...	SPECIMEN LABEL DETAILS...	NOTES.
7.	Oval *Bufonite* / Stunsfield / Pickwick	*Bufonite* Sp. 1 oval / a. Stunsfield / b. Pickwick	Fish tooth: ?*Eomesodon* sp. (splenial bone) P4826 / Stonesfield	
8.	Round Bufonitae / Stunsfield / Pickwick / Didmarton	*Bufonite* Sp. 2 circular / a. Stunsfield / b. Pickwick / c Didmarton	Fish tooth: ?*Lepidotes* sp. / (not found)	
9.	Fish Palate / Pickwick	Palates of Fish Sp. 1 / Pickwick	Shark palate: ?*Asteracanthus magnus* P4820 / Pickwick	
10.	Cap-formed Palate / Pickwick	Palates of Fish Sp. 3 / Pickwick	Shark palate: *Asteracanthus tenuis* P4823 / Pickwick	
11.	Shark's Teeth / Stunsfield / Pickwick / Farley Castle	Teeth, oblong, pointed Sp. 3 / a. Stunsfield / b. Pickwick	Fish teeth: probably 1 × shark, 1 × bony fish / (not found)	
16. (152/3)	**CLAY OVER THE UPPER OOLITE.**	**CLAY OVER THE UPPER OOLITE.**	**MIDDLE JURASSIC, BRADFORD CLAY.**	
1.	*Pear Encrinite* / Bradford / Berfield	*Pear Encrinus* / Bradford Heads & vertebrae / Farley vertebrae / Hinton vertebrae / Winsley vertebrae / Pickwick vertebrae	Crinoid: *Apiocrinites elegans* E559 / Bradford	Fig. 1 captioned *Pear Encrinite* in text and as *Pear Encrinus* on the plate and in SS.
2.	The Vertebrae ditto / Farley Castle / Hinton	ditto		On the plate of SIOF Smith calls it The Clavical.
3.	Stems ditto / Winsley / Pickwick	ditto	E559	On the plate of SIOF Smith calls it The Roots and Stems.
4.	*Tubipora* / Broadfield Farm / Farley Castle	*Tubipora* / Broadfield Farm / Farley	Serpulid worm: *Filograna* sp. / (not found)	
5.	*Millepora* / Hinton / Broadfield Farm / Farley Castle / Pickwick / Westwood / near Bradford	Millepora / Hinton / Broadfield Farm / Farley / Pickwick / Westwood	Bryozoan: *Terebellaria ramosissima* D34537 / Hinton	
6.	*Chama crassa* / Stoford	*Chama* Sp. 1 *Chama crassa* / Stoford	Bivalve: *Praeoxygyra crassa* / (not found)	
7.	*Plagiostoma* / Bradford	*Plagiostoma* Sp. 1 / Bradford	Bivalve: *Plagiostoma cardiiformis* (L1605 – substitute) / (Combhay) (Combhay I 3)	The photographed specimen is not the same specimen, nor is it from the same location as the figured one in SIOF but it is the same species, albeit somewhat smaller. The location tallies with *Plagiostoma* Sp.2 Variety in SS (p. 80).
8.	*Avicula costata* / Bradford / Hinton / Winsley	*Avicula costata* / a. Bradford / b. Hinton / c. Winsley	Bivalve: *Oxytoma costata* a) 43258 b) L1611 / a) Bradford / b) Hinton (N 1 b)	
9.	*Terebratula digona*: the long variety / Farley Castle / Stoford / Bradford / Winsley / Pickwick	*Terebratula* not plicated Sp. 1 / *Terebratula digona* Min. Conch. / a. Farley / b. Stoford / c. Bradford / d. Winsley / e Pickwick	Brachiopod: *Digonella digona* B1424 / Farley (O 1 a)	
10.	*Terebratula reticulata* / Farley Castle / Bradford / Stoford / Hinton / Winsley / Pickwick	*Terebratula* plicated Sp. 5 / *Terebratula reticulata* / a. Farley / Bradford / Stoford / b. Hinton / c. Winsley / d. Pickwick	Brachiopod: *Dictyothyris coarctata* B1430 / Farley (O 3 a)	Smith's code 'O 3 a' is written on the specimen which seems to match the figured specimen well. We believe Smith made a mistake and it should be 'O 5 a'.
17. (154/5)	**UPPER OOLITE.**	**UPPER OOLITE.**	**MIDDLE JURASSIC, GREAT OOLITE.**	Smith does not indicate in SS which species are figured so photographed specimens have been matched to identifications (where shown), descriptions and locations.
1.	*Tubipora* / Broadfield Farm / Combe Down / Westwood	*Madrepora* Sp.7? / This may perhaps be a *Tubipora* / a. Broadfield Farm / b. Combe Down	Coral: *Lochmaeosmilia radiata* / (not found)	
2.	*Tubipora* / Combe Down	? Ditto	Coral: *Lochmaeosmilia radiata* / (not found)	Only Combe Down is mentioned in SIOF.
3.	*Madrepora turbinata* / Farley Castle / Broadfield Farm	*Madrepora* Sp. 1 / *Madrepora turbinata* Linn / Farley	Coral: *Montlivaltia* sp. / (not found)	
4.	*Madrepora porpites* / Broadfield Farm	(Button coral not listed)	Coral: *Chomatoseris porpites* / (not found)	This specimen does not appear to be listed in SS so may not have come to BM.
5.	*Madrepora flexuosa* / Castle Combe	*Madrepora* Sp. 6 / *Madrepora flexuosa* Linn / Castle Combe	Coral: *Cladophyllia babeana* B55 / Castle Coombe	
18. (154/5)	**FULLER'S EARTH ROCK.**	**FULLER'S EARTH ROCK.**	**MIDDLE JURASSIC, FULLER'S EARTH.**	
1.	*Nautilus* / Lansdown	*Nautilus* / Lansdown	Nautiloid: indet. / (not found)	
2.	*Ammonites modiolaris* / Dundry / Rowley Bottom	*Ammonites* Sp.1 *Ammonites modiolaris* / Dundry / Rowley Bottom	Ammonite: *Morrisiceras morrisi* / (not found)	
3.	*Modiola anatina* / Avoncliff	*Modiola* Sp.4 *Modiola anatina* / Ancliff	Bivalve: *Modiolus anatinus* 66930 / Avoncliff	
4.	*Cardita* / Hardington / Grip Wood	*Cardita* Sp.1 / b. Hardington / a. Grip Wood	Bivalve: *Ceratomya* aff. *striata* L53451 / Hardington (Hardington)	
5.	*Cardium* / Charlton Horethorn / near Gagenwell / near Redlynch	*Cardium*. Cornbrash / a. Charlton Horethorn / b. near Gagenwell / c. near Redlynch	Bivalve: *Pholadomya* aff. *lirata* L1685 / Charlton Horethorn (K 1)	
6.	*Tellina* / Avoncliff / Hardington	*Tellina* / a. Ancliff / b. Hardington	Bivalve: *Cercomya* aff. *pinguis* L1689 / Avoncliff (G 3 a)	There is only one species of *Tellina* (labelled Fig. 6) so it is possible that the code should read G 1 a (not G 3 a).
7.	*Ostrea acuminata* / Orchardleigh / Avoncliff / Below Combe Down / Caisson / North of Stamford	(not listed in SS)	Bivalve: *Praeexogyra acuminata* / (not found)	As Fig. 7 is not listed in SS the specimen may not have come to BM. However, at a later date Smith notes beside Fig. 9 [Fig. 7 numerous].
8.	*Ostrea Marshii* / Cotswold Hills / Monkton Combe	*Ostrea* Sp.1 *Ostrea Marshii* Min. Conch. / b. Cotswold Hills / a. Monkton Combe	Bivalve: *Actinostreon marshii* LL40856 / Cotswold Hills (P 1 b)	
9.	*Terebratula media* / near Bath / Charlton Horethorn / Orchardleigh	*Terebratula* plicated Sp. 3 / *Terebratula media* Min. Conch. / a. near Bath / b. Charlton Horethorn / c. Orchardleigh	Brachiopod: *Rhynchonelloidella media* (B1488) / no location	This species is also known as *Rhynchonelloidella smithi*.
19. (8)	**(NOT INCLUDED).**	**UNDER OOLITE FIRST PLATE.**	**MIDDLE JURASSIC, INFERIOR OOLITE (1).**	Proposed plate species listed in SS p. 110.
1.		*Melania* Sp. 2 / b. Tucking Mill / a. Coal Canal / c. near Bath	Gastropod: *Pseudomelania* sp. G1644 / Tucking Mill (B 2 b)	
2.		*Trochus* Sp. 1 / near Bath	Gastropod: *Pyrgotrochus conoideus* G1647 / near Bath (C 1)	
3.		*Trochus* Sp. 6 / a. Coal Canal / b. Tucking Mill / c. Between Cross Hands and Petty France	Gastropod: *Pyrgotrochus* sp. G1653 / Coal Canal (C 6 a)	
4.		*Trochus* Sp. 9 / Sherborn / Bath	Gastropod: *Pleurotomaria granulata* G1655 / Sherborne	
5.		*Turritella* Sp. 1 / a. Smallcombe Bottom / b. Coal Canal / c. Tucking Mill	Gastropod: *Nerinea* sp. G1661 / Smallcombe Bottom (E 1 a)	
6.		*Turritella* Sp. 3 / a. Churchill / b. near Bath	Gastropod: ?*Nerinea* sp. G1663 / Churchill (E 3 a)	
7.		*Ampullaria* Sp. 2 / b. Bath / a. Coal Canal	Gastropod: *Ampullina* sp. G1665 / Bath (F 2 b)	
8.		*Nautilus* Sp. 3 / a. Sherborn / b. Between Sherborn & Yeovil / c. Carlton Horethorn	Nautilus: *Cenoceras excavatus* C666 / Sherborne (Sherborne H 3 a)	
9.		*Ammonites* Sp. 2 *Ammonites calix* / Sherborn	Ammonite: *Teloceras calix* C671 / Sherborn (K 2)	
20. (8)	**(NOT INCLUDED).**	**UNDER OOLITE SECOND PLATE.**	**MIDDLE JURASSIC, INFERIOR OOLITE (2).**	Proposed plate species listed in SS p. 110.
1.		*Madrepora* Sp. 4 / Dundry / Tucking Mill / Crickley Hill	Coral: indet. / (not found)	
2.		*Trigonia* Sp. 1 *Trigonia costata* / a. Cotswold Hills / b. Between Cross Hands & Petty France / c. Cross Hands / d. Mitford / e. Coal Canal / f. Tucking Mill / Little Sodbury	Bivalve: *Trigonia costata* (internal mould) L1694 / Cotswold Hills (N 1 a)	
3.		*Trigonia* Sp. 1 *Trigonia costata* / f. Tucking Mill / a. Cotswold Hills / b. Between Cross Hands & Petty France / c. Cross Hands / d. Mitford / e. Coal Canal / Little Sodbury	Bivalve: *Trigonia costata* L1699 / Tucking Mill (N 1 f)	

PLATE...	STRATA IDENTIFIED...	A STRATIGRAPHICAL SYSTEM...	SPECIMEN LABEL DETAILS...	NOTES.
4.		*Astarte ovata* Oaktree Clay / b. Coal Canal / a. Between Sherborne & Yeovil / c. Tucking Mill / d. Bath / Fullbrook (Q 1 e Fullbrook Hill) / f. Between Cross Hands & Petty France / Mitford Inn / Northwest of Northampton	Bivalve: *Astarte elegans* (internal mould) L1719 / Coal Canal (Q 1 b)	Smith writes 'Oak-tree Clay' after his identification of *Astate ovata* and indeed it is figured on that plate (Fig. 8 – it is the type specimen). This is the same genus but we believe it to be *Astarte elegans*.
5.		*Pecten* Sp. 1 / *Pecten equivalvis* Min. Conch. / Ilmington. Dursley. Dowdswell Hill. Sherborn. / Pecten Sp. 2 / Churchill	Bivalve: *Variamussium* cf. *pumilum* (L1730 – substitute) / Churchill (W 2)	No specimens for Sp. 1 have been found and so Sp. 2 has been substituted.
6.		*Inoceramus* fibrous shell / Bath / Between Cross Hands & Petty France / Monkton Coombe / Tucking Mil	Bivalve: Fragment of large *Trichites ploti* L1733 / Bath	
7.		*Terebratula* plicated Sp. 4 / *Terebratula spinosa* / a. Bath, b. Tucking Mill, Chipping Norton / Sp. 3 *Terebratula obsoleta* / c. Churchill / a. Between Cross Hands & Petty France / b. Tucking Mill / d. Chipping Norton / e. Fullbrook	Brachiopod *Acanthothyris spinosa* (B1501 – substitute) / Churchill (X 3 c)	We believe the photographed specimen to be the same species as the one Smith lists under Sp. 4 but slightly less well preserved. The X 3 c code is marked on the specimen chosen but it is not the *Obsoletirhynchia obsoleta* listed under Sp. 3 in SS.
8.		Clypeus Sp. 1 *Clypeus sinuatus* / f. near Naunton / a. Monkton Coombe / b. Stunsfield / c. Chipping Norton / d. Churchill / e. Fullbrook / g. Stow on the Wold / North-west of Northampton.	Echinoid: *Clypeus ploti* E538 / near Naunton (B 1 f)	
21. (192/3)	**(NOT INCLUDED).**	**SAND & SANDSTONE.**	**LOWER JURASSIC, BRIDPORT SAND.**	Figures detailed in text SS pp. 111–12. Smith's Sand & Sandstone mostly refers to a group of Upper Liassic sandstones now grouped as the Bridport Sand.
1.		*Belemnites* / Tucking Mill	Belemnite: *Belemnopsis* sp. C699a / Tucking Mill (1 b)	
2.		*Ammonites* Sp. 4 / Yeovil	Ammonite: *Pleydellia burtonensis* C684 / Yeovil (B 4)	
3.		*Modiola* / Top of Frocaster Hill	Bivalve: *Inoperna sowerbyana* L1734 / Top of Frocester Hill (C 1 Froster Hill)	
22. (192/3)		**MARLSTONE.**	**LOWER JURASSIC, MARLSTONE.**	Figures are detailed against species in SS pp. 113–18 Smith's Marlstone refers to a number of Middle-Upper silty and calcareous sediments, in part marly members of the Bridport Sand. At Glastonbury and Pennard Hill they constitute the Dyrham Formation. At Churchill the equivalent is the Whitby Mudstone Formation.
1.		*Pentacrinus* Sp. 1 / Churchill / Stone Farm, Yeovil / Kennet & Avon Canal	Crinoid: indet. / (not found)	
2.		*Belemnites* / a. Yeovil / b. Churchill / c. Tucking Mill / d. Enstone	Belemnite: indet. / (not found)	
3.		*Ammonites* Sp. 3 / *Ammonites undulatus* / Coal Canal	Ammonite: *Tragophylloceras undulatum* C33499 / Coal Canal C 3	
4.		*Ammonites* Sp. 4 / a. Coal Canal / b. Tucking Mill / c. Yeovil / d. Churchill	Ammonite: *Zugodactylites* sp. C710 / Coal Canal C 4 a	
5.		*Ammonites* Sp. 10 / c. Penard Hill / a. Coal Canal / b. Tucking Mill / Dundry / Forcester Hill / Bathhampton, foot of inclined plane	Ammonite: *Witchellia* sp. C700 / Penard Hill C 10 c	
6.		*Ammonites* Sp. 13 / *Ammonites Walcotti* Min. Conch. / Glastonbury / Coal Canal	Ammonite: *Grammoceras striatulum* C703 / Glastonbury (Glastonbury C 13 b)	
7.		*Pecten* / Kennet & Avon Canal / Northeast of Newark	Bivalve: *Pseudopecten equivalvis* L1774 / Kennet & Avon Canal	
23. (218/9)	**(NOT INCLUDED).**	**BLUE MARL.**	**LOWER JURASSIC, BLUE MARL.**	SS Part 2. The Orders listed are prefaced with A, B, C etc. Specimens selected to match sketch WSF12-01 courtesy of OUMNH. Smith shows his Blue Marl as a formation above the Blue Lias and indeed there are Blue Marls recognized in the Middle Lias. However, the specimens he lists from his Blue Marl at Marston Magna and Mudford are from the older Blue Lias.
1.		B *Ammonites* Sp. 1 / *Ammonites planicosta* Min. Conch. / Marston Magna	Block with ammonites: *Promicroceras planicosta* C736 / Marston Magna (Marston Magna)	This is the famous Marston Magna stone.
2.		B *Ammonites* Sp. [6] / Marston Magna	Ammonite: *Asteroceras smithi* C737 / (no locality)	Sp. 6 is not specified in SS, only 'a keel between two furrows' following Sp. 5. The fossil was named for Smith by James Sowerby.
3.		D *Tellina* Sp. 1 / a. Mudford / b. South of Bedmister Down	Bivalve: *Cardinia listeri* L1765 / Mudford (D 1 a)	
24. (218/9)	**(NOT INCLUDED).**	**LIAS 1.**	**LOWER JURASSIC, LIAS (1).**	SS Part 2. The genera listed are prefaced with A, B, C etc. Specimens selected to match sketch WSF12-02 courtesy of OUMNH.
1.		N *Plagiostoma* Sp. 1 / *Plagiostoma gigantea* Min. Conch. / b. Bath / a. Topcliff / c. Coal Canal / d. Stony Littleton	Bivalve: *Plagiostoma giganteum* L1780 / near Bath (N 1 b)	
2.		C *Ostrea* Sp. 1 Common grypheus / a. Gloucester & Berkley Canal / b. Pyrton Passage / c. Bath / d. Coal Canal / e. Weston / f. Stony Littleton	Bivalve: *Gryphaea arcuata* L1737 / Gloucester & Berkley Canal (O 1 a)	SS labels *Ostrea* 'C' but it should logically be 'O'.
3.		Q *Terebratula* Sp. 10 Spirifer / a. Keynsham / b. Bath	Brachiopod: *Spiriferina walcotti* B1504 / Keynsham (Q 10 a)	
4.		Bones Sp. 2 vertebrae / a. Charmouth / b. Keynsham	Fish: vertebra indet. / (not found)	
5.		(not listed)	Shark fin spine: *Acrodus curtus* P4841 / near Lyme	No fin spines are listed, nor any locations 'near Lyme' (pp. 19–20).
6.		all the vertebrae are listed under Sp. 2 / (see Fig. 4 above)	Fish: vertebra indet. / (not found)	
25. (9, 220/1)	**(NOT INCLUDED).**	**LIAS 2.**	**LOWER JURASSIC, LIAS (2).**	SS Part 2. The genera listed are prefaced with A, B, C etc. Specimens selected to match sketch WSF12-03 courtesy of OUMNH.
1.		*Pentacrinus* Sp. 1 / b. Gloucester & Berkley Canal / a. Purton Passage / c. Keynsham	Crinoid, disarticulated: *Isocrinus psilonoti* E561 / Gloucester & Berkley Canal	
2.		*Pentacrinus* Sp. 3 / Charmouth	Crinoid, articulated: *Pentacrinites fossilis* E542 / near Lyme Regis	Charmouth is near Lyme Regis.
3.		A *Trochus* Sp. 1 / Trochus serrites (?) Min. Conch. / a. Purton Passage / b. Bath / c. Charmouth	Gastropod: *Pleurotomaria* cf. *cognata* G1670 / Purton Passage (A 1 a)	
4.		E *Ammonites* Sp. 16 / a. Stony Littleton / b. Charmouth	Ammonite: *Euagassiceras sauzeanum* C83396 / Stony Littleton (E16a)	
5.		E *Ammonites* Sp. 9 / *Ammonites communis* / Bath	Ammonite: *Zugodactylites braunianus* C723 / Bath (E 9)	
6.		G ?*Unio* Sp. 2 / Weston	Bivalve: *Pleuromya* aff. *uniformis* L1769 / (no location)	The image was inadvertently omitted from the plate in *William Smith's Fossils Reunited*.
26. (220/1)	**(NOT INCLUDED).**	**REDLAND LIMESTONE / [MOUNTAIN LIMESTONE].**	**CARBONIFEROUS LIMESTONE.**	SS Part 2. Specimens selected to match sketch WSF12-04 courtesy of OUMNH. Smith's failure to publish this plate saved him from emphasizing one of his most serious mistakes, because all the specimens shown on Smith's sketch are from the Carboniferous Limestone (his Mountain/Derbyshire Limestone in the Bristol-Mendip area and not from his Redland Limestone ('Magnesium Limestone') Cox, 1942, p. 55) Smith showed this mistaken correlation of the Mendip Limestones with his Redland Limestone in Yorkshire on his 1815 map. However, by the time his Gloucestershire map was published in 1819 he shows the limestones around Bristol to be older Mountain Limestone suggesting that he had, at least in part, rectified his mistake.
1.		*Madrepora* Sp. 1 *Madrepora turbinata* / a Mendip (?=Leigh) / b. Hotwells / c. Worcestershire / d. Malvern	Coral: *Amplexus coralloides* (R1069 – substitute) / (Leigh) (Leigh C 1)	
2.		*Madrepora* Sp. 6 / a. Hotwells / b. Farley	Coral: *Actinocyathus floriformis* (R1067 – substitute) / (Avon Section, Bristol)	Hotwells is close to the Avon in Bristol.
3.		*Encrinus* Sp. 2 / Mells	Crinoid stems: indet. E548 / Mells (Mells)	

SELECTED WORKS OF WILLIAM SMITH.

- Smith, W., *Prospectus of a Work, entitled Accurate delineations and descriptions of the natural order of the various strata that are found in different parts of England and Wales: with practical observations thereon* (London: printed by B. McMillan, 1801)
- Smith, W., *Observations on the Utility, Form and Management of Water Meadows and the Draining and Irrigating of Peat Bogs: with an account of Prisley Bog and other extraordinary improvements conducted for His Grace the Duke of Bedford, Thomas William Coke, Esq. MP and others* (Norwich: printed by R.M. Bacon, 1806)
- Smith, W., *Description of Norfolk, its soil and substrata* (manuscript, 1807)
- Smith, W., *A Memoir to the Map and Delineation of the Strata of England and Wales, with part of Scotland* (London: J. Cary, 1815)
- Smith, W., 'Mr Smith's observations on his map', *Transactions of the Society, Instituted at London, for the Encouragement of Arts, Manufactures, and Commerce*, 33 (1815), 53–58
- Smith, W., 'Explanation of Mr William Smith's map of strata', *Transactions of the Society, Instituted at London, for the Encouragement of Arts, Manufactures, and Commerce*, 33 (1815), 58–60
- Smith, W., *Strata Identified by Organized Fossils: containing prints on coloured paper of the most characteristic specimens in each stratum* (London: printed by W. Arding, 1816–19)
- Smith, W., *Stratigraphical System of Organized Fossils: with reference to the specimens of the original geological collection in the British Museum: explaining their state of preservation and their use in identifying the British strata* (London: E. Williams, 1817)

SELECTED WORKS OF JOHN PHILLIPS.

- Phillips, J., *Illustrations of the Geology of Yorkshire* (Part I, York: Thomas Wilson & Sons, 1829; Part II, London: John Murray, 1836)
- Phillips, J., *Memoirs of William Smith, LL.D* (London: John Murray, 1844)

INTRODUCTION.

- Conybeare, W. D. and Phillips W., *Outlines of the Geology of England and Wales, with an introductory compendium of the general principles of that science, and comparative views of the structure of foreign countries* (London: William Phillips, 1822)
- Farey, J., 'Observations on the priority of Mr. Smith's investigations of the strata of England', *The Philosophical Magazine and Journal*, 45 (1815), 333–44
- Hudson, J. (ed.), *A Complete Guide to the Lakes, comprising minute directions for the tourist; with Mr. Wordsworth's description of the scenery of the country, etc.: and five letters on the geology of the Lake District, by the Rev. Professor Sedgwick* (Kendal: Hudson and Nicholson, 1842)
- Hutton, J., 'Theory of the Earth; or an investigation of the laws observable in the composition, dissolution, and restoration of land upon the globe', *Transactions of the Royal Society of Edinburgh*, 1, 2 (1788), 209–304
- Jameson, R., *System of Mineralogy* (Edinburgh: Constable & Co., 1804–08)
- Phillips, W., *An Outline of Mineralogy and Geology, intended for the use of those who may desire to become acquainted with the elements of those sciences; especially of young persons* (London: William Phillips, 1815)
- Rudwick, M. J. S., *Georges Cuvier, Fossil Bones and Geological Catastrophes: new translations and interpretations of the primary texts* (Chicago: University of Chicago Press, 1997)
- Sedgwick, A., 'Announcement of the first award of the Wollaston Prize', *Proceedings of the Geological Society of London*, 1, 20 (1831), 270–79
- Smiles, S., *Self-help; with illustrations of character and conduct* (London: John Murray, 1859)
- Strachey, J., *Observations on the Different Strata of Earths and Minerals* (London: Walthoe, 1727)
- Williamson, W. C., 'Reminiscences of a Yorkshire Naturalist', *Good Words*, 18 (1877), 62–66

APPRENTICE.

- Billingsley, J. 'A description of Robert Weldon's Hydrostatik or Caison lock', in *A General View of the Agriculture of Somerset* (London: W. Smith, 1794), 317–18
- Cox, L. R., 'New light on William Smith and his work', *Proceedings of the Yorkshire Geological Society*, 25 (1942), 1–99
- Henry, C. J., 'William Smith: The maps supporting his published maps', *Earth Science History*, 35, 1 (2016), 62–98
- Peach, R. E. M., *Historic Houses in Bath and their Associations* (London: Simpkin and Marshall, 1884)
- Torrens, H. S. (ed.), *Memoirs of William Smith, LL.D., by John Phillips* (Bath: The Bath Royal Literary and Scientific Society, 2003)
- Torrens, H. S., 'The Somersetshire Coal Canal Caisson Lock', *Bristol Industrial Archaeological Journal*, 8 (1975), 4–10
- Torrens, H. S., 'Timeless order: William Smith and the search for Raw Materials, 1800–1820', in *The Age of the Earth: From 4004 BC to AD 2002*, ed. C. L. E. Lewis and S. J. Knell (London: The Geological Society, 2001), 61–83
- Warner, R., *The History of Bath* (Bath: R. Cruttwell, 1801)
- Warner, R., *New Guide Around Bath* (Bath: R. Cruttwell, 1811)

MINERAL PROSPECTOR.

- Eyles, J. M., 'William Smith, Richard Trevithick and Samuel Homfray: Their correspondence on steam engines, 1804–1806', *Transactions of the Newcomen Society*, 43, 1 (1970)
- Farey, J., *General view of the agriculture and minerals of Derbyshire* (London: G. & W. Nicol, 1811)
- Fuller, J. G. C. M., 'William Smith's explanation of the colliers' dial, 1798', *Archives of Natural History*, 19, 1 (1992), 107–11
- Fuller, J. G. C. M., 'The forty-four yard problem: a cross-section by John Strachey annotated by William Smith', *Archives of Natural History*, 21, 2 (1994), 195–99
- Thomas, R., *Report on a survey of the mining district from Cornwall to Chacewater to Camborne* (London: John Cary, 1819)

FIELD WORK.

- Cox, L. R., (1942) *see Apprentice*
- Lewis, C., 'Geologists John Farey and William Smith awarded silver medals for agriculture', *Earth Sciences History*, 37, 2 (2018), 293–308
- Tull, Jethro, *Horse-hoeing Husbandry* (London: A. Millar, 1762)
- Walton, G., 'A note on William Smith's drainage works near Churchill', *Earth Sciences History*, 35, 1 (2016), 218–27

CARTOGRAPHER.

- Boulger, G. S., 'Dr William Smith's geological maps', *The Geological Magazine*, 4, 8 (1877), 378
- Cox, L. R., (1942) *see Apprentice*
- d'Aubuisson de Voisins, J. F., *Traité de Géognosie, ou exposé des connaissances actuelles sur la constitution physique et minérale du globe terrestre* (Strasbourg; Paris: F.G. Levrault, 1819)
- Davis, A. G., 'William Smith's geological atlas and the later history of the plates', *Journal of the Society for the Bibliography of Natural History*, 2, 9 (1952), 388–95
- Eyles, J. M., 'William Smith (1769–1839): A bibliography of his published writings, maps and geological sections, printed and lithographed', *Journal of the Society for the Bibliography of Natural History*, 5, 2 (1969), 87–109
- Eyles, J. M., 'William Smith, Sir Joseph Banks and the French geologists', in *From Linnaeus to Darwin, commentaries on the history of biology and geology*, eds. A. Wheeler and J. H. Price (London: Society for the History of Natural History, special publication 3, 1985)
- Eyles, V. A. and Eyles, J. M., 'On the different issues of the first geological map of England and Wales', *Annals of Science*, 3, 2 (1938), 190–212
- Greenough, G. B., *Memoir of a Geological Map of England: To which are added, an alphabetical index to the hills, and a list of the hills arranged according to counties* (London: Longman, Hurst, Rees, Orme, and Brown, 1820)
- Herries Davies, G. L., *Whatever is Under the Earth. The Geological Society of London 1807 to 2007* (London: The Geological Society, 2007)
- Rudwick, M. J. S., 'Cuvier and Brongniart, William Smith and the reconstruction of geohistory', *Earth Sciences History*, 15, 1 (1996), 25–36
- Rudwick, M. J. S., *Bursting the Limits of Time. The reconstruction of geohistory in the age of revolution* (Chicago; London: University of Chicago Press, 2005)
- Sedgwick, A., (1831) *see Introduction*
- Sharpe, T., 'The birth of the geological map', *Science*, 347, 6219 (2015), 230–32
- Sharpe, T., 'William Smith's 1815 map, A Delineation of the Strata of England and Wales: its production, distribution, variants and survival', *Earth Sciences History*, 35, 1 (2016), 47–61
- Sharpe, T. and Torrens, H. S., 'Introduction' in *A Memoir to the Map and Delineation of the Strata of England and Wales, with part of Scotland by William Smith*, ed. C. Lewis (London: History of Geology Group of the Geological Society, 2015), 1–26
- Sheppard, T., 'William Smith: his maps and memoirs', *Proceedings of the Yorkshire Geological Society*, 19 (1917), 75–253
- Smiles, S., (1859) *see Introduction*
- Torrens, H. S., 'William Smith (1769–1839): his struggles as a consultant, in both geology and engineering, to simultaneously earn a living and finance his scientific projects, to 1820', *Earth Sciences History*, 35, 1 (2016), 1–46
- Twyman, M., *John Phillips' Lithographic Notebook. Reproduced in facsimile from the original at Oxford University Museum of Natural History* (London: Printing Historical Society, 2016)

FOSSIL COLLECTOR.

- Buckland, W., 'Description of a series of specimens from the plastic clay near Reading, Berks: with observations on the formation to which those beds belong', *Transactions of the Geological Society of London*, 1, 4 (1817), 277–304
- Charig, A., *A New Look at the Dinosaurs* (New York, NY: Mayflower Books, 1979)
- Cox, L. R., (1942) *see Apprentice*
- Morton, J. L., *Strata: How William Smith drew the first map of the earth in 1801 and inspired the science of geology* (Stroud: Tempus Publishing Ltd, 2001)
- Sedgwick, A., 'On the geology of the Isle of Wight etc.', *Annals of Philosophy* 3 (1822), 329–55
- Sowerby, J., *The Mineral Conchology of Great Britain*, Vol. 1. (London: Benjamin Meredith, 1812)
- Torrens, H. S. (ed.), (2003) *see Apprentice*
- Torrens, H. S., (2016) *see Cartographer*
- Ure, A., *A New System of Geology: In which the great revolutions of the earth and animated nature etc.* (Longman, Rees, Orme, Brown & Green, London, 1814)
- Wigley, P. (ed.), *William Smith's Fossils Reunited. Strata Identified by Organized Fossils and A Stratigraphical System of Organized Fossils by William Smith with fossil photographs from his collection at the Natural History Museum* (London: Natural History Museum, 2019)
- Winchester, S., *The Map that Changed the World: The tale of William Smith and the birth of a science* (London: Viking, 2001)

WELL SINKER.

- Farey, J., 'On the means of obtaining water', *Monthly Magazine, or, British Register*, 23 (1807), 211–12
- Farey, J., 'On artesian wells and boreholes', *Monthly Magazine, or, British Register*, 56 (1823), 309
- Kellaway, G. A. (ed.), *Hot Springs of Bath* (Bath: Bath City Council, 1991)
- Mather, J. D., 'William Smith: the natural order of strata and the search for underground water supplies', *Earth Science History*, 35, 1 (2016), 124–44

MENTOR.

- Edmonds J. M., 'The geological lecture-courses given in Yorkshire by William Smith and John Phillips, 1824–25', *Proceedings of the Yorkshire Geological Society*, 40, 3 (1975), 373–412
- Sedgwick, A., 'Announcement of the first award of the Wollaston Prize', *Proceedings of the Geological Society of London*, 1, 20 (1831), 270–79
- Williamson, W. C., 'Reminiscences of a Yorkshire Naturalist', *Good Words*, 18 (1877), 62–66

All images © University of Oxford, Museum of Natural History, unless stated otherwise.

A note on the 1815 base map reproduced throughout this book.

The print of Smith's seminal 1815 map, dating from 23 February 1816, owes its particular vibrancy to the fact that shortly after it was received it was bound in sections into a book and carefully preserved in the library of the Oxford University Natural History Museum. Although opened occasionally for study, it was never put on public display or exposed to direct sunlight.

Every effort has been made to locate and credit copyright holders of the material reproduced in this book. The contributors and publisher are happy to rectify any omissions or errors, which can be corrected in future editions.

a = above
c = centre
b = below
l = left
r = right

8–9 © The Trustees of the Natural History Museum, London; 20 *Strata identified by organized fossils: containing prints on colored paper of the most characteristic specimens in each stratum*, William Smith, Printed by W. Arding, London, 1816; 26 © The Royal Society; 27 Courtesy Dave Williams; 32a *System of Mineralogy*, Robert Jameson, Printed by Neil & Company, Edinburgh, 1816 ; 32b The Natural History Museum/Alamy Stock Photo; 33 Courtesy John Henry; 34l, cl *The Posthumous Works of Robert Hooke*, Robert Hooke, Pub. Richard Waller, London, 1705 ; 34cr, r *La Vana Speculazione Disingannata dal Senso*, Agostino Scilla, Apresso Andrea Colicchia, Naples, 1670; 35l *Principles of Geology*, Charles Lyell, D. Appleton & Co., New York, 1854 ; 35r Carte géognostique des environs de Paris par MM. Cuvier et Brongniart 1810, Bibliothèque nationale de France, département Cartes et plans, GE C-7345; 36 David Rumsey Map Collection, www.davidrumsey.com; 37 Library of Congress, Washington D.C.; 38l Permit Number CP20/034 British Geological Survey © UKRI 2019. All rights reserved; 38ar, br, 39 David Rumsey Map Collection, www.davidrumsey.com; 54–59 Bodleian Libraries, University of Oxford, 188505 a. 4, Plates 1–21 and 21a-c; 62 *Strata identified by organized fossils: containing prints on colored paper of the most characteristic specimens in each stratum*, William Smith, Printed by W. Arding, London, 1816; 63 © The Trustees of the Natural History Museum, London; 64 *Strata identified by organized fossils: containing prints on colored paper of the most characteristic specimens in each stratum*, William Smith, Printed by W. Arding, London, 1816; 65 ©The Trustees of the Natural History Museum, London; 66 *Strata identified by organized fossils: containing prints on colored paper of the most characteristic specimens in each stratum*, William Smith, Printed by W. Arding, London, 1816; 67 © The Trustees of the Natural History Museum, London; 68 *Picturesque views on the River Thames, from its source in Gloucestershire to the Nore*, Samuel Ireland, T. and J. Egerton, London, 1792; 69 Reproduced with kind permission of the South West Heritage Trust, SHC DD/X/ON/2; 71:1 Private Collection; 71:2 Universal History Archive/Getty Images; 71:3 Universal History Archive/Getty Images; 71:4 SSPL/Getty Images; 71:5 Ann Ronan Pictures/Print Collector/Getty Images; 71:6 SSPL/Getty Images; 72 *General View of the Agriculture of the County of Somerset*, John Billingsley, Printed by R. Cruttwell, Bath, 1798; 73l The Geological Society of London; 73r Private Collection; 76r Digital image provided by Bath & North East Somerset Council ; 77 Courtesy Dave Williams; 78al, ar, acl SSPL/Getty Images; 78acr Guildhall Library & Art Gallery/Heritage Images/Getty Images; 78bcl, bcr bl, br SSPL/Getty Images; 77a Oxford Science Archive/Print Collector/Getty Images; 77b SSPL/Getty Images; 85 *Strata identified by organized fossils: containing prints on colored paper of the most characteristic specimens in each stratum*, William Smith, Printed by W. Arding, London, 1816; 90–92 Bodleian Libraries, University of Oxford, 188505 a. 4, Plates 1–21 and 21a–c; 96 *Strata identified by organized fossils: containing prints on colored paper of the most characteristic specimens in each stratum*, William Smith, Printed by W. Arding, London, 1816; 97 © The Trustees of the Natural History Museum, London; 98 *Strata identified by organized fossils: containing prints on colored paper of the most characteristic specimens in each stratum*, William Smith, Printed by W. Arding, London, 1816; 99 © The Trustees of the Natural History Museum, London; 100 *Strata identified by organized fossils: containing prints on colored paper of the most characteristic specimens in each stratum*, William Smith, Printed by W. Arding, London, 1816; 101 © The Trustees of the Natural History Museum, London; 105:1 SSPL/Getty Images; 105:2 SSPL/Getty Images; 105:3 SSPL/Getty Images; 105:4 SSPL/Getty Images; 105:5 Oxford Science Archive/Print Collector/Getty Images; 105:6 Oxford Science Archive/Print Collector/Getty Images; 106al *The Condition and Treatment of the Children employed in the Mines and Colliers of the United Kingdom*, Great Britain Commissioners for Inquiring into the Employment and Condition of Children in Mines and Manufactories, William Strange, London, 1842; 106br *Richard Trevithick, the Engineer and the Man*, Henry Winram Dickinson, Arthur Titley, Cambridge University Press, 1934 ; 110 Bridgeman Images; 111 National Museum of Wales; 112l The Geological Society of London; 114 *General view of the Agriculture and Minerals of Derbyshire*, John Farey, G. & W. Nicol, London, 1811 ; 116 *British Mineralogy*, James Sowerby, Printed by R. Taylor and Co., London, 1802; 117–119 SSPL/Getty Images; 122 *Horse-hoeing Husbandry*, Jethro Tull, Printed for A. Millar, London 1762; 123 Private Collection; 125:1 *The Mansions of England or Picturesque Delineations of the Seats of Noblemen and Gentlemen*, J. P. Neale, M.A. Nattali, London, 1847; 125:2 National Museum of Wales; 125:3 William Watts after W. Tomkins, 1786; 125:4 Kershaw & Son; 125:5 W. Watts, c1782.; 125:6 *The Mansions of England or Picturesque Delineations of the Seats of Noblemen and Gentlemen*, J. P. Neale, M.A. Nattali, London, 1847; 126 *Observations on the Utility, Form and Management of Water Meadows, and the Draining and Irrigating of Peat Bogs*, William Smith, Printed by R. M. Bacon, 1806; 127 Holmes Garden Photos/Alamy Stock Photo; 128 Courtesy Stephen Bartlett; 129 Wellcome Library, London; 130a Oxford Science Archive/Print Collector/Getty Images; 130c SSPL/Getty Images; 130b, 131a Yale Center for British Art, Paul Mellon Collection; 131c Guildhall Library & Art Gallery/Heritage Images/Getty Images; 131b Yale Center for British Art, Paul Mellon Collection; 135 *Strata identified by organized fossils: containing prints on colored paper of the most characteristic specimens in each stratum*, William Smith, Printed by W. Arding, London, 1816; 140–147 Bodleian Libraries, University of Oxford, 188505 a. 4, Plates 1–21 and 21a–c; 150 *Strata identified by organized fossils: containing prints on colored paper of the most characteristic specimens in each stratum*, William Smith, Printed by W. Arding, London, 1816; 151 © The Trustees of the Natural History Museum, London; 152 *Strata identified by organized fossils: containing prints on colored paper of the most characteristic specimens in each stratum*, William Smith, Printed by W. Arding, London, 1816; 153 © The Trustees of the Natural History Museum, London; 154 *Strata identified by organized fossils: containing prints on colored paper of the most characteristic specimens in each stratum*, William Smith, Printed by W. Arding, London, 1816; 155 © The Trustees of the Natural History Museum, London; 156 Courtesy Dave Williams; 157 Courtesy John Henry; 163r University of Nottingham Manuscripts and Special Collections; 166al The Geological Society of London; 166ar Stanford Libraries, Special Collections, Barchas Collection: G5751. C57 1815 S64; 166bl National Museum of Wales; 166br University of Nottingham Manuscripts and Special Collections; 167al Art Collection 3 / Alamy Stock Photo; 167ar National Museum of Wales; 167bl Permit Number CP20/034 British Geological Survey © UKRI 2019. All rights reserved; 167br Courtesy Dave Williams; 168–169 National Museum of Wales; 170 Permit Number CP20/034 British Geological Survey © UKRI 2019. All rights reserved; 171 © The Trustees of the Natural History Museum, London; 180b Bodleian Libraries, University of Oxford, 188505 a. 4, Plates 1–21 and 21a–c; 184 © The Trustees of the Natural History Museum, London; 186 © University of Oxford, Museum of Natural History, Information courtesy Peter Wigley and the UK Onshore Geophysical Library; 188 © The Trustees of the Natural History Museum, London; 189 *Strata identified by organized fossils: containing prints on colored paper of the most characteristic specimens in each stratum*, William Smith, Printed by W. Arding, London, 1816; 190–195 © The Trustees of the Natural History Museum, London; 201 *Strata identified by organized fossils: containing prints on colored paper of the most characteristic specimens in each stratum*, William Smith, Printed by W. Arding, London, 1816; 210–214 Bodleian Libraries, University of Oxford, 188505 a. 4, Plates 1–21 and 21a–c; 215 Reconstructed by Peter Wigley, courtesy Peter Wigley and the UK Onshore Geophysical Library; 219 ©The Trustees of the Natural History Museum, London; 221 ©The Trustees of the Natural History Museum, London; 222 Guildhall Library & Art Gallery/Heritage Images/Getty Images; 223 *Superstitions Anciennes et Modernes*, Jean-Baptiste Thiers, J.F. Bernard, Amsterdam, 1733; 228 *Poetical sketches of Scarborough in 1813*, John Buonarotti Papworth, Frank Fawcett, Driffield, 1893; 229, 233 The Geological Society of London; 234–235 *Illustrations of the Geology of Yorkshire*, John Phillips, Printed for the author by T. Wilson, York, 1829 ; 236 W. Tindall, Printed by C. Hullmandel, Published by C.R. Todd; 237 Lapworth Museum; 239 The Geological Society of London; 256 *Strata identified by organized fossils: containing prints on colored paper of the most characteristic specimens in each stratum*, William Smith, Printed by W. Arding, London, 1816

Illustrations are in *italic*.

A.

B.

C.

D.

E.

F.

G.

H.

I.

J.

K.

L.

ACKNOWLEDGMENTS.

The *publisher* would like to thank Kathleen Diston at the Oxford University Museum of Natural History for her expert guidance and support throughout, Julia Bettinson at Alta Image for cartographic restoration, Peter Wigley for the provision of raw data allowing the comparative area of each stratum to be shown for each geological map, and all the contributors for their expert knowledge, diligence and enthusiasm for William Smith and his groundbreaking work.

With special thanks to Robert Macfarlane for the foreword.

The *contributors* would like to thank Tristan de Lancey, Jane Laing, Phoebe Lindsley, Isabel Jessop, Sarah Vernon-Hunt, Rachel Heley and the wider team at Thames & Hudson for their individual skills, inclusiveness and dedication in producing this wonderful book about an important geologist, especially during the year of COVID-19.

They would also like to acknowledge the several researchers who have gathered together the facts and contextual information to augment John Phillips' biography of his uncle and built up the body of knowledge about William Smith. They are William S. Mitchell (active 1860–70s), Thomas Sheppard (1876–1945), Leslie R. Cox (1897–1965), J. M. Edmonds (active 1950–70s), Joan (1907–1986) and Victor Eyles (1895–1978), and Hugh Torrens (active). They would like to thank Caroline Lam and Wendy Cawthorne at the Geological Society of London, and Kate Diston and Emily Chan at the Oxford University Museum of Natural History, who keep and care for Smith's maps, diaries and letters, for their help over the years with research.

In addition, Jill Darrell and Diana Clements would like to thank many colleagues at the Natural History Museum for help with fossil identifications, and Kevin Webb and Aimee McArdle for photography. They are grateful for assistance from Librarians and Archivists at the Natural History Museum and the British Museum, and for help with text from Hugh Torrens, John Henry, Richard Fortey, Steve Tracey, Brian Rosen and Fiona Fearnhead. John Mather would also like to thank John Henry for his input into captions. Peter Wigley would like to thank Malcolm Butler, Chairman, and the other Trustees of the UK Onshore Geophysical Library for all their support for the Strata-Smith online website, which has brought the achievements of William Smith to the attention of the wider public, and Dave Williams for his knowledge and the encouragement he has given over the years.

ABOUT THE CONTRIBUTORS

Robert Macfarlane is a Reader in Literature and the Geohumanities at the University of Cambridge, and a Fellow of the Royal Society of Literature. His writing about landscape, nature, memory, language and travel has won many prizes internationally, including the E. M. Forster Award for Literature from the American Academy of Arts and Letters in 2017. His books have been widely adapted for TV, film and radio and he has collaborated with artists, film-makers, photographers and musicians, including Johnny Flynn and Jackie Morris.

Peter Wigley worked for ERICO, a geological consulting company from 1973 to 1991, and as an independent consultant and then director of Lynx Information Systems from 1995 to 2015. He is a Board member of AAPG-Datapages and the Director of Datapages DEO-GIS, which provides online maps and figures from all AAPG publications. He is the editor of the William Smith's Maps interactive website (strata-smith.com). He has published articles on William Smith in AAPG and the Geological Society of America and in 2019, in conjunction with the Natural History Museum and the Geological Society, he edited and compiled a publication entitled *William Smith's Fossils Reunited*.

Douglas Palmer is a science writer specializing in fossil evolution and geology. He has written extensively on both subjects and has authored more than twenty books. He currently works as a communications officer for the Sedgwick Museum, University of Cambridge, writing exhibit texts, most recently on the Ice Age (2018) and Geological Maps (2017). Previously, he was a senior lecturer and researcher in palaeontology at Trinity College, University of Dublin.

John Henry headed the remote sensing unit of ARUP Geotechnics for thirty years in London, producing engineering geological maps for a wide range of civil and structural engineering projects worldwide, and continues as an independent consulting geologist. He established Nineteenth Century Geological Maps in 1994 to trade in early geological maps, sections, illustrations and books, see www.geolmaps.com. He is active in the History of Geology Group of the Geological Society of London.

Tom Sharpe was formerly Curator of Palaeontology and Archives at the National Museum of Wales, Geology Tutor at Cardiff University Centre for Lifelong Learning and Chair of the Geological Society's History of Geology Group. For more than twenty years he has been trying to locate all extant copies of William Smith's 1815 map to understand its production, distribution and evolution, as well as pursuing an interest in the lives of geologist Henry De la Beche (1796–1855) and the Lyme Regis fossil collector Mary Anning (1799–1847).

Jill Darrell is the curator of the William Smith collection of fossils and rocks and the fossil, coral and related organisms collections in the Earth Science Department of the Natural History Museum in London. She contributed to *William Smith's Fossils Reunited*, compiled and edited by Peter Wigley, and is a committee member of the History of Geology Group of the Geological Society of London.

Diana Clements works in the Department of Palaeontology in the Natural History Museum as a non-specialist. She has worked with Jill Darrell on the William Smith fossil collection for a number of years and was involved in selecting images for *William Smith's Fossils Reunited*. She is General Secretary of the Geologists' Association and compiled the GA guide to the Geology of London.

John Mather is Professor Emeritus in Geology at Royal Holloway, University of London. Following a career as a hydrogeologist at the British Geological Survey and as Lyell Professor at Royal Holloway, he began work on the previously neglected history of hydrogeology. Particular interests are spas and mineral springs and pre-1900 ideas on the origin, prospecting for and development of groundwater supplies.

Dave Williams is a mining engineer, geochemist and geophysicist. He has worked as an exploration geologist in Zambia, on high-pressure research at the Institute of Geophysics and Planetary Physics, UCLA, USA, and on high-temperature research at Electricity Council labs in Chester, UK. He was a senior lecturer in Earth Sciences at the Open University.

The University of Chicago Press, Chicago 60637

STRATA: William Smith's Geological Maps © 2020
Thames & Hudson Ltd, London

Foreword © 2020 Robert Macfarlane

All rights reserved. No part of this book may be used or reproduced in any manner whatsoever without written permission, except in the case of brief quotations in critical articles and reviews. For more information, contact the University of Chicago Press, 1427 E. 60th St., Chicago, IL 60637.

Published 2020

29 28 27 26 25 24 23 22 21   2 3 4 5

ISBN-13: 978-0-226-75488-8 (cloth)

First published in the United Kingdom in 2020 by Thames & Hudson Ltd, 181A High Holborn, London WC1V 7QX, in partnership with Oxford University Museum of Natural History. Published by arrangement with Thames & Hudson, Ltd, London.

For image copyright information see p. 253.

Library of Congress Control Number: 2020029300

Designed by Daniel Streat, Visual Fields

Printed in China by Reliance Printing (Shenzhen) Co. Ltd.

FSC
www.fsc.org
MIX
Paper from responsible sources
FSC® C102842

FRONT AND BACK COVER Images taken from sections of Sheet X of William Smith's geological map, *A Delineation of the Strata of England and Wales, with part of Scotland* (1815).

BELOW Sowerby's illustration of Smith's fossil specimen of a mastodon tooth from Whitlingham, near Norwich, Norfolk, which is shown on page 188. The illustration features on the frontispiece of *Strata Identified by Organized Fossils* (1816–19).

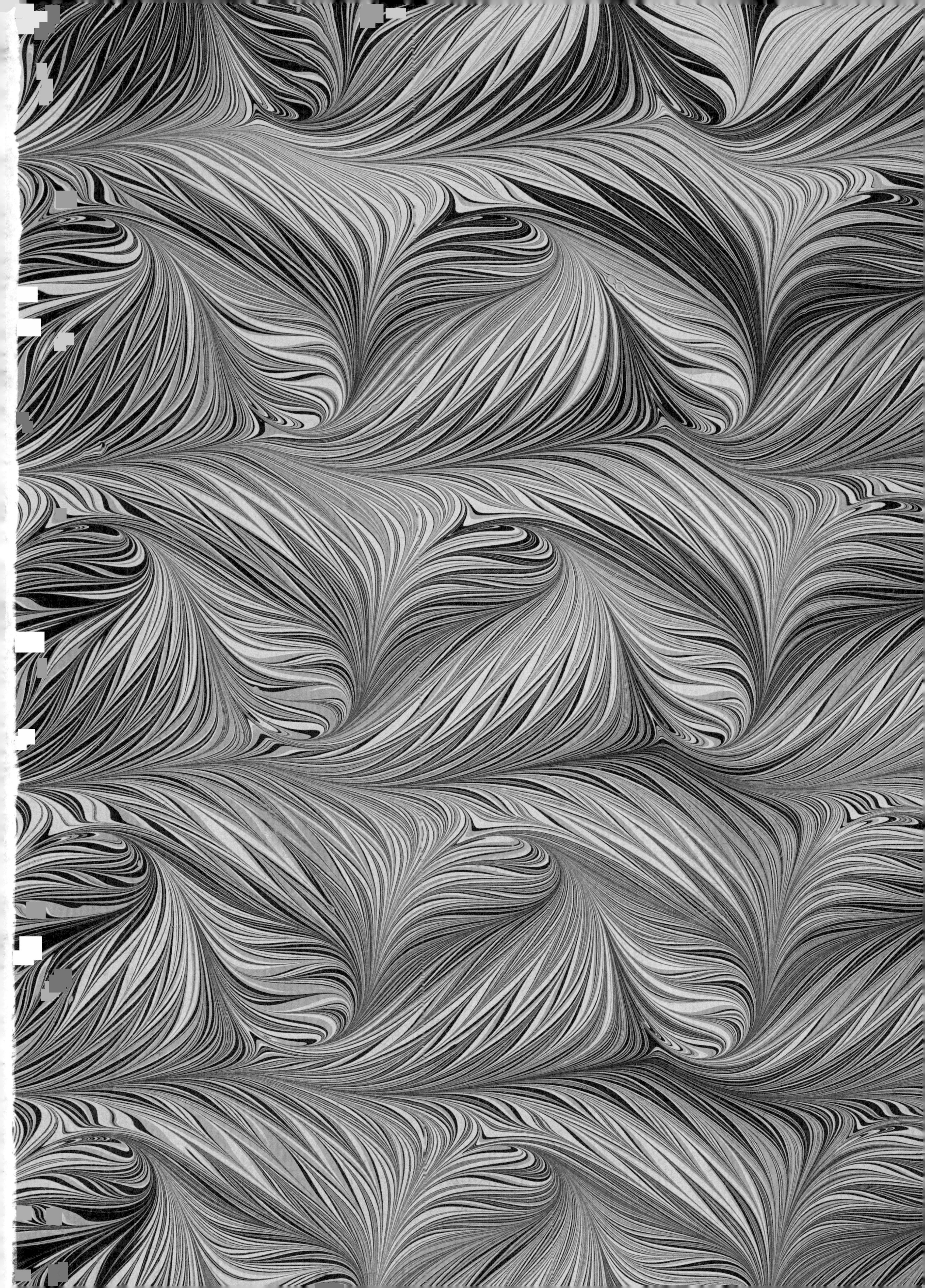

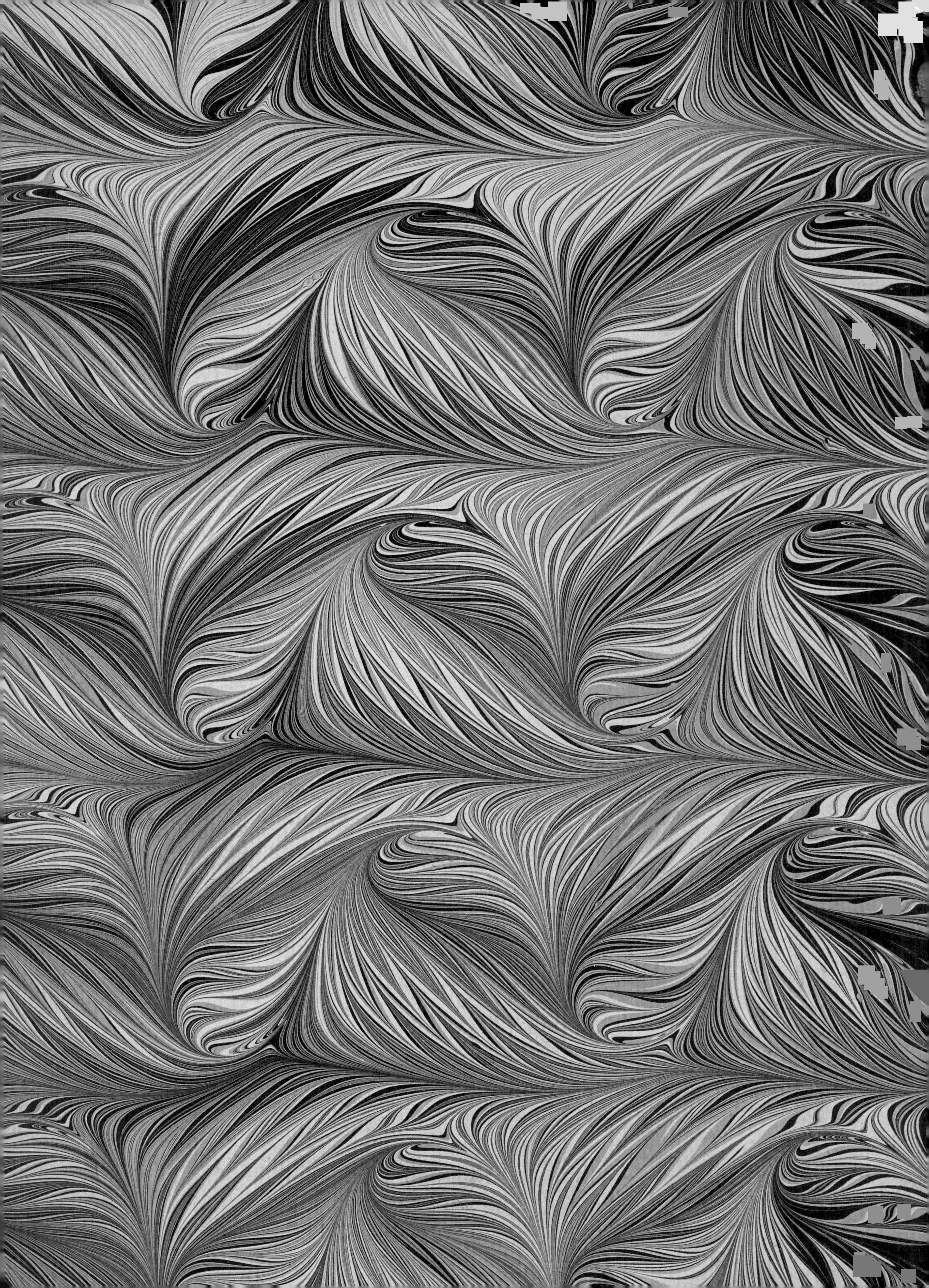